Power Electronics Controller Design Using PSIM and MATLAB

IEEE Press
445 Hoes Lane
Piscataway, NJ 08854

Power Electronics Controller Design Using PSIM and MATLAB

Ilsong Kim
Department of Electrical Engineering
Korea National University of Transportation (KNUT)
Chungju City, Republic of Korea

IEEE Press
WILEY

Published by John Wiley & Sons, Inc., Hoboken, New Jersey.

Library of Congress Cataloging-in-Publication Data applied for:

Hardback ISBN: 9781394410538

Cover Design: Wiley
Cover Image: Ilsong Kim

Set in 9.5/12.5pt STIXTwoText by Straive, Chennai, India
Printed and bound by CPI Group (UK) Ltd, Croydon, CR0 4YY

C9781394410538_021026

Contents

About the Author

ILSONG KIM received the BS degree from Yonsei University and the MS and PhD degrees from Korea Advanced Institute of Science and Technology (KAIST), all in electrical engineering. Born in the Republic of Korea in 1968, he began his career in industry, where he worked on satellite power systems, hybrid electric vehicle battery management systems (BMS), and a wide range of power conversion systems.

After more than a decade of industrial experience, he transitioned to academia in 2007. He is currently a faculty member in the Department of Electrical Engineering at Korea National University of Transportation (KNUT), Republic of Korea. His research interests include power electronics controller design, electric vehicle battery systems, solar power systems, and energy storage systems.

Preface

When graduate students specializing in power electronics or engineers engaged in industrial research and development are asked about their impressions of *controller design*, the most common responses are "difficult," "requires complex mathematics," or "tedious gain tuning." Because controllers are formulated mathematically, their design is often approached through academic literature. However, many of the methods presented in such literature rely on advanced and abstract mathematical theories. This tendency arises from the nature of academic publications, which prioritize theoretical originality over practical, product-oriented implementation.

In addition, a wide variety of controller design methodologies and control techniques have been proposed, making it difficult for engineers and students to determine which approach is most appropriate for a given power conversion system. Control strategies also differ significantly depending on system type, such as DC–DC converters, inverters, grid-connected systems, and motor drives. As a result, it has been challenging to establish a unified and consistent controller design framework applicable across the broad spectrum of power electronic systems.

In practice, however, the fundamental objective of a power conversion system is straightforward: to regulate capacitor voltages, inductor currents, motor speed, or torque. From this viewpoint, the core principle of controller design is simple—generate an appropriate duty command so that the controlled electrical or mechanical variables follow their desired reference values. When controller design is viewed from this fundamental perspective, the same methodology can be applied regardless of system topology.

The primary objective of this book is to present a **unified controller design methodology** that can be consistently applied to a wide range of power conversion systems using only the most fundamental control theory. The proposed approach avoids unnecessary mathematical complexity and advanced theoretical formalisms. Instead, it adopts a single, coherent framework that enables readers to understand and implement controller design in an intuitive and efficient manner. Each chapter follows a structured sequence—**modeling → controller design → verification**. Controller gain calculation, which is often cumbersome and error-prone, is automated using MATLAB SISOTOOL. Verification is performed using PSIM C-block simulations, allowing potential implementation issues to be identified prior to hardware realization. Controller performance is visualized through time-domain waveforms, enabling the validity of the design to be evaluated clearly and intuitively.

This book is intended to serve as a **practical guideline for controller design** for master's-level students in power electronics and engineers working in industrial R&D. It does not aim to cover highly advanced control theories or specialized research topics. Instead, it focuses on applying one or two fundamental theories learned at the undergraduate level to develop controller design and simulation techniques that are directly applicable to commercial products.

The mathematical content presented in this book is deliberately kept simple. Readers who have completed undergraduate courses in circuit theory and power electronics will find the material accessible. Controller design requires only basic undergraduate-level knowledge of control engineering, and familiarity with digital control systems will further facilitate simulation and implementation.

Throughout the book, a unified controller design procedure is consistently applied in every chapter as follows:

1) Derive state-space equations using ON/OFF equivalent circuits.
2) Obtain frequency-domain transfer functions for voltage and current.
3) Identify whether the control target is inductor current or capacitor voltage, and select a single-loop or double-loop control structure.

Up to this point, the design process relies on the engineer's judgment. The subsequent steps are software-assisted:

4) Execute a MATLAB m-file and obtain controller gains using the SISOTOOL library.
5) Verify controller performance through PSIM circuit-level simulations and confirm consistency with the MATLAB-based analytical design.
6) Convert the controller expression into discrete time and implement it in C language for PSIM simulation.

Since the final implementation of a controller is ultimately realized as C code on a microcontroller such as a DSP, most circuits in this book are simulated using PSIM C-blocks. All source codes used in the simulations are fully disclosed. Although these codes are not necessarily optimized for specific industrial products, they represent practical baseline implementations that can be directly adapted for commercial applications.

To ensure consistency in equations and terminology and to maintain coherence between theoretical analysis and simulation, the author personally derived, implemented, and verified all equations and simulation circuits. Consequently, even beginners can reproduce the results presented in this book by studying the theory and executing the simulations to observe the corresponding waveforms.

This book is organized into four major parts: **Introduction, Converters, Inverters, and Special Sessions**. The Introduction presents essential concepts in mathematics, control theory, and electrical components. Although concise, it contains the most fundamental principles underlying controller design and can be regarded as the most important section of the book. The Converter and Inverter sections demonstrate practical controller design based on the methodologies introduced earlier.

The Converter section covers DC/DC converters, solar converters, and DC motor control. DC/DC converters are discussed in Chapters 2–5, including buck, boost, bidirectional, and three-level converters, with both single-loop and double-loop control designs. Solar converters are addressed in Chapters 6–9, covering solar system characteristics and the design of solar boost, buck, and three-level boost converters. Chapter 10 focuses on DC motor control, presenting current and speed controller designs for unipolar and bipolar driving methods using both single-loop and double-loop structures.

The Inverter section includes controller design for single-phase, three-phase, and three-level NPC inverters, as well as solar PCS systems and AC motors. Chapters 11 and 12 present controller designs for single-phase and three-phase inverters using the same unified methodology. Chapter 13 addresses solar PCS design, while Chapter 14 explains single-phase dq inverter control. Chapter 15 introduces PLL design for grid synchronization, presenting algorithms and C code applicable to both single-phase and three-phase systems. Chapter 16 discusses two-stage systems composed of converters and inverters. Chapter 17 explains frequency-response characteristics of converters and inverters. Chapter 18 presents diode-clamped and T-type NPC inverters. Chapter 19 covers PMSM drive controller design, including single-loop and double-loop structures, as well as MTPA and field-weakening control. Chapter 20 addresses the indirect field-oriented control of induction motors.

The Special Sessions present controller design methods for power conversion systems to which state-space averaging cannot be directly applied. Chapters 21–25 cover phase-shift full-bridge converters, dual-active bridge converters, LLC resonant converters, discontinuous-current-mode converters, and modular multilevel converters.

One of the key strengths of the unified design approach presented in this book is that it reveals fundamental equivalences among seemingly different power conversion systems. Specifically, a three-phase inverter can be modeled as three parallel buck converters, while a three-phase AC/DC rectifier can be interpreted as three parallel boost converters. As a result, controller gains designed for buck or boost converters can be directly extended to inverter or rectifier systems, eliminating the need for unnecessarily complex or case-specific design procedures.

If breadth and depth are regarded as complementary dimensions of expertise, the author maintains that the early stage of graduate education should be primarily devoted to strengthening fundamental knowledge. In this context, this book is intended for senior undergraduate students, graduate students, and practicing engineers who seek a systematic and principled introduction to controller design for power conversion systems. Rather than concentrating on isolated techniques or application-specific solutions, the text deliberately addresses a broad range of converter topologies and emphasizes their shared operating principles. Through this unified approach, the author seeks to enable readers to develop a coherent conceptual framework in power electronics control, which will serve as a solid foundation for advanced coursework, independent research, and professional practice.

Although this book is not intended to serve as a guide for producing high-impact academic publications, it is the author's sincere intention that it function as a practical and authoritative reference for controller design in power conversion systems. It is hoped that this book will provide meaningful guidance to students and engineers in the early stages of their professional development, supporting them in building fundamental competence and confidence as they prepare for deeper specialization and innovation in this field.

Winter 2025 *Ilsong Kim*

Postscript

I have tried to present the material in a clear and consistent manner. However, some mathematical errors or differing views on design and analysis may remain. If you find any errors or have suggestions, please email me at ***iskim@ut.ac.kr*** *so they can be reflected in future revisions.*

Thank you.

1

Introduction

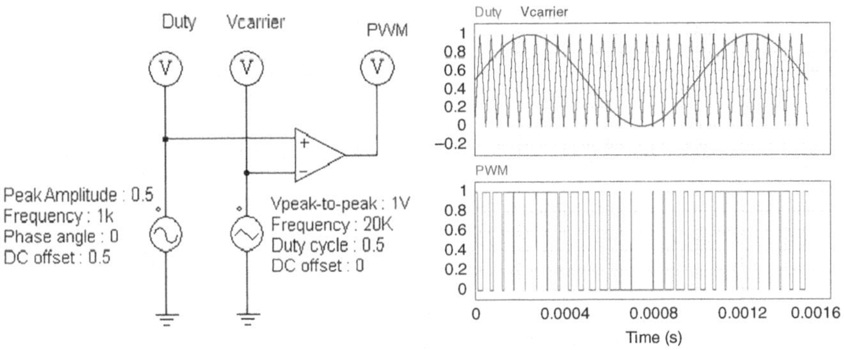

1.1 Design Methodology

1.1.1 Introduction to Controller Design in Power Electronics

Power conversion refers to the use of circuits designed to achieve a desired output voltage or current by utilizing inductors, capacitors, semiconductor switches, and diodes. This process involves comparing the output signal with a reference and adjusting the switching behavior of the semiconductor devices via a controller, typically using pulse width modulation (PWM). By turning the switch ON and OFF repeatedly, the circuit regulates the inductor current or capacitor voltage to reach the target values. Accordingly, the controller is designed to ensure that the waveforms of the inductor current and capacitor voltage meet the desired performance specifications.

Figure 1.1 illustrates a voltage control circuit for a buck converter, implemented using PSIM software. In this example, the capacitor voltage is fed back and

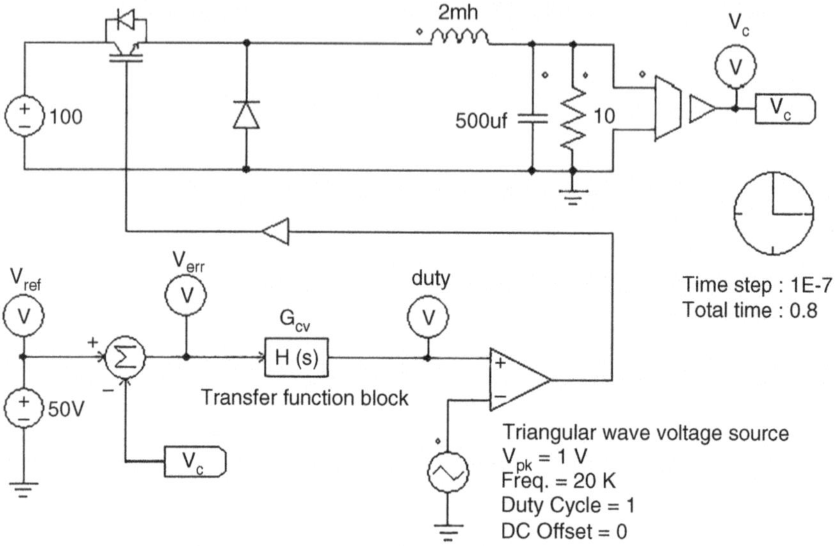

Figure 1.1 Schematic of the capacitor voltage control scheme for a buck converter.

compared to a 50 V reference. The resulting error signal is processed by the voltage controller, denoted as $G_{cv}(s)$, to produce the control duty cycle, which is then applied via PWM. Essentially, the controller $G_{cv}(s)$ ensures that the capacitor voltage waveform conforms to the intended performance criteria.

The voltage controller $G_{cv}(s)$ is designed as a proportional–integral (PI) controller in the frequency domain, with the following transfer function:

$$G_{cv}(s) = \frac{K_p(s + K_i)}{s} \tag{1.1}$$

This transfer function can be implemented using the PSIM library by navigating to: `Elements` → `Control` → `Other Function Block` → `s-domain Transfer Function`. Details on how to enter the coefficients and define the transfer function can be found in the PSIM help documentation.

The primary goal in controller design is to decide suitable values for the proportional gain K_p and the integral gain K_i so that the capacitor voltage waveform satisfies specific performance requirements. These design specifications are as follows:

Design Specifications:

1) Zero overshoot: The voltage waveform must not exceed the target voltage.
2) Zero steady-state error: The output voltage must converge precisely to the desired value with no steady-state error.
3) Rising time: The output voltage should rise from 10% to 90% of the final value (i.e., from 5 V to 45 V) within 0.2 seconds.

The design objective is therefore to determine the PI gains, K_p and K_i, such that the capacitor voltage response exhibits no overshoot, zero steady-state error, and a rise time of 0.2 seconds. To promote hands-on learning, the specific gain values are not prescribed. Instead, an iterative design procedure is recommended: initial gain values are selected, the circuit response is evaluated through simulation, and the controller parameters are progressively tuned until the voltage waveform satisfies the prescribed performance criteria. This trial-and-error process is known as gain tuning, in which various combinations of K_p and K_i are tested in simulation, and the set that best satisfies the requirements is selected. However, this method has several limitations:

- It is time-consuming due to the iterative nature of the tuning process.
- If the circuit parameters or performance specifications change, the tuning must be repeated from scratch.
- The resulting gains may not be optimal, leading to uncertainty regarding their robustness or effectiveness.

Figure 1.2 presents the capacitor voltage waveform obtained using a controller that meets all design specifications. The waveform demonstrates zero steady-state error, no overshoot, and a rise time of 0.2 seconds. The corresponding controller gains were calculated using the mathematical modeling and MATLAB-based approach described in this book. This method offers simplicity and ease of use,

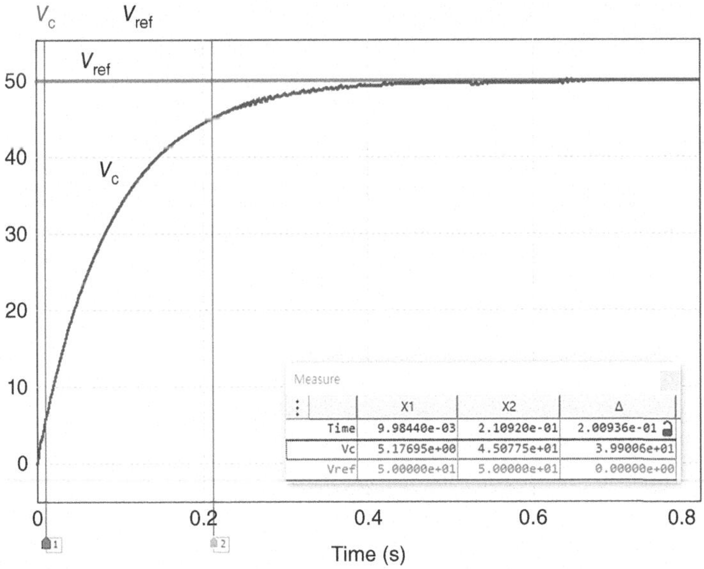

Figure 1.2 Capacitor voltage waveform satisfying the design specifications.

while also allowing for a clear determination of boundary values and stability ranges for the design parameters.

Traditionally, controller design has been viewed as a difficult subject that requires advanced mathematics and strong theoretical knowledge. This perception often makes the topic challenging for students and researchers. Although many research papers and technical resources discuss controller design for power conversion systems, most rely on complex mathematical formulations or focus on specific applications, such as DC–DC converters, inverters, renewable energy systems, and motor drives. As a result, students and practicing engineers often struggle to gain a unified and systematic understanding of controller design, and mastering the subject usually requires considerable time and effort.

This book addresses these challenges by presenting a practical, simple, and systematic approach to controller design. Instead of emphasizing abstract theory, it focuses on intuitive design concepts and efficient techniques. After selecting the controller structure and system transfer function, time-consuming tasks—such as tuning controller gains and verifying circuit performance—are handled using commercial software tools like MATLAB and PSIM. This approach allows readers to design reliable power conversion systems efficiently and with clear, step-by-step procedures.

The proposed methodology is applicable across a wide range of systems, including DC/DC converters, inverters, photovoltaic power conversion systems, grid-connected systems, multilevel inverters, and both DC and AC motor drives. By offering a consistent and unified approach, the methodology enhances design efficiency and repeatability across different platforms, making it a versatile and robust solution for power electronics applications.

1.1.2 Controller Design Process

An example of a single-loop feedback controller configuration is shown in Figure 1.3. As a reference, the capacitor voltage controller for a buck converter is used to illustrate the controller design approach.

The voltage-to-duty transfer function of the buck converter is defined as:

$$G_p(s) = \frac{V_c(s)}{d(s)} \tag{1.2}$$

where $V_c(s)$ represents the capacitor voltage and $d(s)$ denotes the duty cycle. The voltage controller's transfer function is given by:

$$G_{cv}(s) = \frac{d(s)}{V_{\text{err}}(s)} \tag{1.3}$$

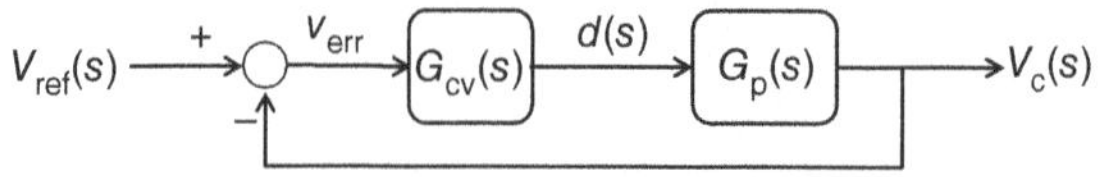

Figure 1.3 Configuration of the voltage controller.

where $V_{\mathrm{err}}(s)$ is the error voltage signal. The closed-loop transfer function T_F of the system can then be expressed as:

$$T_F(s) = \frac{V_c(s)}{V_{\mathrm{ref}}(s)} = \frac{G_{cv}(s)G_p(s)}{1 + G_{cv}(s)G_p(s)} \tag{1.4}$$

The controller design process involves selecting the structure and determining the parameters of the controller such that the closed-loop transfer function T_F satisfies desired performance criteria. These typically include stability, fast response, and minimal steady-state error.

There are two primary approaches to controller design, depending on the type of design method employed:

a) **Time-Domain Design Method**
The time-domain design method utilizes time-domain specifications such as overshoot, rise time, and settling time. In this approach, a step or ramp function is applied to the reference input V_{ref}, and the system's output response—namely, the capacitor voltage—is analyzed in the time domain. The resulting waveform is then evaluated against the design specifications.
However, the time-domain design method has several limitations. It often involves complex theoretical analysis, and precise gain tuning becomes increasingly difficult for higher-order systems (second order or above) due to reliance on approximations. Given these limitations, this book does not adopt the time-domain design approach.

b) **Frequency-Domain Design Method**
The frequency-domain design method is based on prescribed frequency-domain specifications, such as cut-off frequency and phase margin. In this approach, controller design is typically carried out using Bode plots, which enable engineers to assess and shape the gain and phase characteristics of the system. A key advantage of the frequency-domain design method is its strong compatibility with automation, including automatic controller tuning using tools such as *MATLAB Control System Toolbox*. This capability significantly reduces design effort while improving consistency and repeatability in controller synthesis.
Moreover, in more complex configurations such as double-loop control systems—where two controllers are connected in a cascaded structure—time-domain design approaches become impractical. In contrast, frequency-domain methods remain tractable and effective, even for multi-loop systems. For these reasons, this book adopts frequency-domain techniques as the primary methodology for controller design.
In the capacitor voltage control configuration shown in Figure 1.3, it is evident that knowledge of the plant transfer function $G_p(s)$ is essential for applying the frequency-domain design method. In this configuration, the input to the plant

is the duty cycle $d(s)$, and the corresponding output is the capacitor voltage $V_c(s)$. Alternatively, if the control objective were to regulate the inductor current, the plant output of interest would be the inductor current $I_L(s)$, and the transfer function would be defined accordingly.

1.1.3 Frequency-Domain Controller Design Process

The frequency-domain controller design process begins with deriving the plant input–output transfer function, denoted as $G_p(s)$. This requires first defining the plant output according to the control objective.

- If the control target is capacitor voltage, the output is $V_c(s)$.
- If the control target is the inductor current, the output is $I_L(s)$.

In all cases, the plant input is the duty cycle, denoted by $d(s)$.

For clarity and consistency throughout this book, the following naming convention is adopted:

- When the output is the inductor current, the corresponding transfer function is denoted as $G_1(s)$.
- When the output is the capacitor voltage, the corresponding transfer function is denoted as $G_2(s)$.

The next step is to derive the transfer functions $G_1(s)$ and $G_2(s)$. In switching power systems, the equivalent circuit changes depending on whether the power switch is in the ON or OFF state. Consequently, the currents and voltages of the circuit components must, in principle, be determined separately for each switching state.

Although it is theoretically possible to analyze the ON and OFF states independently, solve the corresponding differential equations, and obtain exact solutions, this approach is computationally intensive and often prone to convergence issues in numerical simulations.

To overcome these limitations, the *state space averaging method* is commonly employed. This method simplifies the modeling process by averaging the system dynamics over one switching period. As a result, it significantly reduces computational complexity and improves numerical robustness, albeit at the cost of introducing small modeling approximations. Once the averaged state-space equations are obtained, the corresponding transfer functions can be readily derived using the Laplace transform.

Calculation Using MATLAB Once the input–output transfer functions and the controller structure (e.g., $G_{cv}(s)$ in the previous example) have been defined, *MATLAB* is used to compute the controller gain values that satisfy the prescribed design specifications—typically the cut-off frequency and phase margin.

MATLAB SISOTOOL provides a graphical and semi-automated environment for controller tuning. Using this tool, appropriate gain values are selected while the system's frequency response is evaluated through Bode plots to verify compliance with the design requirements.

In addition to frequency-domain analysis, the time-domain performance is examined by applying a step input to the closed-loop transfer function and observing the resulting output response. This combined frequency- and time-domain evaluation completes the controller design process in *MATLAB*.

Verification Using PSIM The controller design process described thus far is based on mathematical modeling and analytical methods. However, it is essential to verify that the predicted behavior is reproduced in the actual circuit-level implementation.

This verification is carried out through simulation using *PSIM*. The *PSIM* schematic incorporates the power stage, the designed controller block, and the PWM generation module. After the simulation is executed, key waveforms—such as voltages and currents—are examined using *SIMVIEW* and compared with the corresponding results obtained from *MATLAB*.

Agreement between the *PSIM* simulation results and the *MATLAB*-based analysis confirms the consistency between the mathematical model and the circuit-level implementation, thereby validating the correctness of the controller design.

Hardware Implementation In practical applications, the controller is implemented on a hardware platform—such as a digital signal processor (DSP)—using the C programming language. Accordingly, the designed analog controller is discretized and realized in the form of C code. This step represents the final phase of the controller design process, marking the transition from simulation-based validation to physical implementation.

All power conversion systems presented in this book are developed following the design procedure described above. This methodology is broadly applicable to a wide range of power electronic topologies that employ PWM switching techniques. By avoiding unnecessarily complex mathematical formulations, the proposed approach provides an accessible and practical framework for controller design, making it particularly suitable for:

- Graduate students specializing in power electronics.
- Practicing engineers engaged in controller design for power conversion systems

Overview of the Power Conversion System Design Process The design process for a power conversion system can be systematically summarized in the following steps:

a) Modeling and Controller Design
The process begins with the development of a mathematical model of the power conversion system. Based on the desired performance specifications, the controller is then designed using frequency-domain techniques with tools such as *MATLAB SISOTOOL*.

b) Validation Through *PSIM* Simulation
The designed controller is validated through circuit-level simulation using *PSIM*. This step verifies the controller's performance under idealized operating conditions and ensures consistency with the analytical design.

c) Code Generation and Integration
The validated controller is discretized and implemented as *C* code, which is then integrated into the microprocessor development environment, such as *Code Composer Studio* for TI DSP-based platforms.

d) Hardware Implementation and Testing
The compiled code is downloaded to the target hardware. The system is executed, and key voltage and current waveforms are measured and evaluated using laboratory instruments such as an oscilloscope.

e) Design Iteration (If Necessary)
If discrepancies are observed between the simulated results and the experimental waveforms, the controller parameters are adjusted or the controller is redesigned. The process is then repeated starting from Step 2 until satisfactory agreement is achieved.

f) Productization
Once the desired waveforms and overall system performance are confirmed, the design proceeds to the productization and deployment phase.

The laboratory test setup corresponding to this design workflow is illustrated in Figure 1.4.

Figure 1.4 Laboratory test setup.

1.2 Fundamentals: Mathematics and Control

The two most fundamental physical quantities in electrical engineering are voltage (measured in volts V) and current (measured in amperes A).

Voltage can be understood by analogy with water pressure in a dam. A higher dam results in greater water pressure—similarly, a higher voltage represents a stronger electrical potential. Current, on the other hand, corresponds to the flow of water—the amount of water moving through a channel. In electrical terms, current flows naturally from regions of higher voltage (greater potential) to regions of lower voltage.

From these basic quantities, several derived electrical parameters are defined, including resistance Ω, power W, energy Wh, and power factor Φ.

- Resistance is defined as the ratio of voltage to current: $R = V/I$
  It represents the opposition to current flow in a circuit. A higher resistance limits current more significantly.
- Power, defined as the product of voltage and current, quantifies the rate at which electrical energy is used or transferred: $P = V \times I$.
- Energy (measured in watt-hour Wh) represents the total amount of work done over time. Since power is the work done per second (in watts), energy can be seen as the integral (or sum) of power over time. One watt-hour is the amount of energy transferred when one watt of power is applied for one hour (3600 seconds).
- Power factor describes the phase difference between voltage and current in AC systems. It is a dimensionless value between 0 and 1 and plays a crucial role in assessing the efficiency of AC power delivery.

Unit System: MKS (Meter-Kilogram-Second) Electrical engineering commonly uses the MKS (SI) unit system. Below are standard electrical units:

- Voltage: volt V
  - 1 kV = 1000 V
  - 1 V = 1000 mV
- Current: ampere A
  - 1 A = 1000 mA
- Resistance: ohm Ω
  - 1 kΩ = 1000 Ω
- Power: watt W
  - 1 kW = 1000 W

Unit Magnitudes and Prefixes Physical quantities are commonly expressed using standardized metric prefixes to denote powers of ten. These prefixes, summarized

Table 1.1 Unit magnitudes and prefixes.

Prefix	Symbol	Value	Scientific Notation
pico	p	0.000000000001	10^{-12}
nano	n	0.000000001	10^{-9}
micro	µ	0.000001	10^{-6}
mili	m	0.001	10^{-3}
kilo	K	1,000	10^{3}
mega	M	1,000,000	10^{6}
giga	G	1,000,000,000	10^{9}

in Table 1.1, are arranged in ascending order of magnitude to facilitate clear and consistent representation of numerical values across engineering and scientific disciplines.

Relationships Between Units:

$$n = \mu \times m = 10^{-6} \times 10^{-3} = 10^{-9}$$

$$\mu = m \times m = 10^{-3} \times 10^{-3} = 10^{-6}$$

$$M = K \times K = 10^{3} \times 10^{3} = 10^{6}$$

$$C = M \times K = 10^{6} \times 10^{3} = 10^{9}$$

1.2.1 Solving Systems of Differential Equations

Let us begin by solving the following system of linear algebraic equations for the unknown variables x_1, x_2, x_3.

$$12x_1 + 24x_2 - 5x_3 = 9$$

$$3x_1 - 42x_2 + 4x_3 = 40$$

$$2x_1 - 6x_2 - 3x_3 = 4 \tag{1.5}$$

While traditional elimination methods taught in elementary algebra can be applied, they become impractical as the number of equations and variables increases. The most efficient and widely used method in engineering is to express the system in matrix form.

This system can be rewritten as

$$AX = B \tag{1.6}$$

where

$$A = \begin{pmatrix} 12 & 24 & -5 \\ 3 & -42 & 4 \\ 2 & -6 & -3 \end{pmatrix}, \quad X = \begin{pmatrix} x_1 \\ x_2 \\ x_3 \end{pmatrix}, \quad B = \begin{pmatrix} 9 \\ 40 \\ 4 \end{pmatrix} \tag{1.7}$$

Here, matrix A represents the system coefficients, B is the constant input vector, and X is the vector of unknowns. The solution is obtained using the matrix inversion method:

$$X = A^{-1}B = \text{inv}(A) * B \tag{1.8}$$

This approach is general and scalable. *MATLAB* can compute the inverse easily using the inv(A) function, making it especially useful in engineering problems.

Solving Systems of Differential Equations Now consider a system of differential equations, which commonly arises in physical modeling and engineering applications

$$7\frac{dx_1}{dt} + 7x_1 + 14x_2 + 28x_3 = 35$$

$$3\frac{dx_2}{dt} + 9x_1 - 9x_2 - 30x_3 = 60$$

$$\frac{dx_3}{dt} + 3x_1 + 8x_2 + 12x_3 = 14 \tag{1.9}$$

Unlike the previous example, this system includes time derivatives of the state variables. To obtain a standard form, the coefficients of the derivatives are normalized to unity.

$$\frac{dx_1}{dt} + x_1 + 2x_2 + 4x_3 = 5 \Longrightarrow \frac{dx_1}{dt} = -x_1 - 2x_2 - 4x_3 + 5$$

$$\frac{dx_2}{dt} + 3x_1 - 3x_2 - 10x_3 = 20 \Longrightarrow \frac{dx_2}{dt} = -3x_1 + 3x_2 + 10x_3 + 20$$

$$\frac{dx_3}{dt} + 3x_1 + 8x_2 + 12x_3 = 14 \Longrightarrow \frac{dx_3}{dt} = -3x_1 - 8x_2 - 12x_3 + 14 \tag{1.10}$$

This system can be written compactly in matrix form

$$\dot{X} = AX + B \tag{1.11}$$

where

$$\dot{X} = \frac{dX}{dt} = \begin{pmatrix} \frac{dx_1}{dt} \\ \frac{dx_2}{dt} \\ \frac{dx_3}{dt} \end{pmatrix}, \quad X = \begin{pmatrix} x_1 \\ x_2 \\ x_3 \end{pmatrix}, \quad A = \begin{pmatrix} -1 & -2 & -4 \\ -3 & 3 & 10 \\ -3 & -8 & -12 \end{pmatrix}, \quad B = \begin{pmatrix} 5 \\ 20 \\ 14 \end{pmatrix} \tag{1.12}$$

In engineering mathematics, such systems are often solved in the time domain using the state-transition matrix. However, as the system order increases, time-domain solutions become analytically complex and computationally demanding. This naturally leads to the question: What simpler and more general approach can be used instead?

Using the Laplace Transform To simplify the process, the system can be transformed into the frequency domain using the Laplace Transform. This transforms differential equations into algebraic equations, which are easier to solve.

Applying the Laplace Transform to the equation $\dot{X} = AX + B$ yields

$$sX(s) - X(0) = AX(s) + B \Longrightarrow (sI - A)X(s) = X(0) + B \tag{1.13}$$

Solving for $X(s)$

$$X(s) = (sI - A)^{-1}[X(0) + B] \tag{1.14}$$

Finally, the time-domain solution is recovered using the inverse Laplace transform

$$x(t) = L^{-1}[\, X(s) \,] \tag{1.15}$$

This technique significantly reduces complexity, especially for higher-order systems, and is widely used in control systems and circuit analysis.

State-Space Representation The matrix form $\dot{X} = AX + B$ is known as the state-space representation. In this framework, the state variables X are chosen to represent physical quantities described by differential equations.

What can serve as state variables in power conversion systems? In power conversion systems, circuit elements include inductors, capacitors, resistors, diodes, and semiconductor switches. Among these, only inductors and capacitors possess energy storage properties that allow their behavior to be described using differential equations.

- Inductor current as a state variable: In an inductor, the voltage is determined by the rate of change of current. Specifically, when the inductor current increases or decreases linearly, the inductor voltage switches polarity accordingly. This behavior indicates that the inductor current is a continuous and differentiable function, whereas the inductor voltage can be discontinuous and is generally not differentiable.

- Capacitor voltage state variable: In contrast, for a capacitor, the current is determined by the rate of change of voltage. When the slope of the capacitor voltage changes, the capacitor current alternates in direction. Since directly controlling the capacitor current is challenging and often unstable, it is not used as a state variable. Instead, capacitor voltage is chosen as the state variable, as it is the primary stored quantity in the capacitor and behaves in a continuous, differentiable manner.

The number of state variables corresponds to the number of energy storage elements—specifically, the number of inductors and capacitors in the system. If a circuit contains two capacitors and one inductor, the corresponding state variables would be: V_{c1}, V_{c2}, i_{L1}. This results in a total of three state variables.

Establishing State Equations Once the state variables have been identified, the next step is to formulate the state equations. These are first-order differential equations that describe how the state variables evolve over time in response to both their current values and external inputs.

State equations are typically derived using Kirchhoff's current law (KCL) and Kirchhoff's voltage law (KVL). For example, if there are two state variables (SVs), the state equations can be written as

$$\frac{d}{dt}(SV_1) = a_1 \cdot (SV_1) + a_2 \cdot (SV_2) + b_1 \cdot (Input)$$

$$\frac{d}{dt}(SV_2) = a_3 \cdot (SV_1) + a_4 \cdot (SV_2) + b_2 \cdot (\text{Input}) \tag{1.16}$$

Output Equation The output equation expresses the system output as a linear combination of the state variables and inputs

$$\text{Output} = c_1 \cdot (SV_1) + c_2 \cdot (SV_2) + d_1 \cdot (\text{Input}) \tag{1.17}$$

This defines how the system's internal states and inputs contribute to the observable output.

1.2.2 State-Space Representation

This systematic method for modeling dynamic systems is referred to as state-space representation. It enables the structured analysis and design of power conversion systems in a generalizable and scalable manner.

The state-space model is composed of two parts: the state equation, describing the internal system dynamics, and the output equation, specifying how the outputs are obtained from the system states and inputs.

These are expressed in matrix form as

$$\frac{dX}{dt} = AX + Bu$$

$$y = CX + Du \tag{1.18}$$

where

- $\frac{dX}{dt}$: Derivative of the state vector
- A: State matrix, representing the dynamic relationship among state variables
- B : Input matrix, representing how the inputs affect the state variables
- u: Input vector
- y: Output vector
- C: Output matrix, relating the state variables to the outputs
- D: Feedthrough matrix, representing the direct influence of inputs on outputs

This matrix-based representation provides a compact and powerful framework for modeling, simulation, and controller design of dynamic systems, including power electronic converters.

Example 1.1 Converting a Circuit Into State-space Representation
Let us solve an example by converting an electrical circuit into a state-space representation. Refer to Figure 1.5, which shows a simple RLC series circuit. The input is a voltage source $v(t)$, and the circuit consists of a resistor R, an inductor L, and a capacitor C, all connected in series.

The state variables are chosen as

- $i_L(t)$: the current through the inductor
- $V_c(t)$: the voltage across the capacitor

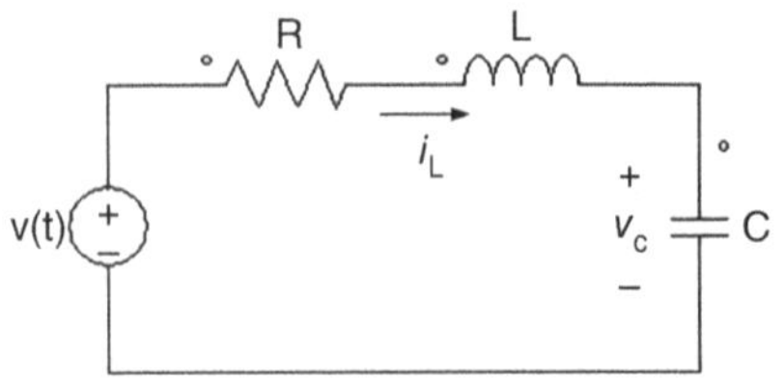

Figure 1.5 Example 1–Series RLC circuit.

Using KCL and KVL, the state equations can be derived as follows:

- Applying KCL at the capacitor

$$i_L = i_c = C\frac{dV_C}{dt} \Rightarrow \frac{dV_C}{dt} = \frac{1}{C}i_L \tag{1.19}$$

- Applying KVL around the loop

$$v(t) = R \cdot i_L + L\frac{di_L}{dt} + V_C \Rightarrow \frac{di_L}{dt} = -\frac{R}{L}i_L - \frac{1}{L}V_C + \frac{1}{L}v(t) \tag{1.20}$$

Assuming the output is the capacitor voltage, i.e., $y = V_C$, the system can now be expressed in state-space form as follows

$$\begin{pmatrix} \frac{di_L}{dt} \\ \frac{dV_C}{dt} \end{pmatrix} = \begin{pmatrix} -\frac{R}{L} & -\frac{1}{L} \\ \frac{1}{C} & 0 \end{pmatrix} \begin{pmatrix} i_L \\ V_C \end{pmatrix} + \begin{pmatrix} \frac{1}{L} \\ 0 \end{pmatrix} v(t)$$

$$y = V_C = \begin{pmatrix} 0 & 1 \end{pmatrix} \begin{pmatrix} i_L \\ V_C \end{pmatrix} + 0 \cdot v(t) \tag{1.21}$$

The corresponding state-space matrices are

$$A = \begin{pmatrix} -\frac{R}{L} & -\frac{1}{L} \\ \frac{1}{C} & 0 \end{pmatrix}, \quad B = \begin{pmatrix} \frac{1}{L} \\ 0 \end{pmatrix}, \quad C = \begin{pmatrix} 0 & 1 \end{pmatrix}, \quad D = 0, \quad X = \begin{pmatrix} i_L \\ V_C \end{pmatrix}, \quad u = v(t) \tag{1.22}$$

This example demonstrates the systematic procedure for deriving the state-space representation of a linear electrical circuit. The same principles can be extended to more complex systems by selecting appropriate state variables and applying KCL and KVL accordingly.

Example 1.2 State-space Representation of a Second-order Circuit
Let us now solve a second example. Figure 1.6 shows a circuit that contains one inductor and one capacitor. Based on this, the state variables are selected as

- $i_L(t)$: the current through the inductor
- $V_c(t)$: the voltage across the capacitor

By applying KCL and KVL, we can derive the two state equations as follows:

Referring to Figure 1.6, the circuit contains one inductor and one capacitor, so the state variables are selected as i_L (inductor current) and V_c (capacitor

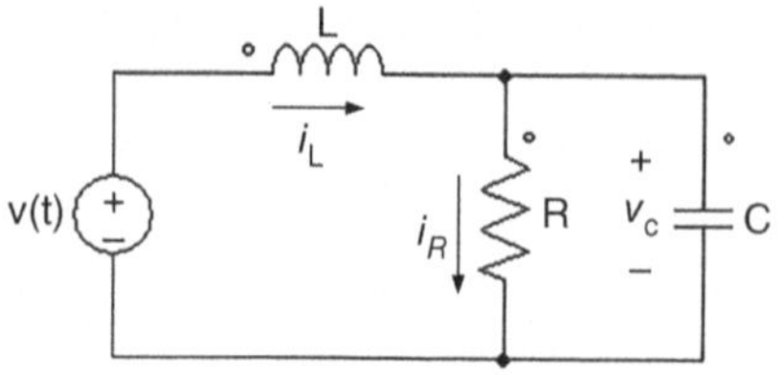

Figure 1.6 Example 2–RLC circuit with output as resistor current.

voltage). By applying KCL and KVL, two state equations can be derived as follows:

- Applying KCL at the top node of the capacitor

$$i_L = i_R + i_c = \frac{V_C}{R} + C\frac{dV_C}{dt} \Rightarrow \frac{dV_C}{dt} = \frac{1}{C}i_L - \frac{1}{RC}V_C \tag{1.23}$$

- Applying KVL around the loop

$$v(t) = L\frac{di_L}{dt} + V_C \Rightarrow \frac{di_L}{dt} = -\frac{1}{L}V_C + \frac{1}{L}v(t) \tag{1.24}$$

Assuming the output is the current through the resistor, $y = i_R$, and using Ohm's law $i_R = \frac{V_C}{R}$, the output equation becomes

$$y = i_R = \frac{V_C}{R} = \begin{pmatrix} 0 & \frac{1}{R} \end{pmatrix} \begin{pmatrix} i_L \\ V_C \end{pmatrix} + 0 \cdot v(t) \tag{1.25}$$

The state-space representation of the system is

$$\begin{pmatrix} \frac{di_L}{dt} \\ \frac{dV_C}{dt} \end{pmatrix} = \begin{pmatrix} 0 & -\frac{1}{L} \\ \frac{1}{C} & -\frac{1}{RC} \end{pmatrix} \begin{pmatrix} i_L \\ V_C \end{pmatrix} + \begin{pmatrix} \frac{1}{L} \\ 0 \end{pmatrix} v(t)$$

$$y = i_R = \begin{pmatrix} 0 & \frac{1}{R} \end{pmatrix} \begin{pmatrix} i_L \\ V_C \end{pmatrix} + 0 \cdot v(t) \tag{1.26}$$

The state-space matrices for this system are

$$A = \begin{pmatrix} 0 & -\frac{1}{L} \\ \frac{1}{C} & -\frac{1}{RC} \end{pmatrix}, \quad B = \begin{pmatrix} \frac{1}{L} \\ 0 \end{pmatrix}, \quad C = \begin{pmatrix} 0 & \frac{1}{R} \end{pmatrix}, \quad D = 0, \quad X = \begin{pmatrix} i_L \\ V_C \end{pmatrix}, \quad u = v(t) \tag{1.27}$$

This example further illustrates how a linear electrical circuit can be modeled using the state-space framework. It highlights how different output choices (e.g., voltage, current) influence the construction of the output matrix C, while the modeling procedure using KVL and KCL remains systematic and generalizable.

Example 1.3 State-space Representation of a Third-order Circuit

Let us now solve a third example. Figure 1.7 shows a circuit that contains two inductors and one capacitor. Based on the energy-storing components, we select the following state variables:

- $i_{L1}(t)$: current through the first inductor
- $i_{L2}(t)$: current through the second inductor
- $V_c(t)$: voltage across the capacitor

Using KCL and KVL, the following state equations can be derived.

- From KCL at the capacitor node

$$i_{L1} = i_C + i_{L2} \Rightarrow \frac{dV_C}{dt} = \frac{1}{C}i_{L1} - \frac{1}{C}i_{L2} \tag{1.28}$$

- From KVL across the second inductor

$$V_C = V_{L2} \Rightarrow \frac{di_{L2}}{dt} = \frac{1}{L_2}V_C \tag{1.29}$$

- From KVL across the first inductor and series resistor

$$v(t) = V_C + L_1\frac{di_{L1}}{dt} + v_R \Rightarrow \frac{di_{L1}}{dt} = -\frac{1}{L_1}V_C - \frac{R}{L_1}i_{L1} + \frac{1}{L_1}v(t) \tag{1.30}$$

To express the system in matrix form, we define the state vector as

$$X = \begin{pmatrix} V_C \\ i_{L1} \\ i_{L2} \end{pmatrix}$$

Then, the state-space equations can be written as

$$\begin{pmatrix} \frac{dV_C}{dt} \\ \frac{di_{L1}}{dt} \\ \frac{di_{L2}}{dt} \end{pmatrix} = \begin{pmatrix} 0 & \frac{1}{C} & -\frac{1}{C} \\ -\frac{1}{L_1} & -\frac{R}{L_1} & 0 \\ \frac{1}{L_2} & 0 & 0 \end{pmatrix} \begin{pmatrix} V_C \\ i_{L1} \\ i_{L2} \end{pmatrix} + \begin{pmatrix} 0 \\ \frac{1}{L} \\ 0 \end{pmatrix} v(t) \tag{1.31}$$

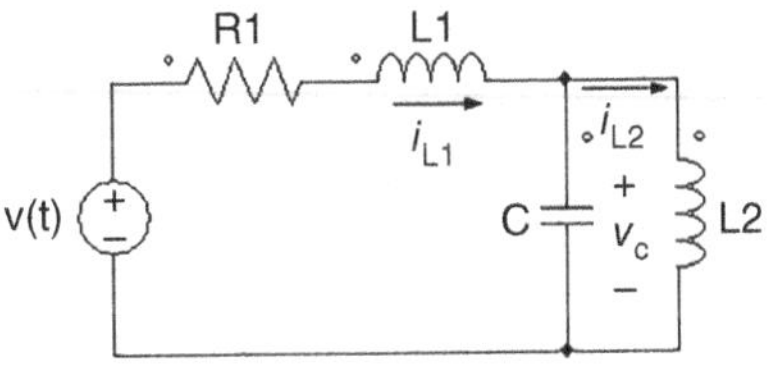

Figure 1.7 Example 3–Two inductors and one capacitor.

The corresponding matrices are

$$A = \begin{pmatrix} 0 & \frac{1}{C} & -\frac{1}{C} \\ -\frac{1}{L_1} & -\frac{R}{L_1} & 0 \\ \frac{1}{L_2} & 0 & 0 \end{pmatrix}, \quad B = \begin{pmatrix} 0 \\ \frac{1}{L} \\ 0 \end{pmatrix}, \quad X = \begin{pmatrix} V_C \\ i_{L1} \\ i_{L2} \end{pmatrix}, \quad u = v(t) \tag{1.32}$$

The output equation is not explicitly defined in this example. This example demonstrates how to extend the state-space modeling process to circuits with multiple inductors and capacitors. The procedure remains systematic: identify energy-storing components, apply KCL/KVL, and arrange the resulting differential equations into matrix form.

1.2.3 State-Averaged Equations of Switching Circuits

In switching circuits, the configuration of the circuit changes depending on the ON/OFF state of the switch within each switching period (T_s). As a result, the system exhibits different state equations during the ON and OFF intervals of the switching cycle. To analyze such systems more conveniently, these time-varying equations are combined into a state-averaged equation, which represents the behavior of the system.

To illustrate the concept of periodic averaging, consider the example shown in Figure 1.8. The capacitor voltage $V_c(t)$ alternates between 100 V for 0.3 seconds and 60 V for 0.9 seconds, repeating periodically with a total cycle time of T = 1.2 seconds.

The average capacitor voltage over one period is calculated using the time-averaging formula

$$V_{C,\text{avg}} = \frac{I}{I+II} \cdot 100 + \frac{II}{I+II} \cdot 60 = \frac{0.3}{1.2} \cdot 100 + \frac{0.9}{1.2} \cdot 60 = 70\ V \tag{1.33}$$

This illustrates how values over subintervals of the switching period contribute to the overall average.

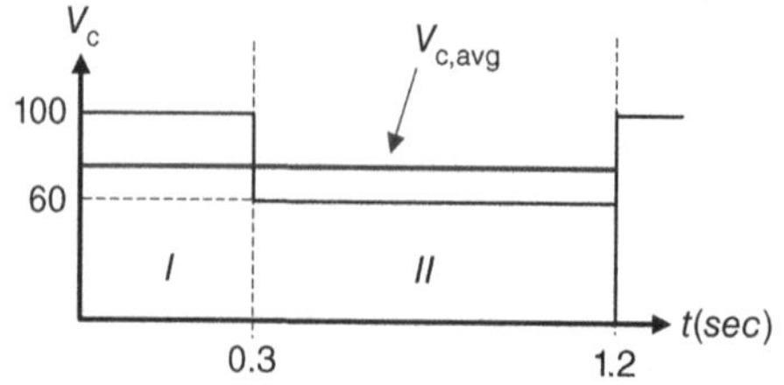

Figure 1.8 Example illustrating the concept of periodic averaging.

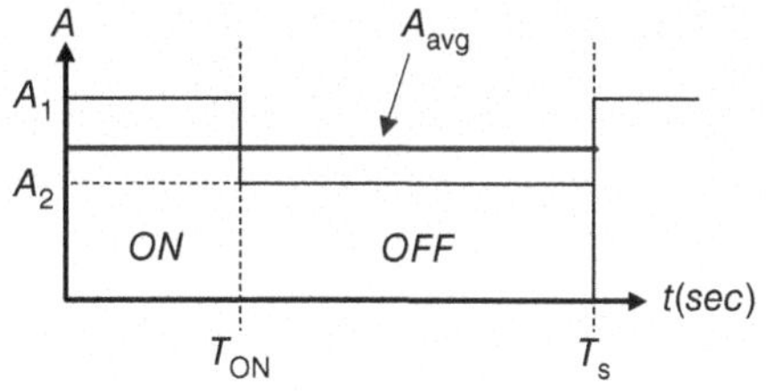

Figure 1.9 Example illustrating the concept of average state-space matrices.

In a similar manner, the average state-space matrices A_{avg} and B_{avg} can be derived for a switching system with multiple operating modes. An illustrative example is presented in Figure 1.9.

The formula for the average state matrix A_{avg} over a switching period T_s, derived from the state-space representations during the ON period (A_1) and the OFF period (A_2), is given by

$$A_{avg} = \frac{T_{on}}{T_s} \cdot A_1 + \frac{T_{off}}{T_s} \cdot A_2 = d \cdot A_1 + (1 - d) \cdot A_2 \tag{1.34}$$

where

$T_s = T_{on} + T_{off}$ is the total switching period
$d = \frac{T_{on}}{T_s}$ is the duty ratio
The same principle applies to the input matrix B_{avg}

$$B_{avg} = d \cdot B_1 + (1 - d) \cdot B_2 \tag{1.35}$$

Thus, the state-averaged equation of the switching circuit can then be expressed in the standard state-space form as follows:

$$\frac{dX}{dt} = A_{avg}X + B_{avg}u$$

$$y = CX + Du \tag{1.36}$$

This modeling approach simplifies the analysis and controller design of switching converters by treating them as linear time-invariant (LTI) systems over one switching period, enabling the use of frequency-domain techniques and linear control methods.

1.2.4 Linearization

Let us now explore the concept of linearization as it applies to functions and matrices.

a) Linearization of a Function

To understand the distinction between linear and nonlinear functions, refer to the graphs shown in Figure 1.10:

When we calculate the slope of the two functions, we observe the following:

- For f_1, the slopes at any two points (e.g., a_1 and a_2) are equal, i.e., $a_1 = a_2$, which indicates that the function is linear.
- For f_2, the slopes differ at different points, i.e., $a_1 \neq a_2$, which means the function is nonlinear.

A linear function is defined as a function that satisfies the properties of additivity and proportionality. Mathematically, a function is linear if it satisfies the following rules.

Additivity: $f(x+y) = f(x) + f(y)$
Proportionality: $f(k \cdot x) = k \cdot f(x)$

Linearizing a Nonlinear Function Now consider the process of linearizing a nonlinear function, such as f_2 shown in Figure 1.10 (*b*). Since the slope of a nonlinear function varies with the input x, it cannot be linearized over the entire input range. Instead, we approximate the nonlinear function locally, around a specific point x_o, where the function behaves approximately linearly.

This is done by approximating the nonlinear function with a linear function (a straight line) that has slope a_1 near x_o, as shown in Figure 1.11(a). To include

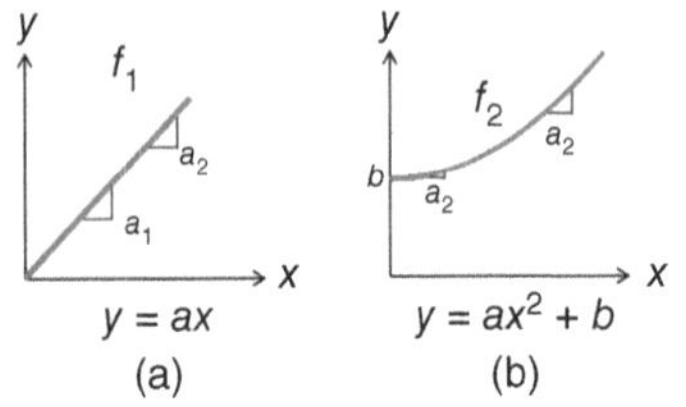

Figure 1.10 Linear function f_1 and nonlinear function f_2. (a) function f_1 (b) function f_2.

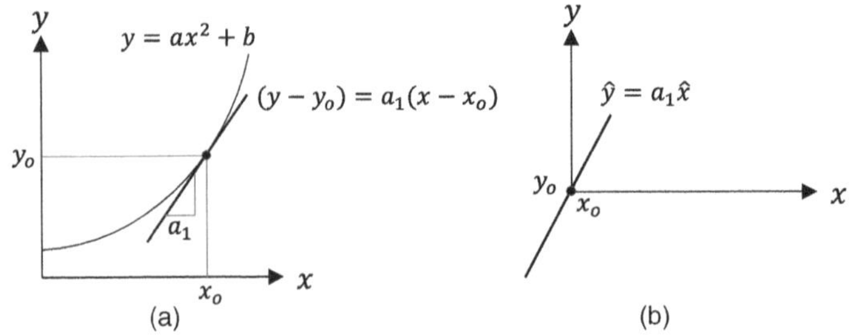

Figure 1.11 Small-signal linearization of the nonlinear function (f_2).

an offset (x_o, y_o), the linearized function becomes:

$$\hat{y} = a_1\hat{x}$$

where $\hat{x}$ is the variable adjusted by the DC offset x_o.

This process is called small-signal linearization. It represents a linearized model around a large signal or DC operating point (x_o, y_o). Figure 1.11(b) illustrates this linearized function.

In summary, small-signal linearization is an effective technique that simplifies the analysis of nonlinear systems by approximating their behavior around a specific operating point. This localized linear model allows the use of linear system analysis tools—such as Bode plots and root locus—on systems that are inherently nonlinear.

Another Example of Linearization Let us now examine another case: a first-order function with an offset, as shown in Figure 1.12.

$$y = ax + b$$

At first sight, this function appears to be linear, since it is expressed as a straight-line equation with slope a. However, to determine whether it is truly a linear function in the mathematical sense, we must apply the formal definitions of additivity and proportionality.

(a) Additivity

- For input x_1, the output is $y_1 = ax_1 + b$
- For input x_2, the output is $y_2 = ax_2 + b$

Now consider the combined input $x_1 + x_2$, the output is

$$y_3 - a(x_1 + x_2) + b - ax_1 + ax_2 + b$$

But the sum of the individual outputs is

$$y_1 + y_2 = (ax_1 + b) + (ax_2 + b) = (ax_1 + ax_2 + 2b)$$

Since $y_1 + y_2 \neq y_3$, the function violates the property of additivity.

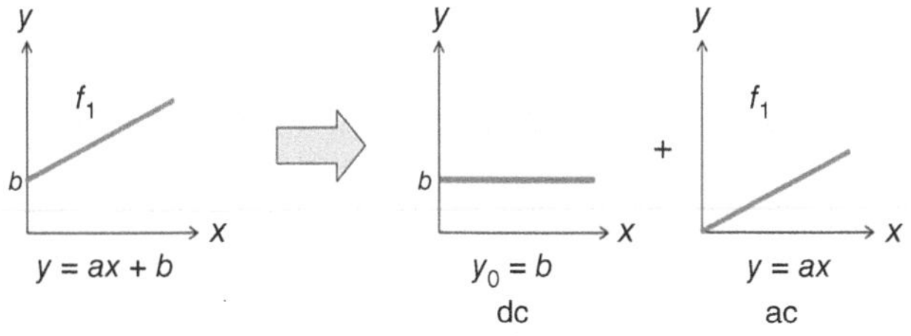

Figure 1.12 Linearize a first-order function with an offset.

(b) Proportionality

- For input kx_1, the output is: $y_1 = a(kx_1) + b$
- Scaling the output for input x_1 by k: $y_2 = k(ax_1 + b) = kax_1 + kb$

Again, since $y_1 \neq y_2$, the function violates the property of proportionality.

Thus, the function $y = ax + b$ does not satisfy the mathematical conditions for linearity (additivity and proportionality) and is therefore classified as a nonlinear function in the context of system analysis.

So, how can we linearize a first-order function with an offset? The key idea is to separate the function into two components

- The DC component, represented by the constant offset b, which defines the y-intercept
- The AC component, represented by the slope a, which forms the line $y = ax$, which passes through the origin.

In this separation, the AC component

$$y = ax.$$

This approach is consistent with the small-signal linearization technique used in control systems and circuit analysis. This method allows us to simplify the analysis of systems by modeling them as linear systems near a specific operating point, even if the overall system is nonlinear.

b) **Linearization of Matrix Equations**

Matrix-based dynamic systems are typically expressed in the following state-space form

$$\dot{X} = AX + Bu$$
$$y = CX + Du \tag{1.37}$$

If the matrices A and B are time-invariant constants, the system is classified as a linear system. In such systems, the output y is linearly proportional to the input u, making it possible to apply classical linear control strategies, such as PI controllers.

However, if the system does not conform to this linear form -i.e., if the relationships between states and inputs are nonlinear—it is considered a nonlinear system.

Because linear controllers are generally not applicable to nonlinear systems without modification, two main approaches exist

1) Design a nonlinear controller, which typically involves advanced control theory and increased complexity
2) Linearize the nonlinear system around a specific operating point, enabling the use of linear controllers within a limited range.

This book focuses on the second approach: approximating nonlinear systems with linear models over a limited operating range. This is a practical and widely used method in power electronics and control engineering.

Common linearization techniques include

- Small-signal linearization
- Feed-forward linearization

Small-Signal Linearization In power electronics systems, circuits such as the boost converter are inherently nonlinear. In such systems, the state-space matrices A and B are not constants, but functions of the input u and state variables X.

To analyze and control these systems using linear methods, small-signal linearization is employed. This technique involves separating each variable or matrix into two components:

- DC component: Represents the large-signal model, corresponding to the steady-state or operating point.
- AC component: Represents the small-signal model, which describes small deviations around the operating point.

This separation yields two models:

- Large-Signal Model: Derived from the DC components and centered around the operating point (x_o, y_o).
- Small-Signal Model: Derived from the AC components, and mathematically analogous to the linearized function $\hat{y} = a_1\hat{x}$.

The specific numerical methods for linearizing boost converters and deriving their models are detailed in Chapter 3: Boost converter. By applying this approach, even nonlinear systems can be analyzed and controlled using linear methods within a limited operating range.

Feedforward Linearization Now, a second technique: Feedforward linearization. Consider a system represented by the following differential equation

$$\dot{X} = AX + Bu + F \tag{1.38}$$

Here

- A, B are time-invariant constant matrices.
- F is an additional nonlinear term, which prevents the system from being linear, even if the rest of the structure appears linear.

To linearize the system, the input u is decomposed into two components

$$u = u_{fb} + u_{ff} \tag{1.39}$$

where

- u_{fb}: Feedback control input, designed to stabilize the system
- u_{ff}: Feedforward input, used to cancel the nonlinear term F

By substituting $u = u_{fb} + u_{ff}$ into the equation and solving for u_{ff} to cancel out F, we get

$$u_{ff} = -\frac{F}{B} \tag{1.40}$$

Substituting this into the original equation gives

$$\dot{X} = AX + Bu_{fb} \tag{1.41}$$

Now, the system has been transformed into a linear form, suitable for applying linear control strategies such as PI control. In this form

- u_{ff} is the feedforward input, computed specifically to cancel the nonlinear effect.
- For feedforward linearization to be feasible, F must be known and predictable, and u_{ff} must be implementable (preferably as a constant or a measurable function).

This approach transforms a nonlinear system into a linear one, enabling the use of linear control strategies such as PI controllers. Feedforward linearization is especially useful when the nonlinear term F is known and predictable. Chapter 11: Single-phase inverter provides a detailed explanation of the numerical methods used for feedforward linearization and model derivation of inverter systems.

The key points regarding the linearization of matrix equations can be summarized as follows:

- Small-signal linearization is suitable when analyzing systems near a known steady-state operating point
- Feedforward linearization is effective when a known nonlinearity can be actively canceled
- Both methods transform nonlinear systems into forms that are compatible with linear control design, thus extending the applicability of simple and robust control techniques to more complex systems.

The block diagram of a typical feedback control system is shown in Figure 1.13. The system is designed with a single-loop configuration in which the output is fed back into the system to regulate its behavior.

a) Feedback Control System
   This is the most used structure in control systems. In this configuration, the output of the controller $d = d_{fb}$ is applied as the input to the plant transfer function G. The feedback loop continuously adjusts the control input to minimize the error between the reference signal and the system output.

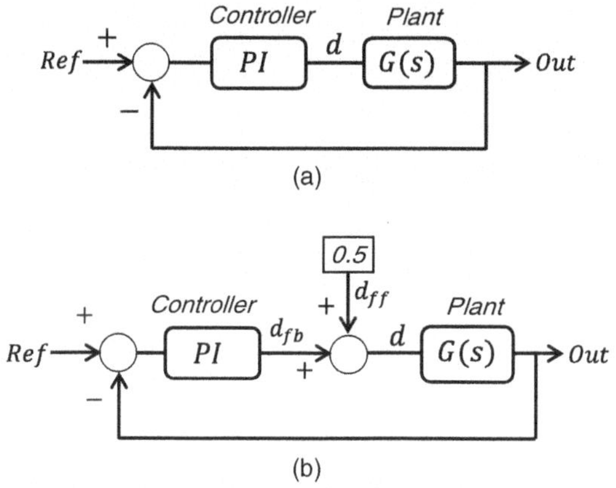

Figure 1.13 Control system configurations: feedback and feedforward structures. (a) Feedback control configuration (b) Feedforward control configuration.

b) Feedforward Control System

In a feedforward control structure, the input to the plant is composed of two parts: the feedback controller output d_{fb} and a feedforward compensation term d_{ff}. The total control input to the plant becomes $u = u_{fb} + u_{ff}$.

For example, the feedforward input d_{ff} may be set to a constant value, such as 0.5, to compensate for predictable disturbances or steady-state offsets.

This dual-input architecture allows the feedforward control system to proactively address known system nonlinearities or external disturbances, while the feedback loop ensures stability and accuracy in the system's dynamic response.

Continuous Conduction Mode Continuous conduction mode (CCM) occurs when the minimum current through the inductor (I_{min}) remains greater than zero throughout the switching period. If there is a period during which the inductor current falls to zero, the circuit is said to operate in discontinuous conduction mode (DCM). The current waveforms for both modes are illustrated in Figure 1.14 (a) and (b).

In CCM, each switching cycle consists of two intervals: T_{on} and T_{off}. These intervals repeat consistently, forming a predictable switching pattern. Under this condition, it is possible to derive a state-averaged model of the converter that is linear, which enables the effective design of PI controllers.

In contrast, DCM introduces an additional zero-current interval. A full switching cycle consists of T_{on}, T_{off1} and T_{off2}, where the inductor current drops to zero during part of the cycle. This discontinuity results in a system whose behavior depends on switching boundaries and operating conditions, making the state-space equations nonlinear. As a result, conventional PI controller design becomes impractical.

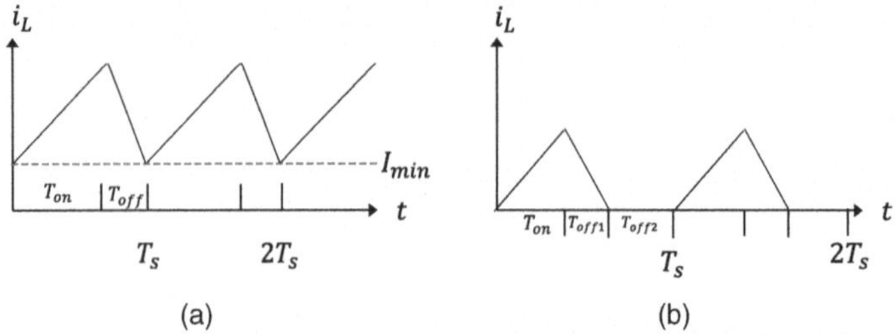

Figure 1.14 Current waveform comparison between CCM and DCM. (a) CCM Mode (b) DCM Mode.

Accordingly, this textbook is primarily devoted to controller design under CCM operation. The design of controllers for DCM is discussed separately in Chapter 24.

Furthermore, this book focuses exclusively on fixed-frequency PWM control techniques. Variable-frequency control strategies, including those applied in resonant converters, are treated separately in Chapter 23. The exclusion of variable-frequency methods in the main text arises from the nonlinear timing characteristics introduced by pulse-frequency modulation (PFM), which significantly complicate the control and hinder automated controller synthesis.

1.2.5 Controller Design Steps

A control system is typically structured with a controller and a plant connected in series, as shown in Figure 1.15. The system output is fed back and compared with the reference input, and the resulting error signal serves as the input to the controller.

Let us now explore the structured steps involved in designing a controller.

- Step 1: Establish a Mathematical Model of the Plant
  a) Time Domain Model: Using the state-space averaging method, the system is represented by the following equations:

$$\dot{X} = AX + Bu$$

$$y = CX + Du$$

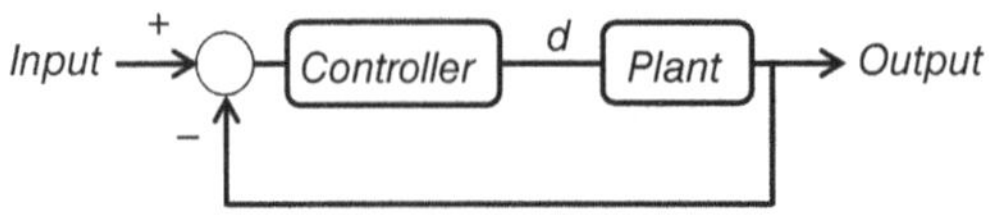

Figure 1.15 Block diagram of the feedback control system.

b) Frequency Domain Model: The system's transfer function is derived from the Laplace-transformed state equations:

$$X(s) = (sI - A)^{-1}BU \Rightarrow \frac{Y(s)}{U(s)} = C(sI - A)^{-1}B$$

- Step 2: Design a Linear Controller
  A commonly used controller in power electronics is the PI controller. Its transfer function is given by:

$$G_c(s) = \frac{K_p(s + K_i)}{s}$$

  The controller design process involves selecting the gain values K_p and K_i to satisfy the desired performance specifications. While time-domain specifications (e.g., rise time, overshoot) may be used, it is generally more practical to design the controller based on frequency-domain criteria, such as cut-off frequency and phase margin. MATLAB's SISOTOOL provides a graphical interface and auto-tuning features to help determine optimal gain values that meet these design specifications.
- Step 3: Verification Process
  Once the controller has been designed, its performance must be verified through simulation and implementation. This involves the following steps:
  Convert the physical circuit into a mathematical model and then derive the corresponding frequency-domain transfer function.
  1) Select and apply appropriate controller gains.
  2) Implement the controller in the simulation environment and observe the system's time-domain response.
     Simulation is carried out using PSIM, where the controller can be implemented either as an analog block or as a C-block. If the simulation results meet the desired specifications, the C-block code is then programmed into the DSP's flash memory.

This completes the controller design and verification process, bridging the gap between mathematical modeling and real-time hardware implementation.

1.3 Electrical Components

1.3.1 Passive and Active Elements

Electrical components are devices that function by utilizing electric energy. These components are broadly categorized into passive components and active components.

- Passive Components: Passive components are elements that cannot generate or amplify electrical signals and do not require an external power source to operate. They typically have two terminals and include elements such as resistors, capacitors, and inductors.
- Active Components: Active components are elements that can control the flow of electricity, amplify signals, or convert energy forms through the use of external power sources or control signals. These components generally have three or more terminals. Examples include diodes and semiconductor switches such as bipolar junction transistors (BJTs), field-effect transistors (FETs), and insulated gate bipolar transistors (IGBTs).

Passive Components Let us begin by discussing passive components. The fundamental physical quantities associated with passive elements are voltage and current. A capacitor is a device that stores voltage, while an inductor stores current. As a result, the relationship between voltage and current in these components is described by differential equations. A resistor, on the other hand, is a component that dissipates electrical energy, and the voltage–current relationship for a resistor is a simple proportional expression. Let us first examine the characteristics of an inductor.

- **Inductor**

When an electric current flows through a conductor, it generates a magnetic field, the strength of which is proportional to the magnitude of the current. To enhance the magnetic field without increasing the current, the conductor can be wound into a closed-loop coil. According to Ampère's Law (right-hand thumb rule), the magnetic flux becomes concentrated within the coil, thereby increasing the magnetic field intensity. Furthermore, increasing the number of turns in the coil proportionally increases the total magnetic flux. If a ferromagnetic core is placed at the center of the coil, the magnetic flux becomes even more concentrated. A device that stores energy in the form of concentrated magnetic flux generated by current flow is known as an inductor.

In circuit analysis, let i_L be the current flowing through the inductor and V_L be the voltage across it. The relationship between voltage and current is described by the following differential equation:

$$V_L = L\frac{di_L}{dt} \tag{1.42}$$

Figure 1.16 shows an inductor circuit along with its corresponding current and voltage waveforms. From the waveforms, it is evident that the voltage across the inductor is determined by the rate of change of current, $\frac{di_L}{dt}$. An inductor requires a continuous current path to ensure that current flows without interruption.

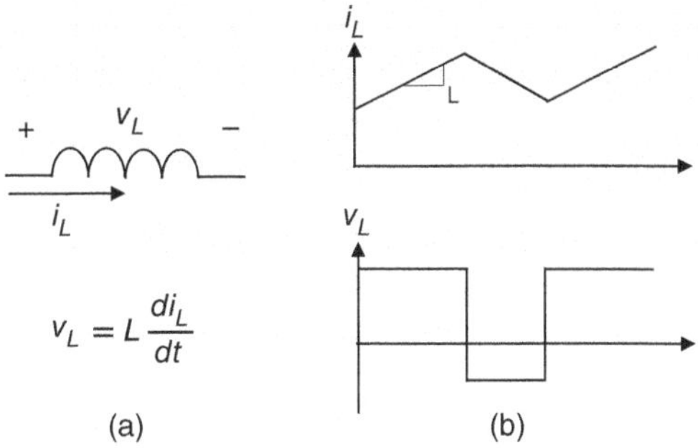

Figure 1.16 Inductor current and voltage characteristics. (a) Inductor circuit (b) Inductor current and voltage waveform.

Key observations from the waveforms

- The current waveform is continuous and varies smoothly over time, increasing and decreasing gradually.
- In contrast, the voltage waveform is discontinuous, with its polarity changing according to the direction of the current slope.

Therefore, in a state-space representation, the appropriate state variable for an inductor is the inductor current i_L.

Capacitor: Opposite Behavior to Inductor Now let us consider a component that exhibits behavior opposite to that of an inductor. When a voltage is applied across a pair of parallel plates, electric charge accumulates on the plates. The amount of stored charge is proportional to the magnitude of the applied voltage.

In circuit analysis, if i_C represents the current through the capacitor and V_C the voltage across it, the voltage–current relationship is described by the following equation:

$$i_C = C\frac{dV_C}{dt} \tag{1.43}$$

Figure 1.17 illustrates a capacitor circuit and its corresponding current and voltage waveforms. The current through the capacitor is governed by the rate of

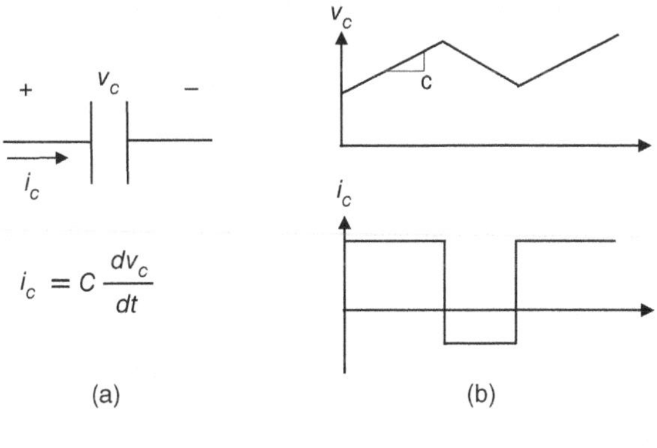

Figure 1.17 Capacitor current and voltage characteristics. (a) Capacitor circuit (b) Capacitor current and voltage waveforms.

change of voltage, $\frac{dV_C}{dt}$. In steady-state conditions, where the voltage is constant, the capacitor behaves as an open circuit, effectively blocking DC current while maintaining the applied voltage.

Key observations from the waveforms

- The voltage is a continuous function that gradually increases and decreases over time.
- The current, in contrast, is a discontinuous function, with polarity switching as the voltage changes.

Therefore, in a state-space representation, the appropriate state variable for a capacitor is the capacitor voltage, V_C.

Resistor: Energy-Consuming Component Unlike inductors and capacitors, which store energy in the form of current and voltage respectively—and are governed by differential relationships—a resistor is an electrical component that consumes energy. The voltage–current relationship for a resistor is described by Ohm's Law and shown in Figure 1.18:

$$V_R = RI_R \tag{1.44}$$

Key characteristics of a resistor

- A resistor does not store energy; instead, it dissipates energy as heat.
- The relationship between voltage (V_R) and current (I_R) is algebraic and proportional, defined by the resistance value R.

Because resistors dissipate energy and do not store it, neither voltage nor current across a resistor can serve as a state variable in the state-space representation. State variables must correspond to quantities that represent stored energy, which do not apply to resistive elements.

Active Components Active components are electronic elements that have the capability to control the flow of electric current using external energy sources. Among the most fundamental active components are diodes and semiconductor switching devices, such as BJTs, MOSFETs, and IGBTs. In controller design or simulation, these devices are typically modeled as switches that turn ON or OFF in response to control signals.

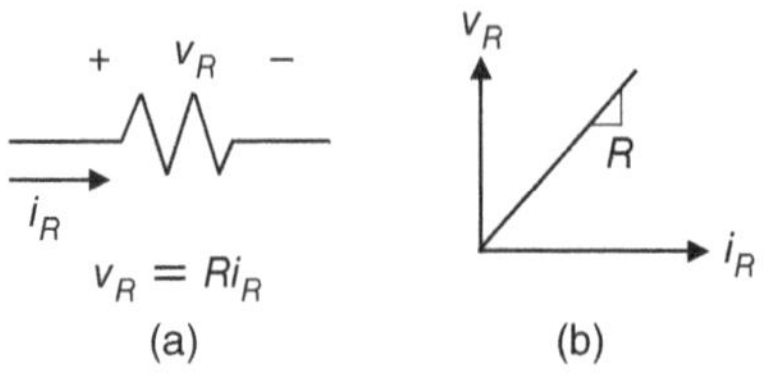

Figure 1.18 Resistor current and voltage characteristics. (a) Resistor circuit (b) Resistor current and voltage waveforms.

Diodes Switching components function by toggling between ON and OFF states in response to control input. One of the most fundamental switching devices is the diode, which acts as an automatic switch whose operation is governed by the voltage applied across its terminals (V_D).

The diode switches between its ON and OFF states depending on the polarity of the applied voltage V_D

- When $V_D > 0$: The diode is forward-biased, the switch closes, and current is allowed to flow.
- When $V_D < 0$: The diode is reverse-biased, the switch opens, and current flow is blocked.

Key characteristics of a diode as a switch

- ON State: In the ideal case, when the diode conducts (switch closed), the voltage drop across it is 0 V. In practice, however, a small forward voltage drop—typically around 0.7 V for silicon diodes or up to 1 V in power applications—is present, contributing to conduction losses.
- OFF State: When reverse-biased (switch open), the diode ideally blocks all current, functioning as an open circuit with negligible leakage current.

Figure 1.19 (b) and (c) show the diode's switching behavior, showing the ON (closed switch) and OFF (open switch) states depending on the polarity of V_D. The diode's inherent ability to switch automatically based on applied voltage makes it an essential component in rectification, freewheeling, and various switching applications.

Switching Devices with Control Terminals Switching devices such as BJT, FET, and IGBT are advanced components derived from diodes by adding control input terminals to enable explicit ON/OFF control. While each device has distinct characteristics, they all share the common feature of being controlled by a dedicated input terminal. The key characteristics of the switching devices are summarized in Table 1.2

Common Characteristics of the three devices are summarized as follows:

- All three devices are semiconductor switches that rely on a control terminal to toggle between conduction (ON) and blocking (OFF) states.

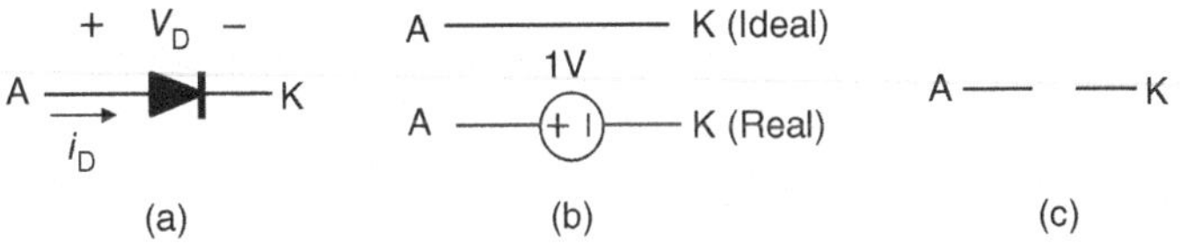

Figure 1.19 Diode and switching characteristics. (a) Diode circuit (b) Switch Closed ($V_D > 0$) (c) Switch Open ($V_D < 0$).

Table 1.2 Key features of BJT, FET, and IGBT.

Feature	BJT	FET	IGBT
Control Type	Current-controlled	Voltage-controlled	Voltage-controlled
Terminals	Base (B), Collector (C), Emitter (E)	Gate (G), Drain (D), Source (S)	Gate (G), Collector (C), Emitter (E)
Input Impedance	Low	High	High
Switching Speed	Moderate	Fast	Moderate to slow
On-State Voltage Drop	Low (typically 0.2–0.5 V)	Low R_{DS}(ON), increasing with current	Higher than MOSFET, ~2–3 V
Drive Power Requirement	Continuous base current needed	Very low gate current (only during switching)	Very low gate current (only during switching)
Power Handling Capability	Moderate	Low to moderate	High
Thermal Stability	Moderate	High	High
Typical Applications	Amplifiers, low-frequency switching	DC–DC converters, digital circuits, motor drivers	Motor drives, inverters, high-voltage power converters

- They differ in control type (current-controlled for BJT, voltage-controlled for FET, and IGBT), switching speed, power handling capabilities, and efficiency.
- The selection of an appropriate device depends on the specific application requirements, such as switching frequency, voltage/current ratings, and thermal performance.

Figure 1.20 shows the symbols and switching characteristics of various switching devices.

1.3.2 R–L–C Circuits Using State-Space Representation

a) **R–C Circuit Analysis**

An R–C circuit, consisting of a resistor and a capacitor connected in series, is illustrated in Figure 1.21. By selecting the capacitor voltage $v_c(t)$ as the state variable, the governing differential equation can be derived.

$$i_c = C\frac{dV_c}{dt} \rightarrow \frac{dV_c}{dt} = i_c = \frac{-V_c}{RC} \tag{1.45}$$

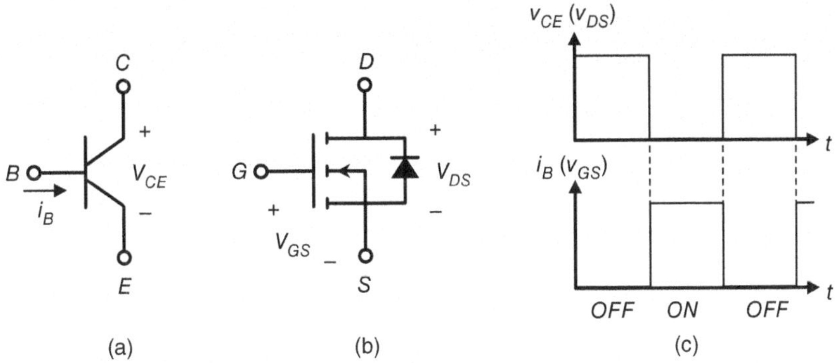

Figure 1.20 Switching devices and their characteristics. (a) Transistor Symbol (b) FET Symbol (c) Switching characteristics.

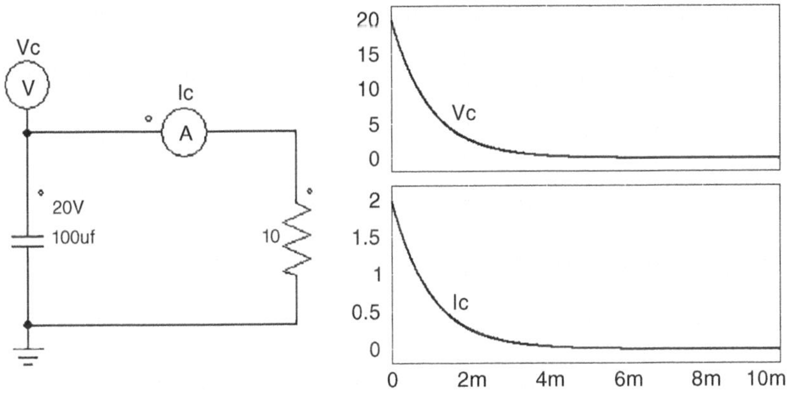

Figure 1.21 R–C circuit and its voltage and current waveforms.

Taking the Laplace transform of both sides yields

$$sV_c(s) - V_c(0) = -\frac{1}{RC}V_c(s) \tag{1.46}$$

Rearranging the equation gives

$$V_c(s) = \frac{V_c(0)}{\left(s + \frac{1}{RC}\right)} \tag{1.47}$$

Applying the inverse Laplace transform, the capacitor voltage in the time domain is obtained as

$$v_c(t) = V_c(0) \cdot e^{-\frac{t}{RC}} \tag{1.48}$$

This solution describes an exponentially decaying voltage, starting from the initial voltage $V_c(0)$ at $t = 0$. The corresponding capacitor current, which is

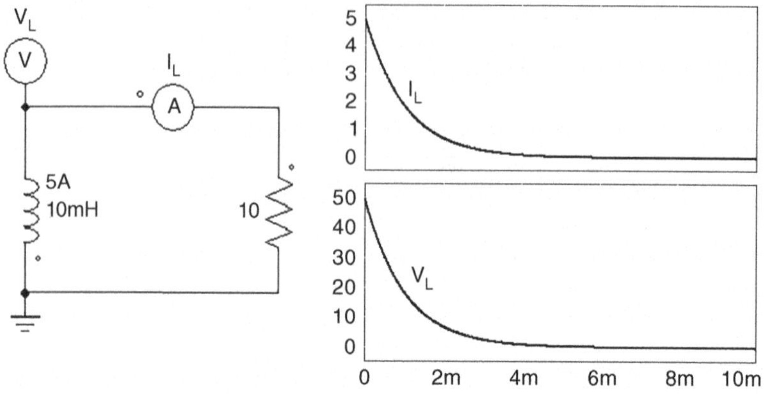

Figure 1.22 R–L circuit and waveforms.

proportional to the derivative of the voltage, also exhibits an exponentially decaying waveform.

b) **R–L Circuit Analysis**

The R–L circuit, composed of a resistor and an inductor in series, can be analyzed in a similar manner. This configuration is shown in Figure 1.22. By selecting the inductor current $i_L(t)$ as the state variable, and applying the inductor voltage–current relationship

$$L\frac{di_L}{dt} + R\,i_L = 0 \tag{1.49}$$

Laplace transform and solution gives the time-domain inductor current $i_L(t)$

$$i_L(t) = I_L(0) \cdot e^{-\frac{Rt}{L}} \tag{1.50}$$

This result describes an exponential decay of the inductor current from its initial value $I_L(0)$, determined by the circuit's time constant $\tau = \frac{L}{R}$.

c) **L–C Circuit Analysis**

An L–C circuit, consisting of an inductor and a capacitor connected in series, is shown in Figure 1.23. For analysis, the capacitor voltage $v_c(t)$ and the inductor current $i_L(t)$ are selected as state variables. From the capacitor and inductor voltage–current relationships

$$i_c = C\frac{dV_c}{dt} = -i_L = -L\frac{di_L}{dt} \rightarrow \frac{dV_c}{dt} = -\frac{1}{C}i_L$$

$$V_c = v_L \rightarrow \frac{di_L}{dt} = \frac{1}{L}V_c \tag{1.51}$$

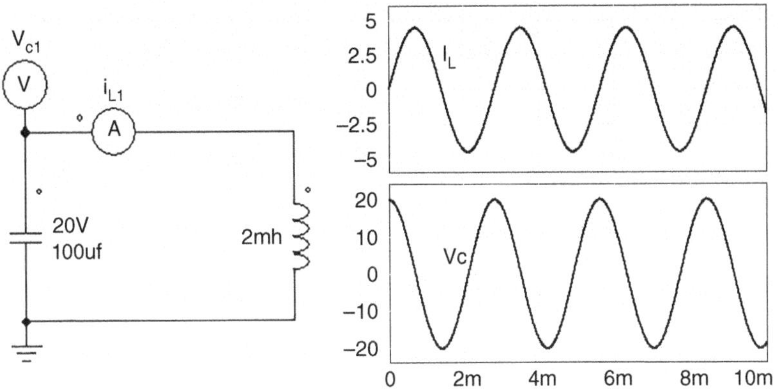

Figure 1.23 L–C circuit and its voltage/current waveforms.

These equations form the state-space representation

$$\begin{pmatrix} \dfrac{di_L}{dt} \\ \dfrac{dV_C}{dt} \end{pmatrix} = \begin{pmatrix} 0 & \dfrac{1}{L} \\ -\dfrac{1}{C} & 0 \end{pmatrix} \begin{pmatrix} i_L \\ V_C \end{pmatrix} \tag{1.52}$$

The Laplace transform is applied to derive

$$sX(s) - X(0) = AX(s) \rightarrow X(s) = (sI - A)^{-1}X(0) \tag{1.53}$$

The inverse of $(sI - A)$ is

$$(sI - A)^{-1} = \begin{pmatrix} s & -1/L \\ 1/C & s \end{pmatrix}^{-1} = \frac{1}{s^2 + 1/LC} \begin{pmatrix} s & 1/L \\ -1/C & s \end{pmatrix}$$

$$X(s) = \begin{pmatrix} I_L(s) \\ V_C(s) \end{pmatrix} = \frac{1}{s^2 + \frac{1}{LC}} \begin{pmatrix} s & \dfrac{1}{L} \\ -\dfrac{1}{C} & s \end{pmatrix} \begin{pmatrix} I_L(0) \\ V_C(0) \end{pmatrix} \tag{1.54}$$

Using initial conditions $I_L(0) = 0$ and $V_C(0) = V_C(0)$

$$I_L(s) = \frac{V_C(0)/L}{s^2 + \frac{1}{LC}},\ V_C(s) = \frac{V_C(0) \cdot s}{s^2 + \frac{1}{LC}} \tag{1.55}$$

Applying the inverse Laplace transform, the time-domain solutions are

$$i_L(t) = V_C(0) \cdot \sqrt{\frac{C}{L}} \cdot \sin \omega t$$

$$v_c(t) = V_C(0) \cdot \cos \omega t,\ \omega = \frac{1}{\sqrt{LC}} \tag{1.56}$$

The inverse Laplace transform formulas for trigonometric functions are as follows:

$$L^{-1}\left(\frac{s}{s^2+\omega^2}\right) = \cos\omega t, \quad L^{-1}\left(\frac{\omega}{s^2+\omega^2}\right) = \sin\omega t \tag{1.57}$$

From the waveform, the oscillation period is observed to be $T = 2.83[ms]$
Substituting $L = 2\,[mH]$ and $C = 100[uF]$ into the resonance frequency formula

$$\omega = 2\pi f = \frac{1}{\sqrt{LC}}, \quad T = 2\pi\sqrt{LC} = 2.83[ms]$$

The computed result matches the observed period precisely, confirming the theoretical model.

d) **R–L–C Circuit Analysis**

In an R–L–C circuit, the presence of a resistor introduces damping, which alters the natural response of the system. This configuration is shown in Figure 1.24. While the inductor and capacitor alone produce sustained oscillations at the natural resonant frequency, the resistor causes the oscillations to decay exponentially over time.

The system's resonant frequency is modified by the damping factor introduced by the resistor, resulting in an under-damped response characterized by oscillations of decreasing amplitude. The governing second-order differential equation reflects this damped harmonic behavior, and the waveform solutions typically exhibit:

Applying KVL to the R–L–C loop gives

$$\begin{pmatrix} \frac{di_L}{dt} \\ \frac{dV_C}{dt} \end{pmatrix} = \begin{pmatrix} -\frac{R}{L} & \frac{1}{L} \\ -\frac{1}{C} & 0 \end{pmatrix} \begin{pmatrix} i_L \\ V_C \end{pmatrix} \tag{1.58}$$

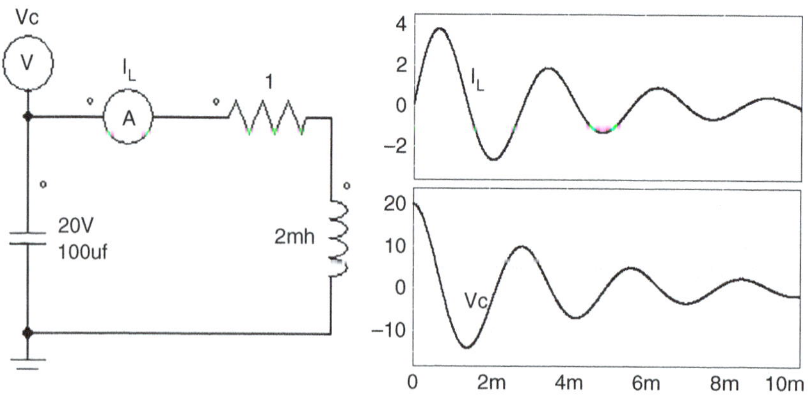

Figure 1.24 R–L–C circuit and corresponding voltage and current waveforms.

Applying the Laplace transform to the state equations and solving yields

$$I_L(s) = \frac{V_C(0)/L}{s^2 + \left(\frac{R}{L}\right)s + \left(\frac{1}{LC}\right)}$$

$$V_C(s) = \frac{V_C(0)\cdot\left(s + \frac{R}{L}\right)}{s^2 + \left(\frac{R}{L}\right)s + \left(\frac{1}{LC}\right)} \tag{1.59}$$

By inverse Laplace transform, the time-domain inductor current and capacitor voltage are

$$i_L(t) = \frac{V_C(0)}{L\omega_d}\cdot e^{-\zeta\omega_n t}\cdot \sin\omega_d t$$

$$v_c(t) = V_C(0)\cdot A\cdot e^{-\zeta\omega_n t}\cdot \sin(\omega_d t - \varphi) \tag{1.60}$$

where the parameters are defined as

$\omega_n = \frac{1}{\sqrt{LC}}$: undamped natural frequency

$\zeta = \frac{R}{2}\sqrt{\frac{C}{L}} = \frac{R}{2L\omega_n}$: damping ratio

$\omega_d = \omega_n\sqrt{1-\zeta^2}$: damped natural frequency

$A = \sqrt{1 + \left(\frac{R}{2L\omega_d}\right)^2}$: amplitude scaling factor

$\varphi = \tan^{-1}\left(\frac{2L\omega_d}{R}\right)$: phase shift

Simulation results confirm this theoretical behavior, displaying damped oscillations with gradually diminishing amplitude. These waveforms show excellent agreement with analytical predictions, validating the effect of damping on the circuit's resonant characteristics and confirming the accuracy of the derived model.

1.4 Linear and PWM Methods

AC Power Conversion Power conversion is the process in which the input voltage and current differ from the output voltage and current, while the overall power remains essentially constant throughout the system. In other words, the form or magnitude of the electrical variables (voltage and current) is altered, but the input power (P_1) and output power (P_2) are effectively equal, assuming negligible losses. The fundamental power relationship can be expressed as:

$$P_1 = V_1\cdot I_1 = V_2\cdot I_2 = P_2 \tag{1.61}$$

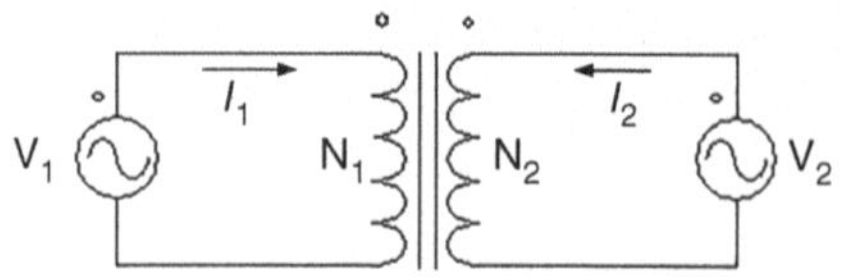

Figure 1.25 AC transformer power relationships.

From this relationship, the proportionality between voltage and current is obtained as:

$$\frac{V_2}{V_1} = \frac{I_1}{I_2}$$

A classic example of power conversion is the AC transformer, illustrated in Figure 1.25. The transformer adjusts voltage and current levels according to the turns ratio between the primary and secondary windings, expressed as

$$\frac{V_2}{V_1} = \frac{I_1}{I_2} = \frac{N_2}{N_1} = n$$

Although a transformer effectively changes voltage and current levels through its winding ratio, it has certain limitations:

- The turns ratio is fixed, making it unsuitable for dynamic or adjustable voltage regulation.
- Transformers operate only with AC signals and are not applicable in DC systems.

DC Power Conversion Unlike AC systems, DC power conversion cannot employ traditional transformers, as transformers depend on time-varying magnetic fields and therefore function only with alternating current (AC). In DC systems, the most basic approach to adjusting the output voltage is the use of a voltage divider circuit, as illustrated in Figure 1.26.

The output voltage is given by:

$$V_2 = \frac{R_2}{R_1 + R_2} V_1 \tag{1.62}$$

By varying R_1 while keeping R_2 constant, the output voltage V_2 can be adjusted to a desired value.

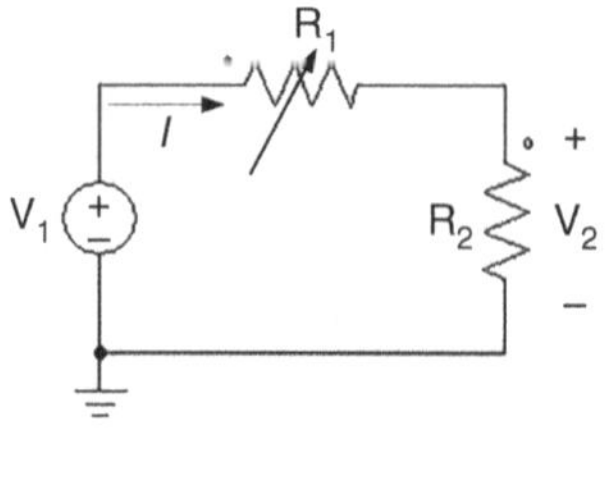

Figure 1.26 Voltage divider.

Table 1.3 Efficiency analysis.

V_1 [V]	R_2 [Ω]	R_1 [Ω]	V_2 [V]	P_{R2} [W]	P_{R1} [W]	Efficiency
100	10	10	50	250	250	50%
100	10	30	25	62.5	187.5	25%
100	10	5	67	444	218	67%

As shown in Table 1.3, adjusting R_1 allows control over the output voltage. However, a large portion of power is dissipated as heat in the resistive elements, resulting in low efficiency. The power delivered to the load is significantly reduced, particularly at lower output voltages.

This method of voltage control using resistive components is known as linear regulation. The term "linear" refers to the use of components that operate in the linear (ohmic) region of their characteristics. Common devices used in linear regulators include:

- Bipolar Junction Transistors (BJT): Conducts between collector and emitter when a base current is applied.
- FET (Field Effect Transistor): Conducts between drain and source when a voltage is applied between gate and source (V_{GS}).

While linear regulators are simple and provide smooth voltage regulation, they suffer from poor efficiency, especially under high current or large voltage-drop conditions. This inefficiency motivates the use of more advanced methods, such as switching regulators, which will be discussed in the following sections.

1.4.1 Linear Regulation Using BJTs and FETs

In power electronics, BJTs and FETs are most commonly used as ON/OFF switches. For a typical FET, the switching behavior is controlled by the gate-to-source voltage V_{GS}. When V_{GS} exceeds a certain threshold voltage (typically above 8 V for many power MOSFETs), the device turns ON, and current flows freely between the drain and source. Conversely, when $V_{GS} < 0$, the FET is OFF, effectively blocking current flow.

However, when a gate voltage in the range of $2 \sim 5$ V is applied—below the full turn-on threshold but above the cutoff region—the FET operates in its linear (or ohmic) region. In this mode, the FET behaves like a voltage-controlled resistor, where the drain-source resistance R_{ds} is inversely proportional to the gate voltage. This property is exploited in linear regulators, where the FET is intentionally biased to act as a variable resistor rather than a binary switch.

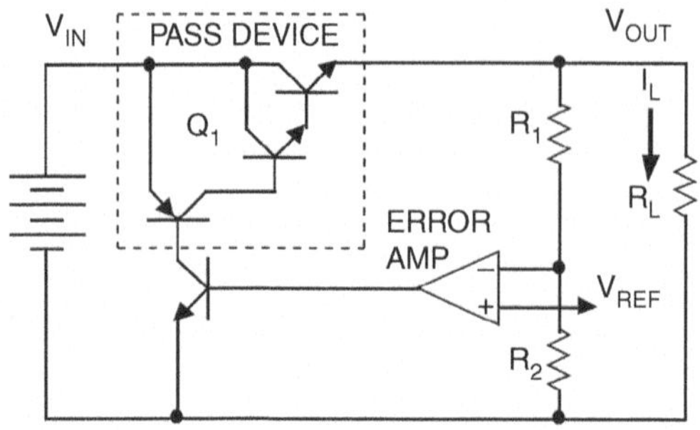

Figure 1.27 Circuit configuration of a typical linear regulator (NPN).

Figure 1.27 illustrates the circuit configuration of a typical linear regulator utilizing an NPN BJT as the pass device. In this structure, the input power is regulated by a pass element operating in its linear region, acting as a variable resistor to control the output voltage. The power is divided between the pass element and the load, and an operational amplifier (Op-Amp) functions as an error amplifier, comparing the output voltage V_o with a reference voltage V_{ref}.

The error signal is amplified and used to drive the base current of the pass transistor via a smaller driver transistor. The pass transistor, which regulates the power flow, is biased to operate in the linear region, allowing for continuous adjustment of output voltage. The pass device in a linear regulator can take various forms—such as NPN, PNP, NMOS, or PMOS transistors. The details are not presented here, as many reference circuits are available in open-source resources.

Figure 1.28 shows a simulation circuit of a linear regulator implemented using BJTs. In this circuit:

- The pass transistor (power BJT) typically has a low current gain β, usually in the range of 10 to 50, due to the high current it handles.
- The driver transistor, responsible for delivering the base current to the pass transistor, typically has a higher β (around 100) to minimize control power requirements.

The Op-Amp is used in place of a typical switch controller to implement a simple voltage feedback control loop. The reference voltage V_{ref} is compared with a scaled version of the output voltage. In this case, the output is scaled by a factor of 0.5, so the regulated output voltage is approximately twice the reference voltage (e.g., for $V_{ref} = 10$ V, the output will be regulated at $V_o = 20$ V).

Despite their low efficiency, linear regulators are still used in certain applications where noise sensitivity, simplicity, or precise analog control is required. One key application is in electronic loads, where power must be dissipated intentionally to simulate resistive behavior. In such cases, linear regulators are often indispensable, as switching regulators cannot operate continuously in a resistive mode.

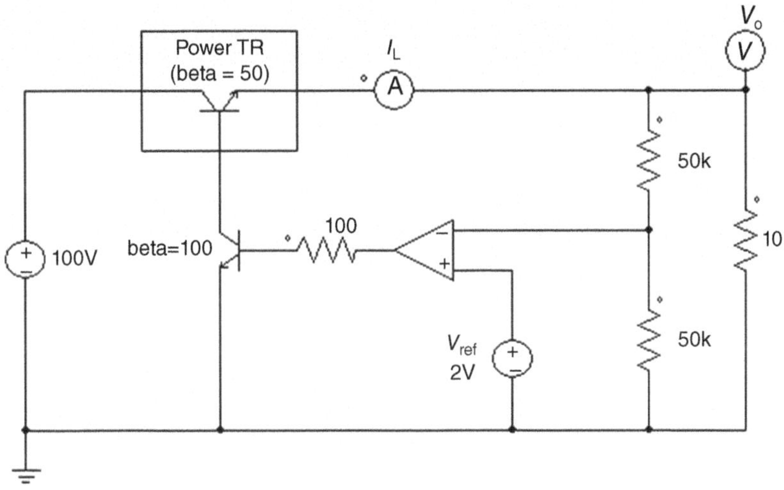

Figure 1.28 Linear regulator circuit using BJT.

For high-current applications, a single pass transistor may not suffice. Typically, multiple power transistors are connected in parallel—often more than 10—to distribute the current and increase the regulator's power-handling capability.

1.4.2 Switching Mode Power Conversion

To overcome the efficiency limitations of linear regulators, switching-mode power conversion offers an effective alternative for DC power regulation. Unlike linear methods, which rely on a pass device operating in the linear region (and thus dissipating power as heat), switching converters operate the pass device—typically a BJT or FET—as a high-speed ON/OFF switch in its saturation region, enabling near-lossless energy transfer.

Assume that the voltage across the switch is denoted by V_{SW}, and the current flowing through it by I_{SW}. The switch is controlled by a digital control input, as illustrated in Figure 1.29. When the control signal is logic high (1), the switch turns ON; when the control signal is logic low (0), the switch turns OFF. The switching behavior is summarized as follows:

- ON State: $V_{SW} = 0\ V$, $I_{SW} = I_{\text{Load}}\ A$, and the power dissipated in the switch is $P_{SW} = 0\ W$. The switch is closed, conducting full load current with negligible voltage drop, thus dissipating minimal power.
- OFF State: $V_{SW} = V_{rev}\ V$, $I_{SW} = 0\ A$, and the power dissipated in the switch is $P_{SW} = 0\ W$. The switch is open, blocking current while sustaining a reverse voltage V_{rev}, again with minimal power dissipation.

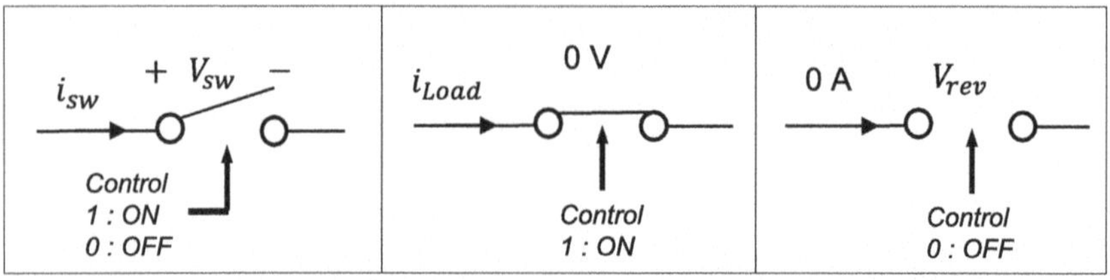

Figure 1.29 Switch equivalent circuit. (a) Basic switch circuit (b) ON state (c) OFF state.

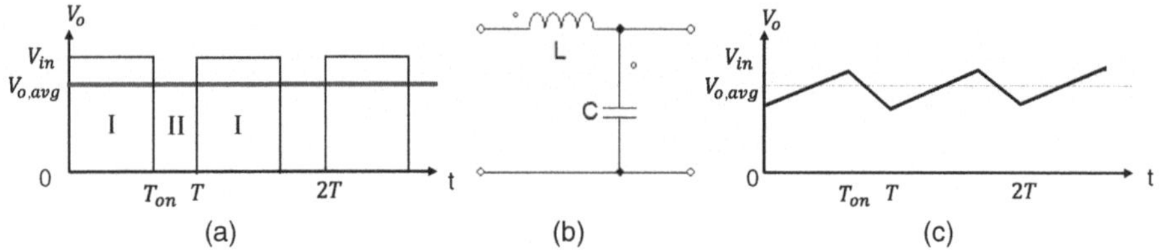

Figure 1.30 Output smoothing using an LC filter. (a) PWM output waveform (b) LC filter circuit (c) Filtered DC output with ripple.

This idealized operation leads to efficient power conversion, as shown in Figure 1.29.

During switching operation, the output voltage rapidly alternates between V_{dc} and 0 V, producing a pulsed waveform, as shown in Figure 1.30(a). This pulse-width modulated (PWM) signal represents an average output voltage that is proportional to the duty ratio of the switching command. To obtain a smooth DC voltage suitable for practical applications, the pulsed waveform is passed through an LC low-pass filter, as illustrated in Figure 1.30(b). The filtered output, shown in Figure 1.30(c), is a DC voltage with superimposed ripple. The ripple magnitude depends on the filter parameters and the load conditions.

By operating the pass device strictly in either the saturation (ON) or cutoff (OFF) region, switching regulators achieve significantly higher efficiency than linear regulators. These regulators form the foundation of modern DC–DC converters, including buck, boost, and buck–boost topologies, which are discussed in the following sections.

Pulse Width Modulation (PWM) The average output voltage $V_{o,avg}$ of the pulsed waveform generated by a switching regulator is given by

$$V_{o,avg} = \frac{T_{on}}{T} \cdot V_{in} + \frac{(T - T_{on})}{T} \cdot 0 = \frac{T_{on}}{T} \cdot V_{in} = d \cdot V_{in} \tag{1.63}$$

where T is the switching period, and the duty ratio d is defined as $d = \frac{T_{on}}{T}$.

This expression demonstrates that the output voltage can be regulated by adjusting the duty ratio, enabling efficient and nearly lossless power conversion using a pulsed voltage waveform. By varying the pulse width, the average output voltage is effectively modulated.

Unlike linear regulators, which continuously vary the resistance of the pass device through an error amplifier, switching regulators employ a saturated error amplifier or comparator to drive the pass device fully ON or OFF. The control action is typically implemented by comparing the output voltage V_o with a reference voltage V_{ref}

- If $V_{ref} > V_o$, the comparator output is HIGH → switch turns ON
- If $V_{ref} > V_o$, the comparator output is LOW → switch turns OFF

To ensure stable switching operation, the system periodically re-enables the switch at a fixed interval T. This control strategy is known as the Constant-ON-Time modulation technique. Although a Constant-OFF-Time method also exists, Constant-ON-Time modulation is more widely used in practical applications.

The simplest implementation of this control logic is achieved using a comparator.

As illustrated in Figure 1.31, PWM waveforms can be generated by applying the error signal $V_{err} = V_{ref} - V_o$ to the non-inverting input (+) of a comparator, while a triangular carrier signal $V_{carrier}$ is applied to the inverting input (−). The comparator output produces a PWM signal whose pulse width increases as V_{err} increases. As the output voltage V_o rises, the error signal V_{err} decreases, resulting in a corresponding reduction in pulse width. This mechanism provides automatic voltage regulation through negative feedback.

Under steady-state conditions, the duty ratio converges to a constant value such that $V_o = V_{ref}$ and $V_{err} = 0$, yielding a fixed duty cycle D.

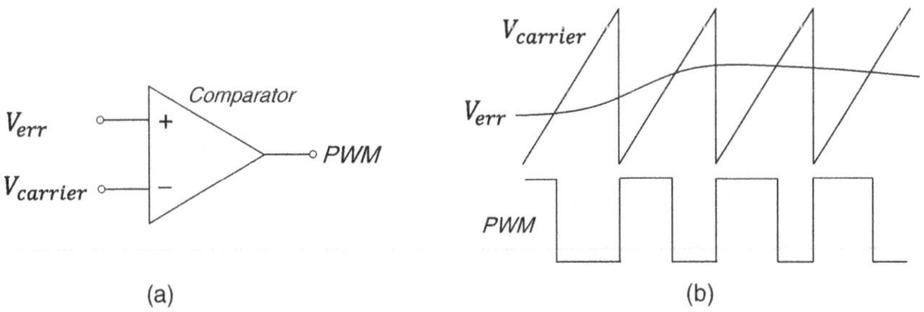

Figure 1.31 Pulse width modulation (PWM). (a) Comparator input signals (b) PWM output waveforms.

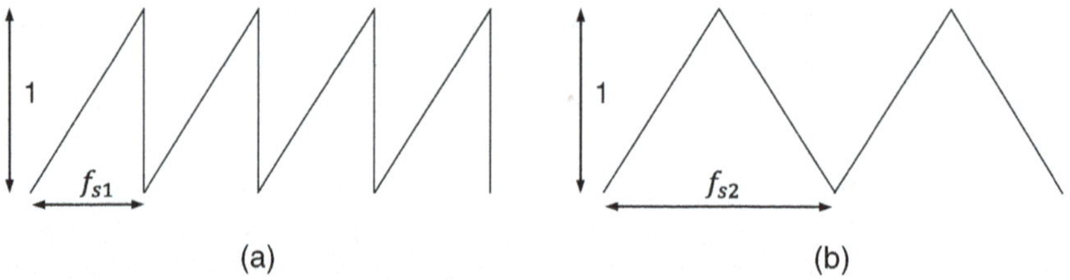

Figure 1.32 Triangular carrier waveforms. (a) Unsymmetrical for converter (b) Symmetrical for inverter.

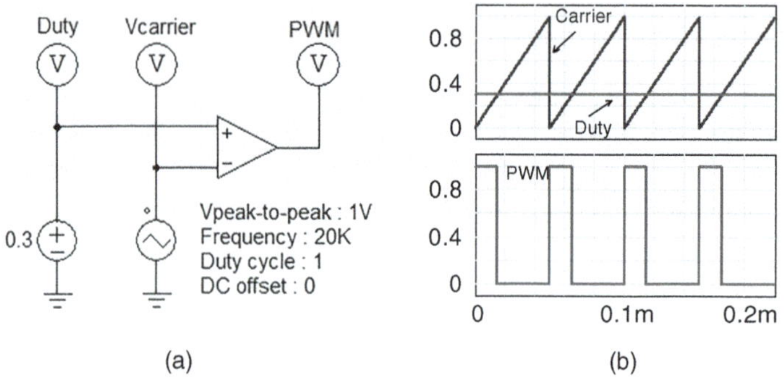

Figure 1.33 PWM circuit and waveform for converter. (a) Schematic diagram (b) Output waveform.

The design of the modulation (carrier) signal depends on the application

- Unipolar modulation (Figure 1.32a) is used for DC–DC converters, where voltages and currents are unidirectional.
- Bipolar modulation (Figure 1.32b) is used for DC–AC inverters, where voltage and current alternate in polarity.

Figure 1.33 illustrates an example of a PWM generation circuit for a converter implemented in *PSIM*. The comparator block can be accessed from the following menu path: `Elements` → `Control` → `Comparator`. The triangular waveform generator used as the carrier signal can be found under: `Elements` → `Source` → `Triangular`.

The output of the comparator generates a PWM waveform whose pulse width is determined by the error voltage. This PWM signal is used to drive switching devices such as MOSFETs or IGBTs in the converter.

A similar PWM configuration for inverter control is shown in Figure 1.34. In this case, symmetrical bipolar modulation is employed to generate AC output waveforms.

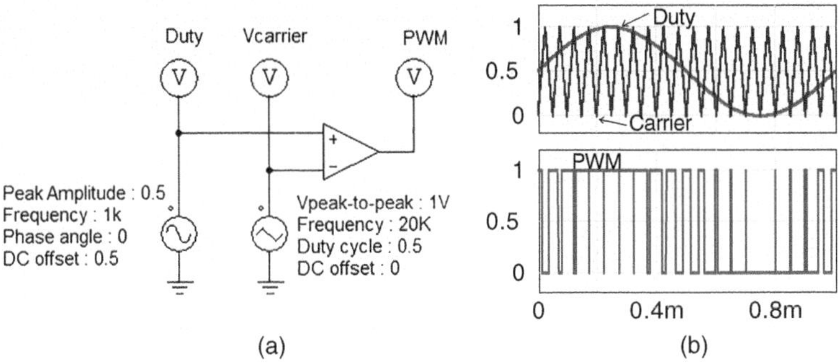

Figure 1.34 PWM Circuit and Corresponding Waveform for Inverter.

Part I

DC Power Conversion

- DC/DC Converter
  2. Buck Converter
  3. Boost Converter
  4. Bi-Directional Converter
  5. Three-Level Boost Converter
- Solar Converter
  6. Solar System
  7. Solar Boost Converter
  8. Solar Buck Converter
  9. Solar Three-Level Boost Converter
- Motor
  10. DC Motor

Part I-A

DC/DC Converter

The types and circuits of DC/DC converters covered in this part are as follows.

2. BUCK CONVERTER
3. BOOST CONVERTER
4. BI-DIRECTIONAL CONVERTER
5. THREE-LEVEL BOOST CONVERTER

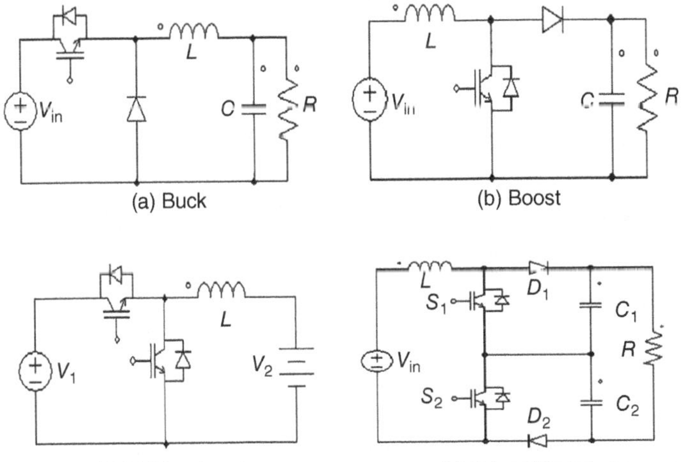

(a) Buck

(b) Boost

(c) Bi-Directional

(d) 3-Level boost

2

Buck Converter

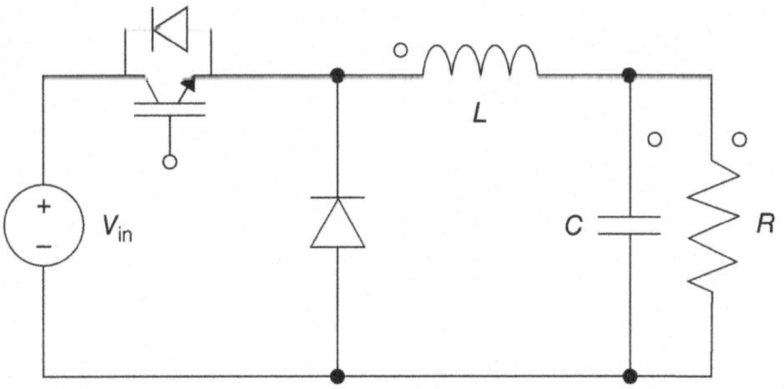

2.1 State-Averaged Equations and Transfer Functions

A buck converter steps down the input voltage by switching a transistor ON and OFF at high frequency. During the ON state, energy is stored in the inductor while supplying the load. During the OFF state, the inductor releases energy to the load through a freewheeling diode. The output voltage is regulated by adjusting the duty ratio of the switch. Figure 2.1 illustrates the circuit diagram of a buck converter.

The equivalent circuits of the buck converter during the switching states – ON and OFF – are shown in Figure 2.2.

The state-space representations for each switching interval are summarized in Table 2.1.

Let T_S denote the switching period, with ON-time T_{ON} and duty ratio $d = T_{ON}/T_S$. The corresponding OFF-time is $T_{OFF} = T_S(1 - d)$. The state-averaged

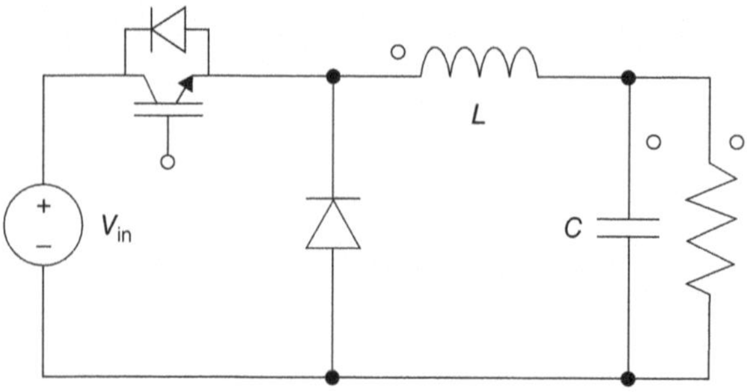

Figure 2.1 Buck converter circuit.

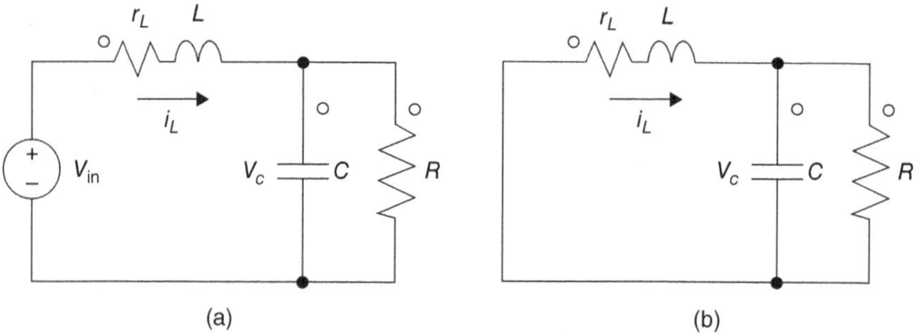

Figure 2.2 Equivalent circuit of the buck converter. (a) ON state and (b) OFF state.

Table 2.1 State-space equations for ON, OFF period.

ON-state equation	OFF-state equation
$V_{\text{in}} = r_L i_L + L\frac{di_L}{dt} + V_C$	$0 = r_L i_L + L\frac{di_L}{dt} + V_C$
$i_L = C\frac{dV_C}{dt} + \frac{V_C}{R}$	$i_L = C\frac{dV_C}{dt} + \frac{V_C}{R}$
$\begin{pmatrix} \frac{di_L}{dt} \\ \frac{dV_C}{dt} \end{pmatrix} = \begin{pmatrix} -\frac{r_L}{L} & -\frac{1}{L} \\ \frac{1}{C} & -\frac{1}{RC} \end{pmatrix} \begin{pmatrix} i_L \\ V_C \end{pmatrix} + \begin{pmatrix} \frac{1}{L} \\ 0 \end{pmatrix} V_{\text{in}}$	$\begin{pmatrix} \frac{di_L}{dt} \\ \frac{dV_C}{dt} \end{pmatrix} = \begin{pmatrix} -\frac{r_L}{L} & -\frac{1}{L} \\ \frac{1}{C} & -\frac{1}{RC} \end{pmatrix} \begin{pmatrix} i_L \\ V_C \end{pmatrix} + \begin{pmatrix} 0 \\ 0 \end{pmatrix} \cdot V_{\text{in}}$
$= A_1 \begin{pmatrix} i_L \\ V_C \end{pmatrix} + B_1 \cdot V_{\text{in}}$	$= A_2 \begin{pmatrix} i_L \\ V_c \end{pmatrix} + B_2 \cdot V_{\text{in}}$ (2.1)

matrices are computed using time-weighted averaging:

$$A = A_{\text{avg}} = \frac{(A_1 \cdot T_{\text{ON}} + A_2 \cdot T_{\text{OFF}})}{T_S} = A_1 \cdot d + A_2 \cdot (1 - d)$$

$$= \begin{pmatrix} -\frac{r_L}{L} & -\frac{1}{L} \\ C & -\frac{1}{RC} \end{pmatrix} \cdot d + \begin{pmatrix} -\frac{r_L}{L} & -\frac{1}{L} \\ C & -\frac{1}{RC} \end{pmatrix} \cdot (1 - d) = \begin{pmatrix} -\frac{r_L}{L} & -\frac{1}{L} \\ C & -\frac{1}{RC} \end{pmatrix}$$

$$B = B_{\text{avg}} = \frac{(B_1 \cdot T_{\text{ON}} + B_2 \cdot T_{\text{OFF}})}{T_S} = B_1 \cdot d + B_2 \cdot (1 - d)$$

$$= \begin{pmatrix} \frac{1}{L} \\ 0 \end{pmatrix} d + \begin{pmatrix} 0 \\ 0 \end{pmatrix} \cdot (1 - d) = \begin{pmatrix} \frac{1}{L} \\ 0 \end{pmatrix} \cdot d \tag{2.2}$$

The state-averaged equations for the buck converter are given as follows.

$$\begin{pmatrix} \frac{di_L}{dt} \\ \frac{dV_C}{dt} \end{pmatrix} = \begin{pmatrix} -\frac{r_L}{L} & -\frac{1}{L} \\ \frac{1}{C} & -\frac{1}{RC} \end{pmatrix} \begin{pmatrix} i_L \\ V_C \end{pmatrix} + \begin{pmatrix} \frac{V_{\text{in}}}{L} \\ 0 \end{pmatrix} d \tag{2.3}$$

In matrix form, the state-averaged equations can be expressed as:

$$\dot{X} = AX + Bu$$

$$X = \begin{pmatrix} \frac{di_L}{dt} \\ \frac{dV_C}{dt} \end{pmatrix}, \quad \text{A} = \begin{pmatrix} -\frac{r_L}{L} & -\frac{1}{L} \\ \frac{1}{C} & -\frac{1}{RC} \end{pmatrix}, \quad \text{B} = \begin{pmatrix} \frac{V_{\text{in}}}{L} \\ 0 \end{pmatrix}, \quad \text{u} = d \tag{2.4}$$

Time-domain analysis can illustrate waveforms over time but cannot reveal the frequency-domain response characteristics. Therefore, frequency-domain analysis is performed using the transfer function. By precomputing the inverse of the system matrix, the following can be achieved:

$$(sI - A)^{-1} = \begin{pmatrix} s + \frac{r_L}{L} & \frac{1}{L} \\ -\frac{1}{C} & s + {}^1\!/_{RC} \end{pmatrix}^{-1} \cong \frac{\begin{pmatrix} s + {}^1\!/_{RC} & -\frac{1}{L} \\ \frac{1}{C} & s + \frac{r_L}{L} \end{pmatrix}}{s^2 + \left(\frac{1}{RC} + \frac{r_L}{L}\right) s + \frac{1}{LC}} \tag{2.5}$$

- When the output variable is the inductor current i_L, the output matrix is defined as $C = [1\ 0]$. The corresponding transfer function is derived as follows:

$$\frac{I_L(s)}{d(s)} = G_1(s) = \frac{\frac{V_{\text{in}}}{L}\left(s + \frac{1}{RC}\right)}{s^2 + \left(\frac{1}{RC} + \frac{r_L}{L}\right) s + \frac{1}{LC}} \tag{2.6}$$

- When the output variable is the capacitor voltage V_C, the output matrix is defined as $C = [0\ 1]$. The corresponding transfer function is derived as follows:

$$\frac{V_C(s)}{d(s)} = G_2(s) = \frac{\frac{V_{in}}{LC}}{s^2 + \left(\frac{1}{RC} + \frac{r_L}{L}\right)s + \frac{1}{LC}} \tag{2.7}$$

- The transfer function between the inductor current and the capacitor voltage is derived from the previously established state-space equations. It is expressed as:

$$\frac{V_C(s)}{I_L(s)} = G_3(s) = \frac{\frac{1}{C}}{s + \left(\frac{1}{RC}\right)} \tag{2.8}$$

The transfer function can also be obtained using *MATLAB*'s built-in function `ss2tf`, which converts a given state-space representation into its equivalent transfer function form. The corresponding *MATLAB* script developed for this purpose is presented in Table 2.2.

From the transfer functions, the controller configuration is shown in Figure 2.3.

Control refers to the technique of feeding back the output to match a given reference value. Based on this principle, two fundamental control architectures are

Table 2.2 MATLAB m-file for deriving transfer function.

```
%% Power Circuit Parameters %%
s=tf('s');
Vin  = 100;   % Input voltage [V]
rL   = 0.01;  % 10 mohm
L  = 2e-3;    % L=2mh  [H]
C  = 500e-6;  % C=500uf  [F]
R  = 10;      % Load resistance [Ohm]

A  =  [-rL/L,     -1/L;
        1/C,      -1/(R*C)];
B  =  [Vin/L;
       0];
C1  =  [1,0]; C2  =  [0,1]; Dx  =  0;

[num1,den1]  =  ss2tf(A,B,C1,Dx);
[num2,den2]  =  ss2tf(A,B,C2,Dx);
G1  =  tf(num1, den1);
G2  =  tf(num2, den2);
G3  =  tf(num2,num1);
```

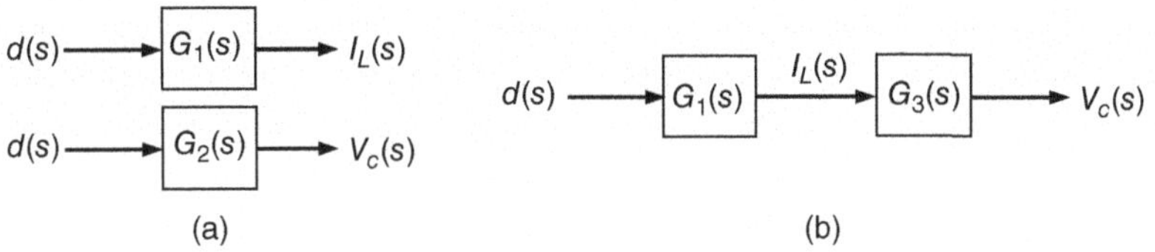

Figure 2.3 Controller configurations based on transfer functions. (a) Single-loop configuration and (b) Double-loop configuration.

commonly employed in power conversion systems: single-loop and double-loop configurations.

In single-loop control, the error is defined as the difference between the reference and the measured output, and negative feedback is used to minimize this error. If the control target is the output voltage, the method is called voltage mode control. If the target is the inductor current, it is referred to as current mode control. These control strategies are illustrated in Figure 2.4.

The double-loop control architecture is depicted in Figure 2.5. In this configuration, the outer loop regulates the output voltage through a voltage controller, while the inner loop regulates the inductor current via a current controller. The output of the voltage controller acts as the reference for the current controller, forming a cascaded control structure. The outer voltage loop is designed with slower dynamics, whereas the inner current loop responds more rapidly.

To minimize interference between the two loops, it is essential to ensure a significant difference in their response speeds. While exact numerical guidelines are difficult to define, a commonly used heuristic is to set the bandwidth of

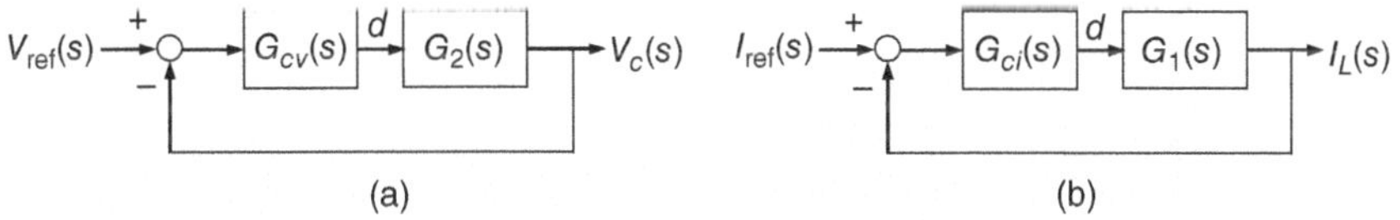

Figure 2.4 Single-loop control architecture. (a) Voltage mode control ($V_{ref} = V_C$) and (b) Current mode control ($I_{ref} = I_L$)

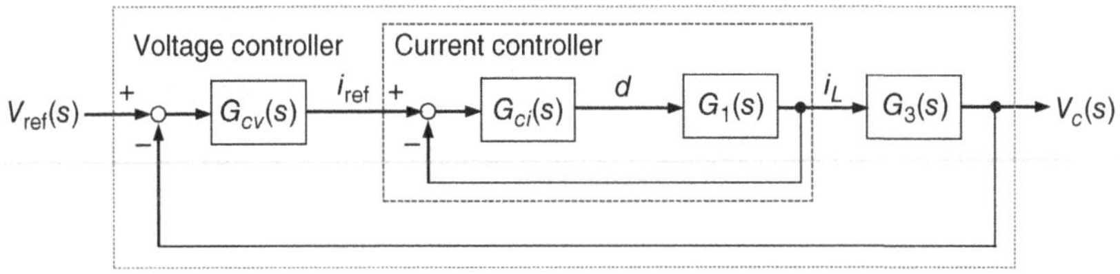

Figure 2.5 Double-loop control architecture.

the current loop approximately 10 times higher than that of the voltage loop. This separation allows the loops to function independently without significant interaction. At steady state, the control objectives must satisfy the following conditions:

$$(V_{\text{ref}} = V_C) \quad \text{and} \quad (I_{\text{ref}} = I_L)$$

Determining the appropriate controller gains is a non-trivial task. Gains are often selected through iterative tuning, though such approaches do not guarantee optimal performance. In the case of *PI* controllers applied in a double-loop structure, four gain parameters (two for each loop) must be determined. This process can be complex and inefficient. Therefore, a systematic and general controller design methodology is necessary to ensure stability, performance, and ease of implementation.

This book adopts a frequency-domain stability analysis approach based on Bode plots, using phase margin and cutoff frequency as primary design parameters. The phase margin is associated with time-domain overshoot, while the cutoff frequency corresponds to the system's rise time. The frequency-domain method is selected due to its compatibility with MATLAB's Auto Tuning library, which streamlines the controller design process. Specifically, MATLAB's SISOTOOL enables automatic computation of controller gains that satisfy the specified design criteria.

How to Use SISOTOOL The usage of SISOTOOL is well documented in MATLAB's official help resources and widely covered in online tutorials, including those on *YouTube*. The initial step involves defining the control system architecture appropriate for your application.

Steps:

- Launch SISOTOOL in MATLAB.
- Navigate to Architecture Editing
  - Select Control Architecture from the menu.
  - Choose one of the six available architectures.

Each architecture corresponds to a specific control system configuration, enabling tailored design according to system requirements. For comprehensive guidance, users are encouraged to consult MATLAB's documentation or video tutorials.

In SISOTOOL, the single-loop control structure corresponds to architecture 1, while the double-loop structure is represented by architecture 6, as illustrated in Figure 2.6. These can be specified using the following commands:

```
sisoinit(1); % Single-loop structure
sisoinit(6); % Double-loop structure
```

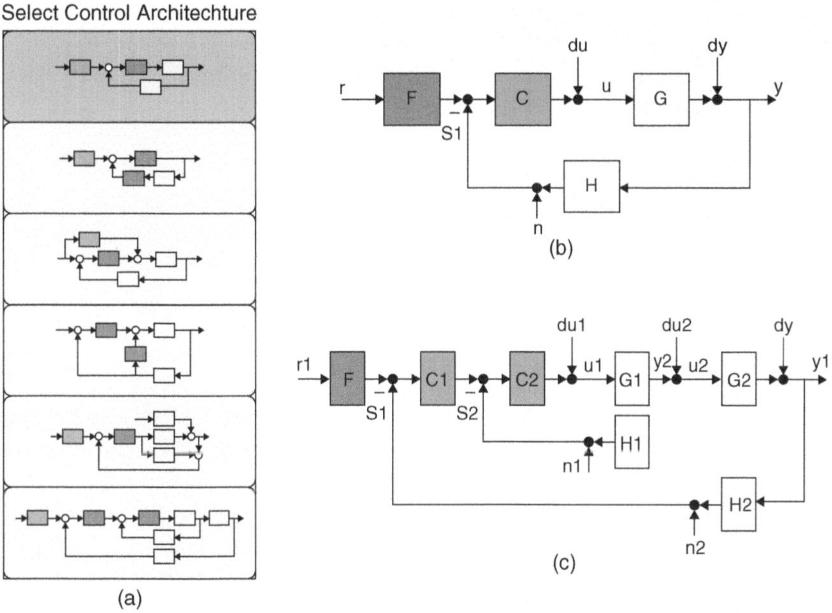

Figure 2.6 Control architectures for use with `sisoinit`. (a) Control architecture, (b) Single-loop configuration (1st architecture) and (c) Double-loop configuration (6th architecture).

These commands initialize the SISOTOOL environment with the desired control architecture, providing a visual and interactive platform for controller tuning and frequency-domain analysis.

The next step is to assign the appropriate transfer functions to each block. For example, in the case of a single-loop controller, the blocks are designated as follows:

1) **F**: Forward Gain
2) **C**: Controller Transfer Function (e.g., $G_c(s)$)
3) **G**: Plant Transfer Function (e.g., $G_p(s)$)
4) **H**: Feedback gain (default value of 1, requiring no additional configuration)

These definitions ensure that each component of the control loop is correctly specified for analysis and design within SISOTOOL.

Example: Automatic Controller Gain Calculation Using SISOTOOL
This example demonstrates how to automatically compute controller gains (auto-tuning) using MATLAB's SISOTOOL. The plant and *PI* (Proportional-Integral)

Table 2.3 MATLAB m-file for SISOTOOL example.

```
s=tf('s');
num=[4167 138900];
den=[0.06 1 3750];
Gp=tf(num,den);
Kp=1; Ki=1; % Initial gain set
Gc = Kp*(s+Ki)/s;
```

```
%sisotool single loop control
T=sisoinit(1);
T.F.Value = 11.0;
T.G.Value = Gp;
T.C.Value = Gc;
sisotool(T);
```

controller transfer functions are defined as follows:

$$G_p(s) = \frac{4167s + 138900}{0.06s^2 + s + 3750}, \quad G_c(s) = \frac{K_p(s + K_i)}{s}$$

The MATLAB M-file presented in Table 2.3 is used to set up the system. The controller gains K_p and K_i are initially set to unity and will be subsequently tuned using the SISOTOOL interface.

In SISOTOOL, the controller format is expressed using the Zero/Pole/Gain representation. Since SISOTOOL supports various controller formats, users must explicitly select the appropriate type. Upon launching SISOTOOL, click *Preferences*, then *Options*, and a settings window will appear. Under the compensator format section, choose *Zero/Pole/Gain*. This process is illustrated in Figure 2.7.

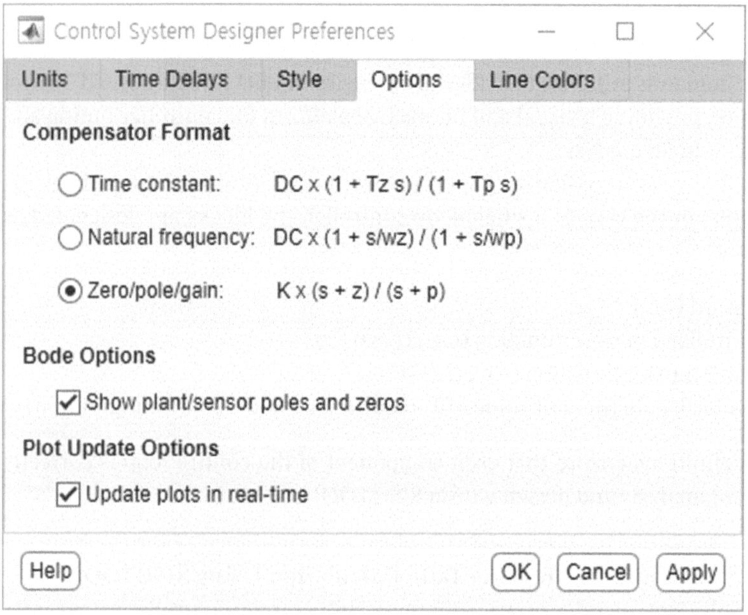

Figure 2.7 Compensator format setting in SISOTOOL.

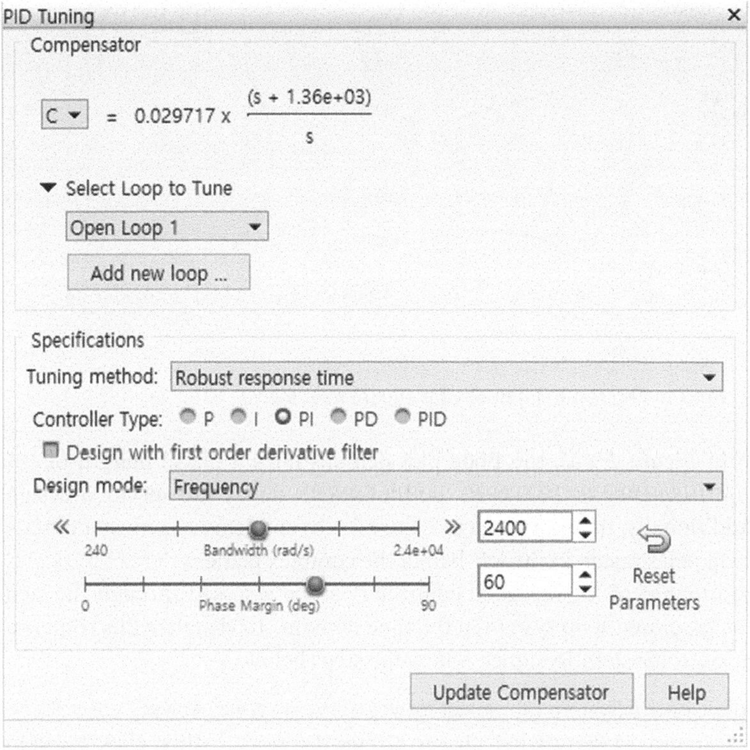

Figure 2.8 PID auto-tuning process.

The PID auto-tuning procedure is illustrated in Figure 2.8, and the sequence is outlined as follows:

1) From the *Tuning Methods* icon, select *PID Tuning* under the *Automated Tuning* menu bar.
2) In the Specifications section, set the Tuning Method to *Robust Response Time.*
3) For Controller Type, choose *PI*.
4) Under Design Mode, select *Frequency*.
5) Specify the desired Bandwidth (in rad/s) and Phase Margin (in degrees). For example, set the bandwidth to 2400 and the phase margin to 60, then click *Update Compensator*.
6) The compensator obtained from the auto-tuning is displayed in the *PID Tuning* tab.

The outcome of the compensator design is presented in the Control System Designer window. The performance characteristics of the designed compensator are evaluated through the open-loop Bode plot and the root locus diagram.

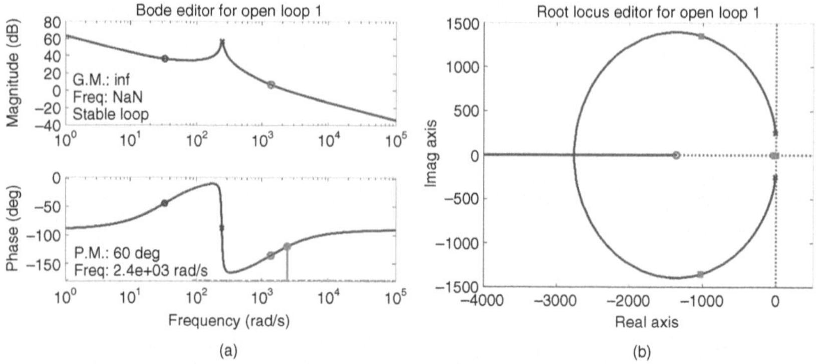

Figure 2.9 Auto-tuning result. (a) Bode plot and (b) Root locus.

As shown in Figure 2.9(a), the Bode plot demonstrates a phase margin of 60° and a cutoff frequency of 2400 rad/s, both of which satisfy the specified design criteria. Additionally, the root locus in Figure 2.9(b) confirms system stability, as all open-loop poles reside in the left-half of the complex plane.

The performance of the designed controller can be assessed through the step response of the closed-loop system in the time domain. To visualize this response using the Control System Designer, follow the steps below.

1) From the Control System tab, select `New Plot` → `New Step`.
2) In the Response to Plot dialog, choose `Closed-Loop 1`, then click `Plot` to generate the step response graph.

The resulting graph is shown in Figure 2.10.

Determining Gains for a Double-Loop Controller This section explains how to determine the controller gains for a double-loop control structure, as illustrated in Figure 2.6(c). To configure this architecture in the Control System Designer, use the following command: `sisoinit(6)`. The corresponding structure for *sisoinit(6)* is reproduced in Figure 2.6(c) for clarity.

Enter the appropriate transfer functions for each block using the naming convention below

1) F: Forward Gain
2) C_1: Voltage Controller ($G_{cv}(s)$)
3) C_2: Current Controller ($G_{ci}(s)$)
4) G_1: Transfer Function from Control Input to Inductor Current $G_1(s)$

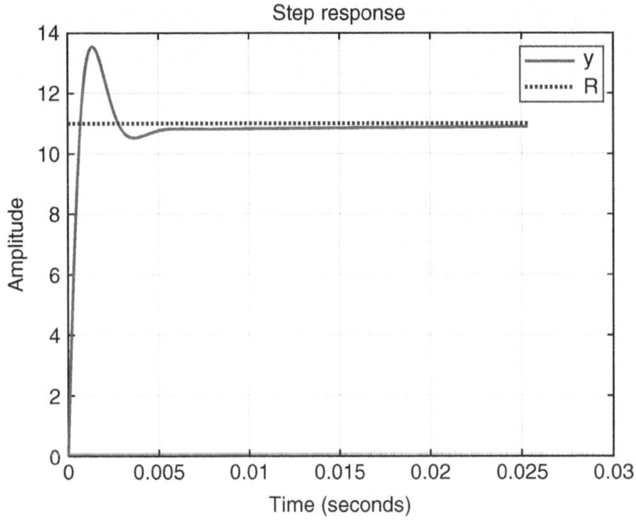

Figure 2.10 Step response of the closed-loop system.

5) G_2: Transfer Function from Inductor Current to Capacitor Voltage $G_3(s)$
6) $H_1(s)$: Current Feedback Gain (default: 1; no configuration required)
7) $H_2(s)$: Voltage Feedback Gain (default: 1; no configuration required)

PI Gain Adjustment Procedure (Auto-Tuning) The *PI* gain tuning process for the double-loop controller follows the same procedure as that of the single-loop controller. Begin by tuning the current controller C_2, followed by the voltage controller C_1.

Since the gain of the voltage controller can influence the dynamics of the inner current loop, it is recommended to repeat this tuning sequence two or three times. After the iterative tuning process, verify that the cutoff frequencies and phase margins of both controllers satisfy the design specifications by inspecting the corresponding Bode plots.

2.2 Controller Design

This chapter presents the design methodology for a buck converter controller. It begins with the design of single-loop controllers—specifically, the voltage and current controllers—and proceeds to a step-by-step explanation of how to implement a double-loop control structure.

2.2.1 Voltage Controller (Single Loop)

The configuration of a single-loop voltage controller for regulating the capacitor voltage V_C is illustrated in Figure 2.11.

To eliminate steady-state error and control the transient response, a *PI* controller is employed for voltage regulation. The controller is expressed as

$$G_{cv}(s) = \frac{K_{pv}(s + K_{iv})}{s} \tag{2.9}$$

The proportional and integral gains, K_{pv} and K_{iv}, are determined using MATLAB's SISOTOOL based on the following frequency-domain design specifications.

1) Bandwidth: 10 rad/sec
2) Phase Margin: 90 deg.

The MATLAB m-file for voltage controller is presented in Table 2.4.

Using the Automated Tuning feature in SISOTOOL, the optimized PI controller is obtained as

$$G_{cv}(s) = \frac{0.00002(s + 5000)}{s} \tag{2.10}$$

The design result is shown in Figure 2.12. As illustrated in Figure 2.12(a), the Bode plot confirms that the phase margin is 90° and the cutoff frequency is 10 rad/sec, exactly matching the specified design criteria. Additionally, the root locus in Figure 2.12(b) verifies system stability, with all open-loop poles located in the left-half of the complex plane.

Step Response of Capacitor Voltage The step response of the closed-loop system in the time domain is presented in Figure 2.13. When the phase margin is set to 90°, the system exhibits no overshoot, confirming that the transient response is well-damped.

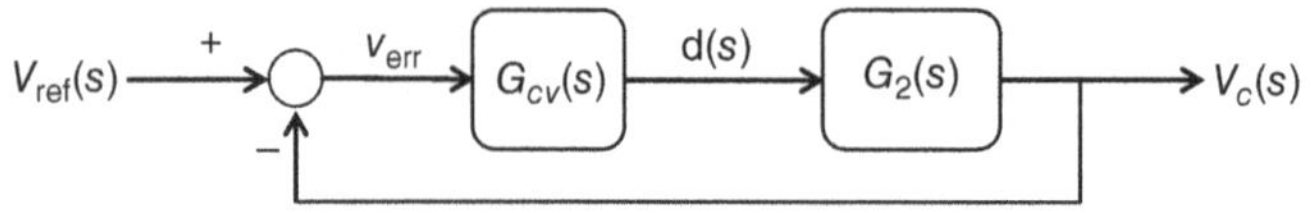

Figure 2.11 Voltage controller configuration (single loop).

Table 2.4 MATLAB m-file for voltage controller.

```
s=tf('s');
Vin = 100;   % Input voltage [V]
L = 2e-3;    % L=2mh [H]
C = 500e-6;  % C=500uf [F]
R = 10;      % Load resistance [Ohm]

G1 = (Vin/L)*(s+1/(R*C)) / (s^2 + s/(R*C) + 1/(L*C) );
G2 = Vin/(L*C) / (s^2 + s/(R*C) + 1/(L*C) );
G3 = (1/C) / (s + 1/(R*C));

Kpv = 1; Kiv=1;  % Initial PI gains
Gcv = Kpv*(s+Kiv)/s;

Vref = 40;
T=sisoinit(1);
T.F.Value = Vref;
T.G.Value = G2;
T.C.Value = Gcv;
sisotool(T);
```

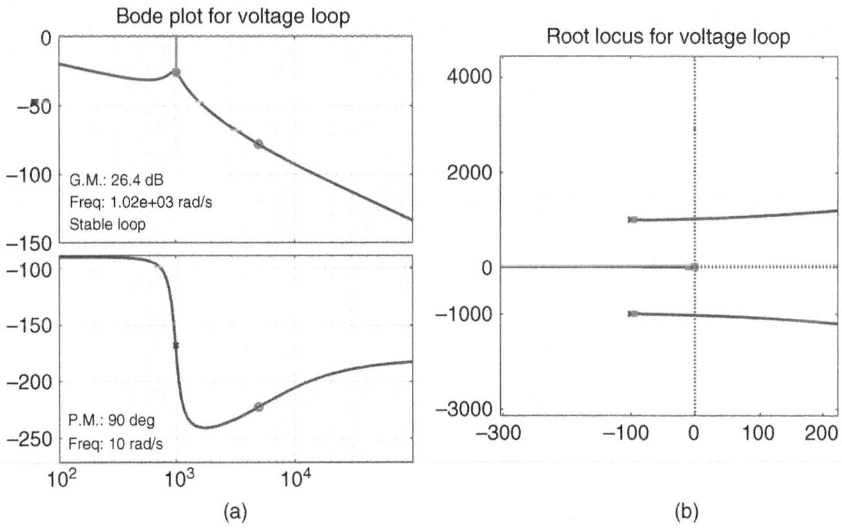

Figure 2.12 SISOTOOL-based voltage controller design (single loop). (a) Bode plot and (b) Root locus.

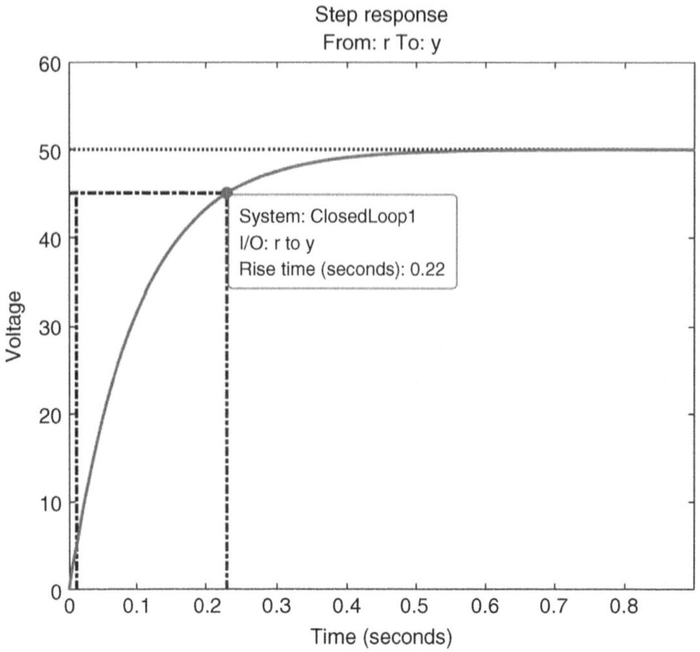

Figure 2.13 Step response of capacitor voltage.

Correlation Between Frequency Design Specifications and Time Response Characteristics Frequency-domain design specifications are typically defined by the cutoff frequency (in rad/sec) and phase margin (in degrees), while time-domain performance is characterized by metrics such as overshoot and rise time. The relationship between these domains can be summarized as follows:

1) Phase Margin → Influences Overshoot
2) Cutoff Frequency → Determines Rise Time

For instance, when the phase margin is set to 90°, the system exhibits zero overshoot, resulting in a critically damped or well-damped response. The mathematical relationship between cutoff frequency and rise time is approximated by

$$t_r = \frac{0.35}{f_{\text{cut}}[Hz]} = \frac{0.35}{\frac{10}{2\pi}} = 0.2199[\text{sec}] \tag{2.11}$$

Therefore, for a cutoff frequency of 10 rad/s, the expected rise time is approximately 0.22 sec, which aligns closely with the step response observed in MATLAB, confirming the accuracy of the frequency-domain design.

Validation of the Designed Voltage Controller The designed voltage controller is derived from a mathematical model based on differential equations. However, the buck converter is a physical system composed of practical components such as inductors, capacitors, switches, and diodes. Therefore, it is essential to validate whether the controller obtained from the theoretical model exhibits the same dynamic response when applied to the actual circuit.

To perform this validation, PSIM—a circuit-level simulation tool widely used for power electronics—is utilized. Using the same component values as those used in MATLAB, the system was modeled in PSIM. The controller $G_{cv}(s)$, originally designed in MATLAB, was implemented in PSIM as a transfer function block $H(s)$ using the same mathematical expression.

Figure 2.14 shows both the PSIM simulation schematic and the resulting step response waveform. In Figure 2.14(a), the simulation circuit includes a voltage source, a switching element, an inductor, a capacitor, a load resistor, and the voltage controller block implemented as a continuous-time transfer function. The reference voltage step input is applied to evaluate the controller's time-domain response.

In Figure 2.14(b), the output voltage waveform clearly demonstrates the dynamic response of the closed-loop system. It exhibits a smooth rise with no overshoot, settling at the reference voltage—exactly as predicted by the MATLAB step response. The observed rise time and steady-state accuracy closely align with the design specifications, confirming the validity of the controller.

This consistency between the MATLAB-based design and the PSIM circuit-level simulation verifies that the voltage controller performs as intended in a realistic environment. It ensures that the controller is not only theoretically sound but also practically implementable.

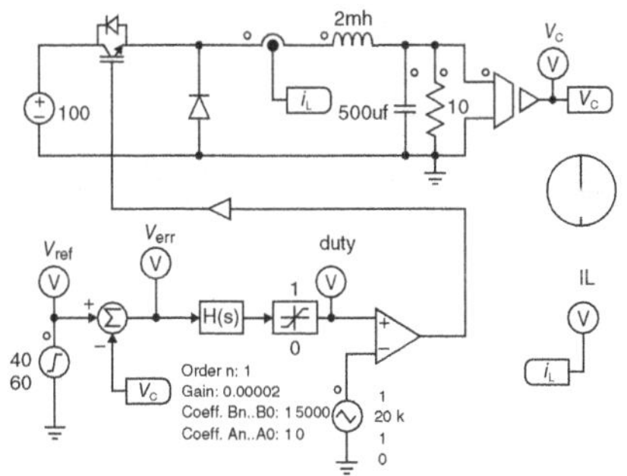

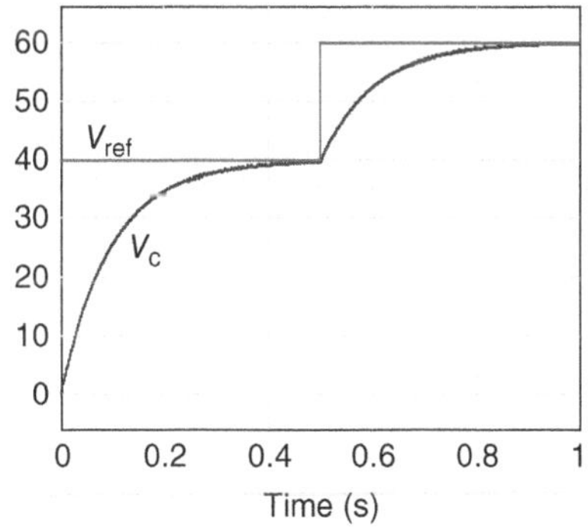

Figure 2.14 Circuit-level validation of the voltage controller using PSIM. (a) PSIM simulation schematic and (b) Step response waveform.

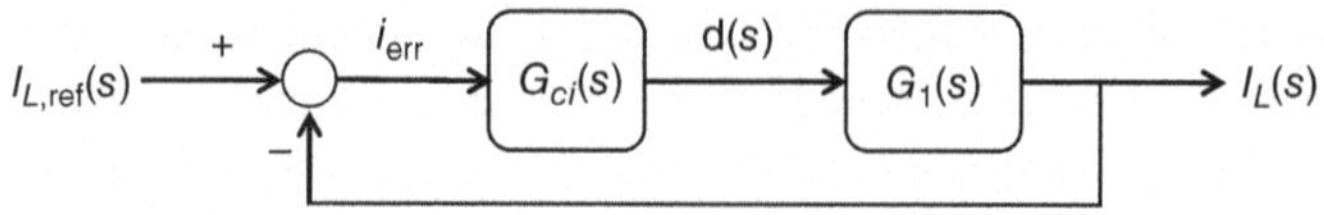

Figure 2.15 Current controller configuration (single loop).

2.2.2 Current Controller (Single Loop)

The configuration of a single-loop current controller for regulating the inductor current i_L is illustrated in Figure 2.15.

To eliminate the steady-state error and control the transient state, the current controller is designed as a *PI* controller, represented by the following equation:

$$G_{ci}(s) = \frac{K_{pi}(s + K_{ii})}{s} \tag{2.12}$$

The gains K_{pi} and K_{ii} of the current controller are determined using MATLAB's SISOTOOL. The MATLAB M-file presented in Table 2.5 is used to set up the system. The frequency design specifications for the current controller are as follows:

Table 2.5 MATLAB m-file for current controller.

```
s=tf('s');
Vin = 100;   % Input voltage [V]
L = 2e-3;    % L=2mh [H]
C = 500e-6;  % C=500uf [F]
R = 10;      % Load resistance [Ohm]

G1 = (Vin/L)*(s+1/(R*C)) / (s^2 + s/(R*C) + 1/(L*C) );
G2 = Vin/(L*C) / (s^2 + s/(R*C) + 1/(L*C) );
G3 = (1/C) / (s + 1/(R*C));

Kpi = 1; Kii=1; % Initial PI gains
Gci = Kpi*(s+Kii)/s;

Iref = 10;
T=sisoinit(1);
T.F.Value = Iref;
T.G.Value = G1;
T.C.Value = Gci;

sisotool(T);
```

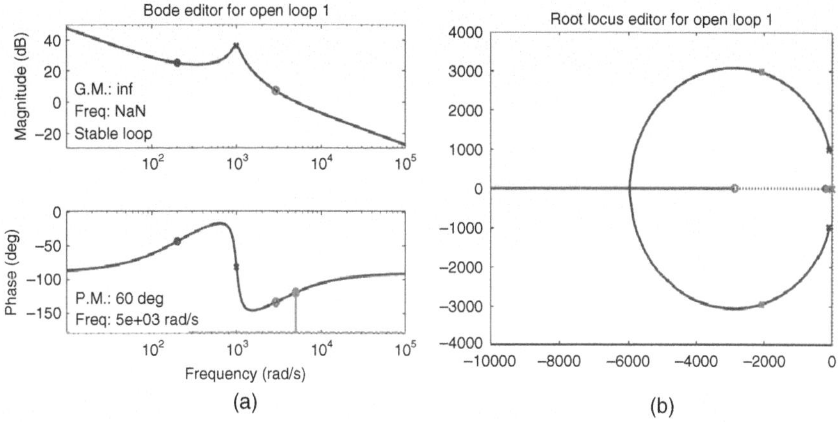

Figure 2.16 SISOTOOL-based current controller design (single loop). (a) Bode plot and (b) Root locus.

a) Bandwidth: 5000 [rad/sec]
b) Phase Margin: 60 [deg]

Using the automated tuning feature in SISOTOOL, the optimized *PI* controller is obtained as

$$G_{ci}(s) = \frac{K_{pi}(s + K_{ii})}{s} = \frac{0.083(s + 2900)}{s} \tag{2.13}$$

The design result is shown in Figure 2.16. As illustrated in Figure 2.16(a), the Bode plot confirms that the phase margin is 60° and the cutoff frequency is 5000 rad/sec, exactly matching the specified design criteria. Additionally, the root locus in Figure 2.16(b) verifies system stability, with all open-loop poles located in the left-half of the complex plane.

Step Response of Inductor Current The time-domain step response of the closed-loop system is illustrated in Figure 2.17.

The closed loop transfer function is given by

$$T_F(s) = \frac{I_L(s)}{I_{\text{ref}}(s)} = \frac{G_{ci}(s)G_1(s)}{1 + G_{ci}(s)G_1(s)} \tag{2.14}$$

Substitute the expressions

$$T_F(s) = \frac{0.083 \cdot 50{,}000 \cdot (s + 2900)(s + 200)}{s(s^2 + 200s + 1{,}000{,}000) + 0.083 \cdot 50{,}000 \cdot (s + 2900)(s + 200)} \tag{2.15}$$

The poles of the resulting closed-loop system are located at

$$s = -183, \quad s = -2083 \pm j2959$$

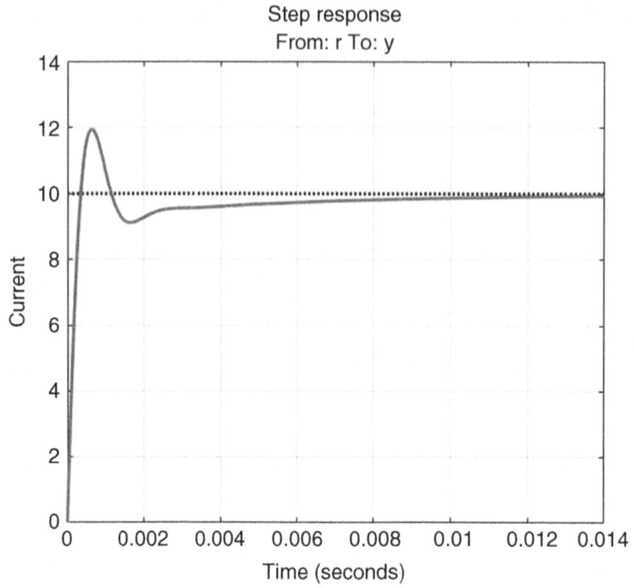

Figure 2.17 Step response of inductor current.

Among these, the real-axis pole at $s = -183$ lies much closer to the origin compared to the complex-conjugate poles. Therefore, the dominant pole approximation is not applicable, as it assumes a single dominant pole significantly closer to the origin than the others. In this case, the step response characteristics—such as overshoot, rise time, and settling time—are evaluated using MATLAB's `stepinfo` function. The closed-loop system can be simulated using the MATLAB code presented in Table 2.6.

The resultant values are given as

1) Settling time: 0.262 ms
2) Overshoot: 19.3 [%]

Table 2.6 MATLAB m-file for step characteristics.

```
G1 = tf([50000, 50000*200], [1, 200, 1e6]);
Gci = tf([0.083, 0.083*2900], [1, 0]);
T = feedback(Gci*G1, 1);

step(T)                    % View step response
S = stepinfo(T)            % View overshoot, rise time, etc.
```

A comparison with Figure 2.17 reveals that the obtained values yield consistent results.

Validation of the Designed Current Controller To validate the performance of the designed current controller, a PSIM simulation was conducted using the same circuit parameters employed in MATLAB. The controller $G_{ci}(s)$ was implemented in PSIM as a transfer function $H(s)$, with an identical expression to that used in MATLAB.

Figure 2.18 presents both the simulation schematics and the resulting step response of the inductor current. The response characteristics observed in PSIM closely match those obtained from the MATLAB simulation, demonstrating strong agreement between the two platforms. This confirms that the current controller designed in MATLAB performs as intended and satisfies the required specifications when applied to the actual circuit model.

2.2.3 Double-Loop Controller Design

A double-loop control scheme is designed to regulate both the output voltage and the inductor current simultaneously. Each loop employs a *PI* controller. The transfer functions corresponding to the control blocks are defined in MATLAB using an m-file, as shown in Table 2.7. Since the initial setup of the m-file is identical to that of the single-loop design, only the modified sections relevant to the double-loop configuration are presented here.

The block-wise assignment of each transfer function is illustrated in Figure 2.19.

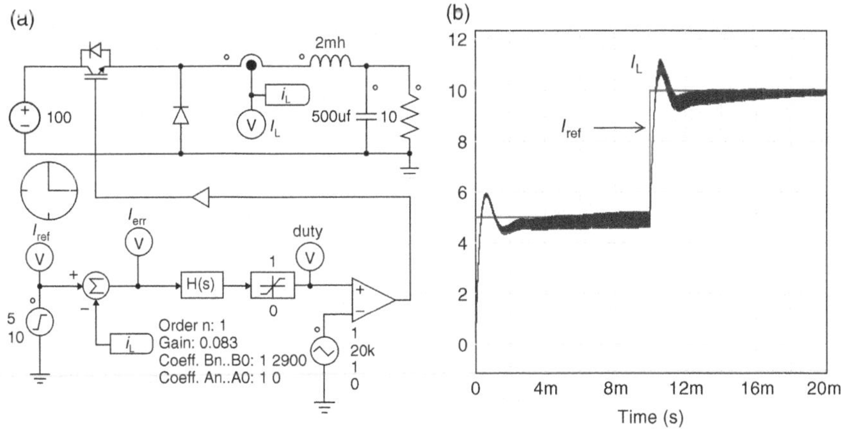

Figure 2.18 Current controller simulation result. (a) Simulation schematics and (b) Simulation waveform.

Table 2.7 MATLAB m-file for double-loop controller.

```
T=sisoinit(6);
T.F.Value = Vref;
T.G1.Value = G1;        % current loop G1= (iL/d)
T.C2.Value = Gci;       % current controller PI = (d/ie)
%T.H1.Value = 1;        % depends on Matlab versions

T.G2.Value = G3;        % voltage loop tf. = (Vc/iL)
T.C1.Value = Gcv;       % voltage controller PI =Vref/Ve)
%T.H2.Value = 1;        % depends on Matlab versions
sisotool(T);
```

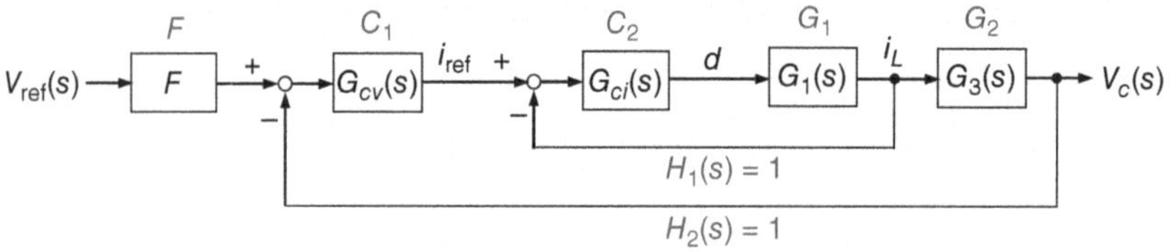

Figure 2.19 Transfer function assignment for each block.

Table 2.8 Frequency domain design specification for double-loop controller.

Controller type	Bandwidth [rad/s]	Phase margin [deg]
Voltage Controller (C_1)	10	90
Current Controller (C_2)	5000	60

The frequency-domain design specifications for the voltage and current controllers are summarized in Table 2.8.

The controllers were tuned using MATLAB's Auto Tuning feature, and the resulting transfer functions are as follows:

$$C_1 = G_{cv}(s) = \frac{0.0054(s+185)}{s}$$

$$C_2 = G_{ci}(s) = \frac{0.0831(s+2880)}{s} \tag{2.16}$$

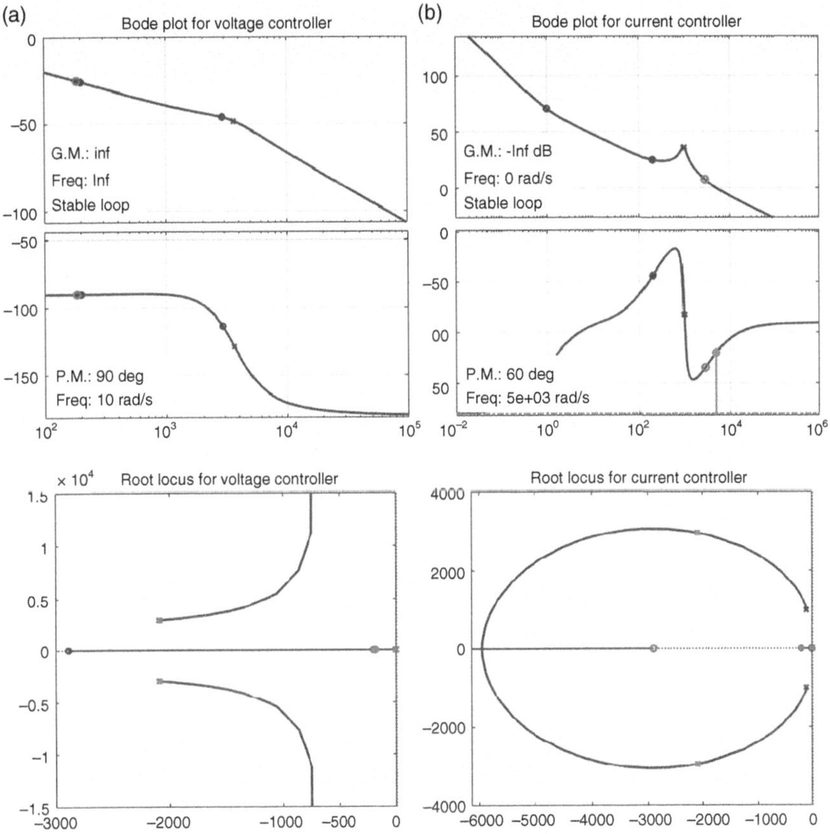

Figure 2.20 SISOTOOL-based double-loop controller design. (a) Voltage controller and (b) Current controller.

The design results are presented in Figure 2.20. The Bode plot and root locus of the voltage controller are shown in Figure 2.20(a), while those of the current controller are displayed in Figure 2.20(b). As illustrated in Figure 2.20(a), the Bode plot confirms a phase margin of 90° and a cutoff frequency of 10 rad/sec, which precisely satisfy the specified design requirements. The root locus further verifies system stability by demonstrating that all open-loop poles reside in the left-half of the complex plane.

Similarly, Figure 2.20(b) shows that the current controller achieves a phase margin of 60° and a cutoff frequency of 5000 rad/sec, also in exact accordance with the design specifications. The root locus confirms system stability, with all open-loop poles located in the left-half plane.

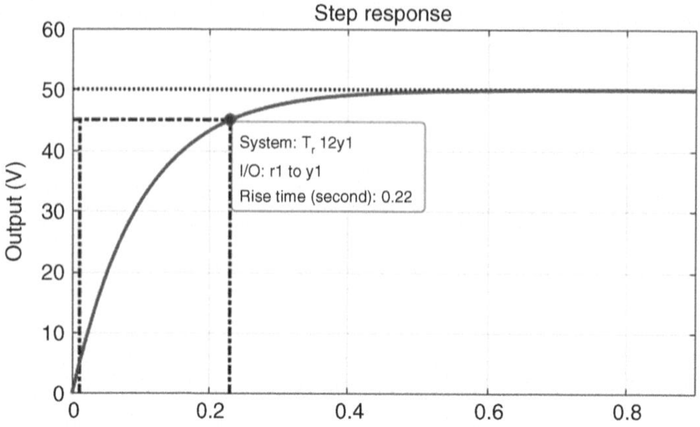

Figure 2.21 Step response of capacitor voltage.

Figure 2.21 illustrates the time-domain step response of the closed-loop system obtained using SISOTOOL.

Validation of the Designed Double-Loop Controller To validate the performance of the designed double-loop controller, a PSIM simulation was conducted. The simulation circuit schematics and the corresponding voltage step response waveform are shown in Figures 2.22 and 2.23, respectively.

As illustrated, the voltage response obtained from the PSIM simulation closely matches the step response generated in MATLAB, confirming the consistency and accuracy of the controller design. Furthermore, the current waveform demonstrates that the inductor current is effectively regulated to track the reference current, indicating proper operation of the inner current control loop.

2.2.4 Digital Conversion

From Analog PI Controllers to Digital PI Controllers PI controllers discussed thus far have been implemented in analog form. A typical analog controller can be expressed as

$$G_C(s) = \frac{2 \cdot 10^{-5}(s + 5 \cdot 10^3)}{s}$$

Analog controllers were commonly used until the 1990s, typically realized with operational amplifiers (op-amps), resistors (R), and capacitors (C). With the advent and widespread adoption of digital microprocessors, however, modern power conversion systems now rely predominantly on digital control techniques.

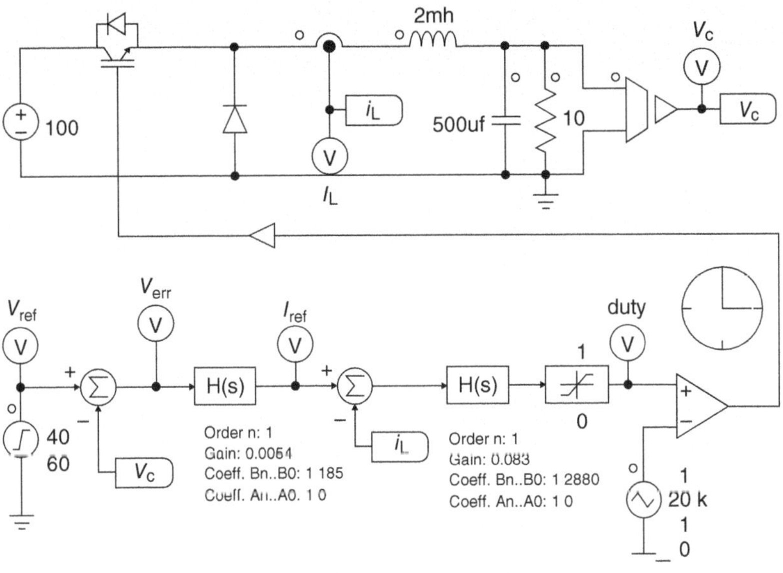

Figure 2.22 Schematics of the double-loop controller.

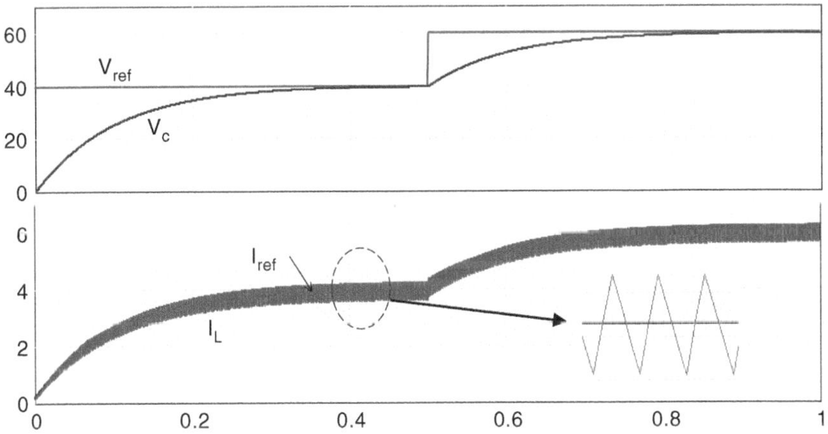

Figure 2.23 Simulation waveforms of the double-loop controller.

To illustrate the analog implementation, consider the design of the analog *PI* controller circuit shown in Figure 2.24. The transfer function of this circuit is given by

$$G_C(s) = \frac{V_o}{V_{in}} = -\frac{R_f}{R_i}\left(1 + \frac{1}{sC_fR_f}\right) = \frac{2 \cdot 10^{-5}(s + 5 \cdot 10^3)}{s} \tag{2.17}$$

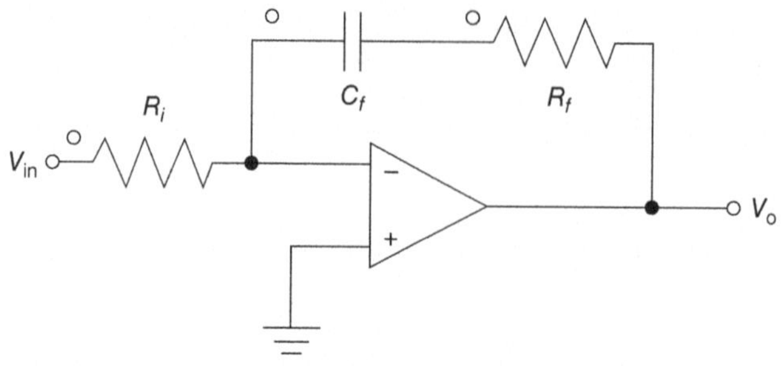

Figure 2.24 PI analog controller circuit.

From this expression, the controller parameters are identified as

$$K_P = \frac{R_f}{R_i}, \quad \omega_Z = \frac{1}{C_f R_f}, \quad K_i = \frac{1}{C_f R_i} \tag{2.18}$$

Practical component values are selected to satisfy $K_P = 2 \cdot 10^{-5}$ and $\omega_Z = 5000$ rad/sec. Let $R_f = 100$ ohm; then $R_i = 5\ M$ohm and $C_f = 2\ uF$.

These choices yield the desired controller transfer function. The negative sign reflects the inverting topology of the op-amp configuration. If a non-inverted response is required, an additional inverter stage may be added, or the sensor/error polarity can be adjusted accordingly.

The fundamental distinction between analog and digital control systems lies in the incorporation of a sampling interval T_s. A digital controller is generally derived by discretizing an already-designed analog controller using methods such as zero-order hold (ZOH) or Tustin's approximation. The resulting digital controller is then implemented in C-code and embedded in the microprocessor's flash memory.

Although the digital conversion process can be performed analytically using standard digital control references, MATLAB's SISOTOOL provides a highly efficient and automated environment for controller discretization. The procedure for implementing a switching frequency of 20 kHz is summarized in Table 2.9.

Example: Digital Conversion Using ZOH

Consider a digital conversion using the ZOH method, with a switching frequency of 20 kHz. The converted digital controller takes the form

$$G_C(z) = \frac{2 \cdot 10^{-5}(z - 0.75)}{z - 1}$$

In this expression, the proportional gain remains $K_p = 2 \cdot 10^{-5}$, but the integral component is modified to $K_{iz} = 0.75$. Note that the continuous-time pole at the origin ($s = 0$) maps to the discrete-time pole at $z = 1$.

Table 2.9 The Procedure for sample(discrete) time conversion.

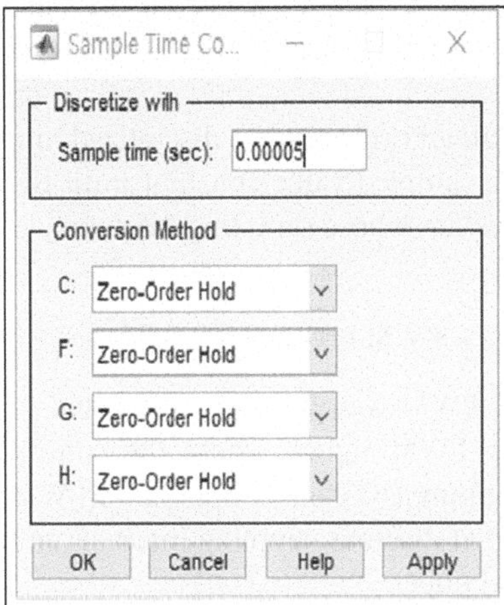	*SISOTOOL* ↓ *Edit Architecture* ↓ *Sample time conversion* ↓ *Sample time(sec): 0.00005* ↓ *Conversion method: all ZOH* ↓ *OK*

Conversion to a Difference Equation for Digital Implementation For practical implementation in embedded software, the discrete-time transfer function is often rewritten as a difference equation. Starting from

$$G_{CV}(z) = \frac{d(z)}{V_{\text{err}}(z)} = K_P \frac{(z - K_I)}{z - 1} \tag{2.19}$$

Rewriting and cross-multiplying

$$z \cdot d(z) - d(z) = z \cdot K_P \cdot V_{\text{err}}(z) - K_P K_I \cdot V_{\text{err}}(z) \tag{2.20}$$

Converting to the time domain yields

$$d(k) - d(k-1) = K_P \cdot V_{err}(k) - K_P K_I \cdot V_{\text{err}}(k-1) \tag{2.21}$$

Thus, the final difference equation suitable for implementation is

$$d(k) = d(k-1) + K_P \cdot V_{\text{err}}(k) - K_P \cdot K_I \cdot V_{\text{err}}(k-1) \tag{2.22}$$

This recursive form offers a straightforward and efficient approach for implementing the PI controller on a digital processor using C or any other programming language.

C-Code Representation of the PI Controller The previously derived difference equation can be directly implemented in C as follows:

```
Verr = Vref-Vc;
duty = duty_old + Kp*Verr - Kp*Ki*Verr_old;
duty_old=duty;
Verr_old=Verr;
```

In a typical digital control system, sampling is synchronized with the PWM cycle—occurring once per cycle. The sampled input remains constant throughout each switching period. Consequently, the sampling time, PWM period, and switching period are all equivalent.

When the ZOH method is employed, the analog PI controller is discretized into a digital controller. The ZOH block holds the sampled value constant until the next sampling instant, resulting in a digital system whose sampling frequency is identical to the PWM switching frequency.

$$\text{ZOH Frequency} = \text{Sampling Frequency} = \text{PWM Frequency}$$

The difference between analog and digital PWM waveforms is illustrated in Figure 2.25. Subfigure (a) shows the analog PWM signal, where the control voltage continuously varies in real time. Subfigure (b) depicts the digital PWM signal, in which the control voltage is updated at discrete sampling intervals and held constant between updates due to the ZOH effect.

To ensure simulation results are consistent with actual hardware behavior, it is recommended to avoid using PSIM's built-in control blocks when implementing digital control logic. Instead, the control algorithm should be written directly in C. This approach not only increases fidelity between simulation and real-world operation but also allows the same C-code to be downloaded into the DSP or microcontroller's flash memory for deployment, significantly streamlining the development process.

Single-Loop Voltage Controller Implementation Using a C-Block The analog voltage controller PSIM circuit depicted in Figure 2.14 can be converted into a digital controller. The original analog controller is defined by the transfer function in Eq. (2.11), reproduced below for reference

$$G_{cv}(s) = \frac{0.00002(s + 5000)}{s}$$

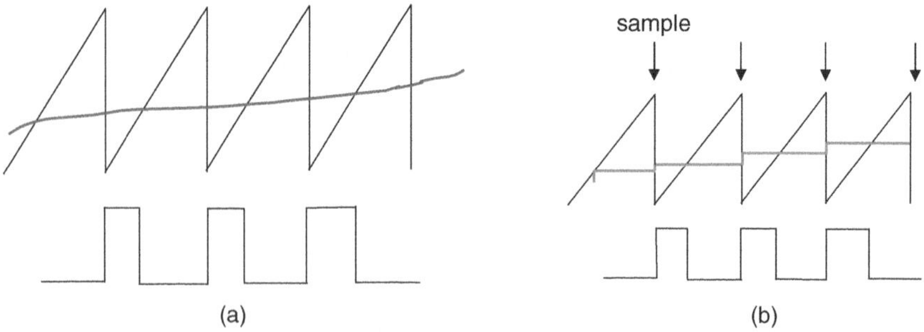

Figure 2.25 Digital conversion using ZOH. (a) Analog PWM and (b) Digital PWM.

Table 2.10 C-block code: Voltage_Controller.

```
double Vc, IL, Vref, duty, Verr;
static double duty_old=0.0,Verr_old=0.0;

Vc   = in[0];
IL   = in[1];
if (t < 0.5) Vref = 40;
else         Vref = 60;
Verr = Vref - Vc;

duty = duty_old + 0.00002*Verr - 0.00002*0.75*Verr_old;
duty_old=duty;
Verr_old=Verr;

out[0]=duty;
out[1]=Vref;
```

When sampled at a switching frequency of 20 kHz, the analog transfer function is discretized into the following digital controller

$$G_{CV}(z) = \frac{d(z)}{V_{err}(z)} = \frac{2 \times 10^{-5}(z - 0.75)}{z - 1} \tag{2.23}$$

The corresponding difference equation, derived from the digital transfer function, is

$$d(k) = d(k-1) + 0.00002 \cdot V_{err}(k) - 0.00002 \cdot 0.75 \cdot V_{err}(k-1) \tag{2.24}$$

This difference equation can be implemented using a C-block in the simulation environment. The C-code presented in Table 2.10 demonstrates the implementation.

This implementation assumes that the sampling interval matches the PWM period, consistent with the 20 kHz switching frequency. The output duty represents the control signal to be applied to the PWM modulator, and V_{ref} is output for monitoring purposes.

The schematic for implementing a single-loop voltage controller using a C-block is shown in Figure 2.26.

By examining the simulation results presented in Figure 2.27, it can be observed that the digital controller $G_C(z)$ produces a response that closely matches that of the analog controller $G_C(s)$, validating the effectiveness of the digital implementation.

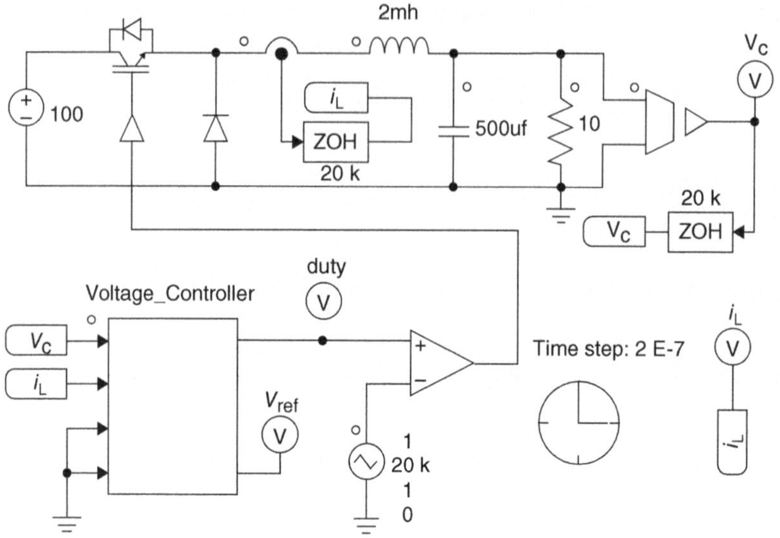

Figure 2.26 Schematic of the C-block implementation for a voltage controller.

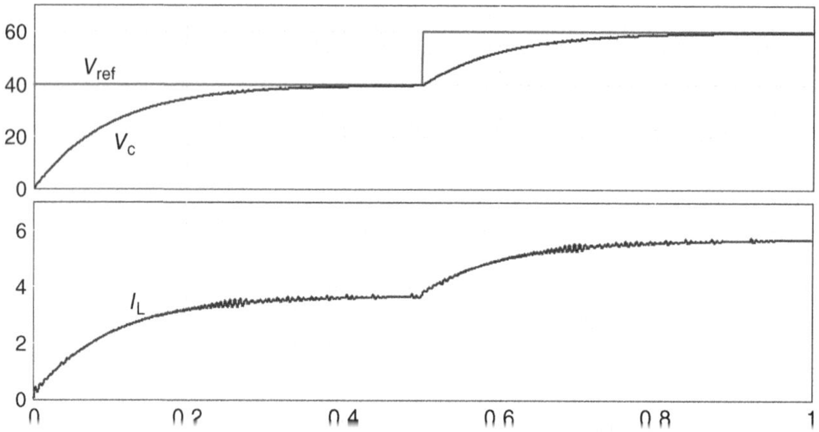

Figure 2.27 Simulation waveforms of the voltage controller implemented using a C-block.

Designing a Double-Loop Controller Using a C-Block When the analog controller given in Eq. (2.16) is converted to a digital form with a switching frequency of 20 kHz, the resulting discrete-time controllers are as follows:

$$C_1 = G_{cv}(z) = \frac{0.00558(z - 0.991)}{z - 1}$$

$$C_2 = G_{ci}(z) = \frac{0.09298(z - 0.9)}{z - 1} \tag{2.25}$$

Table 2.11 C-block code: Double_Loop.

```
double Vc, IL, Vref, duty, Verr,
IL_err, IL_ref;
static double IL_ref_old=0.0,
duty_old=0.0, Verr_old=0.0;
static double IL_err_old=0.0;

Vc = in[0];
IL = in[1];

if (t < 0.5) Vref = 40;
else         Vref = 60;

// Voltage control loop
Verr = Vref - Vc;
IL_ref = IL_ref_old + 0.00558*Verr -
0.00558*0.991*Verr_old;
IL_ref_old=IL_ref;
Verr_old=Verr;

// Current control loop
IL_err = IL_ref - IL;
duty = duty_old + 0.09298*IL_err -
0.09298*0.9*IL_err_old;
duty_old=duty;
IL_err_old=IL_err;

out[0]=duty;
out[1]=Vref;
out[2]=IL_ref;
```

These two digital controllers correspond to the outer voltage loop (C_1) and the inner current loop (C_2), respectively.

The implementation of this double-loop controller C-block is shown in Table 2.11.

The schematic for implementing the double-loop controller is nearly identical to that shown in Figure 2.26, except for the inclusion of inductor current sensing and the modified C-block code. Therefore, it is not shown separately.

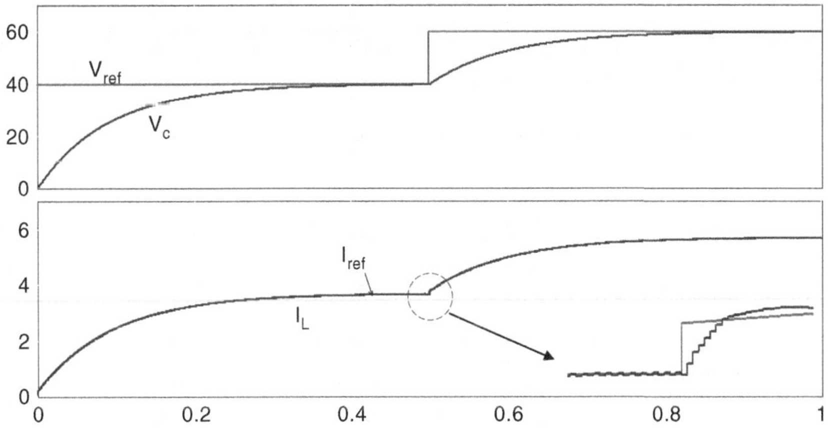

Figure 2.28 Simulation waveforms of the double-loop controller using a C-block.

By analyzing the simulation results in Figure 2.28, it can be confirmed that the digital double-loop controller provides a dynamic response closely matching that of its analog counterpart, thereby validating the effectiveness of the digital implementation.

For practical hardware deployment, the digital double-loop controller is implemented in C as part of an Interrupt Service Routine (ISR) on a high-speed processor such as a digital signal processor (DSP). This ensures real-time operation and accurate control performance under actual system conditions.

2.2.5 Practice: Double-Loop Controller Design and Verification

Objective Design both analog and digital double-loop controllers for a given power converter circuit using MATLAB. Validate the performance of the controllers through simulation in PSIM, particularly focusing on transient behavior under load change conditions.

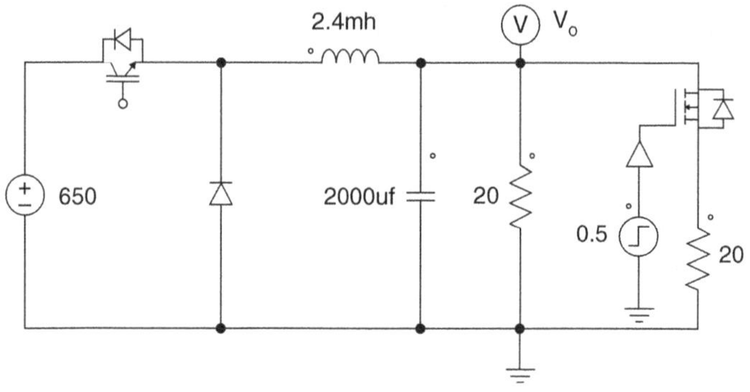

Instructions

1) **Design Requirements**
   - The system features a load disturbance occurring at 0.5 sec.
   - Implement a double-loop control architecture consisting of an inner current loop and an outer voltage loop.
   - Design the analog controllers in MATLAB to meet the specified phase margin (PM) and crossover frequency (f_c) requirements.
   - Convert the analog controllers into their digital equivalents using appropriate discretization methods (e.g., ZOH or Tustin).
   - Implement the digital controller using a C-block in PSIM.

2) **System Specifications**

Switching frequency	Nominal duty	Current controller		Voltage controller	
		f_c(Hz)	PM	f_c(Hz)	PM
10 k	0.4	1500	60	10	85
12 k	0.5	1800	65	15	87
15 k	0.6	2000	60	15	89
18 k	0.4	2500	65	20	90

3) **Design Steps**
   - Derive the transfer functions of the plant for both current and voltage loops.
   - Use MATLAB's SISOTOOL or custom scripts to design compensators meeting the cutoff frequency and phase margin requirements.
   - Verify stability and performance using Bode plots and step response analyses.
   - Convert the analog controllers into digital form.
   - Implement the digital controller in PSIM using a C-block and simulate the system response under the specified load change condition.
4) **Verification Criteria**
   - Validate that the designed controller maintains voltage regulation and current tracking across all switching frequencies.
   - Confirm that the transient response meets performance expectations in terms of overshoot, settling time, and steady-state error.
   - Compare the behavior of the analog and digital implementations to ensure consistency.
5) **Deliverables**
   - MATLAB code (M-file for analog controllers).
   - Plots: Root locus, Bode plots, and step responses.
   - PSIM simulation circuit diagrams.
   - Waveforms for both analog and digital controllers.
   - Digital controller design process and C-code for implementation.

3

Boost Converter

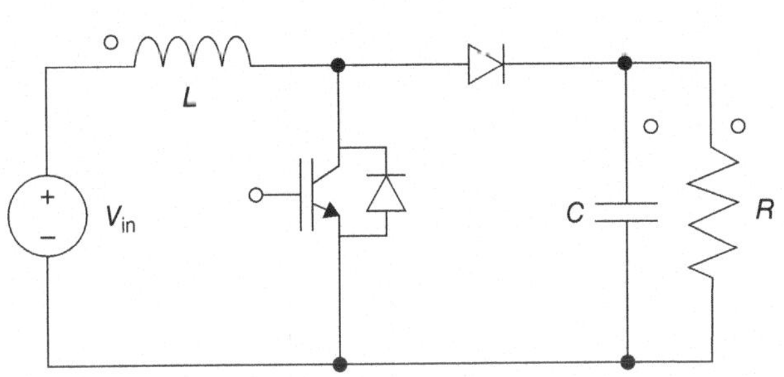

3.1 State-Averaged Equations and Transfer Function

A boost converter is a non-isolated DC–DC power converter that increases the input voltage to a higher output voltage by periodically switching a semiconductor device at a high frequency. Energy transfer and voltage regulation are achieved through controlled operation of an inductor, diode, switch, and output capacitor. The operating principle of the boost converter can be described by two switching intervals within one switching period:

- **ON state:** When the switch is turned ON, the input voltage is applied directly across the inductor, causing the inductor current to increase and energy to be stored in the magnetic field. During this interval, the diode is reverse-biased, and the load is supplied solely by the output capacitor.
- **OFF state:** When the switch is turned OFF, the inductor current cannot change instantaneously and is forced to flow through the diode to the output. The inductor releases its stored energy to both the load and the output capacitor, resulting in an output voltage that is higher than the input voltage.

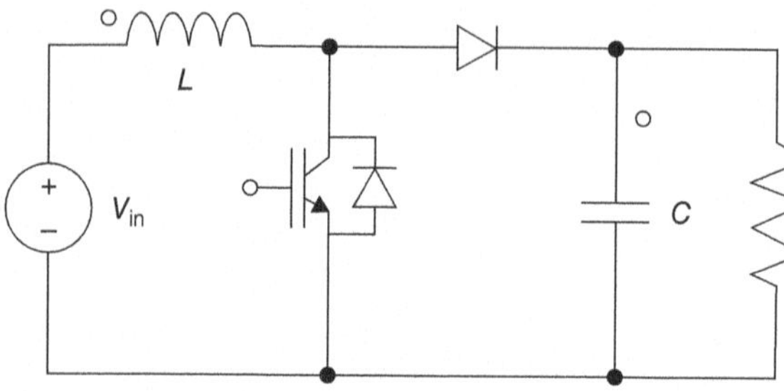

Figure 3.1 Boost converter circuit.

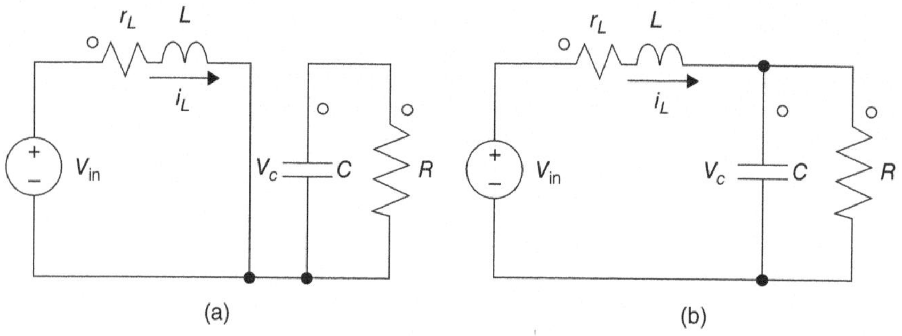

Figure 3.2 Equivalent circuits of the boost converter during ON and OFF states. (a) ON time and (b) OFF time.

The output voltage of the boost converter is regulated by controlling the duty ratio, defined as the fraction of the switching period during which the switch remains in the ON state. By adjusting the duty ratio, the amount of energy transferred from the input to the output can be precisely controlled. Figure 3.1 illustrates the basic circuit configuration of the boost converter.

Equivalent Circuit and State-Space Modeling of the Boost Converter To derive the dynamic model of the boost converter, the circuit is analyzed separately for the two switching states. The corresponding equivalent circuits are shown in Figure 3.2.

State-Space Averaging of the Boost Converter The state-space representations of the boost converter for each switching interval are summarized in Table 3.1. The state variables are chosen as the inductor current i_L and the output capacitor voltage v_C.

Using state-space averaging, the average system matrices over one switching period T_S are:

Table 3.1 State space equations for ON, OFF period.

ON state	OFF state
$V_{\text{in}} = r_L i_L + L\frac{di_L}{dt}$	$V_{\text{in}} = r_L i_L + L\frac{di_L}{dt} + V_C$
$0 = C\frac{dv_C}{dt} + \frac{v_C}{R}$	$i_L = C\frac{dv_C}{dt} + \frac{v_C}{R}$
$\begin{pmatrix} \frac{di_L}{dt} \\ \frac{dv_C}{dt} \end{pmatrix} = \begin{pmatrix} -\frac{r_L}{L} & 0 \\ 0 & -\frac{1}{RC} \end{pmatrix} \begin{pmatrix} i_L \\ v_C \end{pmatrix} + \begin{pmatrix} \frac{1}{L} \\ 0 \end{pmatrix} V_{\text{in}}$	$\begin{pmatrix} \frac{di_L}{dt} \\ \frac{dv_C}{dt} \end{pmatrix} = \begin{pmatrix} -\frac{r_L}{L} & -\frac{1}{L} \\ \frac{1}{C} & -\frac{1}{RC} \end{pmatrix} \begin{pmatrix} i_L \\ v_C \end{pmatrix} + \begin{pmatrix} \frac{1}{L} \\ 0 \end{pmatrix} \cdot V_{\text{in}}$
$= A_1 \begin{pmatrix} i_L \\ v_C \end{pmatrix} + B_1 \cdot V_{\text{in}}$	$= A_2 \begin{pmatrix} i_L \\ v_c \end{pmatrix} + B_2 \cdot V_{\text{in}}$ (3.1)

$$A = A_{\text{avg}} = \frac{(A_1 \cdot T_{\text{ON}} + A_2 \cdot T_{\text{OFF}})}{T_S} = A_1 \cdot d + A_2 \cdot (1-d)$$
$$= \begin{pmatrix} -\frac{r_L}{L} & 0 \\ 0 & -\frac{1}{RC} \end{pmatrix} \cdot d + \begin{pmatrix} -\frac{r_L}{L} & -\frac{1}{L} \\ \frac{1}{C} & -\frac{1}{RC} \end{pmatrix} \cdot (1-d) = \begin{pmatrix} -\frac{r_L}{L} & -\frac{(1-d)}{L} \\ \frac{(1-d)}{C} & -\frac{1}{RC} \end{pmatrix}$$
$$\text{B} = B_{\text{avg}} = \frac{(B_1 \cdot T_{\text{ON}} + B_2 \cdot T_{\text{OFF}})}{T_S} = B_1 \cdot d + B_2 \cdot (1-d)$$
$$= \begin{pmatrix} \frac{1}{L} \\ 0 \end{pmatrix} d + \begin{pmatrix} \frac{1}{L} \\ 0 \end{pmatrix} \cdot (1-d) = \begin{pmatrix} \frac{1}{L} \\ 0 \end{pmatrix} \tag{3.2}$$

The state-averaged equations of the boost converter are therefore expressed as

$$\begin{pmatrix} \frac{di_L}{dt} \\ \frac{dv_C}{dt} \end{pmatrix} = \begin{pmatrix} -\frac{r_L}{L} & -\frac{(1-d)}{L} \\ \frac{(1-d)}{C} & -\frac{1}{RC} \end{pmatrix} \begin{pmatrix} i_L \\ v_C \end{pmatrix} + \begin{pmatrix} \frac{V_{\text{in}}}{L} \\ 0 \end{pmatrix} \tag{3.3}$$

For compact representation, the model can be written in matrix form as

$$\dot{X} = A(d) + B$$

where

$$\dot{X} = \begin{pmatrix} \frac{di_L}{dt} \\ \frac{dv_C}{dt} \end{pmatrix}, \quad \text{A(d)} = \begin{pmatrix} -\frac{r_L}{L} & -\frac{(1-d)}{L} \\ \frac{(1-d)}{C} & -\frac{1}{RC} \end{pmatrix}, \quad \text{B} = \begin{pmatrix} \frac{V_{\text{in}}}{L} \\ 0 \end{pmatrix} \tag{3.4}$$

Since the system matrix A explicitly depends on the duty ratio d, the boost converter is inherently *nonlinear*. This nonlinearity implies that the input–output relationship is not strictly proportional, which complicates controller design. Consequently, conventional linear controllers such as PI controllers may not provide satisfactory performance over a wide operating range.

To address nonlinear behavior, two general approaches are available. One approach is to design a nonlinear controller, such as input–output linearization or adaptive control. Alternatively, the nonlinear system can be linearized around a nominal operating point or within a limited operating range, enabling the application of conventional linear control techniques.

A widely adopted method for this purpose is *small-signal modeling*, which approximates the system dynamics in the vicinity of a steady-state operating point. This approach forms the basis for transfer-function derivation and frequency-domain controller design. A detailed discussion of nonlinear system linearization using small-signal perturbation is provided in Section 1.2.4 of the previous chapter.

Small-Signal Model of the Boost Converter To linearize the state-averaged model of the boost converter, a perturbation approach is used. Each variable is expressed as the sum of a steady-state (DC) value and a small perturbation (AC) component.

$$d = D + \hat{d}, \quad i_L = I_L + \hat{i}_L, \quad v_C = V_C + \hat{v}_C \tag{3.5}$$

Inductor Current Equation Starting from the nonlinear state-averaged model

$$\frac{\mathrm{d}}{\mathrm{d}t}(I_L + \hat{i}_L) = -\frac{r_L}{L(I_L + \hat{i}_L)} - \frac{1}{L}(1 - D - \hat{d})(V_C + \hat{v}_C) + \frac{1}{L}(V_{\text{in}}) \tag{3.6}$$

Separating the DC and AC components

1) DC (steady-state) component:

$$\frac{\mathrm{d}}{\mathrm{d}t}(I_L) = -\frac{r_L}{L}I_L - \frac{1}{L}(1 - D)\, V_C + \frac{1}{L}V_{\text{in}} \tag{3.7}$$

2) AC (perturbation) component:

$$\frac{d}{\mathrm{d}t}(\hat{i}_L) = -\frac{r_L}{L}\hat{i}_L - \frac{1}{L}(1 - D)\,\hat{v}_C + \frac{1}{L}V_C\,\hat{d} \tag{3.8}$$

Capacitor Voltage Equation

$$\frac{\mathrm{d}}{\mathrm{d}t}(V_C + \hat{v}_C) = \frac{1}{C}(1 - D - \hat{d})(I_L + \hat{i}_L) - \frac{1}{RC}(V_C + \hat{v}_C) \tag{3.9}$$

Separating the DC and AC components:

1) DC (steady-state) component:

$$\frac{\mathrm{d}}{dt}(V_C) = \frac{1}{C}(1-D)\,I_L - \frac{1}{RC}V_C \tag{3.10}$$

2) AC (perturbation) component:

$$\frac{\mathrm{d}}{\mathrm{d}t}(\hat{v}_C) = \frac{1}{C}(1-D)\,\hat{i}_L - \frac{1}{C}I_L\,\hat{d} - \frac{1}{RC}\hat{v}_c \tag{3.11}$$

Higher-order nonlinear perturbation terms like $\hat{d}\,\hat{v}_C$, $\hat{d}\,\hat{i}_L$ are neglected under the small-signal assumption.

Matrix Representation The small-signal model can be expressed in state-space form as

$$\begin{pmatrix} \dfrac{\mathrm{d}(\hat{i}_L)}{\mathrm{d}t} \\ \dfrac{\mathrm{d}(\hat{v}_C)}{\mathrm{d}t} \end{pmatrix} = \begin{pmatrix} -\dfrac{r_L}{L} & -\dfrac{1-D}{L} \\ \dfrac{1-D}{C} & -\dfrac{1}{RC} \end{pmatrix} \begin{pmatrix} \hat{i}_L \\ \hat{v}_C \end{pmatrix} + \begin{pmatrix} \dfrac{V_C}{L} \\ -\dfrac{I_L}{C} \end{pmatrix} \hat{d} \tag{3.12}$$

Or in compact form

$$\dot{X} = AX + Bu$$

$$X = \begin{pmatrix} \hat{i}_L \\ \hat{v}_C \end{pmatrix}, \quad A = \begin{pmatrix} -\dfrac{r_L}{L} & -\dfrac{1-D}{L} \\ \dfrac{1-D}{C} & -\dfrac{1}{RC} \end{pmatrix}, \quad B = \begin{pmatrix} \dfrac{V_C}{L} \\ -\dfrac{I_L}{C} \end{pmatrix}, \quad u = \hat{d} \tag{3.13}$$

where

$$V_C = \frac{V_{\text{in}}}{(1-D)}, \quad I_L = \frac{V_{\text{in}}}{R(1-D)^2}$$

This small-signal model captures the linearized dynamics of the boost converter around a steady-state operating point.

Small-Signal Transfer Function Derivation Let us now compute the transfer functions of the boost converter based on the small-signal state-space model. Starting from the Laplace-transformed state-space equation

$$(sI - A)X(s) - X(0) = BU \quad \Rightarrow \quad X(s) = (sI - A)^{-1}[X(0) + BU(s)] \tag{3.14}$$

Given the initial state vector

$$X(0) = \begin{pmatrix} I_L(0) \\ V_C(0) \end{pmatrix} = \begin{pmatrix} 0 \\ V_{\text{in}} \end{pmatrix}$$

The inverse of $(sI - A)$ is calculated as

$$(sI - A)^{-1} = \begin{pmatrix} s + \frac{r_L}{L} & \frac{1-D}{L} \\ -\frac{1-D}{C} & s + \frac{1}{RC} \end{pmatrix}^{-1} = \frac{1}{\Delta(s)} \begin{pmatrix} s + \frac{1}{RC} & -\frac{1-D}{L} \\ \frac{1-D}{C} & s + \frac{r_L}{L} \end{pmatrix}$$

Where the characteristic polynomial (determinant) is

$$\Delta(s) = s^2 + \left(\frac{1}{RC} + \frac{r_L}{L}\right) s + \frac{1}{LC\left(\frac{r_L}{R} + (1-D)^2\right)} \tag{3.15}$$

Thus, the state vector in the Laplace domain becomes

$$X(s) = \frac{1}{\Delta(s)} \begin{pmatrix} s + \frac{1}{RC} & -\frac{1-D}{L} \\ \frac{1-D}{C} & s + \frac{r_L}{L} \end{pmatrix} \left[\begin{pmatrix} \frac{V_C}{L} \\ -\frac{I_L}{C} \end{pmatrix} \hat{d}(s) + \begin{pmatrix} 0 \\ V_{in} \end{pmatrix} \right] \tag{3.16}$$

Using the steady-state values

$$V_C = \frac{V_{in}}{(1-D)}, \quad I_L = \frac{V_{in}}{R(1-D)^2} \tag{3.17}$$

We now compute the small-signal transfer functions.

Derived Transfer Functions

1) Inductor Current to Duty ratio

$$G_1(s) = \frac{\hat{i}_L(s)}{\hat{d}(s)} = \frac{\frac{V_{in}}{L(1-D)}\left(s + \frac{2}{RC}\right)}{S^2 + \frac{1}{RC}s + \frac{(1-D)^2}{LC}} \tag{3.18}$$

2) Output Voltage to Duty ratio

$$G_2(s) = \frac{\hat{v}_C(s)}{\hat{d}(s)} = \frac{\frac{V_{in}}{LC}\left(1 - \frac{Ls}{R(1-D)^2}\right)}{S^2 + \frac{1}{RC}S + \frac{(1-D)^2}{LC}} \tag{3.19}$$

3) Output Voltage to Inductor Current

$$G_3(s) = \frac{\hat{v}_C(s)}{\hat{i}_L(s)} = \frac{\left(\frac{1-D}{C}\right)\left(1 - \frac{Ls}{R(1-D)^2}\right)}{\left(s + \frac{2}{RC}\right)} \tag{3.20}$$

The transfer function can also be obtained using MATLAB's built-in function `ss2tf`, which converts a given state-space representation into its equivalent transfer function form. The corresponding MATLAB script developed for this purpose is presented in Table 3.2.

Table 3.2 MATLAB m-file for deriving transfer function.

```
%-- Boost Converter transfer functions--%

s=tf('s');
Vin  = 100;
rL   = 0.01;    % 10 mohm
L  = 2.4e-3;    % L=2.4mh
C  = 2400e-6;   % C= 2400uf
R  = 25;
D  = 0.4;
Vc  = Vin/(1-D);
IL  = Vc/(R*(1-D));

A  = [-rL/L,     -(1-D)/L;
      (1-D)/C,   -1/(R*C)];
B  = [Vc/L;
      -IL/C];
C1  = [1,0]; C2  = [0,1]; Dx  = 0;

[num1,den1]  = ss2tf(A,B,C1,Dx);
[num2,den2]  = ss2tf(A,B,C2,Dx);
G1  = tf(num1, den1);
G2  = tf(num2, den2);
G3  = tf(num2,num1);
```

Stability Analysis of the Current Loop Transfer Function $G_1(s)$ To examine the dynamic characteristics of the current control loop, we begin by analyzing the root locus of the small-signal transfer function $G_1(s)$. The circuit parameters are defined as follows:

$$V_{\text{in}} = 100\,V,\ \ L = 2.4\,\text{mH},\ \ C = 2400\,\text{uF},\ \ R = 25\,\Omega,\ \ D = 0.4$$

The steady-state operating points are calculated as

$$V_C = \frac{V_{\text{in}}}{(1-D)} = 166\,[V], \qquad I_L = \frac{V_{\text{in}}}{R(1-D)^2} = 11.1[A].$$

The small-signal transfer functions are presented in Table 3.3.

To examine the location of the poles–zeros of $G_1(s)$, we execute *pole*($G1$) and *zero*($G1$). The pole–zero plot shows one zero located on the real axis at −33, and two poles located in the complex plane at $(-8.33 \pm j250)$. These results indicate

Table 3.3 MATLAB m-file for boost converter transfer functions.

```
%%%%%%% Boost Converter %%%%%%%%%%%
s=tf('s');

Vin = 100;
L = 2.4e-3;     % L=2.4mh
C = 2400e-6;    % C= 2400uf
R = 25;
D = 0.4;
Vc = Vin/(1-D);
IL = Vc/(R*(1-D));

G1=(Vc/L)*(s+1/(R*C)+(1-D)*IL/(Vc*C))/(s^2+s/(R*C)+(1-D)^2/(L*C));
G2=(IL/C)*(-s+(1-D)*Vc/(L*IL))/(s^2+s/(R*C)+(1-D)^2/(L*C));
G3=(IL/Vc)*(L/C)*(-s+(1-D)*Vc/(L*IL))/(s+1/(R*C)+(1-D)*IL/(Vc*C));
```

that the system is inherently stable, as all poles lie in the left-half plane (LHP). To further investigate the closed-loop behavior, the root locus of $G_1(s)$ is plotted using the *rlocus*(*G*1) command. As shown in Figure 3.3, all root-locus trajectories remain within the LHP, confirming that the current control loop exhibits stable dynamics across a wide range of gain values.

For frequency-domain analysis, the Bode plot of $G_1(s)$ is generated by executing *bode*(*G*1) and *margin*(*G*1), with the results shown in Figure 3.4. The margin function returns both the gain margin and phase margin, which are found to be

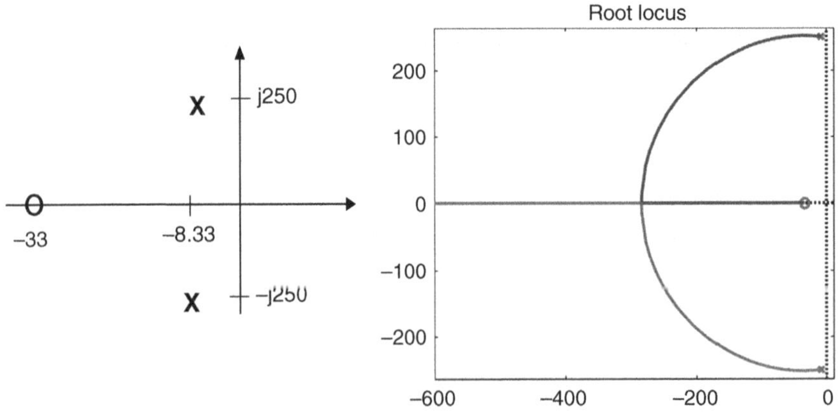

Figure 3.3 Pole-zero diagram and root locus of $G_1(s)$.

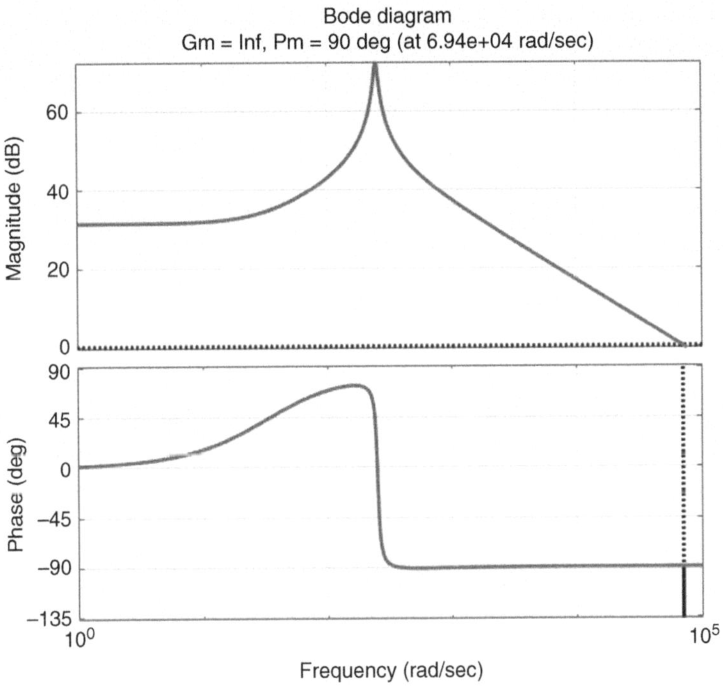

Figure 3.4 Bode plot and stability margins of $G_1(s)$.

positive. This confirms that $G_1(s)$ is not only stable but also robust to variations in gain and phase.

Stability Analysis of the Voltage Loop Transfer Function $G_2(s)$ To assess the stability characteristics of the voltage loop, we first analyze the pole-zero locations of the transfer function $G_2(s)$. Executing the commands *pole*(*G*2) and *zero*(*G*2) in MATLAB yields the pole-zero diagram shown in Figure 3.5. The results indicate that the system has two complex conjugate poles located in the LHP at ($s = -8.33 \pm j250$), and one real zero located in the right-half plane (RHP) at ($s = +3750$).

A system with a zero in the RHP is classified as a *non-minimum phase system*, which introduces a fundamental limitation in control design due to phase lag and inverse response.

To further investigate closed-loop behavior, the root locus of $G_2(s)$ is generated using the *rlocus*(*G*2) command. The resulting plot, shown in Figure 3.6, illustrates the movement of the closed-loop poles as the feedback gain K increases. Although the poles initially lie in the LHP, they migrate toward the RHP beyond a certain gain value, resulting in instability. Specifically:

- The system remains stable for $0 < K < 0.0036$
- The system becomes unstable for $K > 0.0036$

This behavior indicates that the transfer function from duty cycle to output voltage, $\hat{v}_C(s)/\hat{d}(s)$, is *conditionally stable*, with stability dependent on the selected feedback gain.

To evaluate stability in the frequency domain, the Bode plot of $G_2(s)$ is generated using *bode*(*G*2), and the gain and phase margins are obtained using *margin*(*G*2). The results are shown in Figure 3.7.

The analysis reveals that both the gain margin and phase margin are negative, confirming that $G_2(s)$ is an unstable system under proportional control. As a result, direct control of the output voltage using a simple proportional (*P*) controller is inadequate for the boost converter.

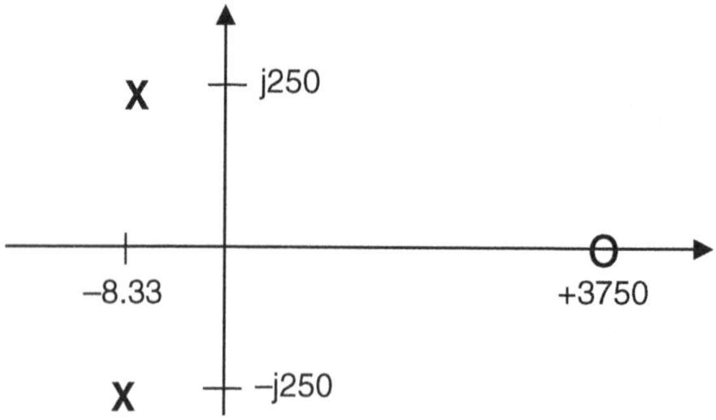

Figure 3.5 Pole-zero diagram of the $G_2(s)$.

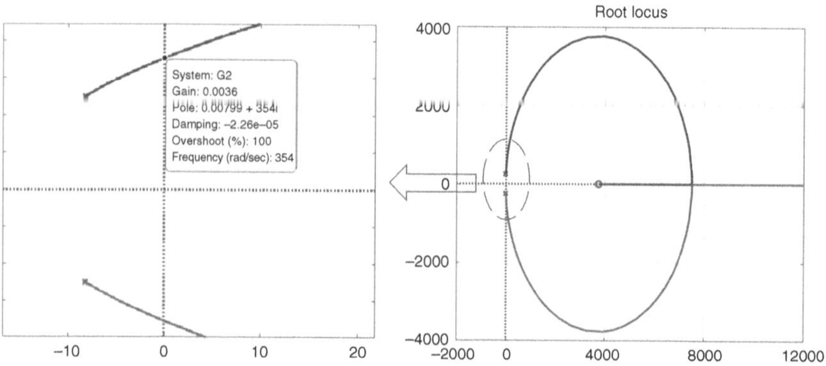

Figure 3.6 Root locus of $G_2(s)$.

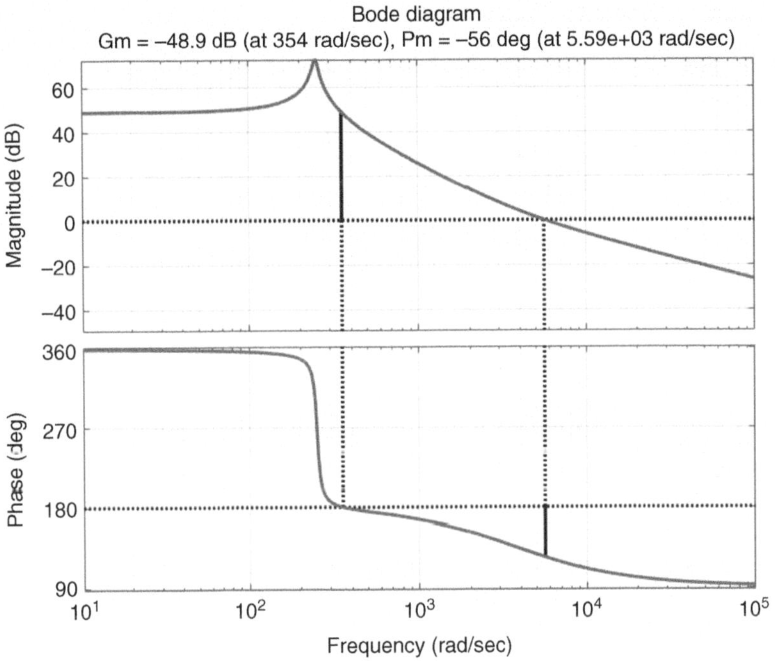

Figure 3.7 Bode plot and stability margins of $G_2(s)$.

To achieve stable output voltage regulation, it is necessary to implement a proportional-integral (PI) controller by introducing an integrator (i.e. a pole at the origin) or alternatively, to adopt a double-loop control structure that separates voltage and current regulation.

3.2 Controller Design

Based on the derived transfer functions, the controller can be implemented using the configuration shown in Figure 3.8.

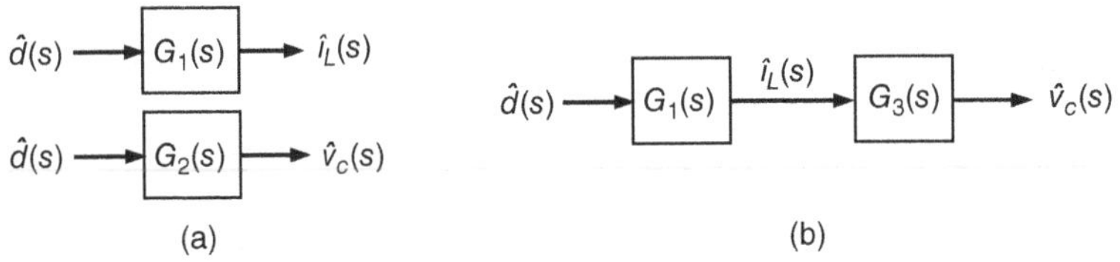

Figure 3.8 Controller configuration using transfer functions. (a) Single-loop configuration and (b) Double-loop configuration.

3.2.1 Current Controller (Single Loop)

The single-loop current controller is designed based on the following performance specifications.

- Bandwidth: $\omega_c = 2400$ [rad/sec]
- Phase Margin: P.M. = 60 [deg]

The MATLAB m-file for current controller is presented in Table 3.4.

Using SISOTOOL's auto-tuning feature, a controller that meets the specified design conditions was obtained.

$$G_{ci}(s) = \frac{K_{pi}(s + K_{ii})}{s} = \frac{0.02972(s + 1364)}{s} \tag{3.21}$$

The resulting Bode plot and the step response of the closed-loop system are shown in Figure 3.9.

Table 3.4 MATLAB m-file for current controller.

```
/********** Boost Converter ************/
s=tf('s');
Vin = 100;
L = 2.4e-3;    % Inductance=2.4mh
C = 2400e-6;   % C= 2400uf
R = 25;
D = 0.4;
Vc = Vin/(1-D);
IL = Vc/(R*(1-D));

G1=(Vc/L)*(s+1/(R*C)+(1-D)*IL/(Vc*C))/(s^2+s/(R*C)+(1-D)^2/(L*C));
G2=-Vin/(L*C)*(-1+L*s/(R*(1-D)^2))/(s^2+s/(R*C)+(1-D)^2/(L*C));
G3=(IL/Vc)*(L/C)*(-s+(1-D)*Vc/(L*IL))/(s+1/(R*C)+(1-D)*IL/(Vc*C));

Kpi = 1; Kii = 1;   % Initial gains
Gci = Kpi*(s+Kii)/s;

%sisotool Current Single loop
T = sisoinit(1);
T.F.Value = 11.0;
T.G.Value = G1;
T.C.Value = Gci;
sisotool(T);
```

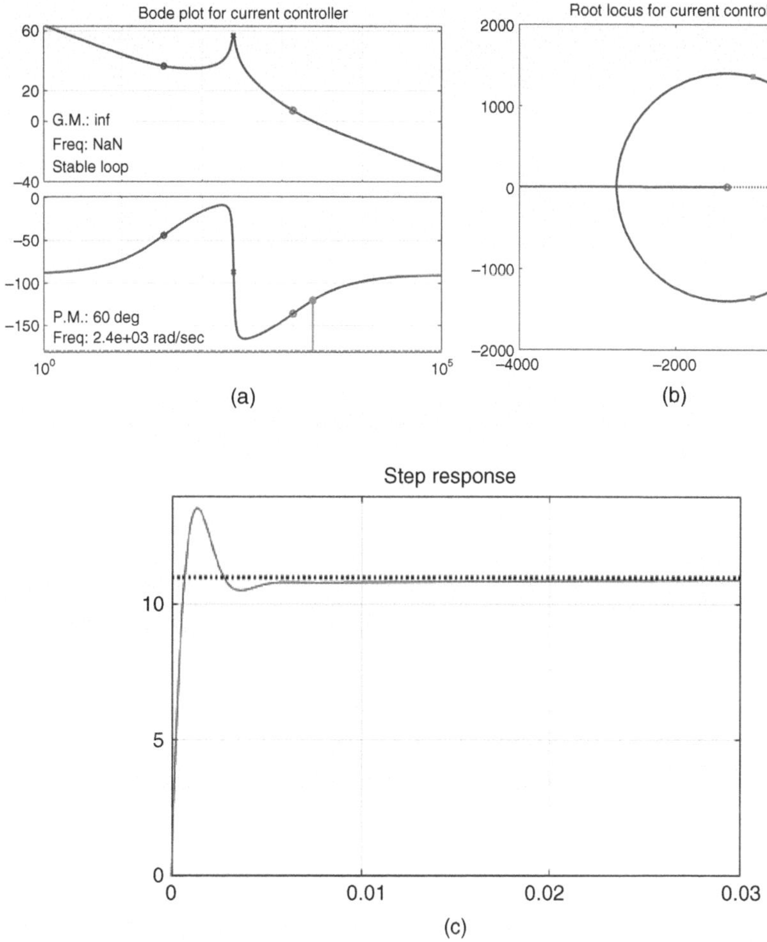

Figure 3.9 SISOTOOL current response of the boost converter. (a) Bode diagram, (b) Root Locus and (c) Closed-loop step response.

This controller was then applied to the current control of a Boost converter, and its behavior was validated through simulation. Figure 3.10 presents the PSIM schematic and waveform, confirming that the simulation results closely match the MATLAB response.

To investigate the impact of changing the design conditions, the crossover frequency was increased to $\omega_c = 4000$ rad/sec while maintaining the phase margin at 60 degrees. The corresponding simulation results are shown in Figure 3.11.

Next, the phase margin was increased to 90 degrees while keeping $\omega_c = 2400$ rad/sec. As shown in Figure 3.12, the resulting controller achieved a bandwidth of

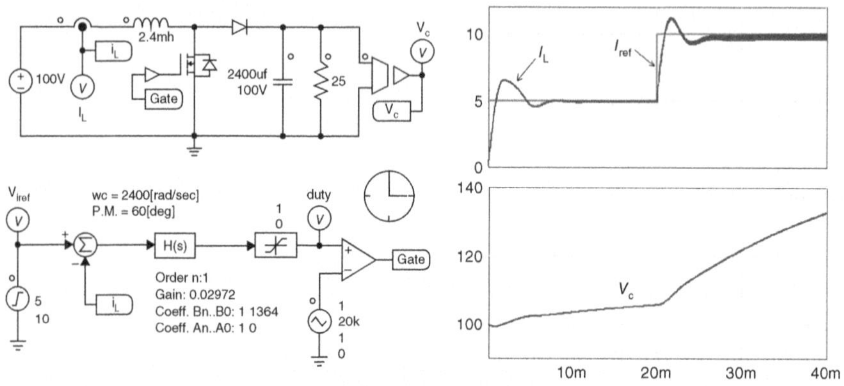

Figure 3.10 PSIM schematic and simulation waveforms of the current controller.

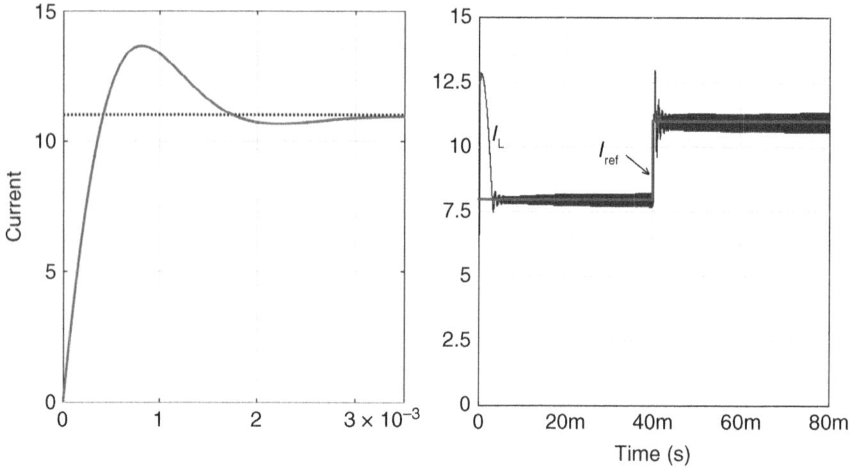

Figure 3.11 MATLAB and PSIM current waveforms at $\omega_c = 4000$ rad/sec.

2400 rad/sec and a phase margin of 88.6 degrees—close but not precisely meeting the design specifications. Nonetheless, the MATLAB and PSIM current waveforms are in good agreement.

Finally, the design was updated to both $\omega_c = 4000$ rad/sec and PM = 90 degrees. As shown in Figure 3.13, although the transient responses are similar, a discrepancy is observed in the steady-state values between MATLAB and PSIM results. It means that the designed phase margin of 88.8° indicates that the design conditions cannot be satisfied.

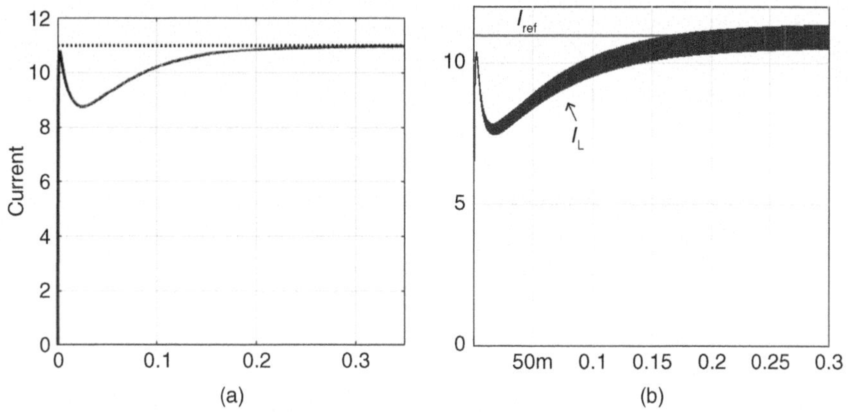

Figure 3.12 Current waveforms for PM = 90° (Design specification not fully satisfied). (a) MATLAB waveform and (b) PSIM waveform.

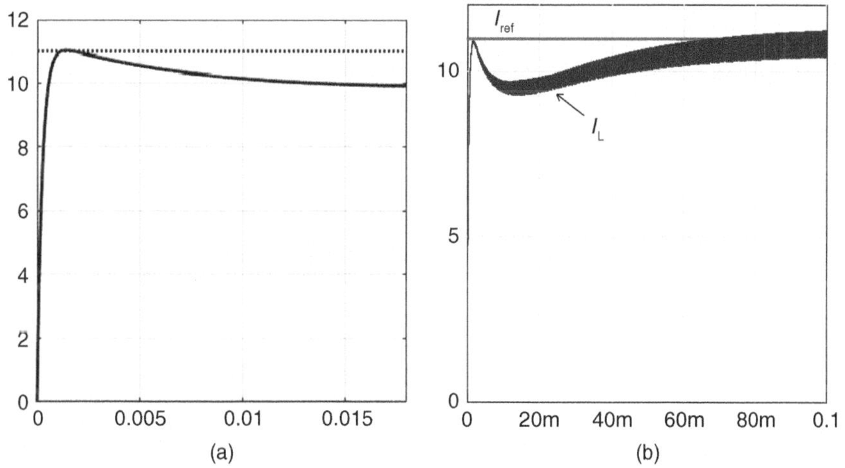

Figure 3.13 Current waveforms for $\omega_c = 4000$ rad/sec and PM = 90°. (a) MATLAB waveform and (b) PSIM waveform.

3.2.2 Voltage Controller (Single Loop)

We now proceed with the design of the voltage controller. According to the results obtained from auto-tuning, a stable system cannot be achieved when the crossover frequency exceeds 16 rad/sec under the given circuit parameters. This limitation stems from the presence of a RHP zero, which imposes a constraint on increasing the loop gain due to phase lag and potential instability. To ensure stable operation, the controller is designed using the following specifications.

- Bandwidth : $\omega_c = 10$ [rad/sec]
- Phase margin : P.M. = 90 [deg]

The structure of the MATLAB m-file for voltage control is similar to that of the current controller, except the plant transfer function is replaced with G_2. The MATLAB m-file for voltage controller is presented in Table 3.5.

Using SISOTOOL's auto-tuning, the PI controller satisfying the design criteria is obtained as

$$G_{cv} = \frac{K_{pi}(s + K_{ii})}{s} = \frac{0.00001918\,(s + 1870)}{s} \tag{3.22}$$

The resulting Bode plot and closed-loop step response are presented in Figure 3.14, confirming that the design specifications are satisfied.

Table 3.5 MATLAB m-file for voltage controller.

```
Kpv = 1; Kiv=1;  % Initial gains
Gcv = Kpv*(s+Kiv)/s;
%sisotool Voltage Single loop
T=sisoinit(1);
T.F.Value = 170;
T.G.Value = G2;
T.C.Value = Gcv;
sisotool(T);
```

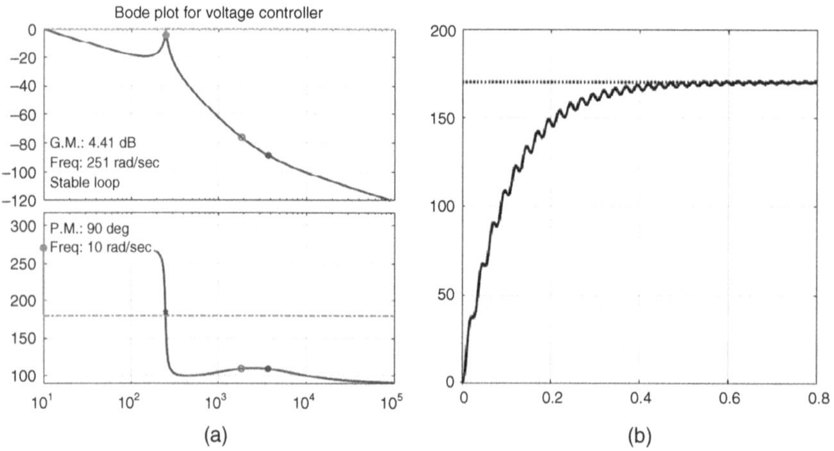

Figure 3.14 SISOTOOL voltage response of the boost converter. (a) Bode diagram and (b) Closed-loop step response.

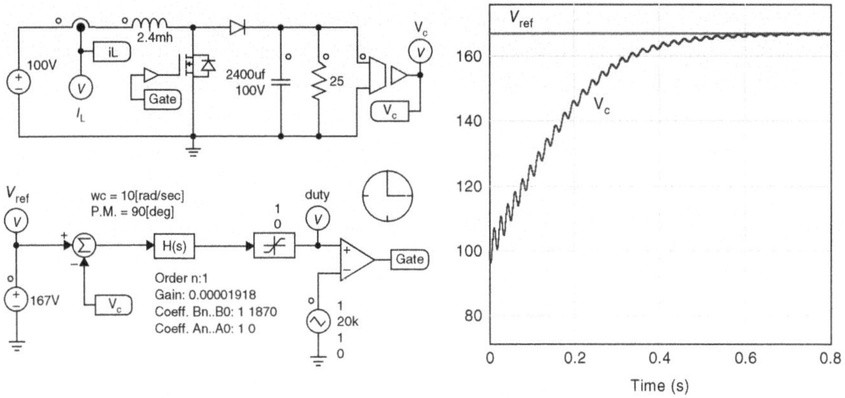

Figure 3.15 Simulation schematic and waveforms of the voltage controller.

To verify the performance in a practical setting, a PSIM simulation was conducted. The circuit schematics and simulation waveforms are shown in Figure 3.15. The results demonstrate close agreement between MATLAB and PSIM, validating the accuracy of the control model and the transfer function-based design.

To assess controller performance under varying operating conditions, the reference voltage was changed from 170 to 250 V. The resulting waveforms are shown in Figure 3.16.

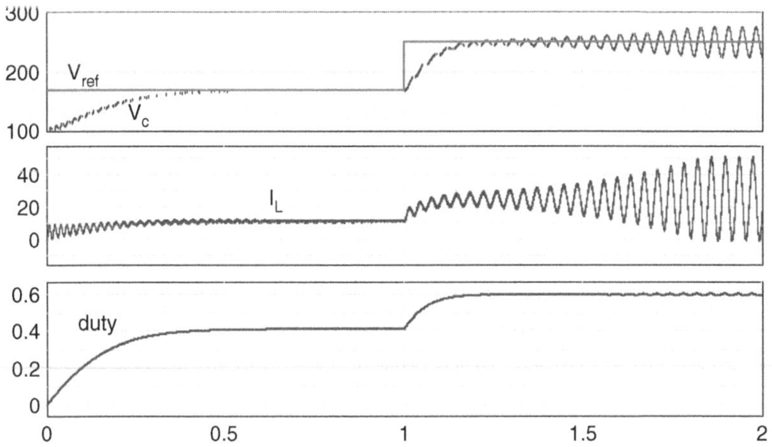

Figure 3.16 Simulation waveforms of the voltage controller.

1) At the nominal reference voltage of 170 V, the controller maintains stability and achieves the desired output regulation. However, at 250 V, the output exhibits divergent behavior, indicating a failure in maintaining regulation.
2) This instability arises because the controller was designed based on a small-signal linear model around the 170 V operating point, which becomes invalid at 250 V. This highlights the limitation of single-point linearization in wide-range operating conditions.

To further investigate the effect of controller parameters, the design condition was modified to $\omega_c = 1$ rad/sec while maintaining the phase margin at 90°. The resulting controller has

- A proportional gain was reduced by a factor of 10, due to the decrease in crossover frequency.

An unchanged integral gain, as the integrator provides low-frequency compensation independent of ω_c.

As illustrated by the MATLAB and PSIM waveforms in Figure 3.17, the system response becomes approximately 10 times slower when the crossover frequency is reduced to 1 rad/sec. This behavior aligns with the well-known inverse relationship between crossover frequency and settling time. From this observation, the following conclusions can be drawn:

- Reducing ω_c enhances system robustness and decreases sensitivity to parameter variations and disturbances. However, this improvement comes at the cost of a significantly slower dynamic response.

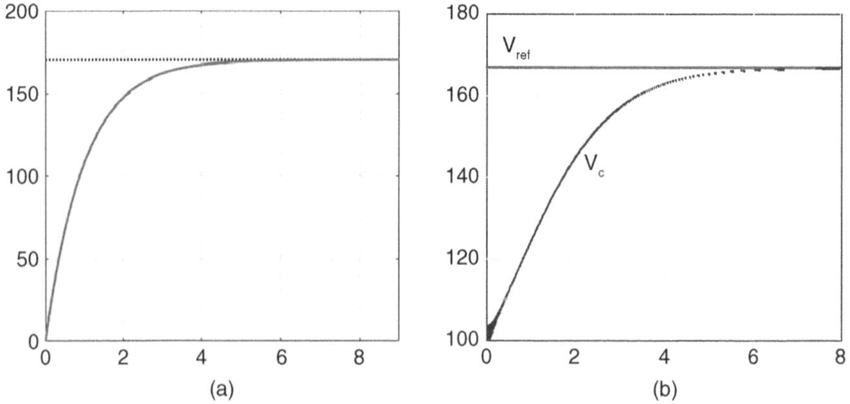

Figure 3.17 Voltage response for $\omega_c = 1$ rad/sec in MATLAB and PSIM. (a) MATLAB waveform and (b) PSIM waveform.

The consistency between MATLAB and PSIM results validates the proposed design methodology across different simulation platforms. It also demonstrates the effectiveness of systematically tuning the controller bandwidth to achieve an appropriate trade-off between performance and stability in voltage loop control.

3.2.3 Double-Loop Controller Design

Based on the transfer functions derived in the previous chapter, the implementation of the double-loop controller is provided in the MATLAB m-file presented in Table 3.6.

The transfer function assignments for each control block in the double-loop structure are illustrated in Figure 3.18.

In contrast to the single-loop voltage controller—where the maximum achievable cutoff frequency was limited to 15 rad/sec due to the RHP zero—the double-loop architecture enables a substantial increase in the allowable cutoff frequency, up to 100 rad/sec for the voltage loop. This results from the improved dynamic separation and enhanced flexibility provided by the cascaded control structure.

The frequency-domain design specifications for the voltage and current controllers are summarized in Table 3.7.

Table 3.6 MATLAB m-file for the double-loop controller.

```
T=sisoinit(6);
T.F.Value  = Vref;
T.G1.Value  = G1;       % current loop G1= (iL/d)
T.C2.Value  = Gci;      % current controller PI = (d/ie)

T.G2.Value  = G3;       % voltage loop tf.  = (Vc/iL)
T.C1.Value  = Gcv;      % voltage controller PI =Vref/Ve)
sisotool(T);
```

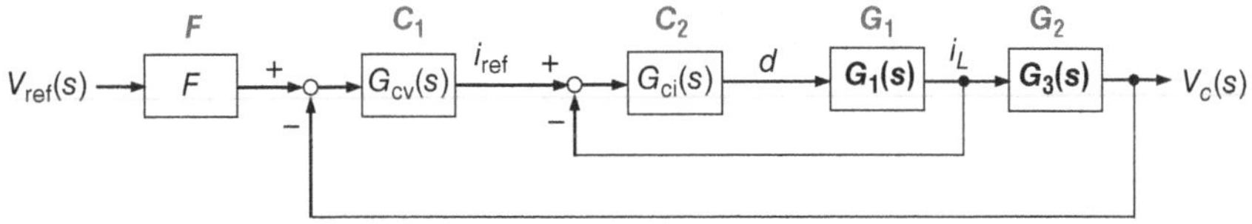

Figure 3.18 Transfer function block assignment for double-loop controller.

Table 3.7 Frequency domain design specification for double-loop controller.

Controller type	Bandwidth [rad/sec]	Phase margin [deg]
Voltage controller (C_1)	100	90
Current controller (C_2)	2400	60

By applying SISOTOOL's auto-tuning feature, the optimal controller transfer functions are obtained as:

$$C_1 = G_{cv}(s) = \frac{0.4111(s+30)}{s}$$

$$C_2 = G_{ci}(s) = \frac{0.0312(s+1230)}{s} \tag{3.23}$$

The corresponding Bode plot of the overall double-loop system is shown in Figure 3.19, clearly verifying that both the voltage and current loops satisfy the desired frequency-domain specifications.

The time-domain step responses of the individual voltage and current loops, simulated in MATLAB, are presented in Figure 3.20.

To validate the performance under practical conditions, a PSIM simulation was conducted. The schematic of the circuit implementation is shown in Figure 3.21, and the corresponding output waveforms are shown in Figure 3.22.

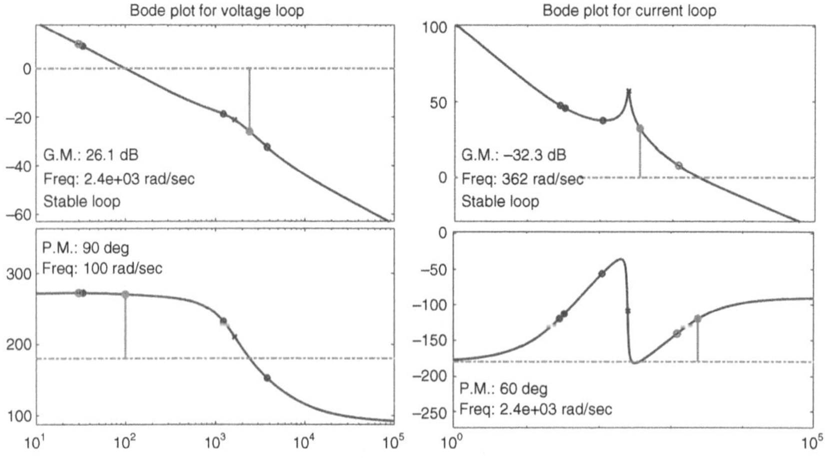

Figure 3.19 Bode diagram of the double loop controller.

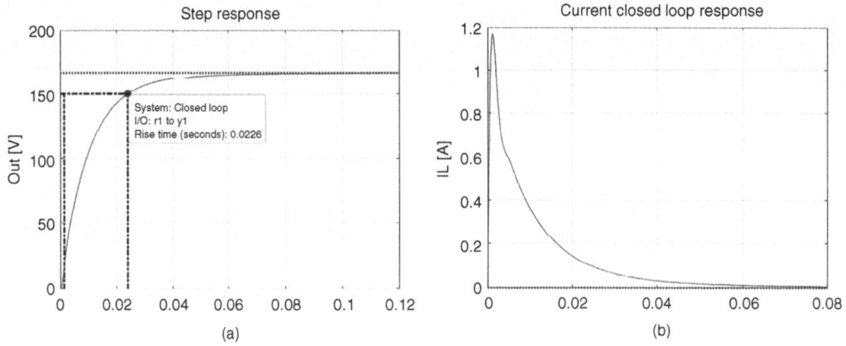

Figure 3.20 Step response characteristics of the double-loop controller. (a) Voltage closed-loop response and (b) Current closed-loop response.

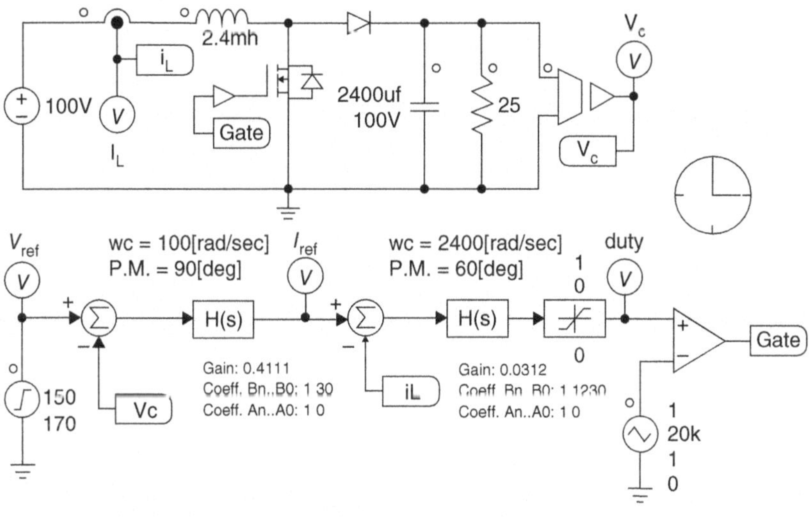

Figure 3.21 PSIM implementation of the double-loop controller.

A comparison between the MATLAB and PSIM simulation results reveals excellent agreement, thereby confirming the accuracy and robustness of the proposed design methodology.

The double-loop controller demonstrates superior performance compared to single-loop implementations, by enabling higher bandwidths and achieving faster and more stable dynamic responses. The design results are validated across both MATLAB and PSIM platforms, confirming that the proposed method is not only theoretically sound but also practical for real-world converter control applications.

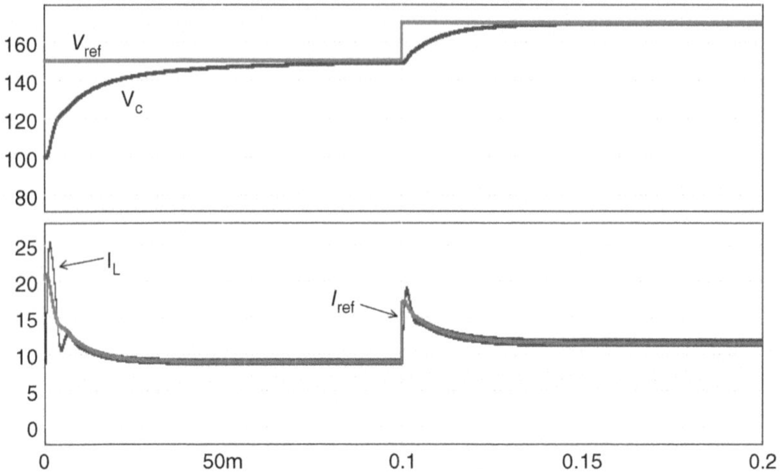

Figure 3.22 Voltage and current waveforms under double-loop control.

3.2.4 Digital Conversion of the Double-Loop Controller

To implement the analog controllers in a digital control system, the continuous-time PI controllers must be converted into discrete-time form. This process involves discretization, difference equation derivation, and code implementation for simulation and hardware deployment:

1) Set the Sampling Time: In MATLAB's *SISOTOOL*, navigate to Architecture > Edit Architecture > Sampling Time Conversion. Set the sampling time to 0.00005 seconds (equivalent to a 20 kHz sampling rate). Use the zero-order hold (ZOH) method.
2) Discrete-Time Controller Expressions: After conversion, the controllers are expressed as:

$$C_1 = G_{cv}(z) = \frac{0.4127(z - 0.999)}{z - 1}$$

$$C_2 = G_{ci}(z) = \frac{0.0331(z - 0.949)}{z - 1} \tag{3.24}$$

3) Difference Equation for C-Implementation: The z-domain transfer function is transformed into difference equation, which is directly implemented in C code, as presented in Table 3.8.
4) C-Code Implementation in PSIM: The following Ccode implements the digital double-loop controller using the derived difference equations. This code is integrated into a PSIM C-block.

Table 3.8 C-block code: Double_Loop.

```
double Vc, IL, Vref, duty, Verr, IL_err, IL_ref;
static double IL_ref_old=0.0, duty_old=0.0, Verr_old=0.0;
static double IL_err_old=0.0;

Vc  = in[0];
IL  = in[1];

if (t < 0.2) Vref = 140;
else         Vref = 170;
Verr = Vref - Vc;

IL_ref = IL_ref_old+0.41273*Verr-0.41273*0.999*Verr_old;
IL_ref_old=IL_ref;
Verr_old=Verr;

IL_err = IL_ref - IL;
duty = duty_old+0.0331*IL_err-0.0331*0.949*IL_err_old;
duty_old=duty;
IL_err_old=IL_err;

out[0]=duty;
out[1]=Vref;
out[2]=IL_ref;
```

PSIM Schematic and Results The simulation schematic and resulting waveforms from the C-block implementation are shown in Figure 3.23. These results demonstrate that the digital controller reproduces the same dynamic behavior as the analog transfer function model, confirming the accuracy of the discretization and implementation process.

Hardware Deployment: For deployment on actual hardware (e.g., TI C2000 DSP), the controller portion of the C code can be directly reused from the PSIM C-block. The code can be ported into Code Composer Studio (CCS) or any similar embedded development environment. This facilitates a seamless transition from simulation to real-time control without requiring changes in controller logic.

By following this structured workflow—from discretization to code implementation and simulation—the designed analog controllers can be successfully transitioned to a real-time digital environment. This approach ensures that the digital controller preserves the performance, stability, and frequency-domain characteristics of its analog counterpart.

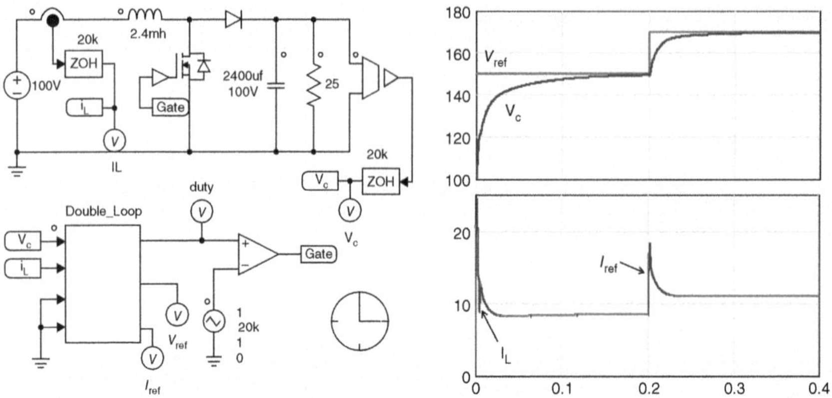

Figure 3.23 PSIM schematic and simulation waveforms of the double-loop C-block controller.

4

Bi-Directional Converter

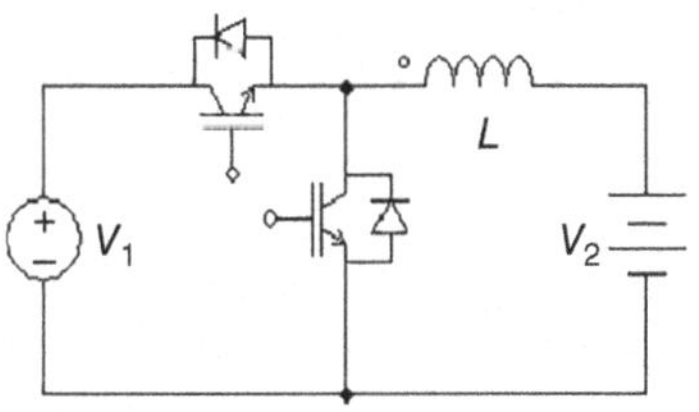

4.1 Structure and Operation

The basic circuit configurations of the buck and boost converters are shown in Figure 4.1. Both converters are inherently unidirectional, meaning that power flows only from the input to the output. This characteristic arises from the use of diodes, which permit current conduction in only one direction and block reverse current flow.

In these conventional configurations

- The input is modeled as a voltage source V_{in}.
- The output voltage V_0 is regulated and stabilized by an output capacitor.
- The load is represented by a resistive element R.

When the load is replaced by a voltage source—such as a battery or a DC-link capacitor connected to an inverter—the converter effectively has voltage sources on both the input and output sides, as illustrated in Figure 4.2. In such a configuration, bidirectional power transfer becomes theoretically possible, depending on the control strategy and circuit topology.

However, bidirectional operation cannot be achieved with conventional diodes, as they inherently block reverse current. To allow current flow in both directions,

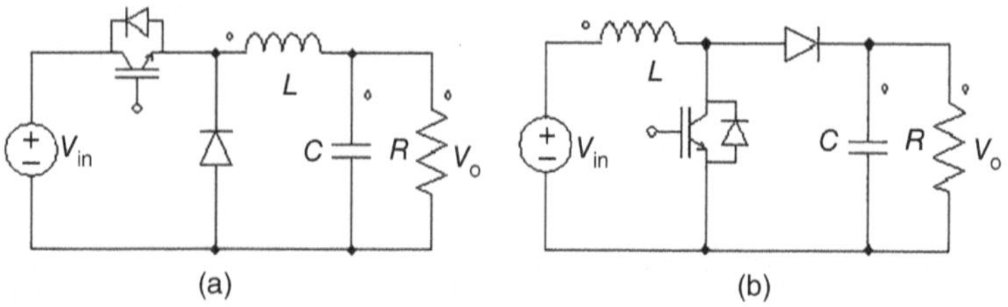

Figure 4.1 Conventional buck and boost converter circuits. (a) Buck converter and (b) Boost converter.

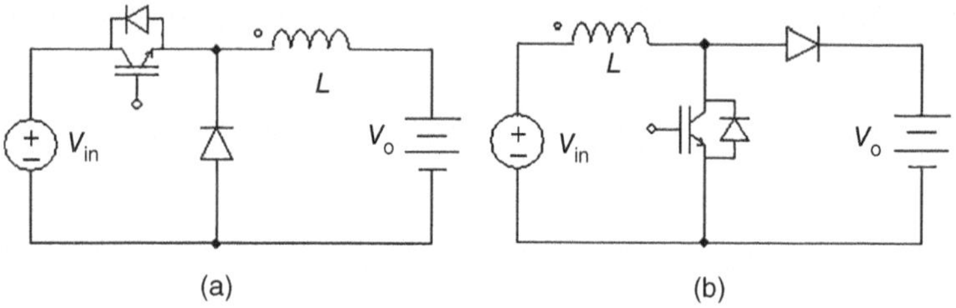

Figure 4.2 Buck and boost converter with output-side voltage source (Battery). (a) Buck converter and (b) Boost converter.

the diodes must be replaced with controllable semiconductor switches. This modification enables active control of current direction.

By replacing the diodes with switches, the converter topology is transformed into the *bi-directional buck–boost configuration*, as shown in Figure 4.3. This structure allows seamless power transfer in either direction, significantly enhancing operational flexibility and expanding the range of possible applications.

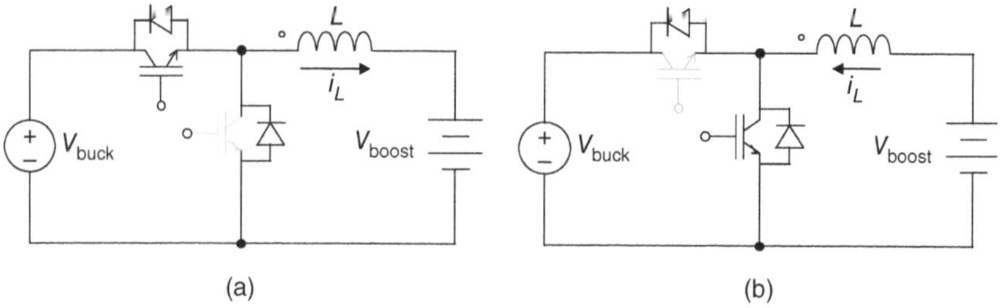

Figure 4.3 Buck–boost type bidirectional converter circuit. (a) Buck mode and (b) Boost mode.

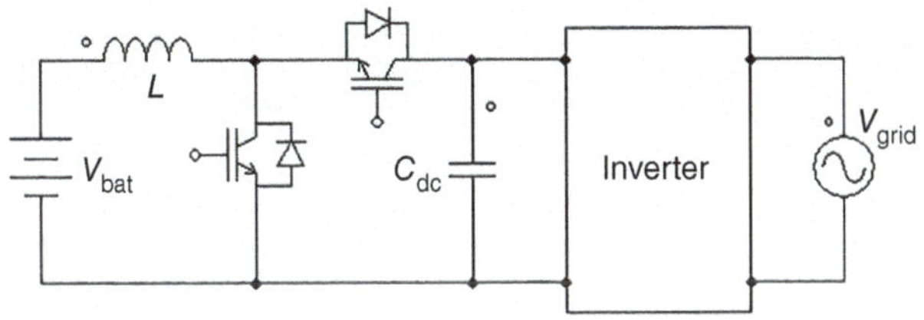

Figure 4.4 Two-stage configuration with bi-directional converter and grid-tied inverter.

A closer examination reveals that, under bidirectional operation, the buck and boost converters become structurally identical. In fact, a boost converter can be interpreted as a buck converter with the roles of the input and output voltage sources reversed. Consequently, replacing the output-side diode with a controllable switch effectively converts the circuit into a bidirectional converter, commonly referred to as a buck–boost type bidirectional converter.

Application Domains of Bidirectional Converters Bidirectional converters are widely employed in power-electronic systems in which both the input and output sides behave as voltage sources, most notably in battery-based applications. Batteries are capable of both delivering and absorbing electrical energy, making them well-suited for bidirectional power conversion.

A representative and widely deployed application is the battery energy storage system (BESS). As illustrated in Figure 4.4, a typical BESS adopts a two-stage power-conversion architecture consisting of:

- a bidirectional DC–DC converter interfacing the battery, and
- a grid-tied inverter connected to the AC grid.

Although a battery can, in principle, be directly connected to the DC link of an inverter, this approach is impractical in low-voltage battery systems. In such cases, a bidirectional DC–DC converter is required to boost the battery voltage to a suitable DC-link level before interfacing with the inverter. In this configuration:

- Battery-to-grid operation (discharging mode): The bidirectional converter operates in boost mode to raise the battery voltage. A double-loop control structure—typically comprising an outer DC-link voltage loop and an inner current loop—is employed to regulate the DC-link voltage supplied to the inverter.
- Grid-to-battery operation (charging mode): The converter operates in buck mode to charge the battery. In this case, the inverter regulates the DC-link voltage, while the bidirectional converter employs a single-loop current controller, since the battery voltage can be regarded as approximately constant over the control bandwidth.

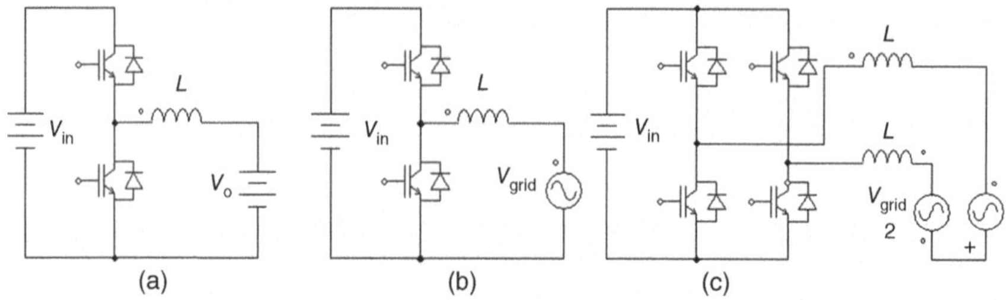

Figure 4.5 Functional equivalence and configurations of grid-tied bi-directional converters. (a) Bidirectional converter, (b) Half-wave inverter operation and (c) Single-phase grid-tied inverter.

Beyond BESS applications, bidirectional converters are also widely used in:

- Photovoltaic (PV) systems with battery energy storage, where power flow must be managed between PV arrays, batteries, and the grid, and
- Regenerative braking systems in electric and hybrid electric vehicles, where kinetic energy is recovered and stored in the battery during deceleration.

Relation to Grid-Tied Inverters The bidirectional DC–DC converter can be redrawn in the form shown in Figure 4.5(a), which reveals that its circuit structure is functionally equivalent to a voltage-source grid-tied inverter. This equivalence provides valuable insight into the control and operational characteristics of bidirectional converters.

- When power is delivered from the battery to the grid, the converter effectively operates in buck mode, and a single-loop current controller is sufficient to regulate the grid current.
- When power flows from the grid to the battery, the converter operates in boost mode, again using a single-loop current controller. This mode of operation is illustrated in Figure 4.5(b).

For single-phase grid-tied inverters, two bi-directional converter legs are operated in parallel with a 180° phase shift, as shown in Figure 4.5(c). In three-phase grid-tied systems, three bidirectional converters are employed, each operating with a 120° phase displacement. Despite the increased number of phases, the fundamental operating principle remains identical to that described for the single-phase case in Figure 4.5(b).

4.2 Controller Design

The bi-directional converter circuit is illustrated in Figure 4.6. The circuit includes two switches: the upper switch operates during buck mode (charge mode), while

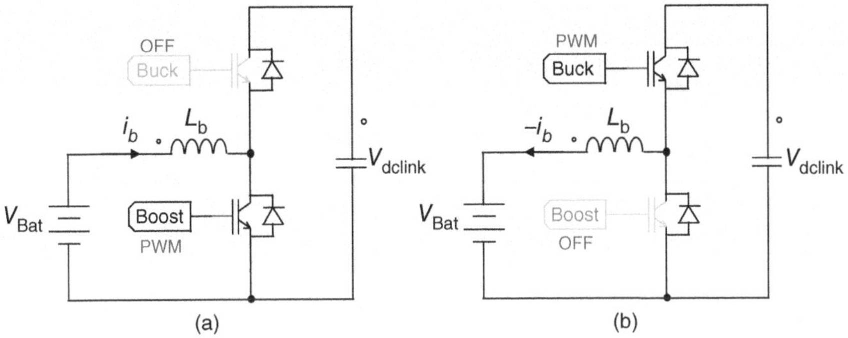

Figure 4.6 The bi-directional converter circuit. (a) Boost mode (discharge mode) and (b) Buck mode (charge mode).

the lower switch is active during boost mode (discharge mode). The switching behavior for each mode is summarized as follows, depending on the direction of power flow.

1) Boost mode (discharge mode)
   - The lower switch (*Boost*) is modulated to regulate current from the battery to the DC-link.
   - The upper switch (*Buck*) remains OFF.
2) Buck mode (charge mode)
   - The upper switch (*Buck*) is modulated to regulate current from the DC-link to the battery.
   - The lower switch (*Boost*) remains OFF.

This mode-dependent operation ensures proper directional control of power flow in the bi-directional converter.

4.2.1 Boost Mode Operation (Discharge Mode)

In discharge mode—corresponding to the boost operation—the controller regulates both the DC-link voltage and the inductor current, forming a double-loop control structure. The corresponding controller block diagram is illustrated in Figure 4.7. Since this double-loop configuration is identical to the one used in conventional boost converter control, its derivation is omitted here. The same *PI* gains as defined in Eq. (3.24) are applied for this mode.

4.2.2 Buck Mode Operation (Charge Mode)

In charge mode, the system operates as a buck converter. The DC-link voltage is assumed to be regulated by the grid-tied inverter, and the battery voltage is

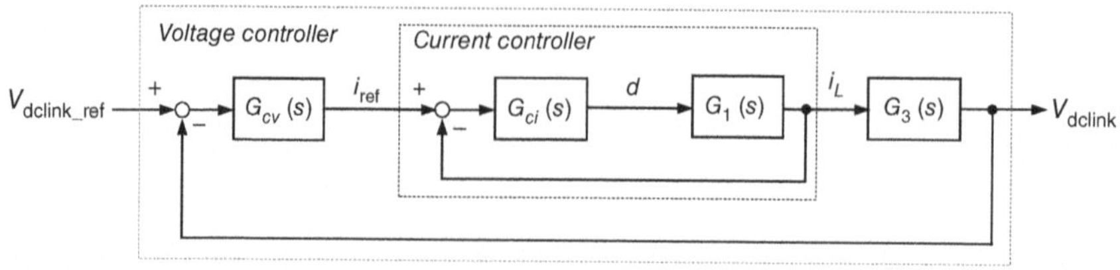

Figure 4.7 Double-loop controller configuration for boost mode.

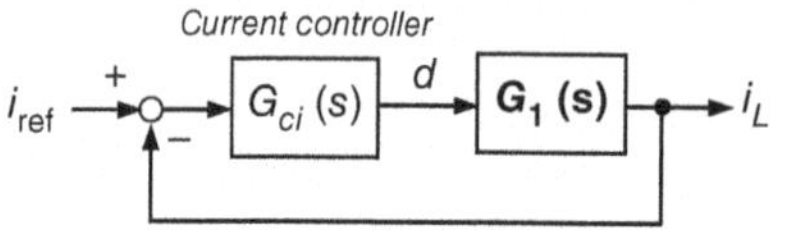

Figure 4.8 Controller configuration in buck mode.

considered constant over a short time interval. Thus, the control objective is limited to regulating the inductor current, resulting in a single-loop control structure, shown in Figure 4.8.

Let us derive the state-averaged equations for the buck mode (charge).

$$\text{ON time } (d): \quad V_{\text{dc-link}} = L\frac{di_L}{dt} + r_L i_L + V_{\text{Bat}} \tag{4.1}$$

$$\text{OFF time } (1-d): \quad 0 = L\frac{di_L}{dt} + r_L i_L + V_{\text{Bat}} \tag{4.2}$$

$$\text{Averaged equation:} \quad L\frac{di_L}{dt} + r_L i_L = V_{\text{dc-link}} \cdot d - V_{\text{Bat}} \tag{4.3}$$

To linearize the system, the duty input is separated into feedforward and feedback terms.

$$d = d_{fb} + d_{ff}, \quad \text{where} \quad d_{ff} = \frac{V_{\text{Bat}}}{V_{\text{dc-link}}} \tag{4.4}$$

Substituting yields the linearized state equation

$$L\frac{di_L}{dt} + r_L i_L = V_{\text{dc-link}} \cdot d_{fb} \tag{4.5}$$

The resulting transfer function is

$$G_1(S) = \frac{I_{L(S)}}{d(s)} = \frac{V_{\text{dc-link}}}{sL + r_L} \tag{4.6}$$

Current Controller Design in Buck Mode The MATLAB m-file for current controller is presented in Table 4.1.

Table 4.1 MATLAB m-file for current controller.

```
%% Buck mode - Current control %%
s=tf('s');

Vbat = 250;       % Input voltage
L = 2.4e-3;       % L=3mh
rL= 0.02;         % rL = 20 mohm
Vdclink = 450; % Dclink voltage

G1 = Vdclink/(s*L+rL);

Kpi =1; Kii =1; %Initial gains
Gci = Kpi*(s+Kii)/s;

IL_ref  = 8;

T=sisoinit(1);
T.F.Value = IL_ref;
T.G.Value = G1;
T.C.Value = Gci;
sisotool(T);
```

The design specifications are as follows:

- BandWidth: 4800 rad/sec
- Phase margin: 60 deg.

The values designed using auto-tuning are as follows.

$$G_{Ci}(s) = 0.0221\frac{(s + 2780)}{s} \tag{4.7}$$

When converted into a digital expression, the results are as follows (f_{sw}=20 [kHz])

$$G_{Ci}(z) = 0.02468\frac{(z - 0.902)}{z - 1} \tag{4.8}$$

The MATLAB step response is shown in Figure 4.9.

PSIM Verification Figure 4.10 presents the PSIM simulation schematic of the bi-directional converter. A mode selector variable is used to toggle between the two modes: mode = 0 is Discharge (boost) and mode = 1 is Charge (buck).

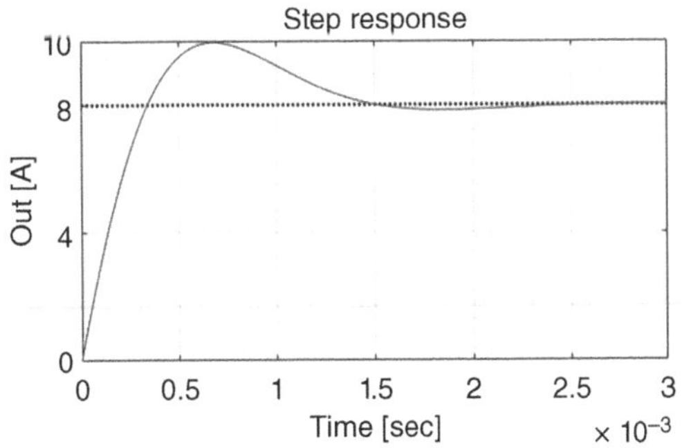

Figure 4.9 Closed-loop step response (MATLAB).

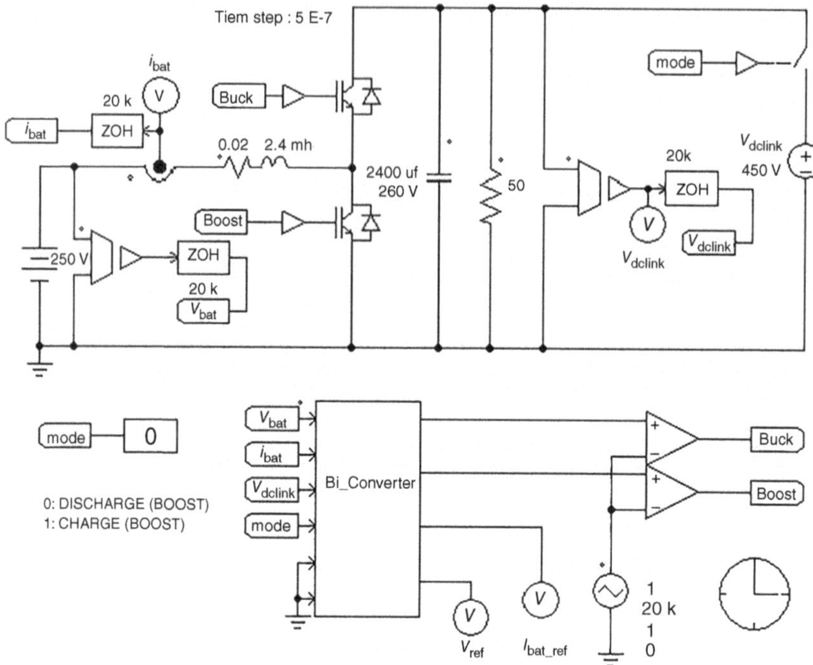

Figure 4.10 PSIM schematic for bi-directional converter simulation.

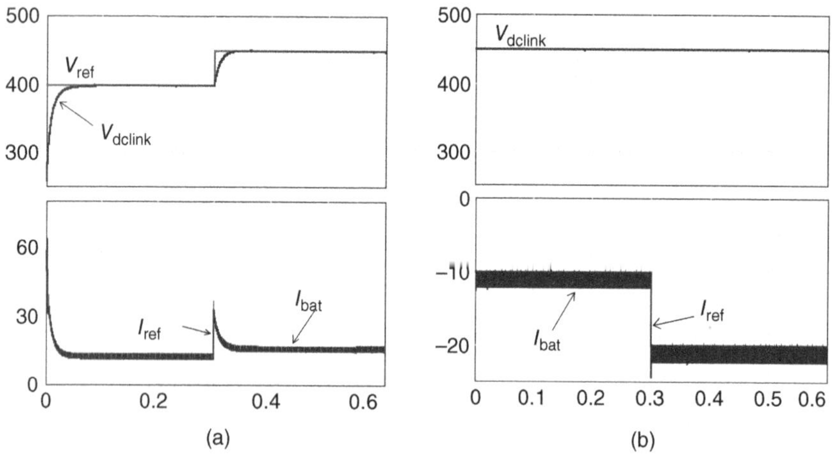

Figure 4.11 Simulation results for bi-directional converter. (a) Discharge mode waveform and (b) Charge mode waveform.

Table 4.2 C-Block code for bi-directional converter: Bi_Converter.

```
double Vbat, Ibat, Vlink, Mode;
double Vdc_ref, Vdc_err, Ichg_ref;
double IL_ref, IL_err, duty;
static double  IL_ref_old=0, duty_old=0,
Vdc_err_old=0, IL_err_old=0 ;

Vbat  = in[0];
Ibat  = in[1];
Vlink = in[2];
Mode  = in[3];

if (Mode == 0)  { // discharge mode
  if (t < 0.3) Vdc_ref = 400;
  else         Vdc_ref = 450;
  Vdc_err = Vdc_ref - Vlink;
  IL_ref=IL_ref_old+0.4127*Vdc_err
     -0.41273*0.999*Vdc_err_old;
  IL_ref_old=IL_ref;
  Vdc_err_old=Vdc_err;

  IL_err = IL_ref - Ibat;
  duty = duty_old +0.033082*IL_err
     -0.033082*0.949*IL_err_old;
  duty_old=duty;
  IL_err_old=IL_err;
  if (duty > 0.9)  duty=0.9;
  if (duty < 0)    duty=0;

  out[0] = 0 ;       // buck
  out[1] = duty;     // boost
  out[2] = IL_ref;
  out[3] = Vdc_ref;
} // end of discharge mode

else if (Mode == 1)  {  // Charge
  if (t < 0.3) Ichg_ref = -10;
  else         Ichg_ref = -20;
  IL_err = Ichg_ref - Ibat;
  duty = duty_old +0.02468*IL_err
     - 0.02468*0.902*IL_err_old;
  duty_old=duty;
  IL_err_old=IL_err;

  out[0] = -duty + (Vbat/Vlink);
  out[1] = 0;    // boost
  out[2] = Ichg_ref;
} // end of charge mode
```

The implementation of the bidirectional converter C-block is presented in Table 4.2. The discharge mode reuses the double-loop control logic from the boost converter design, whereas the charge mode employs a single-loop current controller. It should be noted that in charge mode, the current sensor measures in the opposite direction; therefore, the charge current reference is defined as negative, and a sign inversion is applied to the duty output.

The simulation results are presented in Figure 4.11. The output waveforms from both MATLAB and PSIM are in excellent agreement, verifying the accuracy of both the control design and digital implementation.

5

Three-Level Boost Converter

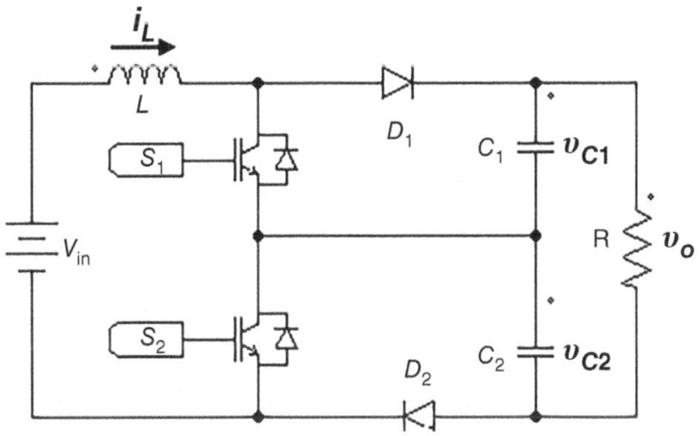

5.1 Structure and Operation

In the preceding chapter, the operation of the conventional boost converter was discussed, including the voltage stress across the switching device. As illustrated in Figure 5.1, the drain-to-source voltage (V_{ds}) of the switch alternates depending on its operating state. When the switch is ON, V_{ds} is ideally 0 V; when the switch is OFF, the entire output voltage (V_c) appears across the device.

Accordingly, the switching device—commonly an Insulated Gate Bipolar Transistor (IGBT)—must have a breakdown voltage rating that exceeds the output voltage. In typical applications, commercially available IGBTs are rated at 600 and 1200 V. For single-phase 220 V AC systems, 600 V IGBTs are commonly used, while three-phase 380 V AC systems generally require 1200 V devices.

With the growing demand for high-capacity (~MW-class) power conversion systems, input voltage levels are continuously increased to improve efficiency. For instance, photovoltaic (PV) inverters, which historically operated below 1000 V,

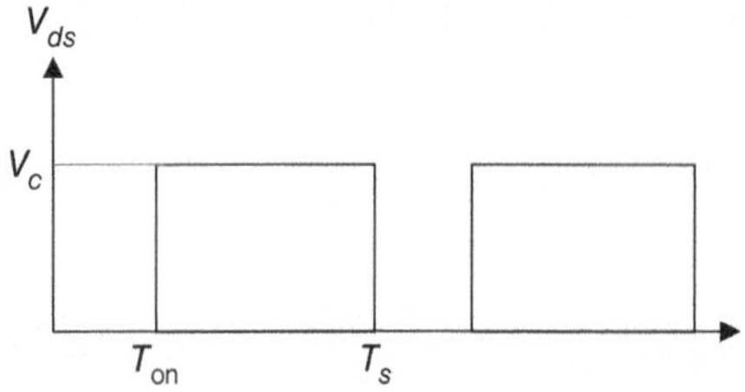

Figure 5.1 Voltage stress across the switch in a conventional boost converter.

are now designed to support PV string voltages up to 1500 V. Similarly, in energy storage systems (ESS), battery voltages have been elevated from around 600 to 1200 V.

Such high-voltage applications necessitate switching devices with voltage ratings exceeding 1200 V. Although IGBTs rated at 1700, 2400 V, and higher are available, they are substantially more expensive. A step increase in voltage rating typically results in a 4 to 6-fold increase in device cost. This cost escalation poses significant challenges for designers targeting cost-effective, mass-producible systems.

To address these challenges, this chapter introduces the *three-level boost converter*, an enhanced topology derived from the conventional two-level configuration. The key innovation of the three-level boost converter lies in its use of two series-connected switches. This structure effectively halves the voltage stress applied to each switch, allowing the use of two 1200 V IGBTs in systems requiring up to 1500 V operation.

Although the number of switching devices is doubled, the total cost is typically lower than that of a single higher-voltage-rated IGBT due to the steep price increase associated with higher voltage classes. As a result, the three-level boost converter presents a cost-efficient solution for high-voltage power conversion systems.

Three-Level Boost Converter Topology Figure 5.2 presents the circuit configurations of the conventional boost converter and the three-level boost converter.

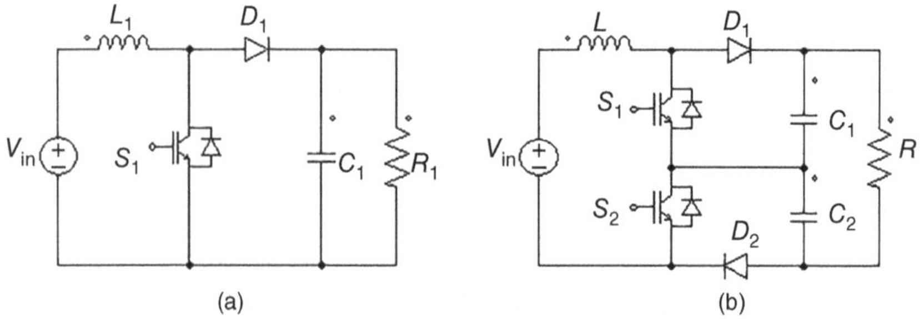

Figure 5.2 Configurations of conventional and three-level boost converter. (a) Conventional boost and (b) Three-level boost.

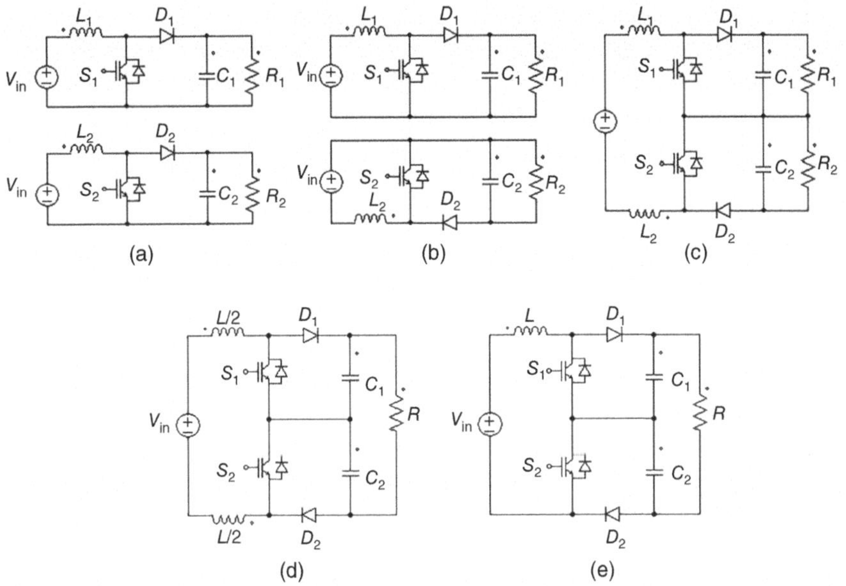

Figure 5.3 Topological development of the three-level boost converter. (a) Parallel boost converters, (b) Relocation of L_2, D_2, (c) Converter merging, (d) Circuit integration and (e) Final three-level boost converter.

In the conventional boost converter (Figure 5.2a), the output voltage alternates between two discrete levels: 0 and V_c. In contrast, the three-level boost converter (Figure 5.2b) introduces three voltage levels: 0, V_c, and $2V_c$. This topology is derived by connecting two conventional boost converters in parallel and then restructuring their components to achieve an integrated multi-level configuration. The topological evolution process is illustrated in Figure 5.3.

Advantages and Disadvantages of the Three-Level Boost Converter

Advantages

1) Higher Voltage Gain: Compared to the conventional boost converter, the three-level boost converter enables a higher voltage gain. In a traditional configuration, the maximum achievable duty cycle is constrained by practical factors such as inductor resistance, switch on-resistance, and parasitic inductance. In contrast, the three-level topology allows each switch to operate with half of the total effective duty cycle. This enables higher cumulative duty ratios and consequently greater voltage amplification.
2) Reduced Inductor Size: The effective switching frequency experienced by the inductor is doubled, which increases the ripple frequency of the inductor current. This frequency doubling allows for a reduction in the size of the inductor while maintaining acceptable ripple levels, thereby contributing to improved power density.

3) Lower Voltage and Current Stress: Each switch in the three-level converter is subjected to only half of the total output voltage. This reduction in voltage stress also translates into lower current stress, improving reliability and allowing the use of devices with lower voltage ratings.

Disadvantages

1) Increased Component Count: The topology requires twice the number of switches and diodes compared to the conventional boost converter, which increases system complexity and component cost.
2) Complex Control Design: The control design is significantly more complicated. The system includes one inductor and two output capacitors, resulting in three state variables—i_L, v_{c1}, and v_{c2}. This yields a third-order state-space representation and corresponding transfer function, complicating the analysis and controller design.
   Additionally, the converter operates in two distinct modes depending on the boundary condition at a duty cycle of $D=0.5$. Each mode is governed by a different equivalent circuit, further complicating control strategies.
3) Capacitor Voltage Imbalance: A potential voltage imbalance between the two output capacitors v_{c1} and v_{c2} may arise during operation. Maintaining voltage balance is critical for stable performance and requires sophisticated control algorithms or auxiliary balancing circuits.
4) Challenges in Mathematical Modeling: Due to the increased number of components, dual-mode operation, and interaction between state variables, developing an accurate mathematical model of the converter is challenging. This complexity makes controller design both critical and non-trivial.

5.2 State-Space Averaged Equation and Transfer Function

To analyze the dynamic behavior of the three-level boost converter, the state-space averaged equations are derived based on three state variables: the inductor current i_L, and the voltages across the two output capacitors, v_{c1} and v_{c2}, as illustrated in Figure 5.4.

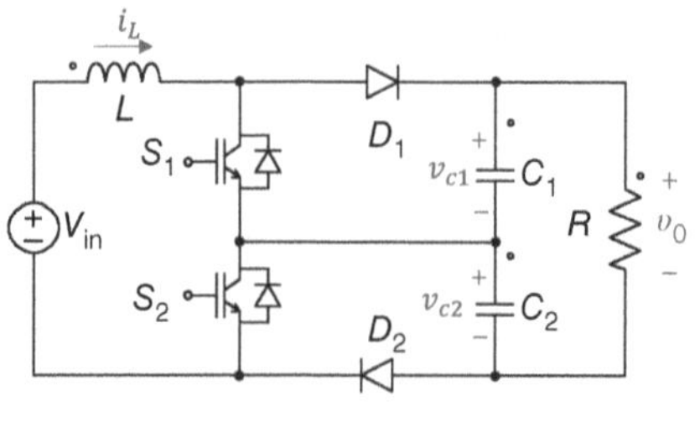

Figure 5.4 State variables in the three-level boost converter.

The operation of the converter is determined by the switching states of S_1 and S_2. These switches form four possible conduction intervals, as summarized in the table below:

S_1	S_2	Interval
1	1	(a)
1	0	(b)
0	1	(c)
0	0	(d)

Switches S_1 and S_2 are operated with identical conduction times but with a 180-degree phase shift. Let D represent the duty cycle of each switch. The following conditions apply:

- For intervals (b), (c), and (d), the effective duty cycle satisfies $0 < D < 0.5$
- For interval (a), both switches are ON simultaneously, resulting in an overlap condition where $D > 0.5$

Due to the varying conduction states, the current flow paths differ depending on whether the system operates in the low-duty ($D < 0.5$) or high-duty ($D > 0.5$) region. Figure 5.5 illustrates the corresponding switching patterns based on the duty cycle boundary, and the equivalent circuits for each switching combination are depicted in Figure 5.6.

To derive the state-space averaged model, we analyze each switching interval separately, constructing its equivalent circuit and formulating the state equations accordingly. These equations are then averaged over one switching period based on the proportion of time spent in each interval. This process allows for the development of a unified, time-continuous model that captures the average dynamics of the three-level boost converter under both low and high duty cycle conditions. The resulting state-space averaged equations serve as the foundation for deriving transfer functions, enabling frequency-domain analysis and controller design.

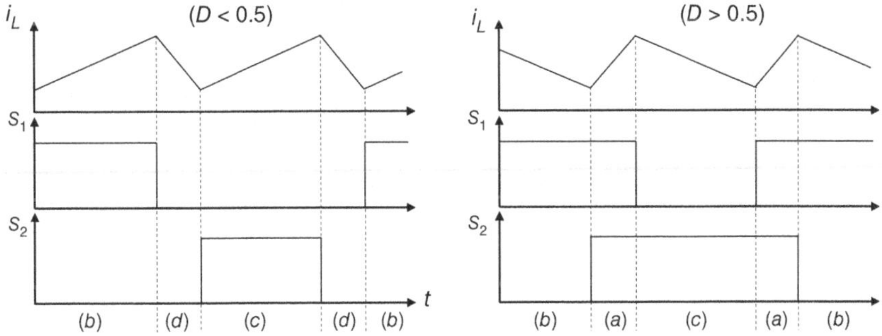

Figure 5.5 Switching pattern according to duty cycle boundary.

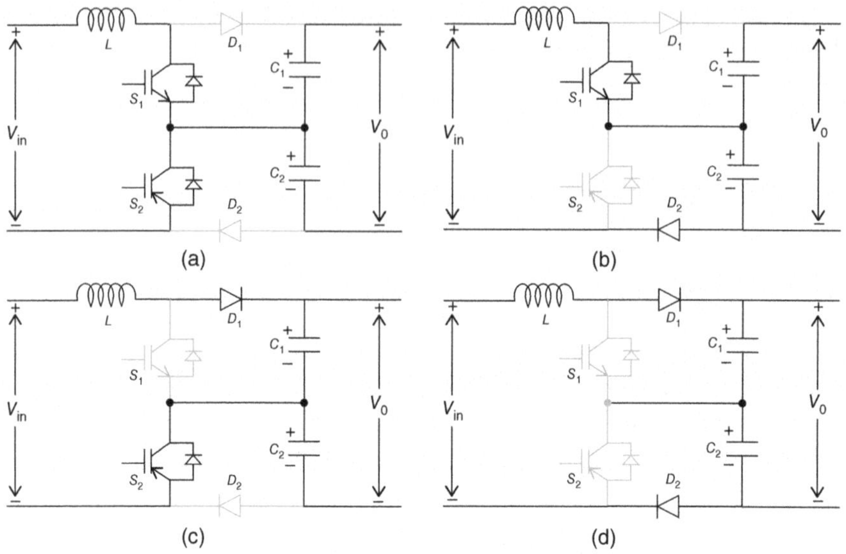

Figure 5.6 Equivalent circuits corresponding to each switching interval.

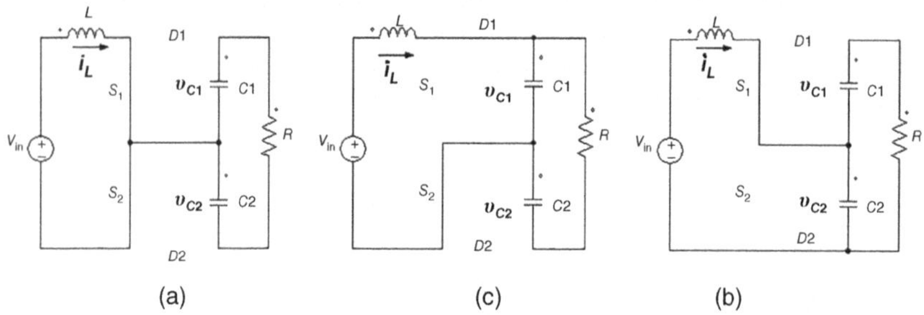

Figure 5.7 Equivalent circuits for each switching interval (a), (c), and (b). (a) Interval, (b) Interval (c) and Interval.

Case 1: Duty Cycle* D > *0.5 From the second diagram in Figure 5.5, one complete switching period consists of four sequential intervals: (a), (c), (a), and (b). The corresponding equivalent circuits for each switching interval are shown in Figure 5.7.

The state-space equations for each interval, based on the state variables i_L, v_{c1}, and v_{c2}, are derived as follows:

- Interval (a): Both switches S_1 and S_2 are ON

$$\frac{di_L}{dt} = \frac{1}{L}V_{in}$$

$$\frac{dv_{c1}}{dt} = -\frac{1}{RC_1}(v_{C1} + v_{C2})$$

$$\frac{dv_{c2}}{dt} = -\frac{1}{RC_2}(v_{C1} + v_{C2})$$

$$\begin{pmatrix} \frac{di_L}{dt} \\ \frac{dv_{c1}}{dt} \\ \frac{dv_{c2}}{dt} \end{pmatrix} = \begin{pmatrix} 0 & 0 & 0 \\ 0 & -\frac{1}{RC_1} & -\frac{1}{RC_1} \\ 0 & -\frac{1}{RC_2} & -\frac{1}{RC_2} \end{pmatrix} \begin{pmatrix} i_L \\ v_{C1} \\ v_{C2} \end{pmatrix} + \begin{pmatrix} \frac{1}{L} \\ 0 \\ 0 \end{pmatrix} V_{in} = A_a X + B_a V_{in} \tag{5.1}$$

Interval (c) : S_1 OFF, S_2 ON

$$\frac{di_L}{dt} = -\frac{1}{L}v_{c1} + \frac{1}{L}V_{in}$$

$$\frac{dv_{c1}}{dt} = \frac{1}{C_1}i_L - \frac{1}{RC_1}(v_{C1} + v_{C2})$$

$$\frac{dv_{c2}}{dt} = -\frac{1}{RC_2}(v_{C1} + v_{C2})$$

$$\begin{pmatrix} \frac{di_L}{dt} \\ \frac{dv_{c1}}{dt} \\ \frac{dv_{c2}}{dt} \end{pmatrix} = \begin{pmatrix} 0 & -\frac{1}{L} & 0 \\ \frac{1}{C_1} & -\frac{1}{RC_1} & -\frac{1}{RC_1} \\ 0 & -\frac{1}{RC_2} & -\frac{1}{RC_2} \end{pmatrix} \begin{pmatrix} i_L \\ v_{C1} \\ v_{C2} \end{pmatrix} + \begin{pmatrix} \frac{1}{L} \\ 0 \\ 0 \end{pmatrix} V_{in} = A_c X + B_c V_{in} \tag{5.2}$$

Interval (b) : S_1 ON, S_2 OFF

$$\frac{di_L}{dt} = -\frac{1}{L}v_{c2} + \frac{1}{L}V_{in}$$

$$\frac{dv_{c1}}{dt} = -\frac{1}{RC_1}(v_{C1} + v_{C2})$$

$$\frac{dv_{c2}}{dt} = \frac{1}{C_2}i_L - \frac{1}{RC_2}(v_{C1} + v_{C2})$$

$$\begin{pmatrix} \frac{di_L}{dt} \\ \frac{dv_{c1}}{dt} \\ \frac{dv_{c2}}{dt} \end{pmatrix} = \begin{pmatrix} 0 & 0 & -\frac{1}{L} \\ 0 & -\frac{1}{RC_1} & -\frac{1}{RC_1} \\ \frac{1}{C_2} & -\frac{1}{RC_2} & -\frac{1}{RC_2} \end{pmatrix} \begin{pmatrix} i_L \\ v_{C1} \\ v_{C2} \end{pmatrix} + \begin{pmatrix} \frac{1}{L} \\ 0 \\ 0 \end{pmatrix} V_{in} = A_b X + B_b V_{in} \tag{5.3}$$

Interval Duration and Duty Cycle Relations Let T_s be the total switching period and d the duty cycle. From the interval sequence *(a)-(c)-(a)-(b)*, the timing relationships are

1) Total period

$$T_s = 2T_a + T_b + T_c$$

Given that $T_b = T_c$, this simplifies to

$$T_s = 2T_a + 2T_b$$

2) From the definition of duty cycle d

$$dT_s = 2T_a + T_b$$

Solving these equations gives

$$T_a = \frac{(2d-1)T_s}{2}, \quad T_b = T_c = (1-d)T_s \tag{5.4}$$

Derivation of Averaged Matrices The state-space averaged matrices over one switching period T_s are defined as

$$A_{\text{avg}} = (A_a \cdot 2T_a + A_b \cdot T_b + A_c \cdot T_c)/T_s$$

Substituting the expressions for durations

$$A_{\text{avg}} = A_a \cdot (2d-1) + (A_b + A_c) \cdot (1-d)$$

The resulting averaged A-matrix is

$$A_{\text{avg}} = \begin{pmatrix} 0 & -\frac{(1-d)}{L} & -\frac{(1-d)}{L} \\ \frac{(1-d)}{C_1} & -\frac{1}{RC_1} & -\frac{1}{RC_1} \\ \frac{(1-d)}{C_2} & -\frac{1}{RC_2} & -\frac{1}{RC_2} \end{pmatrix} \tag{5.5}$$

The corresponding averaged input matrix B is

$$B_{\text{avg}} = (B_a \cdot 2T_a + B_b \cdot T_b + B_c \cdot T_c)/T_s$$

Substituting the expressions for durations

$$B_{\text{avg}} = B_a \cdot (2d-1) + (B_b + B_c) \cdot (1-d)$$

The resulting B-matrix is

$$B_{\text{avg}} = \begin{pmatrix} \frac{1}{L} \\ 0 \\ 0 \end{pmatrix} V_{\text{in}} \tag{5.6}$$

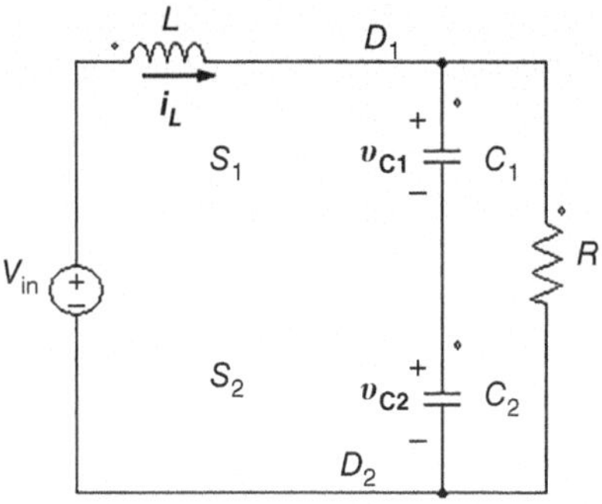

Figure 5.8 Equivalent circuit for interval (d).

These averaged matrices A_{avg} and B_{avg} represent the dynamic model of the three-level boost converter for $D > 0.5$. They serve as the basis for linear system analysis and transfer function derivation for control system design.

Case 2: Duty Cycle* D < *0.5 From the first diagram in Figure 5.5, it can be observed that one complete switching cycle is composed of four intervals: (b), (d), (c), and (d). Since the equivalent circuits for intervals (b) and (c) have already been defined in the $D > 0.5$ case, only the equivalent circuit for interval (d) —where both switches are OFF—requires further analysis. This configuration is illustrated in Figure 5.8.

- Interval (d): Both switches S_1 and S_2 are OFF

The corresponding circuit equations are

$$V_{in} - L\frac{di_L}{dt} + v_{C1} + v_{C2}$$

$$\frac{dv_{c1}}{dt} = \frac{i_L}{C_1} - \frac{1}{RC_1}(v_{C1} + v_{C2})$$

$$\frac{dv_{c2}}{dt} = \frac{i_L}{C_2} - \frac{1}{RC_2}(v_{C1} + v_{C2})$$

Expressed in state-space form

$$\begin{pmatrix} \frac{di_L}{dt} \\ \frac{dv_{c1}}{dt} \\ \frac{dv_{c2}}{dt} \end{pmatrix} = \begin{pmatrix} 0 & -\frac{1}{L} & -\frac{1}{L} \\ \frac{1}{C_1} & -\frac{1}{RC_1} & -\frac{1}{RC_1} \\ \frac{1}{C_2} & -\frac{1}{RC_2} & -\frac{1}{RC_2} \end{pmatrix} \begin{pmatrix} i_L \\ v_{C1} \\ v_{C2} \end{pmatrix} + \begin{pmatrix} \frac{1}{L} \\ 0 \\ 0 \end{pmatrix} V_{in} = A_d X + B_d V_{in} \tag{5.7}$$

Interval Duration and Duty Cycle Relations To express the durations of each interval in terms of the duty cycle d and the switching period T_s, the following relationships apply

1) Total switching period

$$T_s = 2T_d + T_b + T_c$$

Given that $T_b = T_c$, this simplifies

$$T_s = 2T_d + 2T_b$$

2) From the definition of duty cycle d

$$dT_s = T_b$$

Solving these equations yields

$$T_d = \frac{(1-2d)T_s}{2}, \quad T_b = T_c = dT_s \tag{5.8}$$

Derivation of Averaged Matrices The state-space averaged matrices over one switching cycle are given by

$$A_{\text{avg}} = A_d \cdot (1-2d) + (A_b + A_c) \cdot d = \begin{pmatrix} 0 & -\frac{(1-d)}{L} & -\frac{(1-d)}{L} \\ \frac{(1-d)}{C_1} & -\frac{1}{RC_1} & -\frac{1}{RC_1} \\ \frac{(1-d)}{C_2} & -\frac{1}{RC_2} & -\frac{1}{RC_2} \end{pmatrix} \tag{5.9}$$

The resulting B-matrix is

$$B_{\text{avg}} = B_d \cdot (1-2d) + (B_b + B_c) \cdot d = \begin{pmatrix} \frac{1}{L} \\ 0 \\ 0 \end{pmatrix} V_{\text{in}} \tag{5.10}$$

Conclusion: Equivalence of Averaged Models The results corresponding to the $D > 0.5$ case are presented in Eqs (5.5) and (5.6), while those for the $D < 0.5$ case are given in Eqs (5.9) and (5.10). A comparison of these equations clearly reveals that the averaged system matrices A_{avg} and B_{avg} are identical in both cases. This implies that the state-space averaged dynamics of the three-level boost converter are independent of the duty cycle range. Therefore, a unified state-space model can be applied across the entire operating range, simplifying both analysis and controller design.

Comparison with the Conventional Boost Converter When the output capacitors are equal, i.e., $C_1 = C_2 = C$, the system's third-order state-space model can be reduced to a second-order system by assuming a common capacitor voltage $v_c = (v_{c1} + v_{c2})/2$. Under this condition, the averaged matrices A_{avg} and B_{avg} become equivalent to those of the conventional boost converter, as shown in Eq. (3.3).

This observation highlights an important conclusion: the three-level boost converter and the conventional boost converter exhibit identical averaged dynamic behavior when the output capacitance is balanced. However, the three-level topology still offers significant practical advantages, such as reduced voltage stress on switches, improved efficiency, and suitability for high-voltage applications.

5.3 Controller Design

To linearize the state-averaged model of the three-level boost converter, a perturbation approach is used. Each variable is expressed as the sum of a steady-state (DC) value and a small perturbation (AC) component

$$d = D + \hat{d},\ i_L = I_L + \hat{i}_L,\ v_{C1} = V_{C1} + \hat{v}_{C1},\ v_{C2} = V_{C1} + \hat{v}_{C2} \tag{5.11}$$

Inductor Current Equation Starting from the nonlinear state-averaged model

(a) $\frac{d}{dt}(I_L + \hat{i}_L) = -\frac{1}{L}(1 - D - \hat{d})(V_{C1} + \hat{v}_{C1}) + -\frac{1}{L}(1 - D - \hat{d})(V_{C2} + \hat{v}_{C2}) + \frac{1}{L}(V_{IN})$

Separating the DC and AC components

- DC (steady-state) component

$$\frac{d}{dt}(I_L) = \frac{1}{L}[-(1 - D)V_{C1} - (1 - D)V_{C2} + V_{IN}] \tag{5.12}$$

- AC (perturbation) component

$$\frac{d}{dt}(\hat{i}_L) = -\frac{1}{L}(1 - D)\hat{v}_{C1} + \frac{1}{L}V_{C1} \cdot \hat{d} \pm \frac{1}{L}(1 - D)\hat{v}_{C2} + \frac{1}{L}V_{C2} \cdot \hat{d} \tag{5.13}$$

Capacitor Voltage Equation

(b) $\frac{d}{dt}(V_{C1} + \hat{v}_{C1}) = \frac{1}{C_1}(1 - D - \hat{d})(I_L + \hat{i}_L) - \frac{1}{RC_1}(V_{C1} + \hat{v}_{C1}) - \frac{1}{RC_1}(V_{C2} + \hat{v}_{C2})$

Separating the DC and AC components

- DC (steady-state) component

$$\frac{d}{dt}(V_{C1}) = \frac{1}{C_1}(1 - D)I_L - \frac{1}{RC_1}V_{C1} - -\frac{1}{RC_2}V_{C2} \tag{5.14}$$

- AC (perturbation) component

$$\frac{d}{dt}(\hat{v}_{C1}) = \frac{1}{C_1}(1 - D)\hat{i}_L - \frac{1}{C_1}I_L \cdot \hat{d} - \frac{1}{RC_1}\hat{v}_{C1} - \frac{1}{RC_1}\hat{v}_{C2} \tag{5.15}$$

(c) $\frac{\mathrm{d}}{\mathrm{d}t}(V_{C2}+\hat{v}_{C2}) = \frac{1}{C_2}(1-D-\hat{d})(I_L+\hat{i}_L) + -\frac{1}{RC_2}(V_{C1}+\hat{v}_{C1}) - \frac{1}{RC_2}(V_{C2}+\hat{v}_{C2})$

- DC (steady-state) component

$$\frac{\mathrm{d}}{\mathrm{d}t}(V_{C2}) = \frac{1}{C_2}(1-D)I_L - \frac{1}{RC_2}V_{C1} - -\frac{1}{RC_2}V_{C2} \tag{5.16}$$

- AC (perturbation) component

$$\frac{\mathrm{d}}{\mathrm{d}t}(\hat{v}_{C2}) = \frac{1}{C_2}(1-D)\hat{i}_L - \frac{1}{C_2}I_L \cdot \hat{d} - \frac{1}{RC_2}\hat{v}_{C1} - \frac{1}{RC_2}\hat{v}_{C2} \tag{5.17}$$

Higher-order nonlinear perturbation terms like $\hat{d}\cdot\hat{v}_{C1}$, $\hat{d}\cdot\hat{v}_{C2}$, $\hat{d}\,\hat{i}_L$ are neglected under the small-signal assumption.

Matrix Representation The small-signal model can be expressed in state-space form as

$$\begin{pmatrix} \frac{\hat{i}_L}{dt} \\ \frac{\hat{v}_{C1}}{dt} \\ \frac{\hat{v}_{C2}}{dt} \end{pmatrix} = \begin{pmatrix} 0 & -\frac{(1-D)}{L} & -\frac{(1-D)}{L} \\ \frac{(1-D)}{C_1} & -\frac{1}{RC_1} & -\frac{1}{RC_1} \\ \frac{(1-D)}{C_2} & -\frac{1}{RC_2} & -\frac{1}{RC_2} \end{pmatrix} \begin{pmatrix} \hat{i}_L \\ \hat{v}_{C1} \\ \hat{v}_{C2} \end{pmatrix} + \begin{pmatrix} \frac{(V_{C1}+V_{C2})}{L} \\ -\frac{1}{C_1}I_L \\ -\frac{1}{C_2}I_L \end{pmatrix} \hat{d} \tag{5.18}$$

The steady-state values are calculated as

$$I_L = \frac{V_{\mathrm{IN}}}{R(1-D)^2}, \quad V_{C1}+V_{C2} = \frac{V_{\mathrm{IN}}}{(1-D)} \tag{5.19}$$

Thus, the input–output transfer function at steady state (DC) is expressed as

$$\frac{V_{C1}+V_{C2}}{V_{\mathrm{IN}}} = \frac{1}{(1-D)}$$

In matrix form, it is represented as follows

$$\dot{X} = AX + Bu$$

where

$$X = \begin{pmatrix} \hat{i}_L \\ \hat{v}_{C1} \\ \hat{v}_{C2} \end{pmatrix}, \quad A = \begin{pmatrix} 0 & -\frac{(1-D)}{L} & -\frac{(1-D)}{L} \\ \frac{(1-D)}{C_1} & -\frac{1}{RC_1} & -\frac{1}{RC_1} \\ \frac{(1-D)}{C_2} & -\frac{1}{RC_2} & -\frac{1}{RC_2} \end{pmatrix}, \quad B = \begin{pmatrix} \frac{(V_{C1}+V_{C2})}{L} \\ -\frac{1}{C_1}I_L \\ -\frac{1}{C_2}I_L \end{pmatrix}, \quad u = \hat{d} \tag{5.20}$$

Reduction to Second-Order Model When the output capacitors are equal, i.e., $C_1 = C_2 = C$, the small-signal model can be reduced to a second-order system by defining an average output voltage $v_c = (v_{c1} + v_{c2})/2$. Under this condition, the state-space matrices A and B become equivalent to those of the conventional boost converter.

As a result, the controller design methodology—and even the PI gains—developed for the conventional boost converter can be directly applied to the three-level boost converter. This greatly simplifies the control design process while still providing the benefits of lower switch stress and improved efficiency offered by the multi-level structure.

Small-Signal Transfer Function Derivation To derive the frequency-domain small-signal transfer functions of the three-level boost converter, we consider the general case where the output capacitances are unequal ($C_1 \neq C_2$). Starting from the small-signal state-space representation:

Let us consider the general case where C_1 and C_2 are different. To obtain the frequency-domain transfer function, we apply Laplace transform as follows:

$$\dot{X} = AX + Bu$$

Applying the Laplace transform yields:

$$(sI - A)X(s) - X(0) = BU, \quad X(s) = (sI - A)^{-1}[X(0) + BU(s)]$$

Assuming zero initial perturbation in the inductor current and steady-state capacitor voltages initialized as $X(0) = [0\ V_{\text{in}}\ V_{\text{in}}]^T$, the transfer function from control input d to state variables becomes:

$$\frac{X(s)}{U(s)} = (sI - A)^{-1}B$$

Determinant and Matrix Inversion The inverse of matrix can be computed as :

$$(SI - A)^{-1} = \begin{pmatrix} 0 & \frac{(1-D)}{L} & \frac{(1-D)}{L} \\ \frac{(1-D)}{C_1} & -\frac{1}{RC_1} & -\frac{1}{RC_1} \\ \frac{(1-D)}{C_2} & -\frac{1}{RC_2} & -\frac{1}{RC_2} \end{pmatrix}^{-1}$$

The process of calculating the inverse of a 3 × 3 matrix is as follows:

$$sI - A = \begin{pmatrix} s & 0 & 0 \\ 0 & s & 0 \\ 0 & 0 & s \end{pmatrix} - \begin{pmatrix} a_1 & a_2 & a_3 \\ a_4 & a_5 & a_6 \\ a_7 & a_8 & a_9 \end{pmatrix} = \begin{pmatrix} s - a_1 & -a_2 & -a_3 \\ -a_4 & s - a_5 & -a_6 \\ -a_7 & -a_8 & s - a_9 \end{pmatrix}$$

$$\begin{aligned} \Delta &= s - a_1 \begin{vmatrix} s - a_5 & -a_6 \\ -a_8 & s - a_9 \end{vmatrix} + a_4 \begin{vmatrix} -a_2 & -a_3 \\ -a_8 & s - a_9 \end{vmatrix} - a_7 \begin{vmatrix} -a_2 & -a_3 \\ s - a_5 & -a_6 \end{vmatrix} \\ &= (s - a_1)\{(s - a_5)(s - a_9) - a_6 a_8\} + a_4\{-a_2(s - a_9) - a_3 a_8\} \\ &\quad - a_7\{a_2 a_6 + a_3(s - a_5)\} \\ &= s^3 - (a_5 + a_9 + a_1)s^2 + (a_5 a_9 - a_6 a_8 + a_1 a_5 + a_1 a_9 - a_2 a_4 \\ &\quad - a_3 a_7)s + a_1 a_6 a_8 + a_2 a_4 a_9 - a_3 a_4 a_8 - a_2 a_6 a_7 + a_3 a_5 a_7 \\ &\quad - a_1 a_5 a_9 \end{aligned}$$

$$a_1 = 0, \quad a_2 = -\frac{1 - D}{L}, \quad a_3 = -\frac{1 - D}{L}, \quad a_4 = \frac{1 - D}{C_1}, \quad a_5 = \frac{-1}{RC_1}$$

$$a_6 = -\frac{1}{RC_1}, \quad a_7 = \frac{1 - D}{C_2}, \quad a_8 = -\frac{1}{RC_2}, \quad a_9 = -\frac{1}{RC_2}$$

$$(sI - A)^{-1} = \frac{1}{\Delta} \begin{pmatrix} \Delta_1 & \Delta_2 & \Delta_3 \\ \Delta_4 & \Delta_5 & \Delta_6 \\ \Delta_7 & \Delta_8 & \Delta_9 \end{pmatrix}$$

$$\begin{aligned} \Delta_1 &= \begin{vmatrix} s - a_5 & -a_6 \\ -a_8 & s - a_9 \end{vmatrix} = (s - a_5)(s - a_9) - a_6 a_8 \\ &= s^2 - (a_5 + a_9)s + a_5 a_9 - a_6 a_8 \end{aligned}$$

$$\Delta_2 = -\begin{vmatrix} -a_2 & -a_3 \\ -a_8 & s - a_9 \end{vmatrix} = +a_2(s - a_9) + a_3 a_8 = +a_2 s - a_2 a_9 + a_3 a_8$$

$$\Delta_3 = \begin{vmatrix} -a_2 & -a_3 \\ s - a_5 & -a_6 \end{vmatrix} = a_3(s - a_5) + a_2 a_6 = a_3 s - a_3 a_5 + a_2 a_6$$

$$\Delta_4 = -\begin{vmatrix} -a_4 & -a_6 \\ -a_7 & s - a_9 \end{vmatrix} = a_4(s - a_9) + a_6 a_7 = a_4 s - a_4 a_9 + a_6 a_7$$

$$\begin{aligned} \Delta_5 &= \begin{vmatrix} s - a_1 & -a_3 \\ -a_7 & s - a_9 \end{vmatrix} = (s - a_1)(s - a_9) - a_3 a_7 \\ &= s^2 - (a_1 + a_9)s + a_1 a_9 - a_3 a_7 \end{aligned}$$

$$\Delta_6 = -\begin{vmatrix} s - a_1 & -a_3 \\ -a_4 & -a_6 \end{vmatrix} = +a_6(s - a_1) + a_3 a_4 = +a_6 s - a_1 a_6 + a_3 a_4$$

$$\Delta_7 = +\begin{vmatrix} -a_4 & s - a_5 \\ -a_7 & -a_8 \end{vmatrix} = a_4 a_8 + a_7(s - a_5) = a_7 s + a_4 a_8 - a_5 a_7$$

$$\Delta_8 = -\begin{vmatrix} s - a_1 & -a_2 \\ -a_7 & -a_8 \end{vmatrix} = a_8(s - a_1) + a_2 a_7 = a_8 s - a_1 a_8 + a_2 a_7$$

$$\Delta_9 = \begin{vmatrix} s - a_1 & -a_2 \\ -a_4 & s - a_5 \end{vmatrix} = (s - a_1)(s - a_5) - a_2 a_4$$

$$= s^2 - (a_1 + a_5)s + a_1 a_5 - a_2 a_4$$

$$(sI - A)^{-1}B = \frac{1}{\Delta}\begin{pmatrix} \Delta_1 & \Delta_2 & \Delta_3 \\ \Delta_4 & \Delta_5 & \Delta_6 \\ \Delta_7 & \Delta_8 & \Delta_9 \end{pmatrix}\begin{pmatrix} b_1 \\ b_2 \\ b_3 \end{pmatrix} = \begin{pmatrix} g_i \\ g_{v1} \\ g_{v2} \end{pmatrix}$$

$$g_i = \Delta_1 b_1 + \Delta_2 b_2 + \Delta_3 b_3,\ g_{v1} = \Delta_4 b_1 + \Delta_5 b_2 + \Delta_6 b_3,$$

$$g_{v2} = \Delta_7 b_1 + \Delta_8 b_2 + \Delta_9 b_3$$

$$\therefore (sI - A)^{-1}B = \frac{X(S)}{U(S)} = \frac{1}{\Delta}\begin{pmatrix} g_i \\ g_{v1} \\ g_{v2} \end{pmatrix} = \begin{pmatrix} I_L(s) \\ V_{c1}(s) \\ V_{c2}(s) \end{pmatrix}$$

Final Transfer Functions The small-signal transfer functions are

- Inductor Current

$$G_1(s) = \frac{\hat{I}_L(s)}{\hat{d}(s)} = \frac{g_i}{\Delta} \tag{5.21}$$

- Capacitor Voltage $V_{c1}(s)$

$$G_{21}(s) = \frac{\hat{v}_{c1}(s)}{\hat{d}(s)} = \frac{g_{v1}}{\Delta} \tag{5.22}$$

- Capacitor Voltage $V_{c2}(s)$

$$G_{22}(s) = \frac{\hat{v}_{c2}(s)}{\hat{d}(s)} = \frac{g_{v2}}{\Delta} \tag{5.23}$$

Using the same circuit parameters as those applied in the conventional boost converter, the results of executing $bode(G_1)$ and $bode(G_{21})$ in MATLAB are shown in Figure 5.9.

The Bode plot for $G_{22}(s)$ is identical to that of $G_{21}(s)$ and is therefore omitted for brevity. From the simulation results, it is observed that $G_1(s)$ exhibits stable system characteristics, while both $G_{21}(s)$ and $G_{22}(s)$ represent unstable dynamics. These results are consistent with those of the conventional boost converter, confirming that the same control design methodology can be applied to the three-level boost converter. Consequently, the controller gains previously developed for the boost converter can be directly reused without modification.

PWM Generation in PSIM Schematic To implement PWM control for a three-level boost converter in PSIM, it is necessary to generate two complementary gate signals for the switches S_1 and S_2, with a 180-degree phase shift based on a given duty cycle d. The following outlines the key requirements for PWM generation:

- High-Side Switch (S_1) Duty Cycle: The high-side switch is driven with a duty cycle equal to d, determined by the output of the controller.
- Low-Side Switch (S_2) Duty Cycle: The low-side switch operates with a duty cycle of $(1 - d)$, and its PWM signal is phase-shifted by 180 degrees. This is realized by inverting the phase in the PWM generation circuit.
- Carrier Signal Configuration: A symmetric triangular carrier waveform is employed, in contrast to the asymmetric triangular waveform typically used in conventional boost converter implementations.

The PSIM schematic diagram for the PWM generation circuit is shown in Figure 5.10.

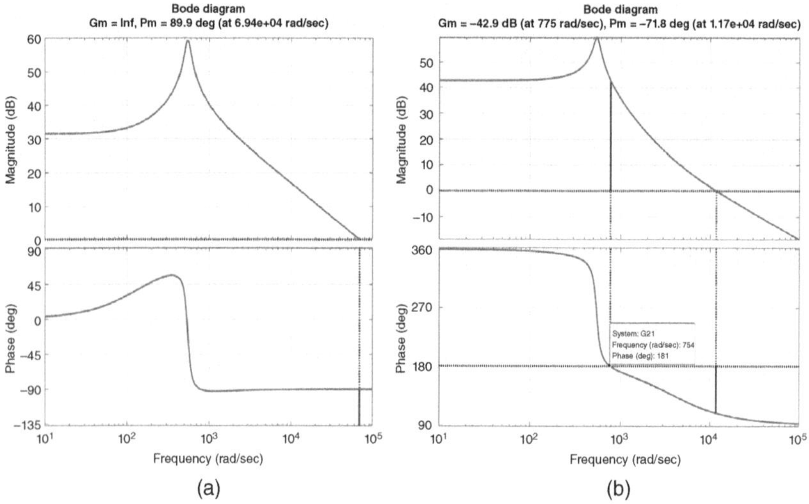

Figure 5.9 Bode diagrams of the inductor current and capacitor voltage transfer functions. (a) Bode plot of $G_1(s)$ and (b) Bode plot of $G_{21}(s)$.

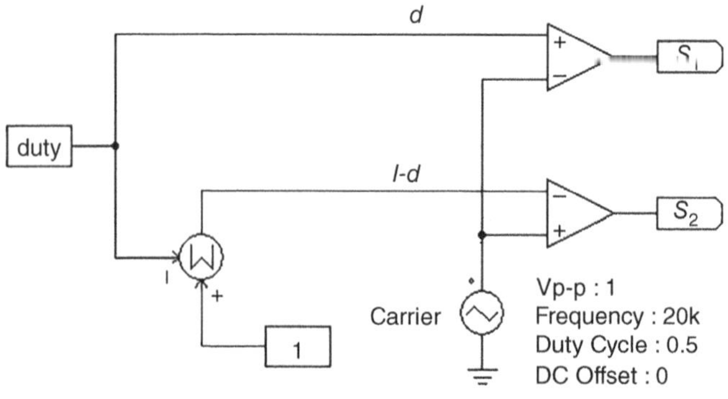

Figure 5.10 PWM generation circuit for three-level boost converter.

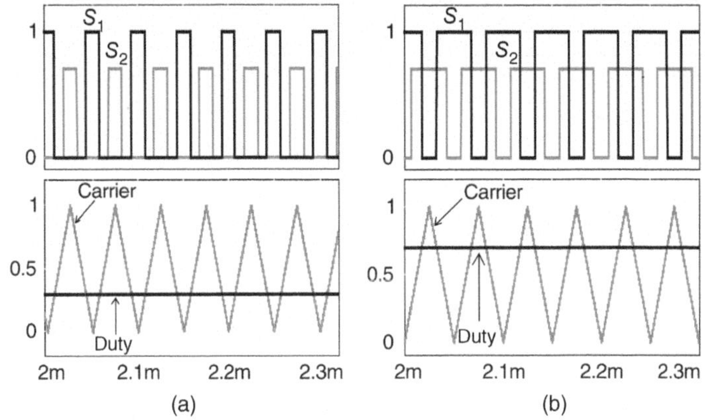

Figure 5.11 Switching pattern under fixed duty conditions. (a) $D < 0.5$ case ($d = 0.3$) and (b) $D > 0.5$ case ($d = 0.7$).

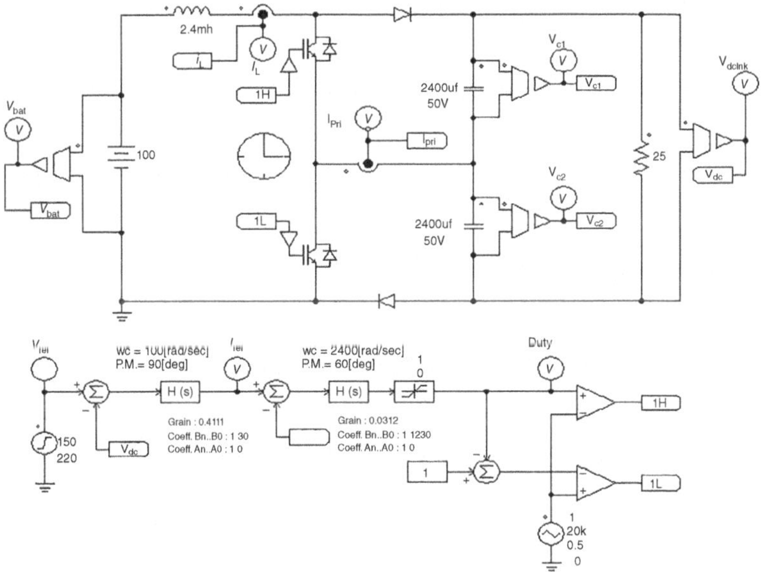

Figure 5.12 PSIM schematic with double-loop control.

The switching waveforms under fixed duty cycle operation are illustrated in Figure 5.11. Figure 5.11(a) shows the case for $d = 0.3$ ($D < 0.5$), while Figure 5.11(b) represents $d = 0.7$ ($D > 0.5$). As observed, switches S_1 and S_2 exhibit identical pulse widths and operate in a complementary manner with a 180-degree phase difference.

Figure 5.12 presents the PSIM simulation schematic using a double-loop control scheme. The controller gains are directly inherited from the conventional boost

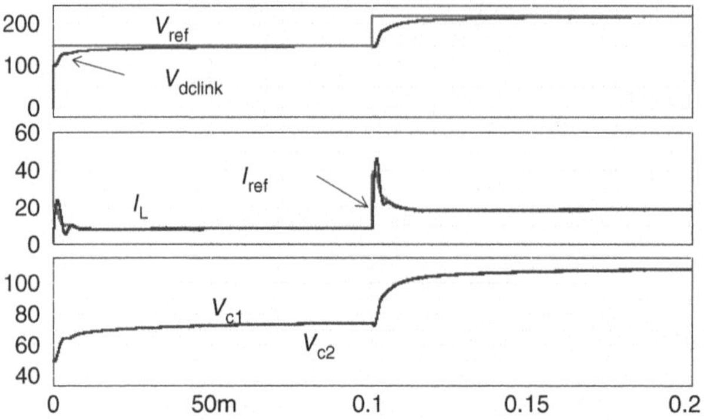

Figure 5.13 PSIM waveforms with double-loop control.

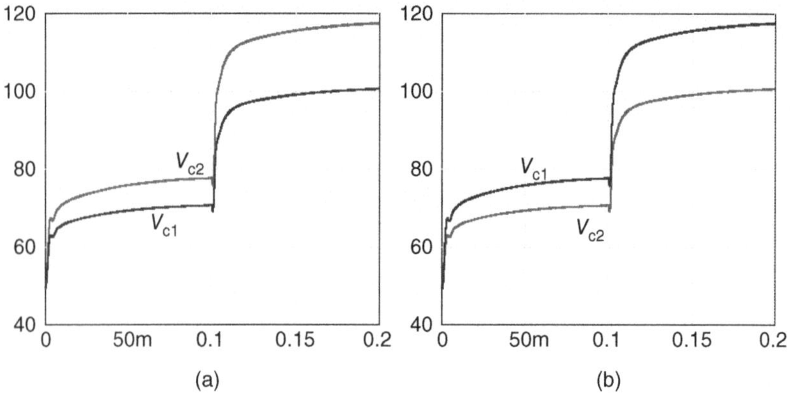

Figure 5.14 Voltage imbalance due to capacitance mismatch. (a) $C_1 = 2400$ uF, $C_2 = 1800$ uF and (b) $C_1 = 1800$ uF, $C_2 = 2400$ uF.

converter. The results are presented in Figure 5.13 and confirm that the output voltage accurately tracks the reference, and the inductor current is well-regulated, demonstrating that the same control gains remain effective in the three-level topology.

In the simulation, the voltages across capacitors C_1 and C_2 remain equal, as both capacitors are set to the same value of 2400 μF. However, in practical applications, electrolytic capacitors generally exhibit tolerance variations of 10%–15%, which leads to inevitable voltage imbalance between them.

Figure 5.14 illustrates the capacitor voltage waveforms for mismatched capacitance cases. Despite the imbalance, the overall DC-link voltage and inductor current remain unaffected and maintain the desired behavior.

In Figure 5.14(a), $V_{C2} > V_{C1}$, while in Figure 5.14(b), $V_{C1} > V_{C2}$. This occurs because the capacitor with smaller capacitance tends to charge to a higher voltage.

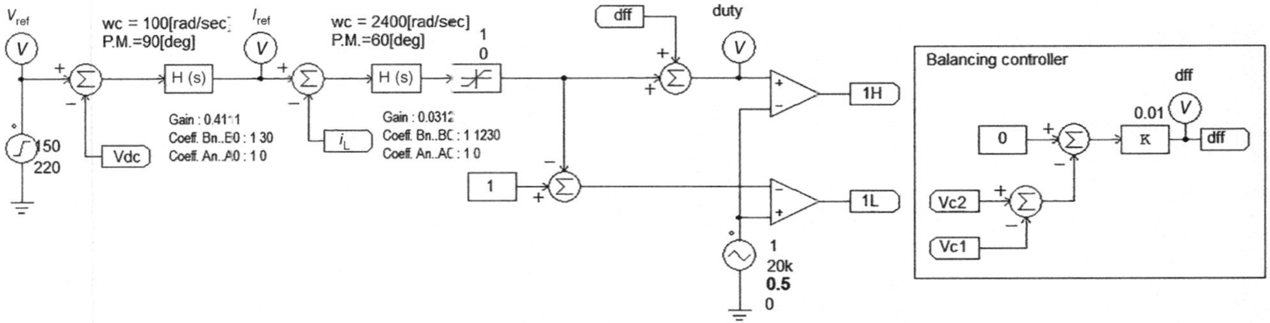

Figure 5.15 Control system configuration with voltage balancing controller.

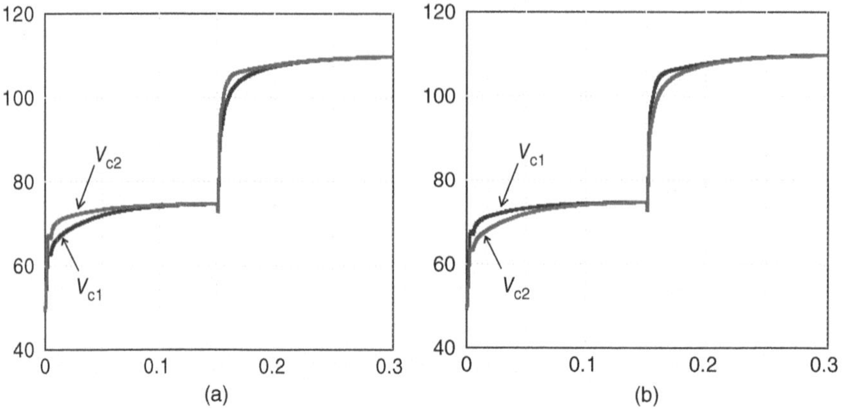

Figure 5.16 Simulation results with balancing controller applied. (a) $C_1 = 2400$ uF, $C_2 = 1800$ uF and (b) $C_1 = 1800$ uF, $C_2 = 2400$ uF.

To address this voltage imbalance, a balancing controller is introduced. The controller uses a feedback loop to minimize the voltage difference between V_{C1} and V_{C2}, implemented using a simple proportional (P) controller. Although a PI controller could be employed, a proportional controller provides a more straightforward solution.

The proportional gain is set such that the voltage error is multiplied by 0.01 and added as a feedforward adjustment to the duty cycle. The configuration of the control system with the integrated balancing function is shown in Figure 5.15.

The balancing controller is integrated into the original control loop, and simulations are conducted for cases where C_1 and C_2 differ by 15%. The results are shown in Figure 5.16.

The simulation demonstrates that, in both scenarios, the capacitor voltages converge to equal values in steady-state operation. This confirms the effectiveness of the proposed balancing controller in mitigating voltage imbalance in practical implementations.

Analogous to the bi-directional converter discussed in the previous chapter, the three-level boost converter can also be extended to a bi-directional configuration by replacing the two diodes with controllable switches. The schematic of the three-level bi-directional converter is presented in Figure 5.17.

The simulation conditions are identical to those used in the previous bi-directional converter case. The controller C-block is reused without modification. The duty cycle assignments for buck and boost modes remain consistent, and the gate-driving logic for upper and lower switches follows the structure of the three-level boost converter. Furthermore, the voltage balancing controller is reused directly.

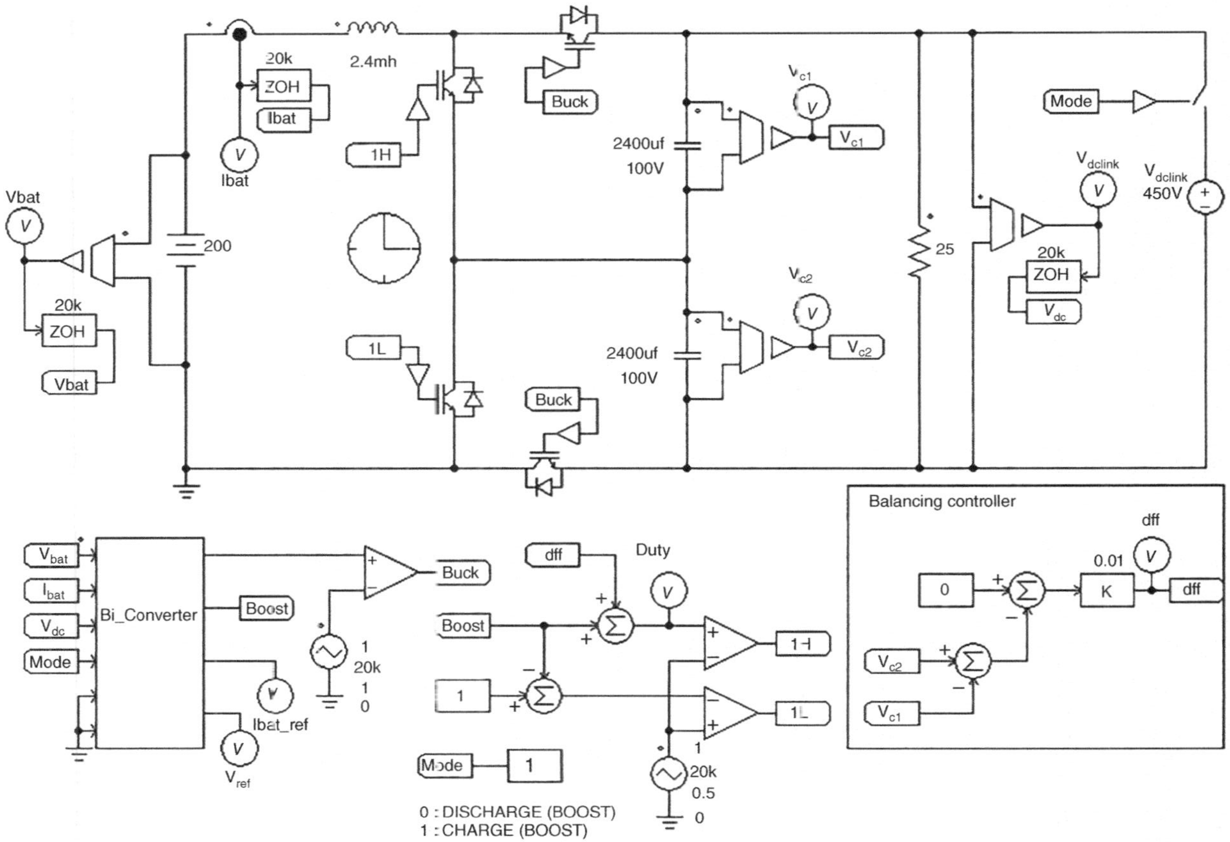

Figure 5.17 Schematic of the thr≥e-level bi-direct onal converter.

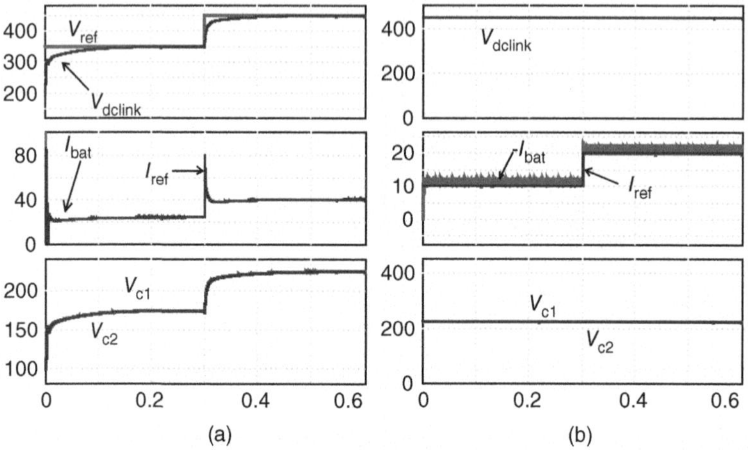

Figure 5.18 Waveforms for charge and discharge modes in the three-level bi-directional converter. (a) Discharge waveform and (b) Charge waveform.

Figure 5.18 shows the simulation results for both discharge and charge modes. The waveforms confirm that the controller, even without gain tuning, ensures stable operation. During discharge, the output voltage and current track their respective references. During charge mode, the battery charging current is well-regulated, verifying the robustness of the control scheme.

Part I-B

Solar Converter

The types and circuits of Solar Converters covered in this part are as follows.

6. SOLAR SYSTEM
7. SOLAR BOOST CONVERTER
8. SOLAR BUCK CONVERTER
9. Three-Level SOLAR BOOST CONVERTER

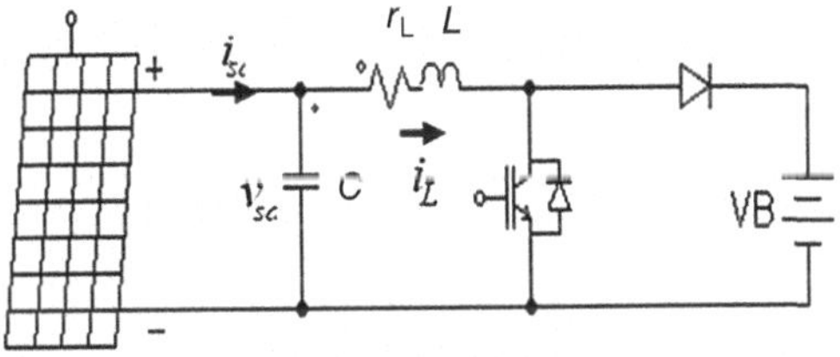

(a) Solar boost converter

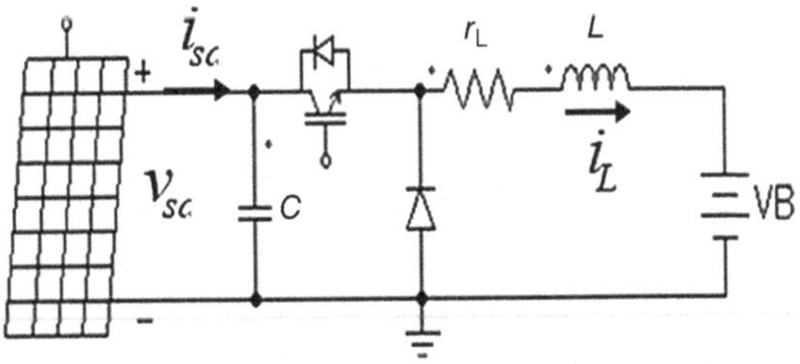

(b) Solar buck converter

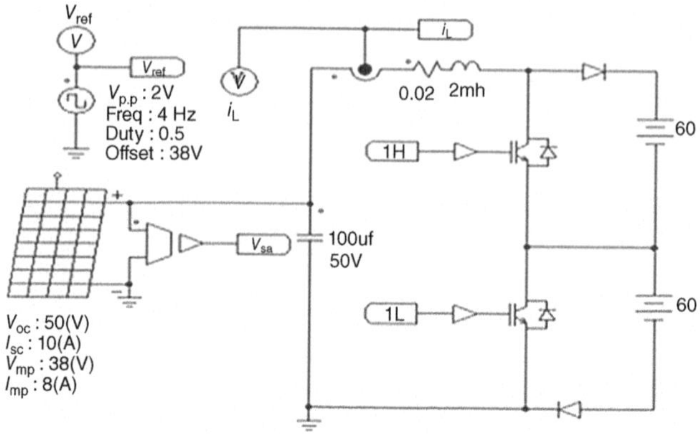

(c) 3-Level solar boost converter

6

Solar System

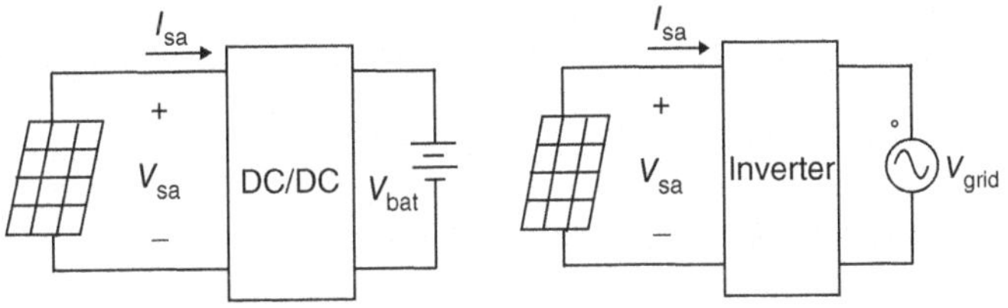

6.1 Understanding Solar Cells

Principle of Electric Power Generation The fundamental principle of electrical power generation can be explained using Maxwell's equations and electromagnetic induction. The core concepts are as follows:

1) When current flows through a conductor, it generates a magnetic field.
2) Direct current (DC) produces a static magnetic field proportional to its magnitude, whereas alternating current (AC) creates a time-varying magnetic field.
3) According to Faraday's Law, a changing magnetic field induces an alternating current: $d\varphi/dt = -i$.
4) A stationary magnetic field does not induce a current.
5) The most efficient way to produce a time-varying magnetic field is through circular motion, which naturally results in sinusoidal (sine and cosine) behavior.
6) Consequently, all conventional electricity generation methods are based on some form of rotational motion.
7) Exceptions to this are solar cells and fuel cells, which can generate electricity without mechanical rotation, naturally producing DC output.

In conventional power-generation systems, electrical energy is produced by establishing a rotating magnetic flux, which induces an alternating voltage in the stator windings in accordance with Faraday's law of electromagnetic induction. This rotating magnetic field is typically generated by a rotor—either a permanent magnet or an electromagnet—mechanically driven by an external energy source. Such an operating principle forms the foundation of rotational generation systems, including hydropower, wind power, thermal power, and nuclear power plants.

In contrast, stationary generation systems operate without any rotating mechanical components. Representative examples include solar cells and fuel cells, in which electrical energy is produced directly through electrochemical or photovoltaic (PV) processes. While rotational generation systems inherently produce AC power, stationary generation systems generate DC power. Among these technologies, solar power generation has emerged as one of the most important and rapidly expanding renewable energy sources, attracting global attention due to its environmental sustainability and long-term energy potential.

The Essence of a Solar Cell A solar cell is, by its nature, a semiconductor device. Although a wide variety of semiconductor devices have been developed for different applications, nearly all of them are fundamentally based on the PN-junction diode structure. The PN junction serves as the most basic building block of modern semiconductor technology.

Advanced semiconductor devices—including bipolar transistors, field-effect transistors (FETs), insulated-gate bipolar transistors (IGBTs), wide-bandgap power switches based on SiC and GaN, high-electron-mobility transistors (HEMTs), MOS-controlled thyristors (MCTs), integrated gate-commutated thyristors (IGCTs), and charge-coupled devices (CCDs)—can all be understood as extensions or functional derivatives of the PN-junction concept. Likewise, the operating principle of a solar cell is rooted in the electrical behavior of the PN junction, with its ability to separate charge carriers and establish an internal electric field being central to PV energy conversion.

As illustrated in Figure 6.1(a), a diode consists of a P-type semiconductor (anode) and an N-type semiconductor (cathode), which together form a depletion layer at the junction. The energy barrier between the P- and N-type materials is typically around 0.6 V, although this value may vary slightly depending on the doping concentration, material properties, and temperature. Notably, as temperature increases, the energy barrier decreases due to enhanced carrier recombination rates. For current to flow from the anode to the cathode, the external voltage must exceed this threshold energy barrier (~0.6 V).

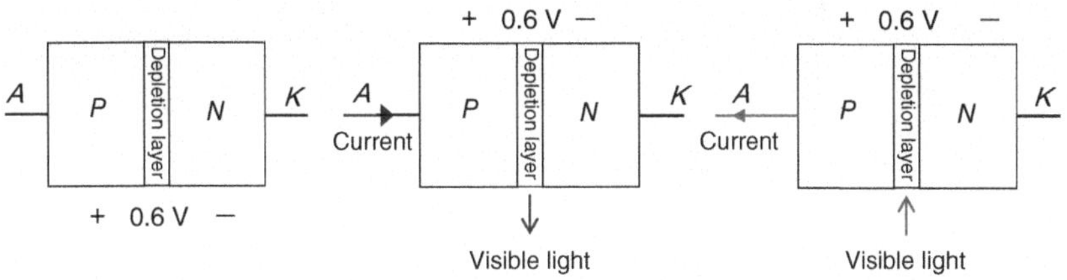

Figure 6.1 Operating principles of the diode, LED, and solar cell. (a) Diode, (b) LED and (c) Solar Cell.

From LED to Solar Cell A well-known application of the PN-junction diode is the light-emitting diode (LED). As illustrated in Figure 6.1(b), when a forward voltage exceeding the diode's threshold is applied, current flows from the P-region to the N-region. During this process, electron–hole recombination occurs in the depletion region, releasing energy in the form of light. Depending on the semiconductor material and doping profile, the emitted radiation may lie in the visible or infrared spectrum. When the emission is in the infrared range and exhibits stimulated emission, the device is commonly referred to as a laser diode (LD).

The operating principle of a solar cell can be understood by applying a contrapositive interpretation of this phenomenon. If *light emission* is a consequence of *current flow* in a forward-biased PN junction, then, conversely, *light absorption* can induce *electrical generation* in the same structure.

As shown in Figure 6.1(c), when photons are incident on the depletion region of a PN junction, their energy generates electron–hole pairs. The built-in electric field within the junction separates these carriers, resulting in the establishment of a voltage across the device terminals. In this manner, a solar cell converts incident light energy directly into electrical energy, operating as the reverse functional counterpart of an LED.

Current Generation in Solar Cells The output current of a solar cell is directly related to the effective area of the depletion region exposed to incident light. A larger illuminated junction area allows more photons to be absorbed, thereby generating a greater number of electron–hole pairs and increasing the resulting photocurrent.

This relationship explains why increasing the surface area of a solar cell leads to a proportional increase in current output. Consequently, solar cell design places strong emphasis on maximizing the light-receiving area and ensuring uniform illumination across the junction. Practical implementations therefore employ large-area cells, surface texturing, and optimized electrode layouts to enhance light absorption while minimizing shading losses, ultimately improving the overall current generation capability of the device.

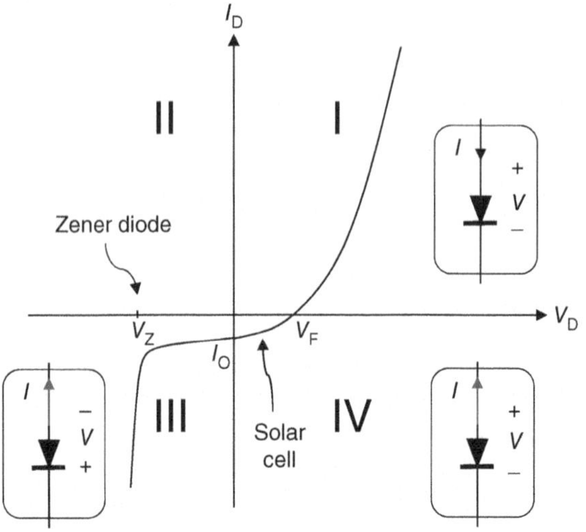

Figure 6.2 Voltage–current (V–I) characteristics and operating regions of a diode.

Voltage–Current Characteristics of a Diode The voltage–current (V–I) characteristics of a diode span the first (*I*), third (*III*), and fourth (*IV*) quadrants, each corresponding to a distinct mode of operation, as illustrated in Figure 6.2.

- First Quadrant (*I*-Quadrant): Forward Bias––*Diode Operation*

In the forward-bias region, the diode conducts when the anode-to-cathode voltage exceeds the forward threshold voltage V_F (typically around 0.6 V for a silicon diode). Beyond this threshold, the current increases exponentially with the applied voltage. This region represents the normal operating mode of conventional diodes in rectification and switching applications.

- Fourth Quadrant (*IV*-Quadrant): Reverse Bias—*Solar Cell Operation*

Under ideal reverse-bias conditions, no current flows when the applied voltage magnitude is below V_F. In practice, however, a small reverse saturation current I_0 exists due to quantum-mechanical effects and the presence of minority carriers. The magnitude of I_0 is strongly influenced by the doping concentrations and material properties of the junction.

In solar cells, the depletion region is intentionally extended toward the surface to maximize exposure to incident light. When photons are absorbed in this region, many electron–hole pairs are generated, significantly increasing the effective reverse current. This light-induced current, commonly referred to as the photocurrent, dominates the device behavior. As a result, the *IV*-quadrant of the diode V–I characteristic corresponds to the operating region of a solar cell, where electrical power is delivered to an external load.

- Third Quadrant (*III*-Quadrant): Breakdown Region––*Zener diode Operation*

In the third quadrant, when the reverse voltage reaches the breakdown voltage V_Z, the diode current increases abruptly due to avalanche or Zener breakdown mechanisms. When properly designed and controlled, this operating region is intentionally utilized in Zener diodes for voltage regulation, owing to their stable and well-defined breakdown voltage characteristics.

Applications to Solar Cells Whereas general-purpose diodes are typically operated in the first quadrant under forward-bias conditions, solar cells operate in the fourth quadrant of the V–I characteristic. In conventional diodes, the reverse saturation current I_0 is extremely small and is generally regarded as negligible.

In a solar cell, however, this current plays a fundamentally different role. When the PN junction is exposed to light, photo-generated carriers dramatically increase the effective reverse current, transforming I_0 into a photocurrent that serves as the primary current source of the device. As a result, electrical power is delivered to an external load while the junction remains reverse biased.

This operating principle demonstrates the structural and functional uniqueness of solar cells compared to other semiconductor devices. Although they share the same PN-junction foundation as ordinary diodes, solar cells are specifically engineered to maximize light absorption and carrier collection, enabling efficient conversion of optical energy into electrical energy.

Power Definition: Load vs. Source In power electronics, the role of a device as either a load or a source is determined by the sign of its instantaneous power, defined as

$$P = V \cdot I$$

According to this definition

- If $P > 0$, the device absorbs energy and operates as a load.
- If $P < 0$, the device delivers energy and operates as a source.

By convention, current is defined as flowing from the anode to the cathode. Based on this sign convention, the behavior of semiconductor devices in different quadrants of the voltage–current (V–I) plane can be interpreted as follows.

Interpretation by V–I Quadrants

- First Quadrant (*I*): $V > 0,\ I > 0\ \Rightarrow P > 0$. This region corresponds to the normal forward-biased operation of a diode, in which the device absorbs electrical energy and functions as a load.
- Third Quadrant (*III*): $V < 0,\ I < 0\ \Rightarrow P > 0$. This region represents Zener or avalanche breakdown operation. The device again absorbs energy and functions as a load.

- Fourth Quadrant (IV): $V > 0, I < 0 \Rightarrow P < 0$. In this region, the device supplies power to the external circuit. This operating mode characterizes a solar cell, which functions as an energy source.

Graphical Representation The voltage–current characteristics of a solar cell are typically plotted such that the saturation current I_0 appears on the positive vertical axis, as shown in Figure 6.3. Under this representation

- Quadrants I and III correspond to conventional diode behavior (forward conduction and Zener breakdown, respectively).
- Quadrant IV represents the operating region of a solar cell, where electrical power is generated and delivered to the load.

This graphical interpretation emphasizes the structural similarity between diodes and solar cells, while clearly illustrating their functional duality, which depends entirely on the operating quadrant.

Effect of Illumination and Temperature on I_0 and V_F The V–I characteristics of a solar cell are strongly influenced by illumination level and temperature:

- Illumination: An increase in incident light intensity raises the effective saturation current I_0, resulting in a higher output current.
- Temperature: At lower temperatures, the semiconductor energy bandgap widens, leading to an increase in the threshold voltage V_F. At higher temperatures, the bandgap narrows, causing V_F to decrease.

In solar cell modeling, these parameters are commonly redefined as

- The open-circuit voltage V_{oc}, which corresponds to the condition $I = 0$.
- The short-circuit current I_{sc}, which corresponds to the condition $V = 0$.

These two quantities serve as key performance indicators and are fundamental to the characterization and modeling of PV devices.

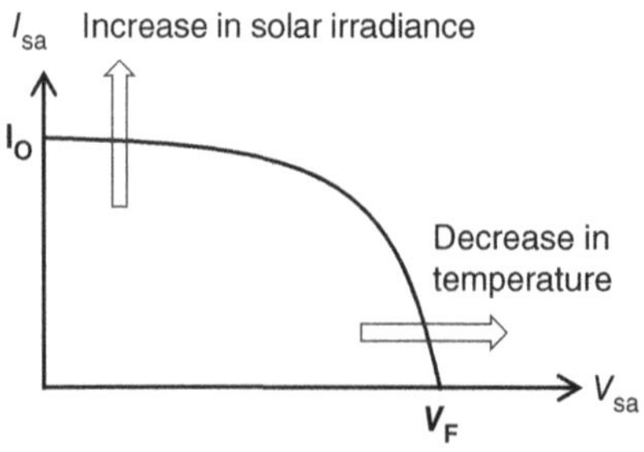

Figure 6.3 V–I characteristics of a solar cell.

Voltage–Current Characteristics for Solar Cell Connections To obtain the desired voltage and current levels, individual solar cells are interconnected in series, parallel, or series–parallel configurations. The resulting V–I characteristics depend directly on the connection method.

1) Series Connection
   In a series connection, as illustrated in Figure 6.4, the output voltages of individual solar cells are additive, while the current remains unchanged.
   - The total voltage increases in proportion to the number of cells.
   - The output current is equal to the current of a single cell.

   For example, when three identical solar cells, each with an open-circuit voltage V_{oc}, are connected in series, the total voltage becomes $3V_{oc}$, while the current remains I_{sc}. This configuration is analogous to connecting three 1.5 V batteries in series to obtain a total voltage of 4.5 V.
2) Parallel Connection
   In a parallel connection, shown in Figure 6.5, the output currents of the solar cells are additive, while the voltage remains constant.
   - The total current increases in proportion to the number of parallel cells.
   - The output voltage is equal to the voltage of a single cell.

   For three identical solar cells connected in parallel, the total current becomes $3I_{sc}$, while the voltage remains V_{oc}.
3) Series–Parallel Configuration
   In practical PV systems, a series–parallel configuration is commonly employed to achieve both higher voltage and higher current, as illustrated in Figure 6.6.
   - The series connection increases the voltage.
   - The parallel connection increases the current.

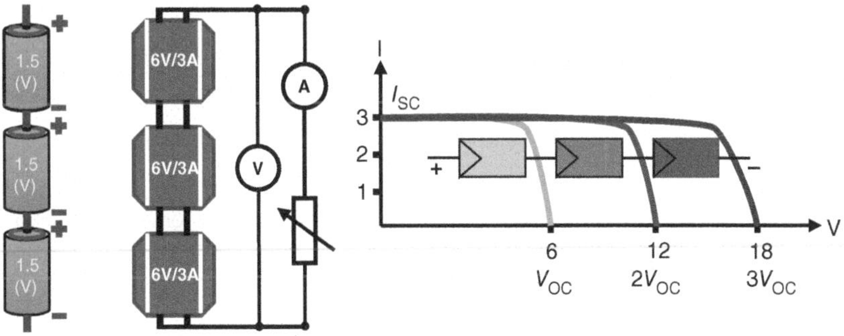

Figure 6.4 Series connection of solar cells.

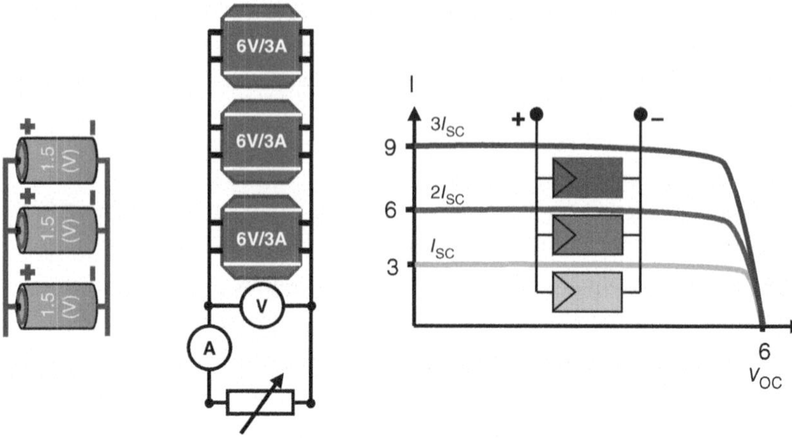

Figure 6.5 Parallel connection of solar cells.

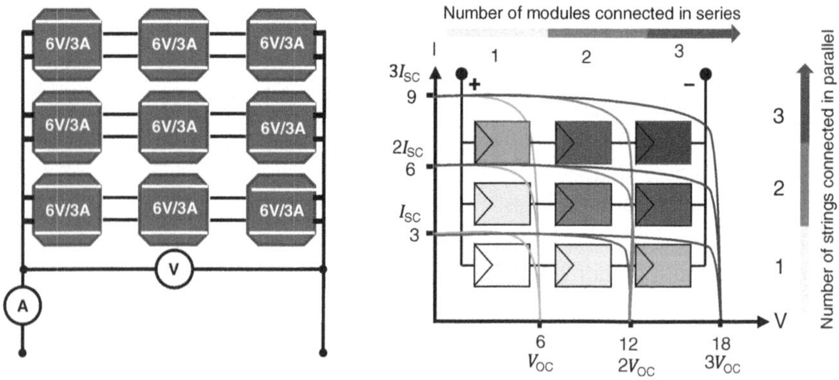

Figure 6.6 Series–parallel connection of solar cells.

For a 3 × 3 configuration, three cells connected in series and three such strings connected in parallel—the overall electrical characteristics are given by

$$V_{\text{total}} = 3V_{\text{oc}},\ \ I_{\text{total}} = 3I_{\text{sc}},\ \ P_{\text{total}} = 9V_{\text{oc}}I_{\text{sc}}.$$

This arrangement provides a scalable means of meeting system-level voltage and power requirements.

6.2 Characteristics of Solar Systems

Solar cell models are broadly categorized into two types: the physical model and the functional model. Each serves distinct purposes in analysis, simulation, and system design.

The physical model represents the actual solar cell by incorporating all the physical characteristics and properties of the materials. The functional model is designed for simulations or system designs by utilizing the voltage–current graph. The physical model requires detailed knowledge of the material parameters and the physical and chemical properties of the applied materials. Generally, these values are not publicly available, so they must be estimated through measurements or reverse engineering, making it difficult to achieve 100% accuracy. This model is typically used in the manufacturing or design process of solar cells.

In contrast, the functional model uses only the voltage–current graph (I–V curve) of the solar cell without requiring physical or chemical parameters. It is easy to understand and convenient to use for designing PV power systems or simulating power conversion devices. Additionally, since manufacturers provide the I–V curve and related constants, this model is easily accessible and widely used. The functional model will be used for controller design and modeling.

a) **Physical model**

A solar cell can be considered a device in which current is generated when light enters the diode. When modeled as a circuit, it is represented by a current source (I_{ph}) generated by light, diode, and series resistance (R_S) and parallel resistance (R_{Sh}) associated with the electrodes. Series (N_S) and parallel (N_p) connections can be similarly represented by a current source ($N_p I_{\mathrm{ph}}$), series (N_S) diodes, and (N_S/N_p) series and parallel resistances. The equivalent circuit representation is shown in Figure 6.7.

The mathematical voltage–current expression for a single cell is as follows:

$$I = I_{\mathrm{ph}} - I_{\mathrm{sat}}\left(e^{\frac{q(V+IR_S)}{AKT}} - 1\right) - \frac{V + IR_S}{R_{\mathrm{sh}}}$$

The mathematical expression for series (N_S) and parallel (N_p) connections is given as follows

$$I_{\mathrm{sa}} = N_p\left(I_{\mathrm{ph}} - I_{\mathrm{sat}}\left(e^{\frac{q}{AKT}\left(\frac{V_{\mathrm{sa}}}{N_s} + \frac{I_{\mathrm{sa}}R_S}{N_p}\right)} - 1\right)\right) - \frac{1}{R_{\mathrm{sh}}}\left(\frac{V_{\mathrm{sa}}}{N_s} + \frac{I_{\mathrm{sa}}R_S}{N_p}\right)$$

Physical and chemical constants are defined in Table 6.1.

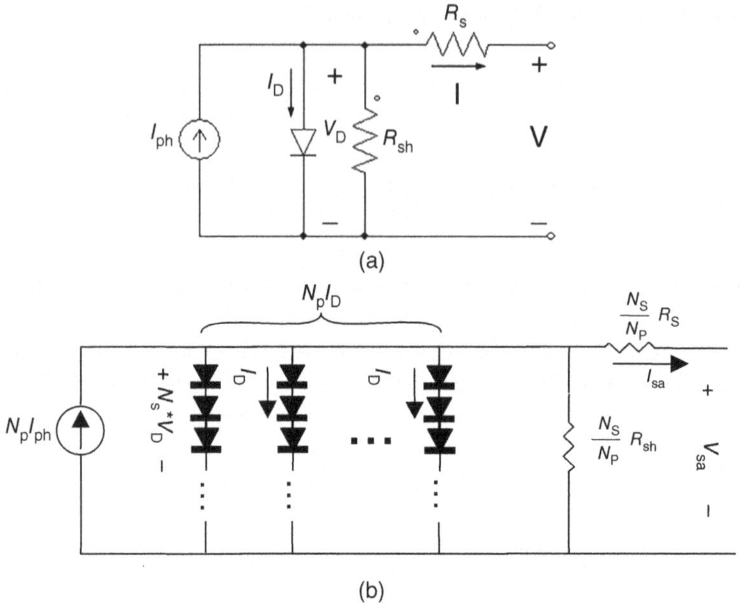

Figure 6.7 Equivalent circuit models for solar cell. (a) Single cell and (b) Series/Parallel connection cells.

Table 6.1 Solar cell physical and chemical constants.

I	Solar cell output current in A
V	Solar cell output voltage in V
I_{ph}	Light-generated current in A
I_{sat}	Solar cell reverse saturation current in A
q	Electronic charge: 1.6022×10^{-19} C
A	Deviation factor from the ideal diode
k	Boltzmann's constant: 1.3807×10^{-23} J/K
T	Cell temperature in K
R_s	Series resistance in Ω
R_{sh}	Shunt resistance in Ω

PSIM provides physical model library. The library can be found under `Elements → Power → Renewable Energy → Solar Module (Physical model)` (Figure 6.8).

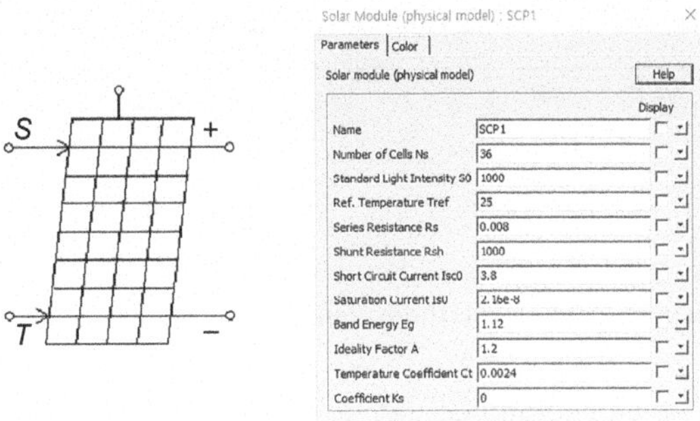

Figure 6.8 PSIM solar cell library for physical model.

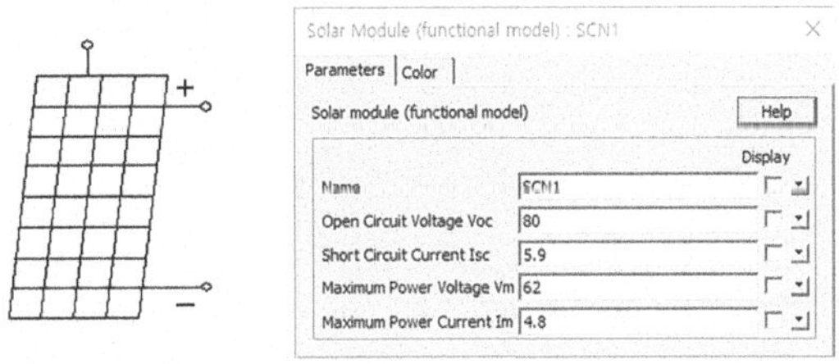

Figure 6.9 PSIM solar cell library for functional model.

b) **Functional model**

This model represents the solar cell using four parameters (three points) on the voltage–current graph. PSIM provides the functional model (Figure 6.9).

The library can be found under `Elements → Power → Renewable Energy → Solar Module (functional model)`. The functional model is used for simulations or controller design.

The definition of parameters and voltage–current graphs is shown in Table 6.2.

The characteristics of the solar cell can be examined using its functional model. The I–V curve, which varies with changes in illumination and temperature, is illustrated in Figure 6.10. When illumination increases, the short-circuit current (I_{sc}) becomes larger. When the temperature decreases, the open-circuit voltage (V_{oc}) increases.

The maximum power point (MPP) voltage (V_{mp}) shifts due to changes in illumination and temperature, and the maximum power (P_{mp}) also changes continuously (Figure 6.10). Figure 6.11 shows the trajectory of the MPP (P_{mp}) over a day based on changes in illumination and temperature. The movement of the (MPP voltage is not linear, so it must be determined using an algorithm.

This is called the maximum power point tracking (MPPT) algorithm, and many methods have been developed for it.

Table 6.2 Parameters of the functional model and I–V graphs.

- Four parameters (three points)

V_{oc}: Open Circuit Voltage

I_{sc}: Short Circuit Current

V_{mp}: Maximum Power Voltage

I_{mp}: Maximum Power Current

Maximum Power $P_{mp} = V_{mp} \cdot I_{mp}$

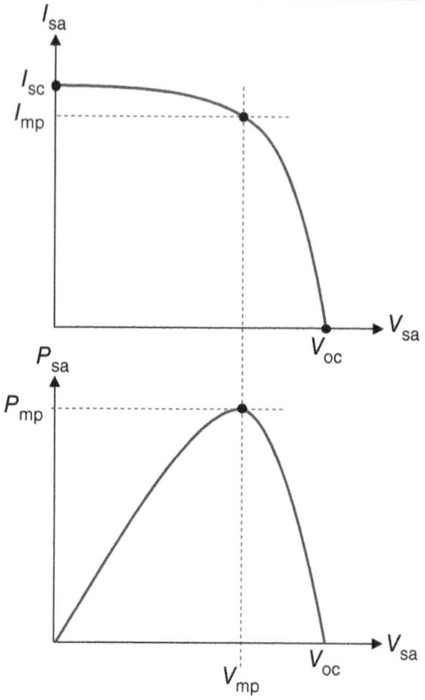

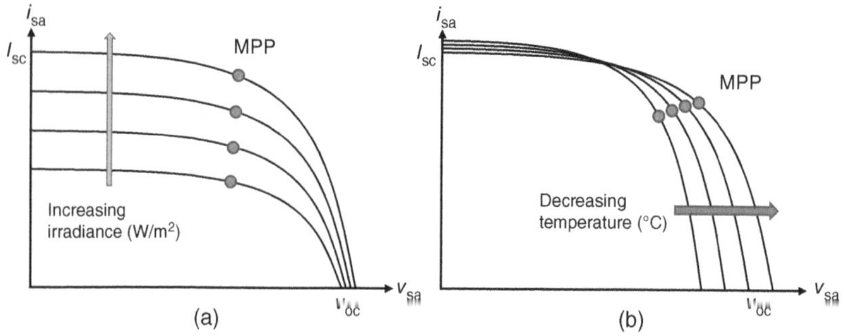

Figure 6.10 I–V characteristics under environmental variations. (a) Illumination change and (b) Temperature change.

Maximum Power Point Tracking (MPPT) Algorithm The process of identifying and operating at the MPP of a solar cell is known as MPPT. Over the years, numerous MPPT algorithms have been developed to optimize power extraction from PV systems under varying environmental conditions.

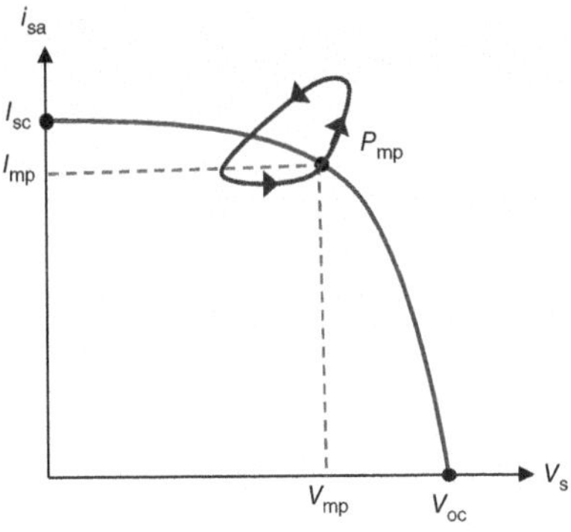

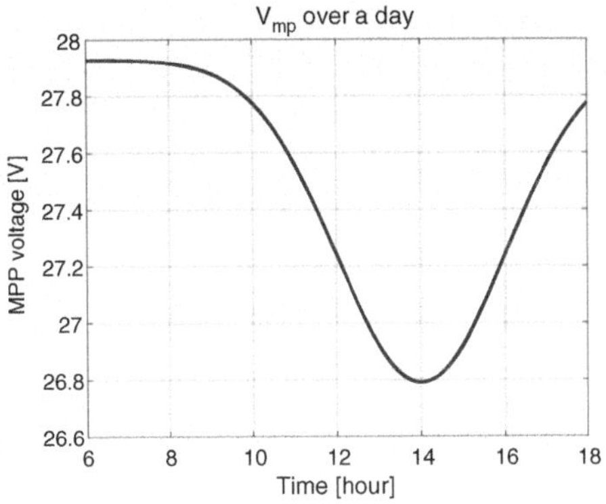

Figure 6.11 Daily movement of the MPP.

The fundamental principle of MPPT is illustrated on the voltage–power (V–P) curve, where the slope of the power with respect to voltage changes sign at the MPP

- To the left of MPP ($V < V_{mp}$): $\frac{dP}{dV} > 0$
- At MPP ($V = V_{mp}$): $\frac{dP}{dV} = 0$
- To the right of MPP ($V > V_{mp}$): $\frac{dP}{dV} < 0$

The core idea of MPPT algorithms is to continuously adjust the reference voltage (V_{ref}) of the voltage controller such that $\frac{dP}{dV} = 0$, thereby operating at the MPP. The algorithm typically oscillates every second between the points where $\frac{dP}{dV} > 0$ and $\frac{dP}{dV} < 0$.

Perturb and Observe (P&O) Method One of the most widely used MPPT methods is the Perturb and Observe (P&O) algorithm. Also referred to as the hill-climbing method, it works by periodically perturbing the system's operating voltage and observing the resulting change in output power. Figure 6.12 provides an explanation of the algorithm.

- If $\frac{dP}{dV} > 0$: Increase V_{ref}
- If $\frac{dP}{dV} < 0$: Decrease V_{ref}
- If $\frac{dP}{dV} = 0$: System is at MPP

At the MPP, the algorithm oscillates around V_{mp}, continuously adjusting the voltage reference to maintain optimal power output.

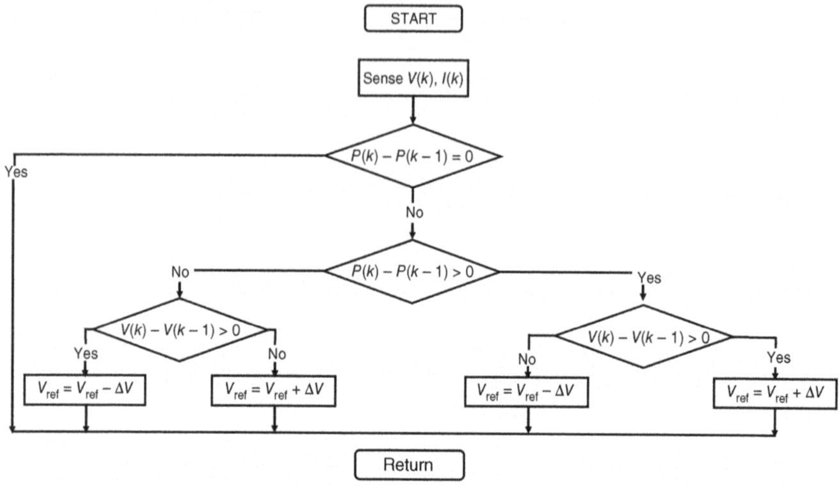

Figure 6.12 P&O MPPT algorithm flowchart.

P&O Algorithm Implementation Steps

1) Measure instantaneous power: $P_{sa} = V_{sa} \cdot I_{sa}$
   Compute the power difference: $dP = P_{sa} - P_{old}$
2) Measure voltage difference: $dV = V_{sa} - V_{old}$
3) Determine the sign of $\frac{dP}{dV}$
   - If positive: V_{ref} increases
   - If negative: V_{ref} decreases
4) Update the reference voltage

$$V_{ref}(k+1) = V_{ref}(k) + M \cdot sgn\left(\frac{dP}{dV}\right)$$

   where M is a predefined step size.
5) Update historical values for next iteration:

$$P_{old} = P_{sa}, \quad V_{old} = V_{sa}$$

6) Repeat the process at regular intervals (typically every one second)
   It is important to note that the *I–V* curve of a solar cell differs significantly from that of conventional batteries or DC sources. While batteries behave as linear voltage sources, a solar cell behaves more like a nonlinear current source, necessitating real-time tracking algorithms such as P&O for efficient operation.

Operating Point As previously discussed, the I–V characteristic of a solar cell is nonlinear, exhibiting a curved profile rather than a straight line. Unlike linear power sources such as DC power supplies or batteries, a solar cell possesses distinct

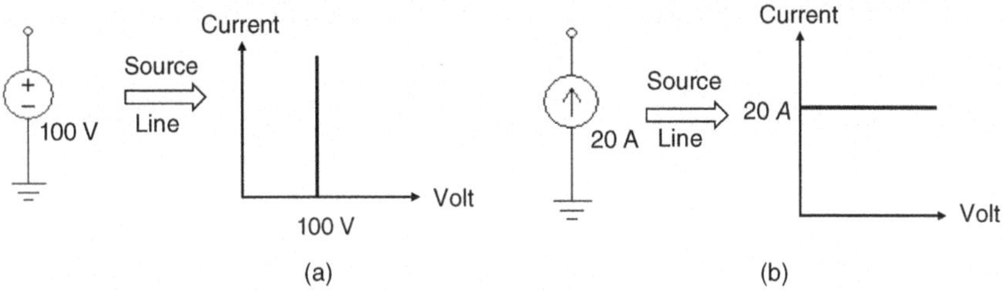

Figure 6.13 Source lines. (a) Voltage source line and (b) Current source line.

electrical behavior. To analyze this behavior effectively, the concept of an *operating point* is introduced, which helps in understanding the voltage and current characteristics of the solar cell.

Traditionally, the term *operating point* refers to the DC bias voltage and current of a transistor. However, in the context of power circuits, it refers to the specific voltage and current at which the source and the load in a circuit are in equilibrium. In other words, the operating point is the intersection where the output characteristics of the power source match the input requirements of the load. For a stable power circuit, this equilibrium is typically represented as a single, well-defined point on the I–V plane.

A *source* is any device that delivers electrical energy, and it can be categorized as either a voltage source or a current source

- A voltage source maintains a constant voltage regardless of the load, and the current is determined by the connected load.
- A current source maintains a constant current, while the voltage across it varies depending on the load.

When voltage and current characteristics of a source are plotted on an X–Y graph, the resulting line is referred to as the source line. Figure 6.13 illustrates two such examples: (a) Voltage source line and (b) Current source line.

From these figures, it is evident that the operating point is not determined solely by the source line itself. Instead, the actual operating point is established at the intersection of the source line and a load line, which represents the behavior of the load.

A *sink* is a device that consumes energy. A resistor, for instance, is a classic example of a sink, defined by a linear voltage-to-current (V/I) relationship. When the voltage and current characteristics of the load are plotted, the resulting graph is known as the *load line*. Figure 6.14 shows the load line of a resistive load.

In certain cases, a source can also function as a load. For instance, a battery (a voltage source) can act as a source during discharge and as a sink during charging. The same principle applies to current sources. Therefore, voltage and current

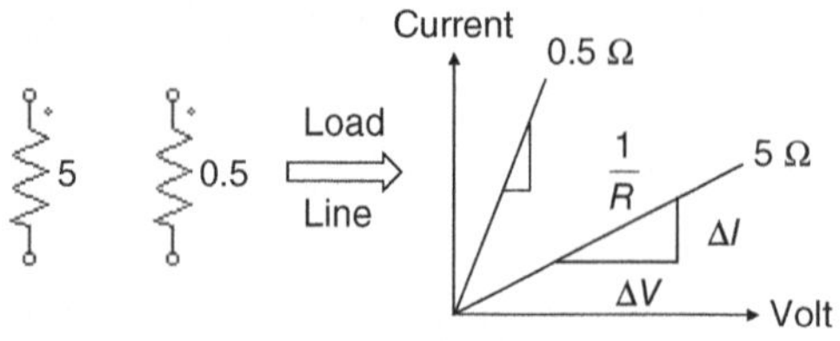

Figure 6.14 Load line of a resistor.

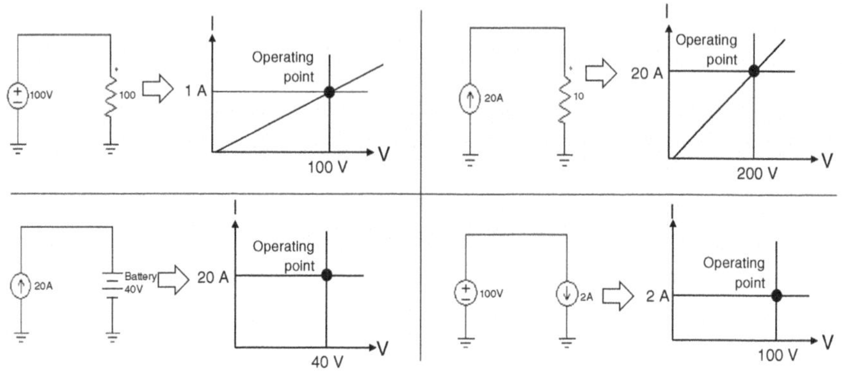

Figure 6.15 Operating point of a linear source and sink.

sources can alternate roles as sources or sinks depending on the operating condition. Regardless, the operating point is always defined at the intersection of the source and load lines.

Figure 6.15 illustrates the operating points of linear sources for various load types.

- When the source is a voltage source, the operating voltage is fixed, and the load determines the current.
- When the source is a current source, the operating current is fixed, and the load determines the voltage.

Consider scenarios where the *load itself* is a source

- If the source is a current source, the load must be a voltage source (e.g., a battery).
- If the source is a voltage source, the load must be a current source (e.g., an inductor).

However, if both the source and the sink are of the same type (e.g., both are voltage sources or both are current sources), the circuit cannot function properly due to the inherent conflict in control variables (voltage or current). To better understand these concepts, it is recommended to simulate each scenario using PSIM and analyze the resulting operating points.

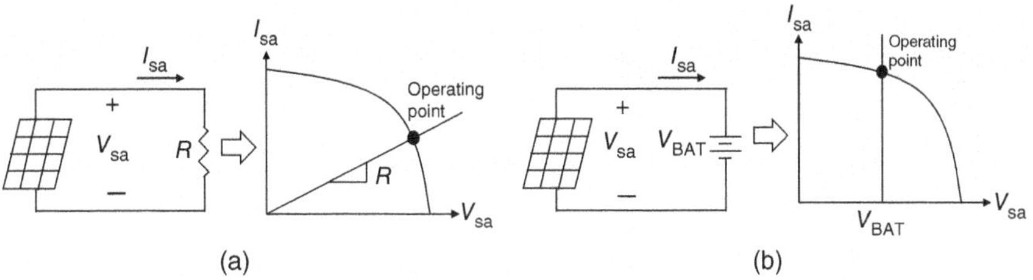

Figure 6.16 Operating point of the solar cell. (a) Solar cell with resistor and (b) Solar cell with battery.

Operating Point of the Solar Cell Next, we examine the operating point of a nonlinear power source—the solar cell. Two representative configurations, analogous to previous examples, are shown in Figure 6.16.

- Figure 6.16(a) illustrates a solar cell connected to a resistive load.
- Figure 6.16(b) shows a solar cell connected to a voltage source (e.g., a battery).

In Figure 6.16(a), when a resistor is connected to the output of the solar cell, the load line is defined by the resistance value. The operating point is established at the intersection of the solar cell's I–V curve and the load line. In this case, both the voltage and current of the solar cell vary depending on the resistance. Conversely, in Figure 6.16(b), when the solar cell is connected to a voltage source such as a battery, the load line becomes vertical, corresponding to a constant voltage. The operating point is then determined by the fixed battery voltage and the resulting current from the solar cell.

To further investigate this concept, Figure 6.17 shows a simple circuit where a solar cell is connected to a capacitor and a resistor. The capacitor voltage represents the solar cell voltage, while the current through the resistor corresponds to the solar cell current. The capacitor is included to model the transient response of the solar voltage and to facilitate voltage-based measurement of the solar cell output. Since the solar cell behaves like a nonlinear current source, the capacitor ensures a realistic response. The rise time of the solar voltage depends on the capacitance value.

Figure 6.17 Simulation circuit of a solar cell with a resistor load.

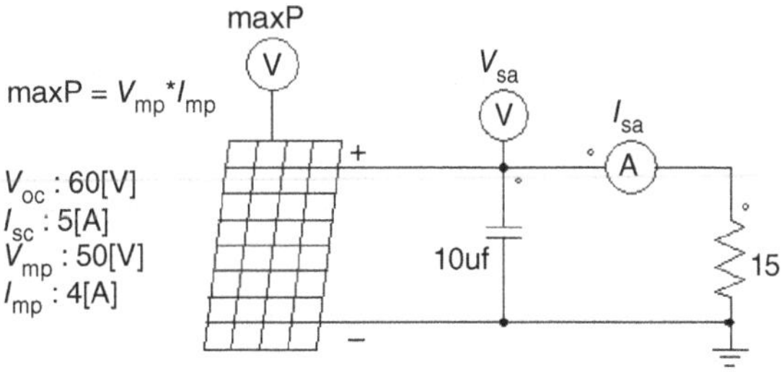

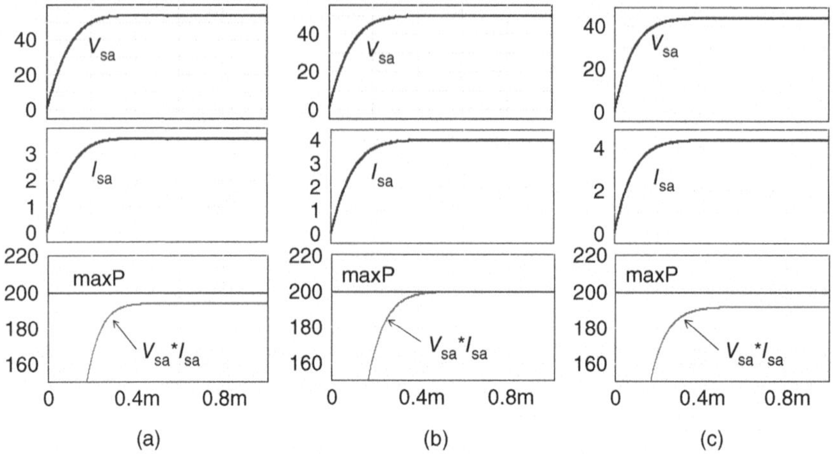

Figure 6.18 Solar cell voltage, current, and power for different resistor loads. (a) $R = 15[\Omega]$, (b) $R = 12.5[\Omega]$ and (c) $R = 10[\Omega]$.

By varying the resistor value, we can observe how the operating point—and thus the output power—changes. Figure 6.18 presents simulation results for three different resistor values.

The solar cell parameters used in the simulation are: $V_{oc} = 60\,\text{V}$, $I_{sc} = 5\,\text{A}$, $V_{mp} = 50\,\text{V}$, $I_{mp} = 4\,\text{A}$. Thus, the maximum theoretical power output is $P_{max} = V_{mp} \cdot I_{mp} = 200\,\text{W}$.

- In Figure 6.18(a), with $R = 15\,\Omega$, the solar cell operate at $V_{sa} = 54\,\text{V}$, $I_{sa} = 3.6\,\text{A}$, yielding $P_{sa} = V_{sa}I_{sa} = 194\,\text{W}$.
- In Figure 6.18(b), with $R = 12.5\,\Omega$, the voltage and current are $V_{sa} = 50\,\text{V}$, $I_{sa} = 4\,\text{A}$, resulting in the maximum power output of $P_{sa} = 200\,\text{W}$.
- In Figure 6.18(c), with $R = 10\,\Omega$, the voltage and current are $V_{sa} = 43.8\,\text{V}$, $I_{sa} = 4.38\,\text{A}$, and $P_{sa} = 192\,\text{W}$.

These results confirm that maximum power is obtained when the load resistance satisfies $=V_{mp}/I_{mp}$. This conclusion aligns with the graphical interpretation using the intersection of the solar cell's source curve and the load line.

Although the solar cell's rated capacity is 200 W, its actual output depends on the operating point, which varies with load conditions. This behavior can be analogized to a high-performance sports car such as a Ferrari, which may have a top speed of 350 km/h. However, this maximum speed can only be reached under optimal conditions, such as on a dedicated race track—not in everyday driving environments. Similarly, a solar cell reaches its maximum power only under specific load conditions; any deviation causes a shift in the operating point, resulting in suboptimal output power.

Using a fixed resistor to extract power from a solar cell is limited in practicality and is rarely employed in real-world applications. However, as demonstrated in the simulation results, varying the resistance from 0 to ∞ Ω allows for the plotting of the solar cell's complete I–V characteristic curve. This method is particularly useful for measurement equipment and experimental purposes.

As a practical exercise, the I–V (current–voltage) and P–V (power–voltage) characteristics of a solar cell will be obtained by varying the load resistance over a wide range, from 0.1 Ω to 1 kΩ. This variation effectively captures the complete nonlinear behavior of the solar cell under diverse load conditions. Table 6.3 presents the exercise for plotting the *I–V* and *P–V* curves of the solar cell.

Next, we replace the resistive load with a battery (voltage source). Figure 6.19 shows the simulation schematic and the resulting waveforms. As previously explained, when a solar cell is connected to a voltage source, the solar cell voltage is fixed at the battery voltage. The corresponding solar cell current is determined by the I–V characteristic, and the output power is calculated as

$$P_{sa} = V_{battery} \cdot I_{sa}$$

To determine the battery voltage that yields maximum power, a simulation can be conducted by sweeping the battery voltage from 0 V to V_{oc} (open-circuit voltage). However, understanding the concept of the operating point allows us to intuitively conclude that maximum power is delivered when the battery voltage equals the MPP voltage V_{mp} of the solar cell.

Similar to the resistor-based approach, varying the battery voltage enables extraction of the solar cell's I–V curve. This method is especially useful for identifying the MPP under controlled voltage conditions.

Table 6.3 Solar cell I–V and P–V curve exercise.

$R[\Omega]$	0.1	1	2	3	5	8	10	20	100	1k
$V_{sa}[V]$										
$I_{sa}[A]$										
$P_{sa}[W]$										

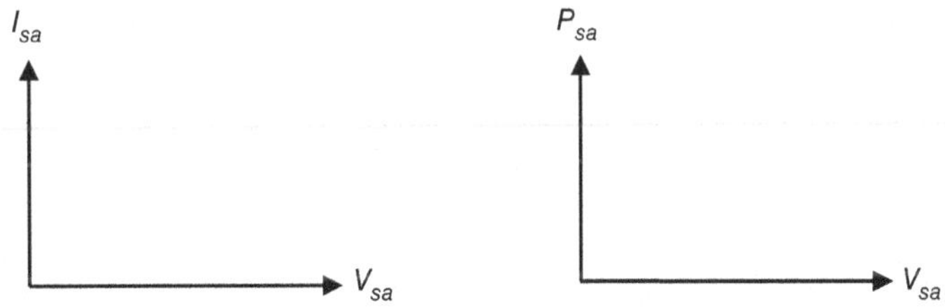

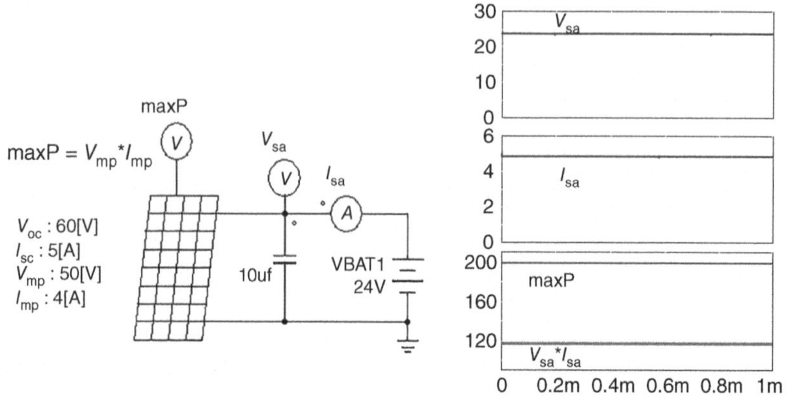

Figure 6.19 Solar voltage, current, and power for a battery load.

Thus far, we have analyzed the voltage, current, and power characteristics of a solar cell using fixed loads—resistors and batteries. While useful for characterization, these fixed-load configurations do not allow for continuous extraction of the solar cell's maximum power under varying environmental conditions (e.g., irradiance and temperature). Because resistors and batteries present fixed load characteristics, they lack the flexibility needed for optimal power tracking.

To address this limitation, power conversion systems that incorporate DC/DC converters are employed. These converters adjust their input impedance dynamically to match the optimal load condition, effectively varying the equivalent resistance presented to the solar cell. As a result, they can continuously track and extract maximum power from the solar cell—a process known as MPPT.

Figure 6.20 presents a circuit in which a constant power load (CPL) is connected to a solar cell. This configuration is realized by interfacing the solar cell with a DC/DC power converter that is controlled to continuously consume 50 W of power. Such power converters inherently exhibit CPL characteristics, as they maintain a fixed power consumption regardless of voltage variations at their input.

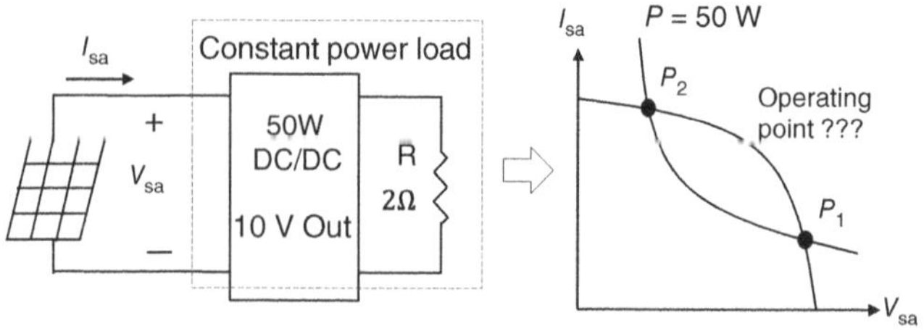

Figure 6.20 Circuit and operating point for a solar cell with a constant power load.

The load characteristic of a CPL can be derived from the fundamental power equation

$$P = VI = K \qquad \text{(constant)}$$

Rearranging the equation yields the current-voltage relationship: $I = \frac{K}{V}$

This inverse relationship results in a hyperbolic or parabolic curve when plotted on the I–V plane, as shown in the figure. When this load line is superimposed on the I–V curve of the solar cell, the intersection points represent potential operating points. In this case, the source curve intersects the load line at two points, labeled P_1 and P_2. This gives rise to a critical question:

Which of these two points represents the valid operating point of the system?

(a) P_1 (b) P_2 (c) P_1 and P_2 (oscillatory behavior)

This issue has historical significance. It first emerged in the 1960s during the development of satellite power systems using solar cells. A phenomenon known as battery latch-up (or battery lock-up) was observed, particularly as satellites exited Earth's shadow (eclipse) while in low Earth orbit. During these transitions, the solar cell voltage would occasionally become fixed at the battery voltage, causing unexpected system behavior. Extensive research and experimentation eventually led to a comprehensive understanding of the underlying dynamics.

The correct answer is: (a) P_1 is the valid operating point.

In systems without active control of the solar cell voltage—where the DC/DC converter regulates only the output voltage—the solar cell will naturally stabilize at P_1, assuming the initial condition begins at or near the open-circuit voltage (V_{oc}) of the solar cell. If, however, the initial voltage were below P_2 (an unlikely real-world scenario), the system could theoretically stabilize at P_2. Nonetheless, no transition between P_1 and P_2 can occur dynamically once the system has settled.

This behavior is explained by nonlinear control theory, particularly through the concept of saddle points. Between P_1 and P_2, a saddle point exists—an inherently unstable equilibrium. As a result, the system cannot traverse from one stable operating point to the other through normal dynamic behavior. Only one stable equilibrium exists under typical starting conditions: P_1.

Configuration of General Power Conversion System The configurations of general power conversion systems—including DC/DC converters and DC/AC inverters—are illustrated in Figures 6.21 and 6.22.

DC/DC Converter is shown in Figure 6.21 as

- Figure 6.21(a): The converter controls the output voltage (V_0) when a resistor is connected to the output.
- Figure 6.21(b): The converter controls the output current (I_0) when a battery is connected to the output.

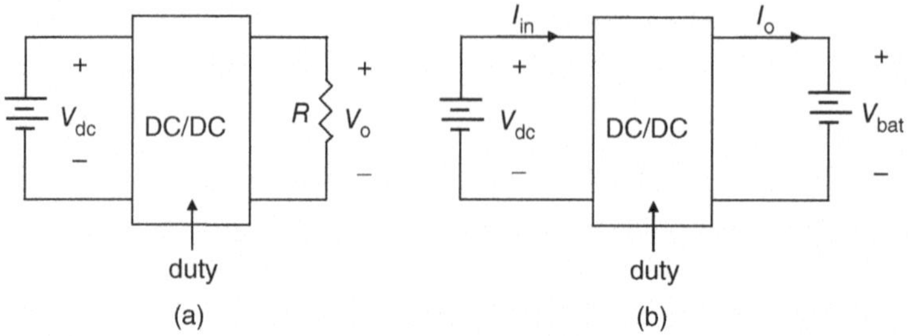

Figure 6.21 DC/DC converter configurations. (a) Output voltage control and (b) Output current control.

In both cases, the input voltage (V_{dc}) is constant and well-regulated. The DC/DC converter regulates either the output voltage or the output current by adjusting the duty cycle of the switching devices. Because the duty cycle is determined by the ratio of the input voltage to the output voltage, stable and effective control requires that either the input or the output be fixed. Consequently, a DC/DC converter is fundamentally a power conversion system with a fixed input and a controlled output.

A similar principle applies to DC/AC inverters, as illustrated in Figure 6.22. In Figure 6.22(a), the inverter operates with a fixed DC input, and the output AC voltage is regulated by adjusting the duty cycle of the switching devices. In contrast, Figure 6.22(b) shows a grid-connected configuration in which the output voltage magnitude and frequency are imposed by the utility grid. Under this condition, the inverter cannot control the output voltage; instead, the only controllable variable is the grid current, which directly determines the power flow.

These configurations demonstrate that, similar to DC/DC converters, DC/AC inverters regulate either the output voltage or the output current by adjusting the duty cycle of the switching devices. Because the duty cycle is determined by the ratio of the input voltage to the output voltage, stable and effective control

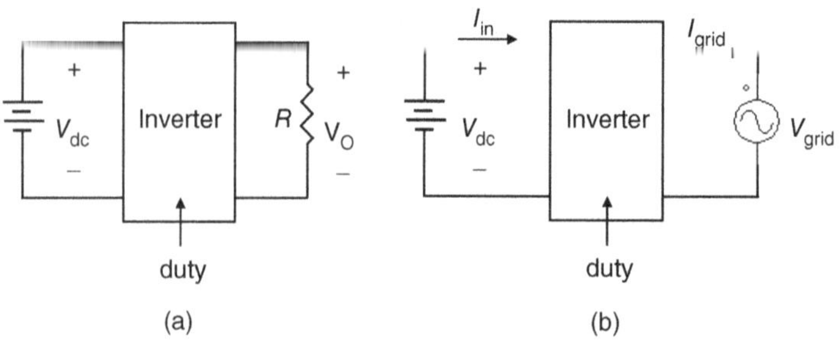

Figure 6.22 DC/AC inverter configurations. (a) Output voltage control and (b) Grid current control.

requires that either the input or the output be fixed. Consequently, a DC/AC inverter is fundamentally a power conversion system with a fixed input and controlled output.

Configuration of Photovoltaic Power Generation Systems Unlike batteries or conventional DC voltage sources, solar cells exhibit an output voltage that varies with the connected load. As discussed earlier, the terminal voltage of a solar cell is not inherently fixed but is determined by its operating point on the I–V characteristic curve. Consequently, to extract power efficiently from a solar cell through a power converter, the converter input voltage—namely, the solar cell voltage—must be actively regulated.

General power converters, including DC–DC converters and DC–AC inverters, employ duty-cycle modulation for control and therefore require a relatively stable input voltage to ensure reliable and efficient operation. In contrast, the voltage generated by a solar cell varies significantly with irradiation, temperature, and load conditions. Since the input voltage cannot be inherently stabilized, voltage regulation must instead be achieved at the output side of the converter.

This requirement necessitates the presence of a voltage sink at the converter output, with a battery being the most common and practical solution. Accordingly, a battery or an AC grid is connected to the output of the power converter in the PV power system configuration shown in Figure 6.23, providing a stable voltage reference and enabling proper power flow control.

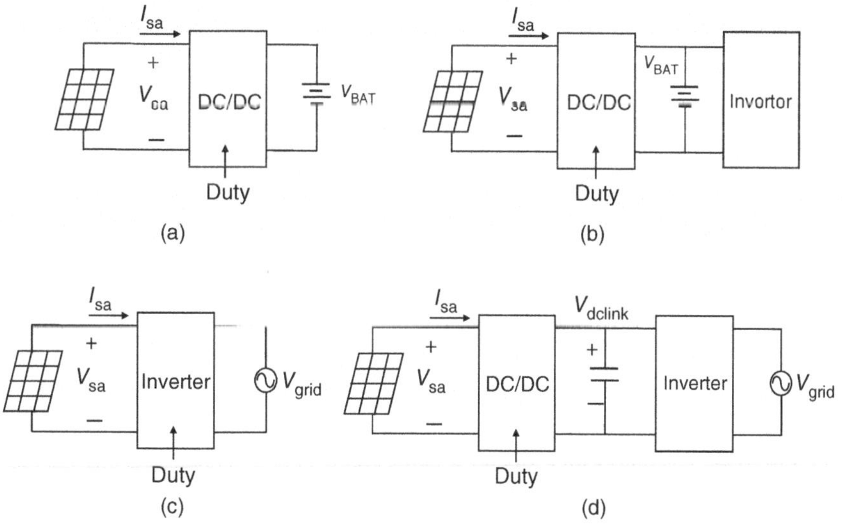

Figure 6.23 Configurations of PV power generation systems. (a) Stand-alone: DC output, (b) Stand-alone: Two-stage, (c) Grid-tied: single-stage and (d) Grid-tied: two-stage.

In two-stage grid-tied PV systems, the DC-link capacitor effectively serves as a voltage sink. The inverter regulates the DC-link voltage to a constant level, causing the capacitor voltage to appear, from the perspective of the upstream DC/DC converter, analogous to a battery voltage. This regulated DC-link voltage provides a stable input condition for the converter, thereby enabling proper control and efficient power transfer.

A similar principle applies to inverters connected to the AC grid. In this case, the grid inherently maintains a fixed voltage magnitude and frequency, effectively functioning as a constant voltage source. This characteristic explains why an AC voltage source—namely, the utility grid—must always be present at the inverter output in grid-tied PV systems. The presence of the grid ensures voltage reference stability and facilitates synchronized power injection.

Figure 6.23 illustrates the typical configurations of PV power generation systems, which can be broadly classified into stand-alone and grid-tied architectures. In this context, *stand-alone* denotes a DC/DC–based output configuration, while *grid-tied* denotes a DC/AC–based configuration interfacing with the utility grid. Figure 6.23(a) and (b) corresponds to stand-alone systems, while Figure 6.23(c) and (d) represents grid-tied systems.

In stand-alone PV systems, a battery is typically connected to the output of the DC/DC converter, serving both as an energy storage element and as a voltage sink. Unlike a simple resistive load, the battery provides a stable voltage reference, which is essential for reliable system operation and effective power regulation. As discussed in this section, a resistive load alone cannot fulfill this role because it does not provide a fixed voltage reference and cannot ensure stable converter operation.

As illustrated in Figure 6.23(b), an inverter may be connected downstream of the battery to supply AC power, thereby operating in a manner analogous to a conventional stand-alone inverter. In this configuration, the DC/DC converter and the inverter are electrically decoupled and operate independently. This separation allows each stage to be controlled separately, simplifying controller design and improving system robustness.

In grid-tied systems, such as the configuration shown in Figure 6.23(c), the inverter directly interfaces the PV system with the utility grid. This raises a fundamental question: why must an AC power source—namely, the grid—always be present during grid-connected operation? This issue has been addressed in detail in this section.

Figure 6.23(d) illustrates a two-stage grid-tied architecture in which a DC-link capacitor is connected at the output of the DC/DC converter, while an inverter interfaces the system with the AC grid. In this configuration, the DC-link voltage $V_{dc,link}$ is regulated by the inverter, which effectively behaves as a constant-voltage sink, similar to a battery. Consequently, from the perspective of the DC/DC converter, the DC-link capacitor voltage appears analogous to a battery voltage V_{bat}. This regulated voltage provides a stable reference that enables proper power transfer and robust control of the converter stage.

Linearized Solar Cell Model Up to this point, the electrical characteristics of solar cells, the configurations of PV power systems, and their operational behavior have been examined. This section focuses on the control-system design of PV power systems, beginning with the mathematical modeling of the solar cell.

Because the physical model of a solar cell results in highly nonlinear equations, it is not well-suited for control-system design and analysis. Instead, a functional model is adopted and linearized in a piecewise manner over the typical operating range. Although this approximation introduces some modeling error, it enables the application of linear control techniques and frequency-domain analysis. Moreover, since the operating point of a PV system is regulated to remain near the MPP—except during the initial tracking phase—the linearized model provides sufficiently accurate representation for most operating conditions.

Piecewise Linear Model of a Solar Array Figure 6.24 (a) illustrates the functional (nonlinear) I–V characteristic of a solar cell, which can be expressed as:

$$i_{sa} = f(V_{sa})$$

where f represents the nonlinear relationship between the solar array current i_{sa} and voltage V_{sa}. To facilitate controller design, this nonlinear characteristic is approximated by a piecewise linear model, as shown in Figure 6.24(b). Straight-line segments are drawn between key operating points and the MPP, denoted by (V_{mp}, I_{mp}). Each segment can be represented by a first-order linear equation of the form

$$i_{sa} = -\alpha \cdot V_{sa} + \beta \tag{6.1}$$

where α is the slope of the linear segment and β is the y-intercept.

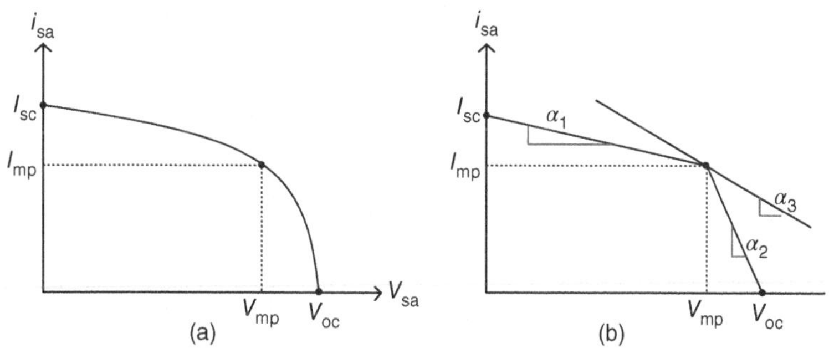

Figure 6.24 Piecewise linear modeling of a solar cell. (a) Functional (nonlinear) model and (b) Piecewise linear approximation.

The operating voltage range of the solar cell is divided into three regions, each described by a different linear approximation:

- Region 1: $V_{sa} < V_{mp}$

$$i_{sa} = -\alpha_1 V_{sa} + \beta_1, \quad \alpha_1 = \frac{I_{sc} - I_{mp}}{V_{mp}}, \quad \beta_1 = I_{sc}$$

- Region 2: $V_{sa} > V_{mp}$

$$i_{sa} = -\alpha_2 V_{sa} + \beta_2, \quad \alpha_2 = \frac{I_{mp}}{V_{oc} - V_{mp}}, \quad \beta_2 = \frac{V_{oc} I_{mp}}{V_{oc} - V_{mp}}$$

- Region 3: $V_{sa} = V_{mp}$

$$i_{sa} = -\alpha_3 V_{sa} + \beta_3, \quad \alpha_3 = \frac{I_{mp}}{V_{mp}}, \quad \beta_3 = 2I_{mp} \tag{6.2}$$

These equations provide a simplified yet sufficiently accurate representation of the nonlinear behavior of a solar cell and are particularly useful for controller design and small-signal analysis. During system start-up, the solar cell voltage typically begins near the open-circuit region ($V_{sa} > V_{mp}$), making the parameters α_2 and β_2 appropriate. Once MPPT is activated, the operating point is maintained near V_{mp}, and the parameters α_3 and β_3 should be used for accurate control modeling.

7

Solar Boost Converter

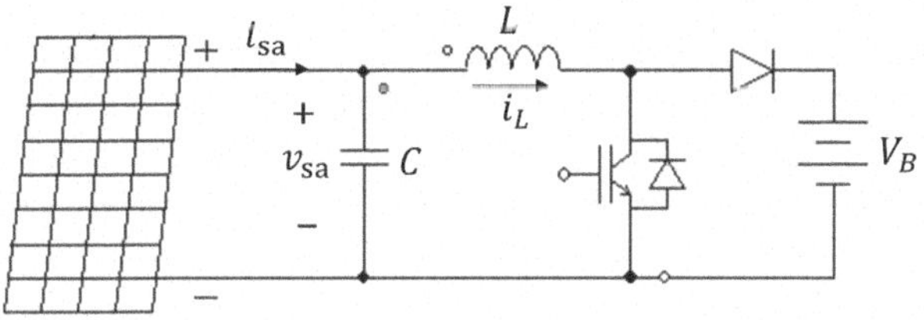

7.1 State-Space Averaged Equation and Transfer Function

Figure 7.1 illustrates the circuit configuration of a solar boost converter, in which a solar cell serves as the input source and a battery is connected as a constant voltage sink at the output. As discussed in the previous chapter, the battery is used to fix the output voltage, enabling control of the solar cell voltage via the converter's duty cycle. This arrangement is essential because, in DC/DC converter operation, either the input or output must remain fixed to facilitate effective control.

In traditional DC/DC converter design, analysis typically begins with a buck converter, as its transfer function is a linear function of the duty cycle d, making modeling and controller design more straightforward. In a buck converter, the input voltage is constant, and the output voltage is regulated by adjusting the duty cycle.

In contrast, when a solar cell is used as the source, the input voltage varies based on environmental conditions, and it becomes more effective to fix the output voltage by connecting a battery. In this configuration, using a boost converter allows the solar cell input voltage to be regulated via duty cycle control. The transfer function of this setup is linear with respect to d, which simplifies system analysis and control design.

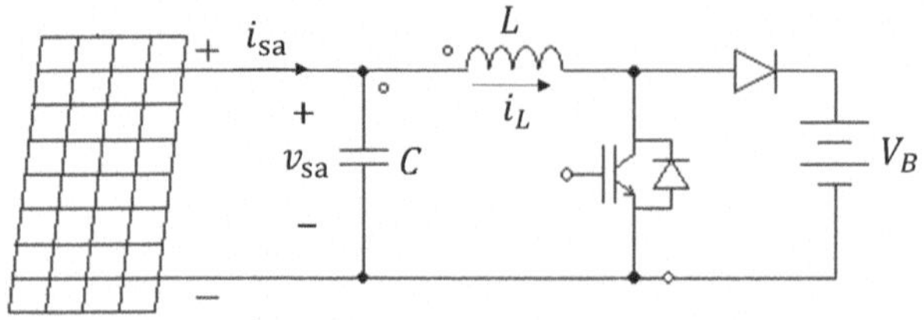

Figure 7.1 Configuration of a solar boost converter.

It is important to note that buck and boost converters exhibit a duality relationship. Conceptually, if the input and output of a buck converter are interchanged, the result is a Boost converter. As such, the transfer function characteristics of a circuit connecting a solar cell to a Boost converter are identical to those of a buck converter fed by a voltage source.

To derive the state-averaged equations, we analyze the system behavior over one switching period by examining the equivalent circuits during the switch ON and OFF intervals, as shown in Figure 7.2. The state variables are defined as the voltage across the capacitor C, which represents the solar cell voltage, and the current through the inductor L. The state equations corresponding to each operating interval are summarized in Table 7.1.

By applying periodic averaging, the following equation can be obtained

$$A = A_{avg} = \frac{(A_1 \cdot T_{ON} + A_2 \cdot T_{OFF})}{T_s} = A_1 \cdot d + A_2 \cdot (1-d)$$

Using this equation to calculate the state-averaged equations

$$A = A_1 d + A_2(1-d) = \begin{pmatrix} -\frac{r_L}{L} & \frac{1}{L} \\ -\frac{1}{C} & -\frac{\alpha}{C} \end{pmatrix}$$

$$B = B_1 d + B_2(1-d) = \begin{pmatrix} \frac{V_B}{L} \\ 0 \end{pmatrix} d + \begin{pmatrix} -\frac{V_B}{L} \\ \frac{\beta}{C} \end{pmatrix}$$

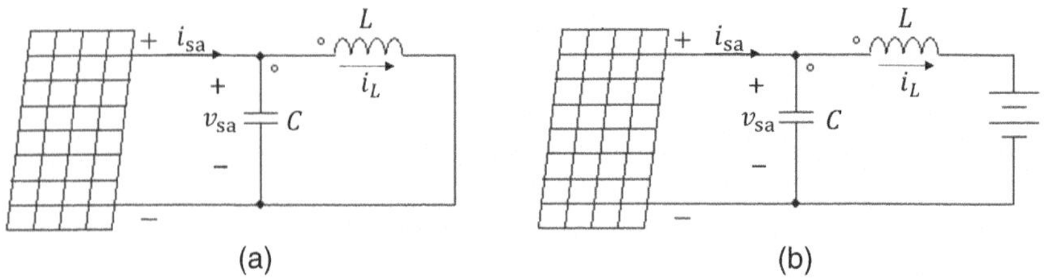

Figure 7.2 Equivalent circuits of the solar boost converter. (a) ON-state equivalent circuit and (b) OFF-state equivalent circuit.

Table 7.1 State equations for each operating interval.

ON-time interval	OFF-time interval
$v_{sa} = L\frac{di_L}{dt} + r_L i_L$	$v_{sa} = r_L i_L + L\frac{di_L}{dt} + V_B$
$i_{sa} = C\frac{dv_{sa}}{dt} + i_L$	$i_{sa} = C\frac{dv_{sa}}{dt} + i_L$
$-\alpha \cdot v_{sa} + \beta = C\frac{dv_{sa}}{dt} + i_L$	$\dot{v}_{sa} = -\frac{1}{C}i_L - \frac{\alpha}{C}v_{sa} + \frac{\beta}{C}$
$\begin{pmatrix} \frac{di_L}{dt} \\ \frac{dv_{sa}}{dt} \end{pmatrix} = \begin{pmatrix} -\frac{r_L}{L} & \frac{1}{L} \\ -\frac{1}{C} & -\frac{\alpha}{C} \end{pmatrix} \begin{pmatrix} i_L \\ v_{sa} \end{pmatrix} + \begin{pmatrix} 0 \\ \frac{\beta}{C} \end{pmatrix}$	$\begin{pmatrix} \frac{di_L}{dt} \\ \frac{dv_{sa}}{dt} \end{pmatrix} = \begin{pmatrix} -\frac{r_L}{L} & \frac{1}{L} \\ -\frac{1}{C} & -\frac{\alpha}{C} \end{pmatrix} \begin{pmatrix} i_L \\ v_{sa} \end{pmatrix} + \begin{pmatrix} -\frac{V_B}{L} \\ \frac{\beta}{C} \end{pmatrix}$
$= A_1 \begin{pmatrix} i_L \\ v_{sa} \end{pmatrix} + B_1$	$= A_2 \begin{pmatrix} i_L \\ v_{sa} \end{pmatrix} + B_2$

Therefore, the state-averaged equation of the solar boost converter is expressed as

$$\begin{pmatrix} \frac{di_L}{dt} \\ \frac{dv_{sa}}{dt} \end{pmatrix} = \begin{pmatrix} -\frac{r_L}{L} & \frac{1}{L} \\ -\frac{1}{C} & -\frac{\alpha}{C} \end{pmatrix} \begin{pmatrix} i_L \\ v_{sa} \end{pmatrix} + \begin{pmatrix} \frac{V_B}{L} \\ 0 \end{pmatrix} d + \begin{pmatrix} -\frac{V_B}{L} \\ \frac{\beta}{C} \end{pmatrix} \tag{7.1}$$

In matrix form, it becomes $\dot{X} = AX + BU + F$.

$$X = \begin{pmatrix} i_L \\ v_{sa} \end{pmatrix}, \quad A = \begin{pmatrix} -\frac{r_L}{L} & \frac{1}{L} \\ -\frac{1}{C} & -\frac{\alpha}{C} \end{pmatrix}, \quad B = \begin{pmatrix} \frac{V_B}{L} \\ 0 \end{pmatrix}, \quad F = \begin{pmatrix} -\frac{V_B}{L} \\ \frac{\beta}{C} \end{pmatrix}, \quad U = d$$

This state equation is a nonlinear matrix equation because of the F matrix. If the input is defined as $= U_{fb} + U_{ff}$, the nonlinear matrix can be canceled using the feed-forward input (U_{ff}). Calculating this gives the feed-forward input $U_{ff} = -\frac{F}{B} = 1$.

To analyze in the frequency domain, using the Laplace transform to find the state variable $X(s)$

$$sX(s) = AX(s) + BU_{fb} \rightarrow X(s) = (sI - A)^{-1}BU_{fb}$$

$$X(s) = \frac{1}{\left(s + \frac{r_L}{L}\right)\left(s + \frac{\alpha}{C}\right) + \frac{1}{LC}} \begin{pmatrix} \frac{V_B}{L}\left(s + \frac{\alpha}{C}\right) \\ -\frac{V_B}{LC} \end{pmatrix} U_{fb}$$

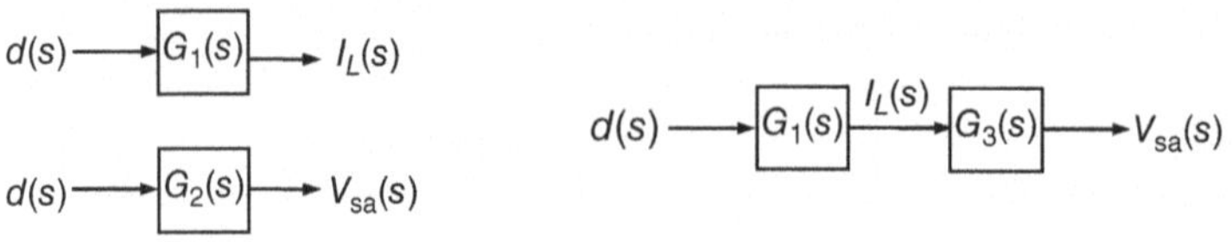

(a) (b)

Figure 7.3 Controller configurations for the solar boost converter. (a) Single-loop and (b) Double-loop.

Transfer Functions Let's calculate the transfer function.

1) Inductor Current Output

$$G_1(s) = \frac{I_{L(S)}}{d(s)} = \frac{\frac{V_B}{L}\left(s + \frac{\alpha}{C}\right)}{s^2 + \left(\frac{r_L}{L} + \frac{\alpha}{C}\right)s + \frac{(r_L \cdot \alpha + 1)}{LC}} \tag{7.2}$$

2) Capacitor Voltage Output

$$G_2(s) = \frac{V_{sa}(s)}{d(s)} = \frac{-\frac{V_B}{LC}}{s^2 + \left(\frac{r_L}{L} + \frac{\alpha}{C}\right)s + \frac{(r_L \cdot \alpha + 1)}{LC}} \tag{7.3}$$

Note: The negative sign in $G_2(S)$ indicates that an increase in duty cycle d results in a **decrease** in the solar cell voltage V_{sa}.

3) Voltage–Current Relationship

$$G_3(s) = \frac{V_{sa}(s)}{I_L(s)} = \frac{-1}{C\left(s + \frac{\alpha}{C}\right)} \tag{7.4}$$

Using these transfer functions, the control system shown in Figure 7.3 can be designed. A single-loop controller is implemented using $G_1(s)$ and $G_2(s)$, whereas a double-loop control structure is obtained by combining $G_1(s)$ and $G_3(s)$.

7.2 Controller Design

7.2.1 Voltage Controller Design

The voltage controller for the solar boost converter is based on a single-loop control structure and utilizes the system transfer function $G_2(s)$. The overall configuration is illustrated in Figure 7.4, where a *PI* (Proportional-Integral) controller, denoted as $G_{CV}(s)$, is employed.

It is important to note the sign convention used for the voltage error v_{err}. Unlike conventional DC–DC converter controllers, which compute the error as the

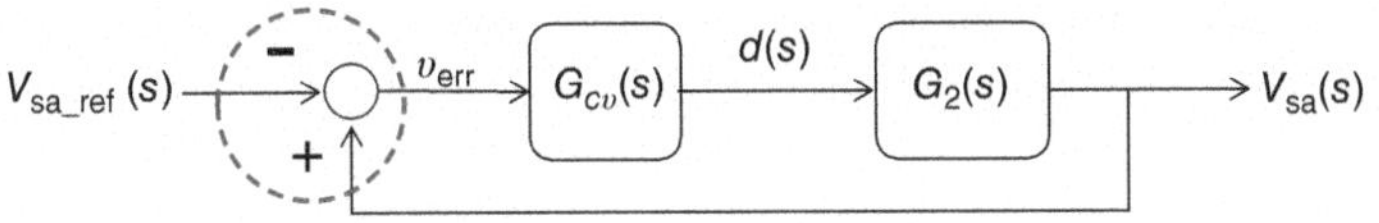

Figure 7.4 Voltage controller configuration for the solar boost converter.

difference between the reference voltage and the output voltage, the solar boost converter uses an inverted sign. This adjustment is necessary because the system transfer function $G_2(s)$ is negative. An increase in the duty cycle d results in a decrease in the solar panel voltage V_{sa}, which is contrary to typical converter behavior.

The closed-loop transfer function is expressed as:

$$\frac{V_{sa}(s)}{V_{sa\_ref}(s)} = \frac{-G_{CV}(s)G_2(s)}{1 + G_{CV}(s)G_2(s)}$$

Although the expression appears to yield a negative result, the overall system response remains positive because $G_2(s)$ is negative. This ensures proper system operation. If the sign of the error were not inverted as shown, the closed-loop response would become negative, leading to instability. Therefore, the voltage error sign must be defined opposite to that of a conventional controller to maintain correct functionality.

For simulation purposes, the following solar cell parameters are used

$$V_{oc}: 50\,\text{V}, \quad I_{sc}: 10\,\text{A}, \quad V_{mp}: 40\,\text{V}, \quad I_{mp}: 8\,\text{A}$$

The boost converter circuit parameters are as follows

$$r_L = 0.02, \quad L = 1\,\text{mH}, \quad C = 100\,\text{uF}, \quad V_B = 70\,\text{V}$$

The linearized solar cell parameters (assuming MPPT operation) are

$$\alpha = \alpha_3 = 0.2, \quad \beta = \beta_3 = 16$$

Using these values, the system transfer function $G_2(s)$ is derived as

$$G_2(s) = \frac{V_{sa}(s)}{d(s)} = \frac{-7 \cdot 10^8}{s^2 + 2020s + 1.004 * 10^7}$$

The *PI* voltage controller is defined as

$$G_{CV}(s) = \frac{K_{pv}(s + K_{iv})}{s}$$

The design objective is to determine the optimal values for K_{pv} and K_{iv} to meet specific control performance criteria. The frequency-domain design specifications are:

- Bandwidth = 50 rad/sec
- Phase margin = 90 deg

Table 7.2 MATLAB m-file for solar boost controllers.

```
% Solar Boost PI Design (Solar_boost1.m) %
Voc=50;
Vmp=40;
Isc=10;
Imp=8;
alpha3 = Imp/Vmp;
beta3 = 2*Imp;

s=tf('s');
rL = 0.02;     % rL=20 mohm
L = 1.0e-3;    % L=1.0mh
C = 100e-6;    % C=100uf
VB=70;
alpha = alpha3;
beta =  beta3;

G1=(VB/L)*(s+alpha/C)/(s^2+s*(rL/L+alpha/C)+
   (rL*alpha+1)/(L*C) ); % IL(s)/d(s)
G2=VB/(L*C) / (s^2+s*(rL/L+alpha/C)+
    (rL*alpha+1)/(L*C));      % Vsa(s)/d(s)
G3=(-1/C)/(s + alpha/C); % Vdq(s)/Idq(s)
```

```
Kpv = 1; Kiv=1; %temporarily
Gcv = Kpv*(s+Kiv)/s;

Kpi =1; Kii =1; %temporarily
Gci = Kpi*(s+Kii)/s;

Vsa_ref = 38;
IL_ref  = 8;
%sisotool Voltage Single loop
 T=sisoinit(1);
 T.F.Value = Vsa_ref;
 T.G.Value = G2;
 T.C.Value = Gcv;
 sisotool(T);

%sisotool Current Single loop
%T=sisoinit(1);
%T.F.Value = IL_ref;
%T.G.Value = G1;
%T.C.Value = Gci;
%sisotool(T);
```

The MATLAB m-file is presented in Table 7.2. Since both voltage and current controllers can be selected, the results corresponding to the voltage controller are shown first.

The following controller satisfies the design requirements

$$G_{CV}(s) = 0.0001443\frac{(s + 4969)}{s}$$

The corresponding discrete-time representation, with a sampling frequency ($f_{sw} = 20\,\text{kHz}$), is

$$G_{CV}(z) = 0.00018\frac{(z - 0.8)}{z - 1}$$

Figure 7.5 presents the Bode plot and root locus. The Bode plot of the voltage controller satisfies the specified design requirements, while the root locus confirms system stability.

Figure 7.6 illustrates the step response, which demonstrates a rise time of approximately 44 ms with no overshoot, thereby validating the effectiveness of the controller.

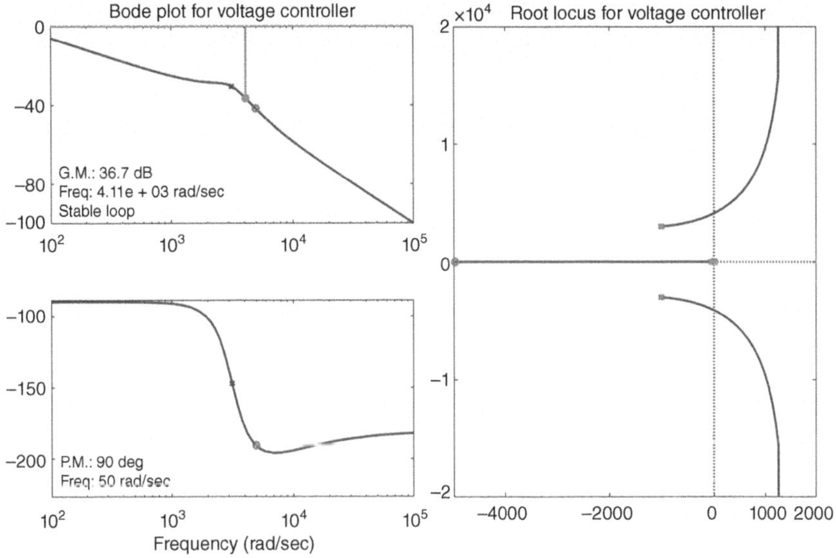

Figure 7.5 Bode plot and root locus of the voltage controller.

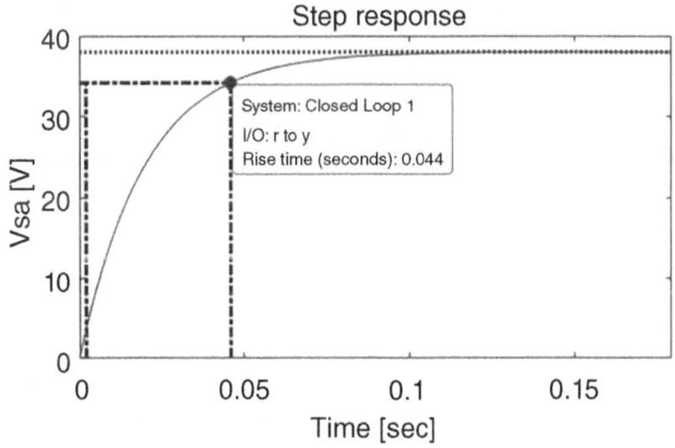

Figure 7.6 Step response of the voltage controller.

When converted into a discrete-time equation, the digital controller becomes

$$G_{CV}(z) = \frac{d(z)}{V_{err}(z)} = K_P \frac{(z - K_I)}{z - 1}$$

Multiplying both sides by the denominator and rearranging terms yields

$$z \cdot d(z) - d(z) = z \cdot K_P \cdot V_{err}(z) - K_P K_I \cdot V_{err}(z)$$

Taking the inverse Z-transform yields the time-domain difference equation

$$d(k) = d(k-1) + K_P \cdot V_{err}(k) - K_P K_I \cdot V_{err}(k-1) \tag{7.5}$$

This equation is implemented in digital control systems to update the duty cycle $d(k)$ at each sampling instant based on the current and previous voltage error values.

The PSIM simulation was performed using the designed PI controller gains. Figure 7.7 illustrates the PSIM schematic of the solar boost converter voltage controller.

The corresponding C-block implementation is provided in Table 7.3.

The simulation results are presented in Figure 7.8. When the reference voltage was step-changed by 4 V—from 30 to 34 V—it is evident that the solar cell voltage is effectively regulated without any overshoot or steady-state error. The third graph illustrates the solar cell power output, calculated as $V_{sa}I_{sa}$, which reaches its maximum value (MaxP) at 38 V, corresponding to the maximum power point voltage V_{mp} specified in the solar cell parameters.

The rightmost graph evaluates rise time when the voltage transitions from 34 to 38 V. After the step input is applied at 0.45 s, rise time is defined as the interval required for the voltage to increase from 10% to 90% of its final value, corresponding to 34.4 up to 37.6 V. The measured rise time is approximately 0.04424 s (44 ms), which agrees exactly with the 44 ms prediction from the MATLAB-based analysis.

MPPT Design Using the Voltage Controller

To implement maximum power point tracking (MPPT), the previously designed voltage controller is utilized. Figure 7.9 illustrates the block diagram of the MPPT

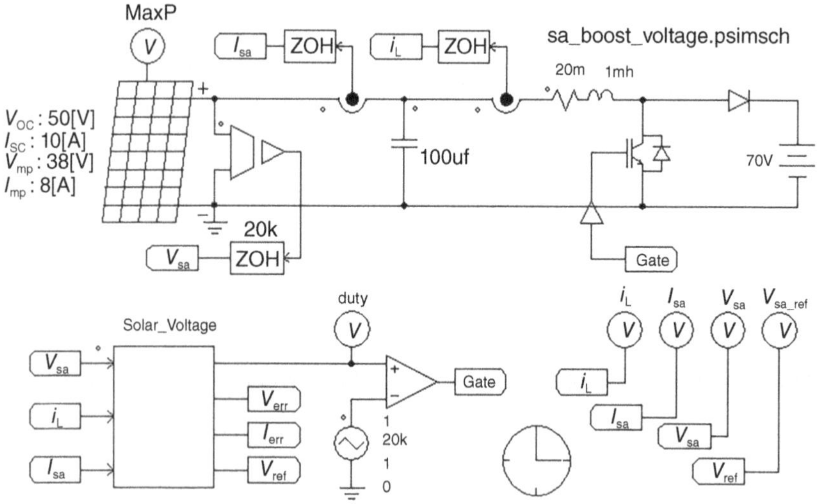

Figure 7.7 PSIM schematic of the solar boost converter voltage controller.

Table 7.3 C-block code for the solar boost voltage controller: Solar_Voltage.

```
double Kp, Ki;
double Vsa, IL, Vref, Isa, Verr;
static double duty_old=0.0;
static double Verr_old=0.0;
double duty;

Vsa  = in[0];
IL   = in[1];
Isa  = in[2];

/**** Voltage controller *****/
if (t <0.3) Vref = 30;
else if (t < 0.45) Vref = 34;
else if (t < 0.6) Vref = 38;
else Vref =42;

Verr = Vsa - Vref;
Kp=0.00018;
Ki=0.8;

duty=duty_old+Kp*Verr-Kp*Ki*Verr_old;
duty_old= duty;
Verr_old= Verr;
if(duty >0.9) {duty= 0.9;}
if(duty <0)   {duty= 0;}

out[0] = duty;
out[1] = Verr;
out[2] = IL_err;
out[3] = Vref;
```

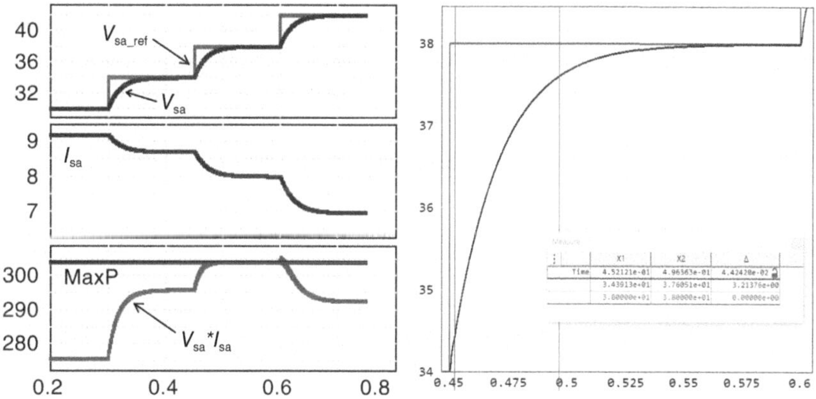

Figure 7.8 Simulation results of the voltage controller.

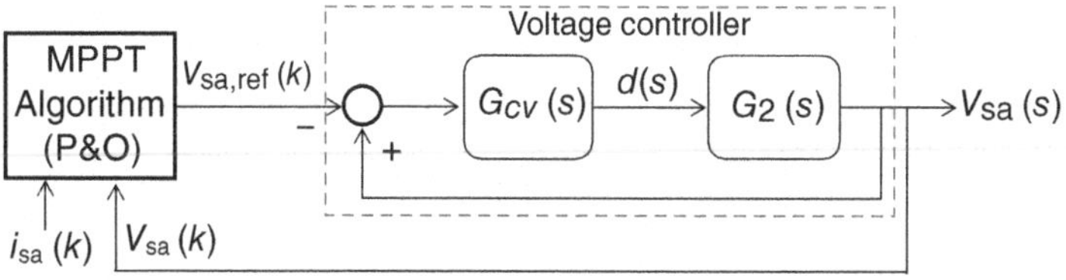

Figure 7.9 Configuration of the MPPT system using voltage control.

control system, where the voltage controller operates in the inner loop, while the MPPT algorithm governs the outer loop. The Perturb and Observe (P&O) algorithm is employed to perform MPPT. The algorithm's output is the reference voltage V_{sa_ref}, which serves as the input to the voltage controller.

A critical design consideration is the separation of bandwidths between the inner and outer loops. Since the outer loop continuously updates the value of V_{sa_ref}, the update interval must be sufficiently long to allow the inner voltage control loop to settle. In other words, the reference voltage should remain constant until the output voltage V_{sa} has stabilized. In commercial implementations, this update interval is typically set to once per second.

The MPPT simulation schematic is shown in Figure 7.10. Within the circuit, the input constant T_{up} defines the MPPT update interval. Although the optimal interval is not explicitly designed in this case, the update frequency is set to ensure it exceeds the settling time of the voltage controller. The parameter DV represents the step size ΔV used to adjust V_{sa_ref}.

The C-block (`MPPT_Voltage`) code for the MPPT logic is shown in Table 7.4. The P&O algorithm is implemented according to the standard flowchart-based control logic. The voltage controller used in the inner loop retains the gain values derived from the previous design.

The simulation results are presented in Figure 7.11. As shown, V_{sa_ref} decreases from the open-circuit voltage and converges toward the maximum power point voltage of 38 V, oscillating within a ±2 V range. The corresponding solar cell

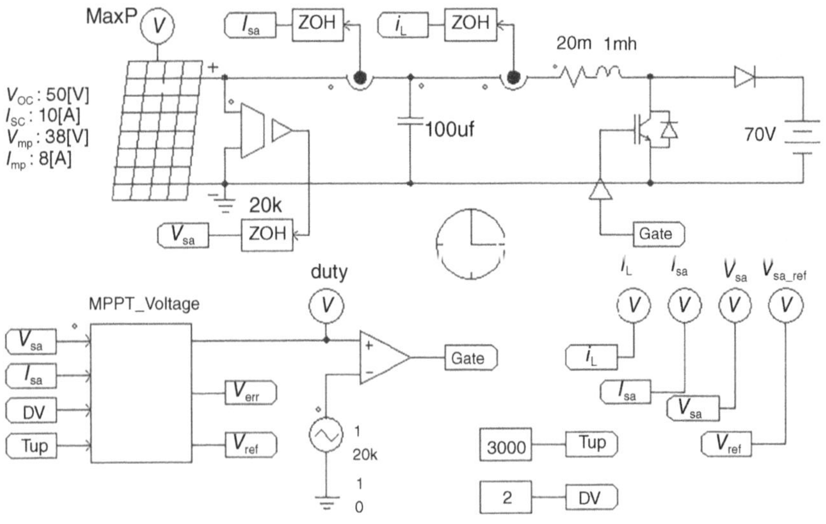

Figure 7.10 Schematic for MPPT voltage control.

Table 7.4 C-block code for the MPPT voltage controller: MPPT_Voltage.

```
#define Kp   0.00018
#define Ki   0.8

double Vsa, Isa, DV, Psa;
double duty, Verr, Vref =50.0;
int TUpdate;

static int i=0;
static double Psa_old=0.0;
static double Vsa_old=51.0;
static double Vcon_old=0.0;
static double Verr_old=0.0;
static double duty_old=0.0;

Vsa= in[0];
Isa = in[1];
DV  = in[2];
TUpdate=in[3];

//** MPPT  P&O Algorithm **//
i++;
if (i % TUpdate == 0) {
 Psa = Vsa*Isa;
 if (Psa >= Psa_old) {
   if (Vsa > Vsa_old) Vref = Vref + DV;
   else               Vref = Vref - DV;
 }
 else {
   if (Vsa > Vsa_old) Vref = Vref - DV;
   else               Vref = Vref + DV;
 }
 Psa_old = Psa;
 Vsa_old = Vsa;
} // end of MPPT
/**** Voltage controller ************/
Verr = Vsa - Vref;
duty = duty_old+Kp*Verr-Kp*Ki*Verr_old;
duty_old=duty;
Verr_old=Verr;
/*****************************/
out[0] = duty;
out[1] = Verr;
out[2] = Vref;
```

power output increases and stabilizes at the maximum power level (MaxP), demonstrating that the P&O algorithm is functioning effectively. The results confirm the expected behavior: when the duty cycle increases, V_{sa} decreases; conversely, when the duty cycle decreases, V_{sa} increases.

To evaluate the performance of the MPPT algorithm, the maximum power point voltage V_{mp} in the circuit was modified from 42 to 44 V. The corresponding simulation results are shown in Figure 7.12.

The MPPT algorithm successfully adjusts the solar cell voltage to track the updated maximum power point. In Figure 7.12(a), when V_{mp}= 42 V, the controller maintains the voltage around this point, ensuring maximum power extraction. In Figure 7.12(b), after V_{mp} is increased to 44 V, the controller adapts accordingly, and the solar cell voltage converges to the new reference. In both cases, the output power aligns with the updated maximum power point, confirming the effectiveness of the tracking algorithm.

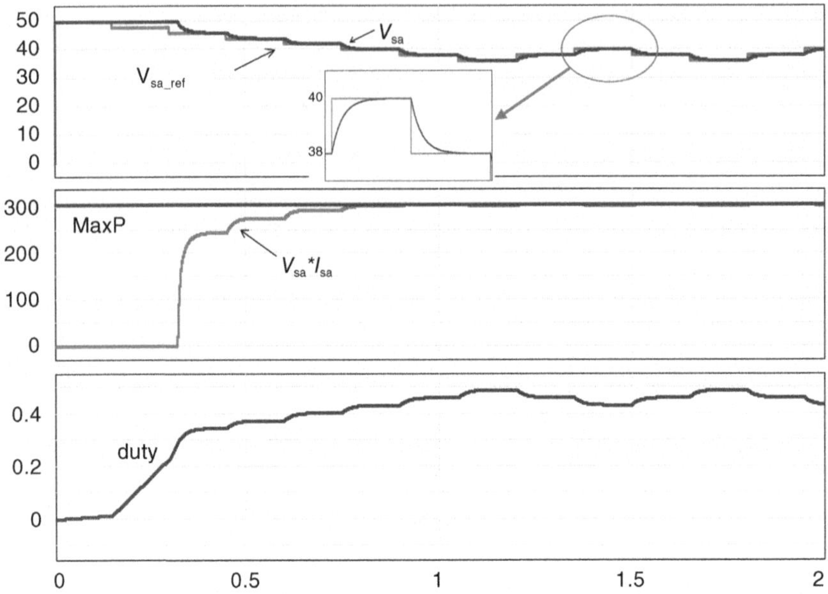

Figure 7.11 Simulation results of the MPPT controller.

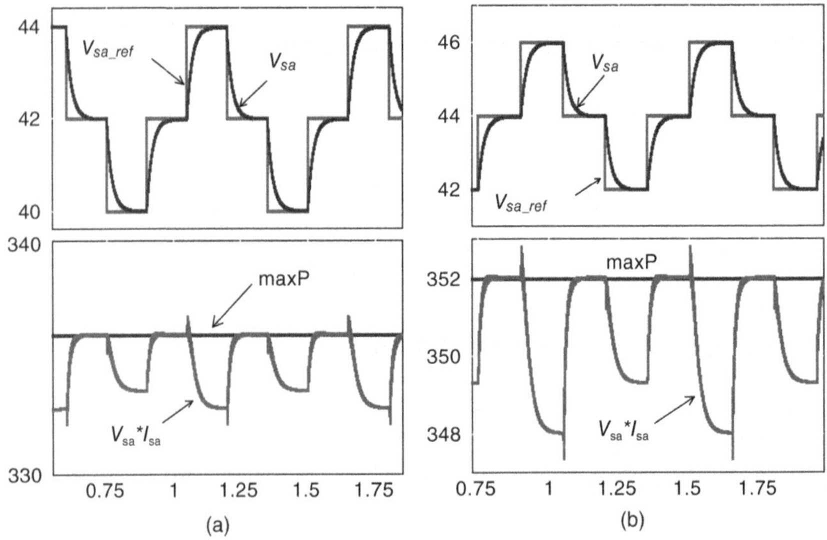

Figure 7.12 MPPT tracking performance for varying voltage references. (a) $V_{mp} = 42$ V and (b) $V_{mp} = 44$ V.

7.2.2 Current Controller Design

The current controller for the solar boost converter regulates the inductor current I_L instead of the solar cell voltage. It is based on a single-loop control structure and uses the system transfer function $G_1(s)$. The configuration of the current controller is illustrated in Figure 7.13. A PI (proportional–integral) controller, denoted $G_{Ci}(s)$, is employed.

Since the transfer function $G_1(s)$ is positive, the controller structure follows a standard configuration. That is, an increase in the duty cycle d results in an increase in the inductor current I_L, consistent with typical boost converter behavior.

By applying the previously defined circuit parameters, the transfer function $G_1(s)$ is calculated as follows:

$$G_1(s) = \frac{I_L(s)}{d(s)} = \frac{\frac{V_B}{L}\left(s + \frac{\alpha}{C}\right)}{s^2 + \left(\frac{r_L}{L} + \frac{\alpha}{C}\right)s + \frac{(r_L \cdot \alpha + 1)}{LC}} = \frac{70000(s + 1.4 \times 10^8)}{s^2 + (2020)s + 1.004 \times 10^7}$$

The pole-zero diagram of $G_1(s)$, shown in Figure 7.14, reveals a classical structure: two complex poles and a single real-axis zero.

To eliminate steady-state error, the controller $G_{Ci}(s)$ must form a *type-1 system*, meaning it includes an integrator (a pole at the origin). Thus, the controller structure must be

$$G_{Ci}(s) = \frac{K_{pi}(s + K_{ii})}{s}$$

The controller was designed using SISOTOOL in MATLAB by activating the masked current controller section within the voltage controller design framework. The following frequency-domain design criteria were applied

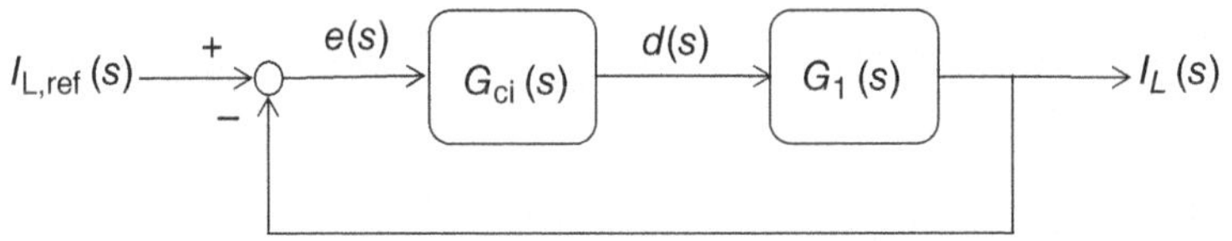

Figure 7.13 Configuration of the current controller for the solar boost converter.

Figure 7.14 Pole-zero diagram of the transfer function $G_1(s)$.

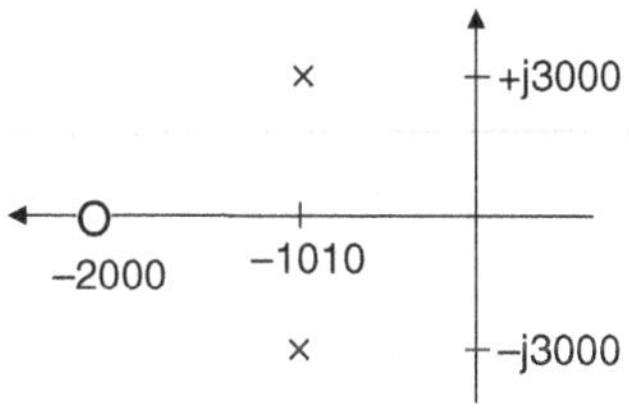

- Bandwidth = 50 rad/sec
- Phase margin = 90 deg

The resulting controller from SISOTOOL is

$$G_{Ci}(s) = \frac{3.58}{s}$$

Converting this to a discrete-time expression using a sampling frequency ($f_{sw} = 20$ kHz)

$$G_{Ci}(z) = 0.0000895\frac{(z+1)}{z-1}$$

Figure 7.15 presents the Bode plot and root locus of the current controller. The Bode plot indicates a phase margin of 90.9°, which is slightly lower than the specified design requirement; nevertheless, the overall control performance remains nearly satisfactory. The root locus analysis further confirms closed-loop stability, as all system poles are located in the left half-plane.

Figure 7.16 illustrates the step response of the current controller. The response exhibits a rise time of approximately 44 ms, thereby validating the effectiveness of the controller design.

The designed PI controller was implemented in PSIM, as illustrated by the simulation schematic in Figure 7.17. Note that the duty-cycle calculation uses a positive sign, consistent with the positive gain of the system transfer function $G_1(s)$.

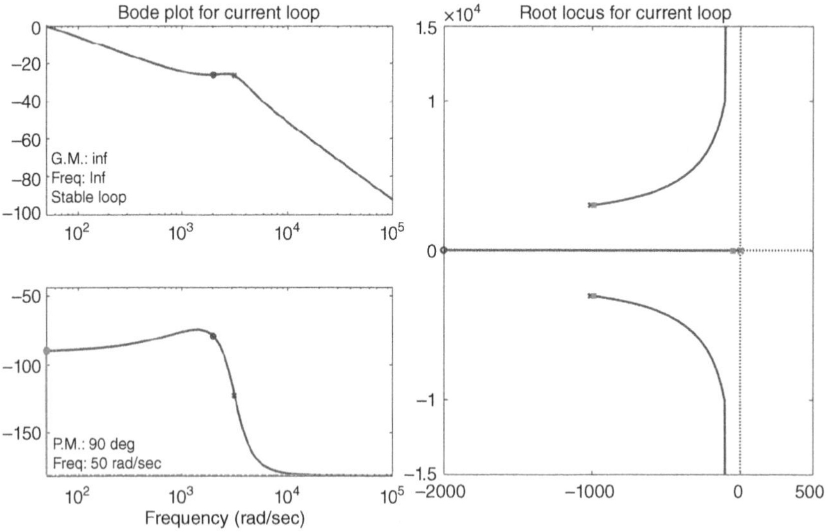

Figure 7.15 Bode plot and root locus of the current controller.

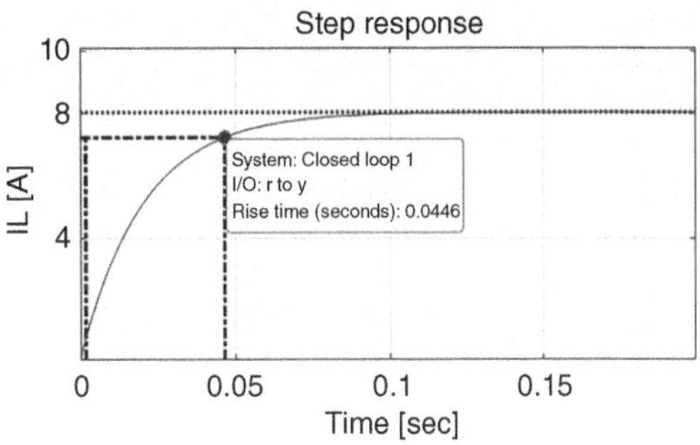

Figure 7.16 Step response of the current controller.

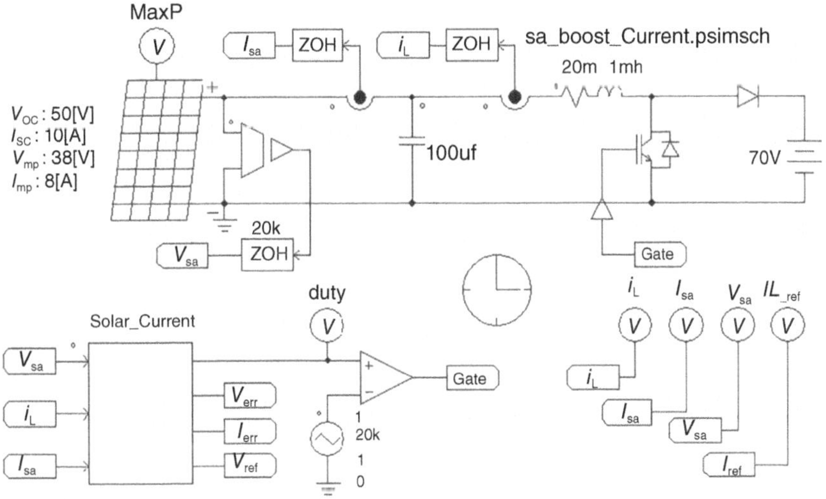

Figure 7.17 PSIM schematic of the solar boost current controller.

The corresponding C-block implementation is presented in Table 7.5.

The simulation results for the current controller are shown in Figure 7.18. When the reference current was step-changed through a sequence of 3 A → 7.5 A → 5 A → 3 A, the solar cell current responded smoothly, reaching its steady-state values without overshoot. The third graph illustrates the solar cell power, calculated as $V_{sa} \cdot I_{sa}$, achieving the maximum power (*MaxP*) when the reference current is set to 7.5 A. According to the solar cell parameters, the maximum power point current I_{mp} is 8 A. Since the inductor current I_L is slightly less than the solar cell current I_{sa}, the reference current was appropriately set to 7.5 A.

The measured settling time was within 100 ms, which closely aligns with the MATLAB simulation result of 79 ms, confirming consistency between simulation environments.

Table 7.5 C-block code for the solar boost current controller: Solar_Current.

```
#define Kp      0.0000895
#define Ki      1
double Vsa, IL, Vref, Isa;
double duty, Verr, IL_err, IL_ref;
static double duty_old=0.0,
static double IL_err_old=0.0;

Vsa   = in[0];
IL    = in[1];
Isa   = in[2];

if (t <0.2)            IL_ref = 3;
else if (t < 0.4)      IL_ref = 7.5;
else if (t < 0.6)      IL_ref=  5;
else                   IL_ref=  3;

IL_err = IL_ref - IL;

duty=duty_old+Kp*IL_err +
Kp*Ki*IL_err_old;
duty_old=duty;
IL_err_old = IL_err;

if(duty>1) {duty=1;}
if(duty<0) {duty=0;}

out[0] = duty;
out[1] = Verr;
out[2] = IL_err;
out[3] = IL_ref;
```

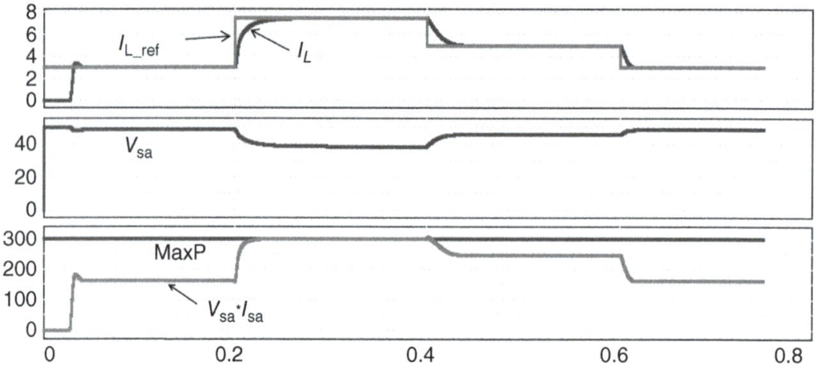

Figure 7.18 Simulation results of the current controller.

MPPT Design Using the Current Controller To implement an MPPT system, the previously designed current controller is integrated into a double-loop control structure. Figure 7.19 illustrates the block diagram of the MPPT system using current control. The current controller operates in the inner loop, while a modified P&O algorithm forms the outer loop. Unlike the traditional P&O method, which is based on voltage changes, this version is adapted to track the rate of change in current. The output of the MPPT algorithm is the reference current I_{L_ref}, which acts as the input to the current controller.

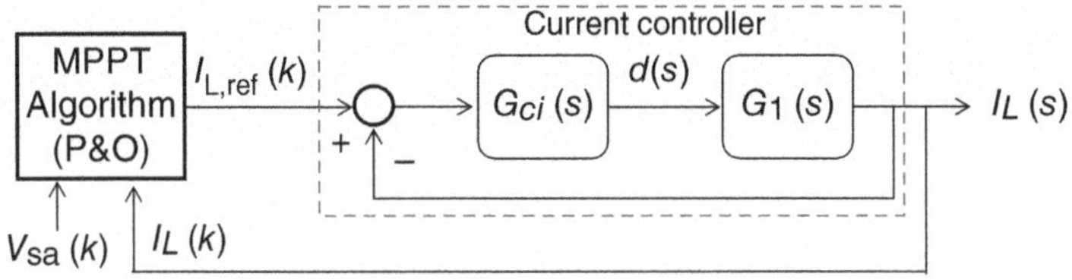

Figure 7.19 Configuration of the MPPT system using current control.

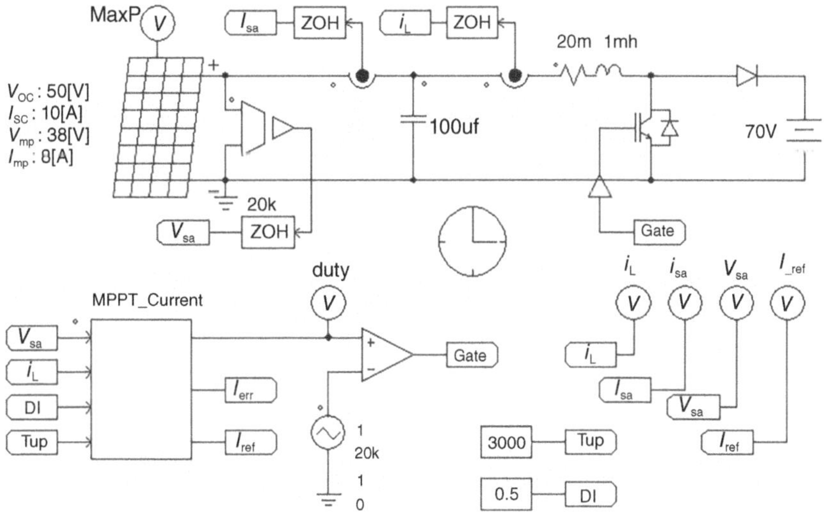

Figure 7.20 Schematic for MPPT current control.

The corresponding MPPT simulation schematic is shown in Figure 7.20. In the circuit, the input constant T_{up} defines the MPPT update interval, while the constant DI specifies the step size for adjusting the reference current I_{L_ref}.

The C-block (`MPPT _Current`) code for the current-based MPPT logic is presented in Table 7.6. It implements the modified P&O algorithm in accordance with the control flowchart. The inner-loop current controller retains the previously designed PI parameters without modification.

The simulation results are presented in Figure 7.21. The reference current I_{L_ref} begins at 0 A and gradually increases, eventually stabilizing around the maximum power point of 8 A within a ±0.5 A range. This fluctuation is attributed to the current step size (DI) being set to 0.5 A. The solar cell power output increases steadily and stabilizes at the maximum power point ($MaxP$), confirming that the current-based P&O algorithm is functioning as intended.

Table 7.6 C-block code for the MPPT current controller: MPPT_Current.

```
#define Kp      0.0000895
#define Ki      1

double Vsa, IL, DI;
int TUpdate;
static int i=0;

static double Psa_old=0.0;
static double duty_old=0;
static double IL_err_old=0.0;
static double IL_old=0,
IL_ref=8.0;
double Psa, IL_err, duty;

Vsa= in[0];
IL  = in[1];
DI  = in[2];
TUpdate=in[3];
//**** MPPT  P&O Algo.****//
i++;
if (i % TUpdate == 0) {
  Psa = Vsa*IL;
  if (Psa >= Psa_old) {
    if (IL>IL_old)  IL_ref=IL_ref + DI;
    else            IL_ref=IL_ref - DI;
  }
  else {
    if (IL>IL_old)  IL_ref=IL_ref - DI;
    else            IL_ref=IL_ref + DI;
  }
  Psa_old = Psa;
  IL_old = IL;
}
/****Current controller ************/
IL_err = IL_ref - IL;
duty=duty_old+Kp*IL_err+Kp*Ki*IL_err_old;
duty_old=duty;
IL_err_old = IL_err;
/*****************************/
out[0] = duty;
out[1] = IL_err;
out[2] = IL_ref;
```

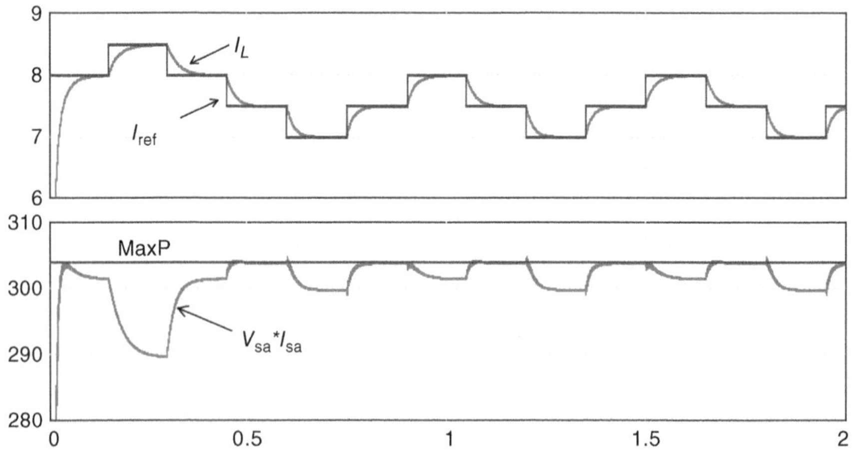

Figure 7.21 Simulation results of the MPPT controller using current reference.

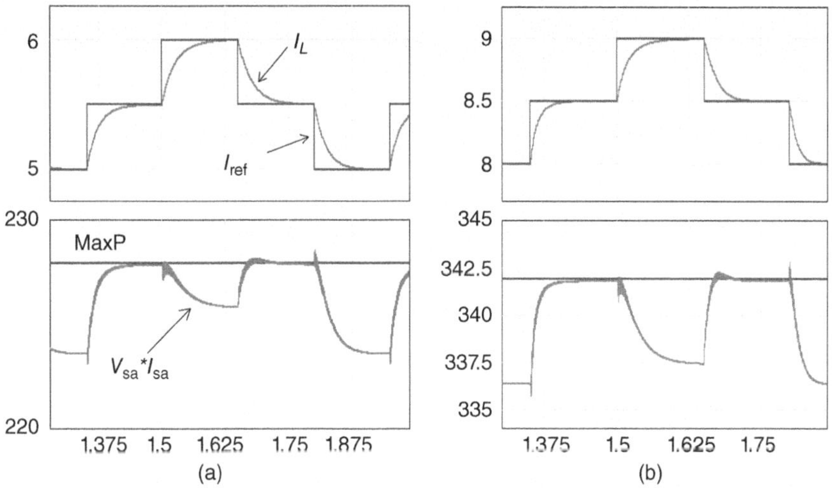

Figure 7.22 MPPT tracking performance for varying current references. (a) $I_{mp} = 6$ A and (b) $I_{mp} = 9$ A.

It is also observed that the inductor current (or solar cell current) increases as the duty cycle increases and decreases when the duty cycle decreases—consistent with the system's positive transfer function. Users are encouraged to simulate and compare the performance of current-based and voltage-based MPPT algorithms to understand their respective advantages and behavioral differences.

To further validate the performance of the current-based MPPT algorithm, the same evaluation method used in the previous chapter is applied. By changing the maximum power point current I_{mp} in the circuit to 6 and 9 A, the system's response is observed. As shown in Figure 7.22, the MPPT algorithm successfully adjusts the solar cell current to track the newly updated maximum power points at $I_{mp} = 6$ A and $I_{mp} = 9$ A. The corresponding power output reaches and maintains the maximum level, demonstrating effective tracking performance.

8

Solar Buck Converter

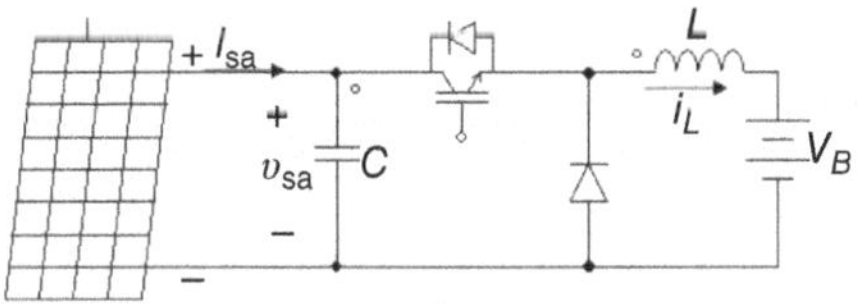

8.1 State-Space Averaged Equation and Transfer Function

The circuit configuration connecting a solar cell to a buck converter is illustrated in Figure 8.1. This topology can be analyzed using the same methodology as the solar boost converter discussed in the previous chapter.

To derive the state-averaged equations, the equivalent circuits for the switch ON and OFF intervals are considered, as shown in the corresponding figures. The state variables are defined as follows:

- v_{sa}: The voltage across the capacitor C, representing the solar cell voltage.
- i_L: The current flowing through the inductor L.

The state equations for each switching interval are presented in Table 8.1.

By applying the state-space averaging method over one switching period, the averaged system dynamics are obtained as follows:

$$A = A_1 d + A_2(1-d) = \begin{pmatrix} -\frac{r_L}{L} & \frac{1}{L}d \\ -\frac{1}{C}d & -\frac{\alpha}{C} \end{pmatrix}$$

$$B = Bd + B(1-d) = \begin{pmatrix} -\frac{V_B}{L} \\ \frac{\beta}{C} \end{pmatrix}$$

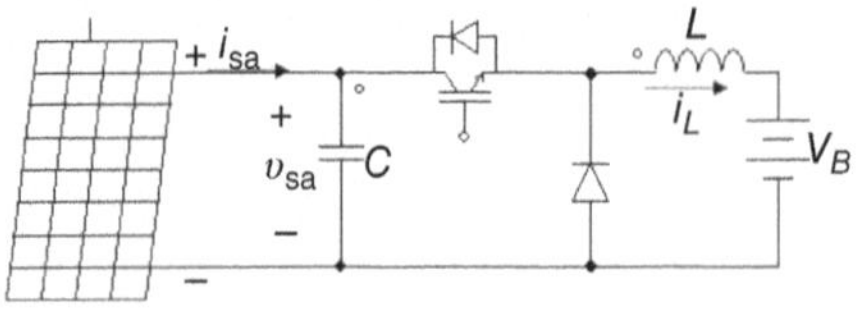

Figure 8.1 Schematic of the solar buck converter.

Table 8.1 State equations for each operating interval.

ON time	OFF time
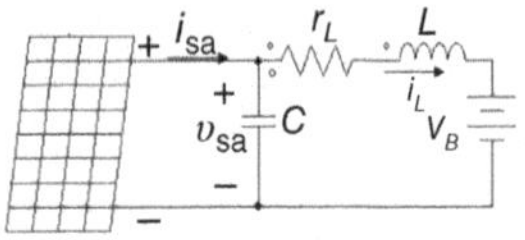	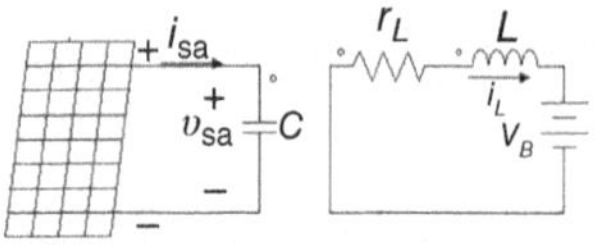
$\frac{di_L}{dt} = -\frac{r_L}{L}i_L + \frac{1}{L}v_{sa} - \frac{1}{L}V_B$	$\frac{di_L}{dt} = -\frac{r_L}{L}i_L - \frac{1}{L}V_B$
$\frac{dv_{sa}}{dt} = -\frac{1}{C}i_L + \frac{1}{C}i_{sa}$	$\frac{dv_{sa}}{dt} = \frac{1}{C}i_{sa}$
$\frac{dv_{sa}}{dt} = -\frac{1}{C}i_L - \frac{\alpha}{C}v_{sa} + \frac{1}{C}\beta$	$\frac{dv_{sa}}{dt} = -\frac{\alpha}{C}v_{sa} + \frac{1}{C}\beta$
$\begin{pmatrix} \frac{di_L}{dt} \\ \frac{dv_{sa}}{dt} \end{pmatrix} = \begin{pmatrix} -\frac{r_L}{L} & \frac{1}{L} \\ -\frac{1}{C} & -\frac{\alpha}{C} \end{pmatrix} \begin{pmatrix} i_L \\ v_{sa} \end{pmatrix} + \begin{pmatrix} -\frac{V_B}{L} \\ \frac{\beta}{C} \end{pmatrix}$	$\begin{pmatrix} \frac{di_L}{dt} \\ \frac{dv_{sa}}{dt} \end{pmatrix} = \begin{pmatrix} -\frac{r_L}{L} & 0 \\ 0 & -\frac{\alpha}{C} \end{pmatrix} \begin{pmatrix} i_L \\ v_{sa} \end{pmatrix} + \begin{pmatrix} -\frac{V_B}{L} \\ \frac{\beta}{C} \end{pmatrix}$
$= A_1 \begin{pmatrix} i_L \\ v_{sa} \end{pmatrix} + B$	$= A_2 \begin{pmatrix} i_L \\ v_{sa} \end{pmatrix} + B$

Therefore, the state-averaged equations of the solar buck converter are expressed as

$$\begin{pmatrix} \frac{di_L}{dt} \\ \frac{dv_{sa}}{dt} \end{pmatrix} = \begin{pmatrix} -\frac{r_L}{L} & \frac{1}{L}d \\ -\frac{1}{C}d & -\frac{\alpha}{C} \end{pmatrix} \begin{pmatrix} i_L \\ v_{sa} \end{pmatrix} + \begin{pmatrix} -\frac{V_B}{L} \\ \frac{\beta}{C} \end{pmatrix} \tag{8.1}$$

Since the matrix $A(d)$ is dependent on the duty ratio d, the system is inherently nonlinear. The dynamic system can be represented in matrix form as

$$\dot{X} = A(d) \cdot X + B$$

To analyze the steady-state behavior, the state vector X can be determined by setting $\dot{X} = 0$ and solving

$$X = -A^{-1}(D)B$$

This yields the steady-state values of the inductor current I_L and the capacitor voltage V_{SA} as

$$\begin{pmatrix} I_L \\ V_{SA} \end{pmatrix} = \begin{pmatrix} \dfrac{I_{SA}}{D} \\ \dfrac{r_L I_L + V_B}{D} \end{pmatrix} = \begin{pmatrix} \dfrac{I_{SA}}{D} \\ \dfrac{r_L I_{SA}}{D^2} + \dfrac{V_B}{D} \end{pmatrix} \tag{8.2}$$

Here, I_{SA} denotes the steady state output current of the solar cell. This formulation illustrates the dependence of the system variables on the duty cycle d, confirming the nonlinear nature of the solar buck converter.

Small-Signal Modeling To analyze the dynamic behavior of the solar buck converter, small-signal linearization is performed around the steady-state operating point. The linearized differential equations are derived from the perturbation of the state-space equations using

$$\frac{d\hat{i}_L}{dt} = -\frac{r_L}{L}(I_L + \hat{i}_L) + \frac{1}{L}(V_{SA} + \hat{v}_{sa})(D + \hat{d}) - \frac{V_B}{L}$$

Expanding and neglecting second-order perturbation terms

$$\frac{d\hat{i}_L}{dt} = -\frac{r_L}{L}\hat{i}_L + \frac{1}{L}(V_{SA}\hat{d} + D\hat{v}_{sa})$$

Similarly, the capacitor voltage dynamics are

$$\frac{d\hat{v}_{sa}}{dt} = -\frac{1}{C}(I_L + \hat{i}_L)(D + \hat{d}) - \frac{\alpha}{C}(V_{SA} + \hat{v}_{sa}) + \frac{\beta}{C}$$

Linearizing this equation yields

$$\frac{d\hat{v}_{sa}}{dt} = -\frac{1}{C}(D\hat{i}_L + I_L\hat{d}) - \frac{\alpha}{C}\hat{v}_{sa}$$

Hence, the linearized small-signal state-space model is

$$\begin{pmatrix} \dfrac{d\hat{i}_L}{dt} \\ \dfrac{d\hat{v}_{sa}}{dt} \end{pmatrix} = \begin{pmatrix} -\dfrac{r_L}{L} & \dfrac{D}{L} \\ -\dfrac{D}{C} & -\dfrac{\alpha}{C} \end{pmatrix} \begin{pmatrix} \hat{i}_L \\ \hat{V}_{sa} \end{pmatrix} + \begin{pmatrix} \dfrac{V_{SA}}{L} \\ -\dfrac{I_L}{C} \end{pmatrix} \hat{d} \tag{8.3}$$

The corresponding steady-state solution is given by

$$\begin{pmatrix} I_L \\ V_{SA} \end{pmatrix} = \begin{pmatrix} \dfrac{I_{SA}}{D} \\ \dfrac{r_L I_L + V_B}{D} \end{pmatrix} = \begin{pmatrix} \dfrac{I_{SA}}{D} \\ \dfrac{r_L I_{SA}}{D^2} + \dfrac{V_B}{D} \end{pmatrix}$$

Frequency Domain Model: Transfer function The system transfer functions are then derived as follows:

a) Control- to-inductor current transfer function

$$G_1(s) = \frac{\hat{i}_L(s)}{\hat{d}(s)} = \frac{\frac{V_{SA}}{L}\left(s + \frac{\alpha}{C}\right) - \frac{DI_L}{LC}}{s^2 + \left(\frac{r_L}{L} + \frac{\alpha}{C}\right)s + \frac{(r_L \cdot \alpha + D^2)}{LC}} \tag{8.4}$$

b) Control-to-solar voltage transfer function

$$G_2(s) = \frac{\hat{v}_{sa}(s)}{\hat{d}(s)} = \frac{-\frac{I_L}{C\left(s + \frac{r_L}{L}\right)} - \frac{DV_{SA}}{LC}}{s^2 + \left(\frac{r_L}{L} + \frac{\alpha}{C}\right)s + \frac{(r_L \cdot \alpha + D^2)}{LC}} \tag{8.5}$$

c) Inductor current–to–solar voltage transfer function

$$G_3(s) = \frac{\hat{v}_{sa}(s)}{\hat{i}_L(s)} = \frac{-L \cdot I_L\left(s + \frac{r_L}{L}\right) - DV_{SA}}{CV_{SA}\left(s + \frac{\alpha}{C}\right) - DI_L} \tag{8.6}$$

8.2 Controller Design

The MATLAB m-file is provided in Table 8.2. Since both voltage and current controllers can be selected, the results corresponding to the voltage controller are presented first. Each section of the code can be activated as needed to perform the respective designs. In this section, the design process begins with the voltage controller.

8.2.1 Voltage Controller Design

The Solar Buck converter employs a single-loop voltage control structure, utilizing the system transfer function $G_2(s)$. The configuration of the voltage controller is illustrated in Figure 8.2. Since the controller architecture is identical to that of the solar boost converter, detailed derivations are omitted for brevity.

The design specifications in the frequency domain are defined as follows

- Bandwidth = 50 rad/sec
- Phase margin = 90 deg

To meet the above specifications, a *PI* controller is designed with the following transfer function:

$$G_{CV}(s) = 8.7785 \times 10^{-6} \frac{(s + 93630)}{s}$$

Table 8.2 MATLAB m-file for solar buck controllers.

```
% Solar Buck PI Design (Sa_buck.m) %
Voc=50;
Vmp=38;
Isc=10;
Imp=8;
alpha= Imp/Vmp;

s=tf('s');
rL = 0.02;     % rL=20 mohm
L = 1.0e-3;    % L=1.0mh
C = 100e-6;    % C=100uf
VB=24;

% Steady state value
D=VB/Vmp;
IL=Imp/D;
VSA=rL*Imp/D^2 + VB/D;

G1=( (VSA/L)*(s+alpha/C)-D*IL/(L*C) ) / (s^2 +
s*(rL/L+alpha/C) + (rL*alpha+D^2)/(L*C));
G2 =( (IL/C)*(s+rL/L)+D*VSA/(L*C) )/ (s^2 +
s*(rL/L + alpha/C) + (rL*alpha+D^2)/(L*C) );
G3=(-1/C)/(s + alpha/C); % Vdq(s)/Idq(s)
```

```
Kpv = 1; Kiv=1; %temporarily
Gcv = Kpv*(s+Kiv)/s;

Kpi =1; Kii =1; %temporarily
Gci = Kpi*(s+Kii)/s;

Vsa_ref = 40;
IL_ref  = 7;

%sisotool Voltage Single loop
 T=sisoinit(1);
 T.F.Value = Vsa_ref;
 T.G.Value = G2;
 T.C.Value = Gcv;
 sisotool(T);

%sisotool Current Single loop
%T=sisoinit(1);
%T.F.Value = IL_ref;
%T.G.Value = G1;
%T.C.Value = Gci;
%sisotool(T);
```

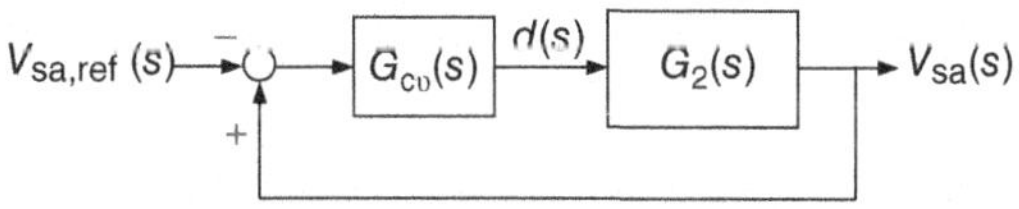

Figure 8.2 Configuration of the voltage controller for the solar buck converter.

Figure 8.3 presents the Bode plot and root locus. The Bode plot of the voltage controller satisfies the specified design requirements, while the root locus confirms system stability.

Figure 8.4 illustrates the step response, which demonstrates a rise time of approximately 44 ms with no overshoot, thereby validating the effectiveness of the controller.

For digital implementation, with a switching frequency $f_{sw} = 20$ kHz and sampling time $T = 0.00005$ sec, the controller is discretized as

$$G_{CV}(z) = 5.0232 \times 10^{-5} \frac{(z - 0.182)}{z - 1}$$

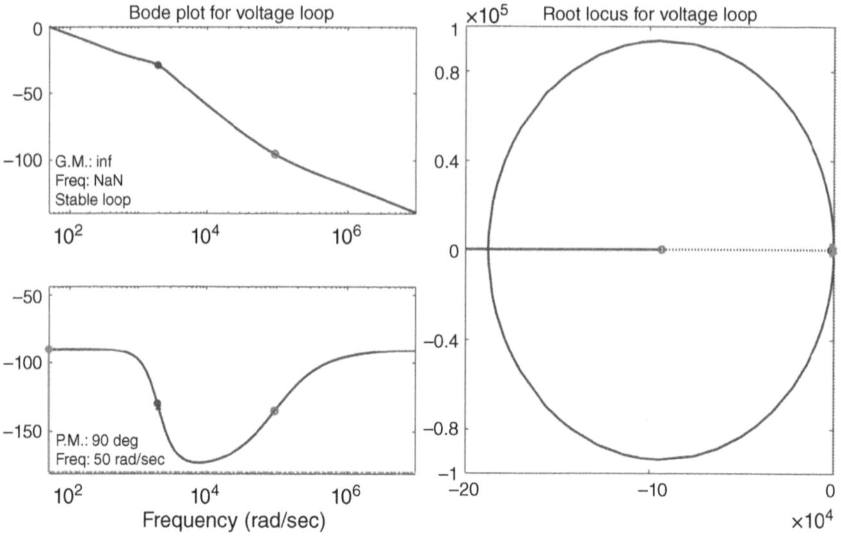

Figure 8.3 Bode plot and root locus of the voltage controller.

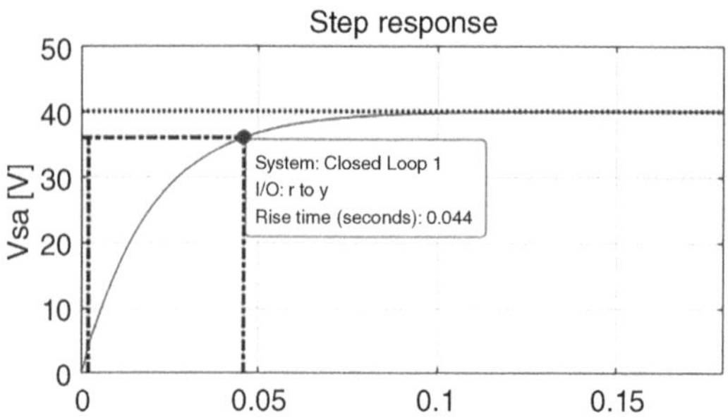

Figure 8.4 Step response of the voltage controller.

The PSIM simulation was performed using the designed PI controller gains. Figure 8.5 illustrates the PSIM schematic of the Solar Buck converter voltage controller.

The corresponding C-block implementation is provided in Table 8.3.

Figure 8.6 presents the simulation waveforms. When the reference voltage is step-changed from 45 V → 40 V → 45 V, it is observed that the solar panel voltage is effectively regulated. The response exhibits no overshoot and achieves zero steady-state error, demonstrating the excellent performance of the designed controller.

The design of the MPPT system employing the developed voltage controller follows the same methodology as that of the solar boost converter discussed in

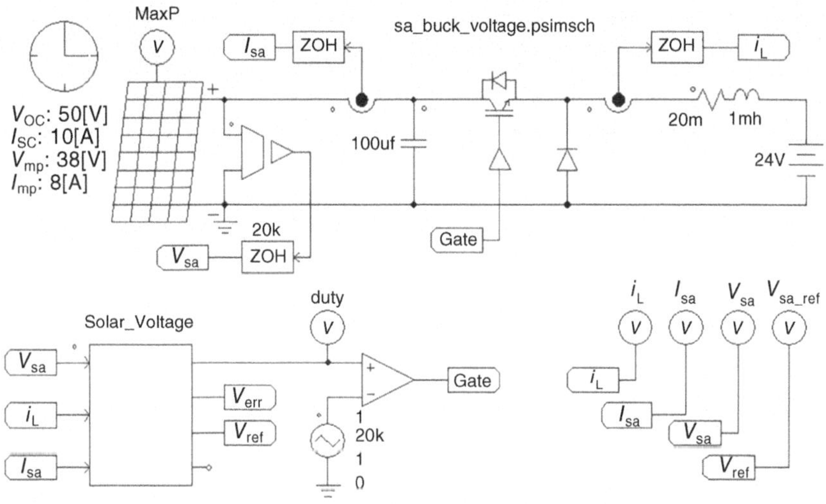

Figure 8.5 PSIM schematic of the solar buck converter voltage controller.

Table 8.3 C-block code for the solar buck voltage controller: Solar_Voltage.

```
#define Kp   0.000050232
#define Ki   0.182

double Vsa, IL, Vref, Isa, Verr;
static double duty_old=0.0;
static double Verr_old=0.0;
double duty;

Vsa  = in[0];
IL   = in[1];
Isa  = in[2];

/**** Voltage controller *****/
if (t<0.3) Vref = 45;
else if (t<0.45) Vref = 40;
else Vref =45;
Verr = Vsa - Vref;

duty=duty_old+Kp*Verr-Kp*Ki*Verr_old;
duty_old= duty;
Verr_old= Verr;

out[0] = duty;
out[1] = Verr;
out[2] = Vref;
```

the previous chapter. Therefore, detailed explanations are omitted here and will instead be assigned as a term project for further investigation.

8.2.2 Current Controller Design

The current controller for the Solar Buck converter regulates the inductor current rather than the solar panel voltage. This controller also adopts a single-loop

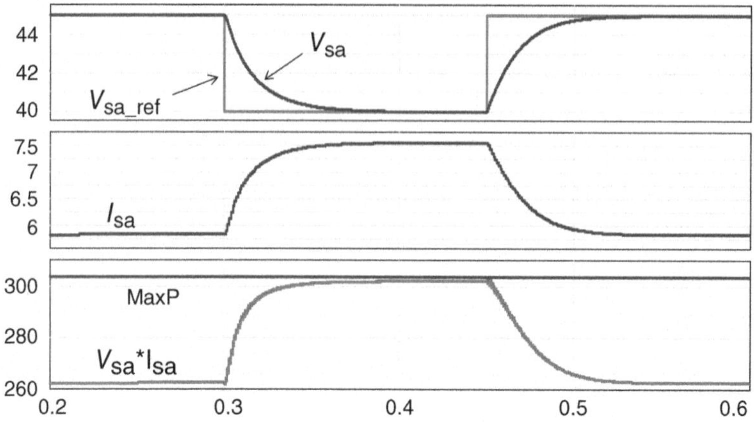

Figure 8.6 Simulation waveforms of the voltage controller.

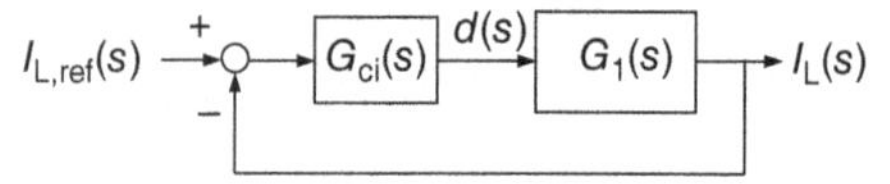

Figure 8.7 Configuration of the current controller for the solar buck converter.

structure and is designed based on the system transfer function $G_1(s)$. The configuration of the current controller is illustrated in Figure 8.7. The controller $G_{Ci}(s)$ is implemented as a proportional–integral (PI) controller. Since the transfer function $G_1(s)$ has a positive sign, the controller follows a conventional structure without requiring polarity inversion.

To design the current controller using SISOTOOL, the same m-file used for voltage controller design can be reused by enabling the current controller section. The frequency-domain design specifications are defined as follows:

- Bandwidth: 2400 rad/sec
- Phase Margin: 90°

Running SISOTOOL with the above conditions yields the following controller

$$G_{\mathrm{Ci}}(s) = \frac{0.019266(s + 6872)}{s}$$

For digital implementation, with a switching frequency $f_{sw} = 20\,\mathrm{kHz}$ and sampling time $T = 0.00005$ sec, the discrete-time form of the controller is

$$G_{\mathrm{Ci}}(z) = 0.02584\frac{(z - 0.749)}{z - 1}$$

Figure 8.8 presents the Bode plot and root locus. The Bode plot of the current controller satisfies the specified design requirements, while the root locus confirms system stability.

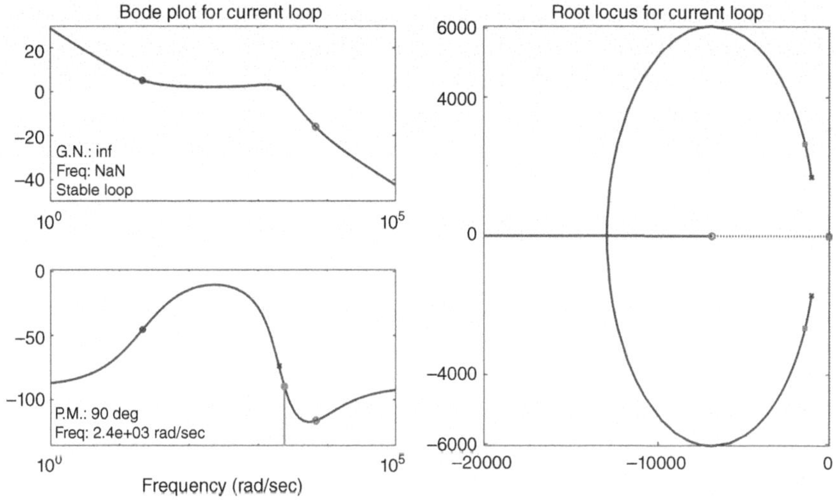

Figure 8.8 Bode plot and root locus of the current controller.

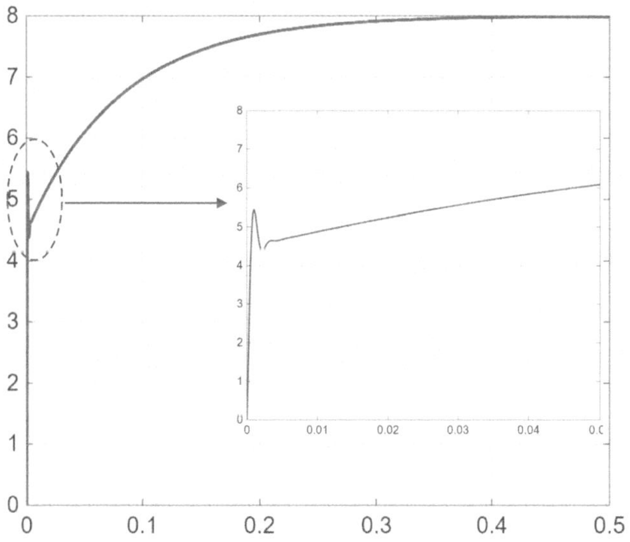

Figure 8.9 Step response of the current controller.

Figure 8.9 illustrates the step response, which demonstrates a rise time of approximately 121 ms with no overshoot, thereby validating the effectiveness of the controller.

The PSIM schematic remains unchanged from the previous design. Only the C-block code needs to be updated to implement the newly designed current controller. The corresponding C-block implementation is provided in Table 8.4.

Figure 8.10 presents the simulation waveforms. When the reference current is step-changed from 7 A → 12 A → 7 A, the inductor current is accurately regulated

Table 8.4 C-block code for the solar buck current controller: Solar_Current.

```
#define Kp 0.02584   //2400 rad/s
#define Ki 0.749     // 90 deg

double Vsa, IL, Iref, Isa, Ierr;
static double duty_old=0.0;
static double Ierr_old=0.0;
double duty;

Vsa  = in[0];
IL   = in[1];
Is   = in[2];

/**** Current controller ****/
if (t<0.3) Iref =7;
else if (t<0.4) Iref =12;
else Iref =7;

Ierr = Iref-IL;

duty=duty_old+Kp*Ierr-Kp*Ki*Ierr_old;
duty_old= duty;
Ierr_old= Ierr;

out[0]  = duty;
out[1]  = Ierr;
out[2]  = Iref;
```

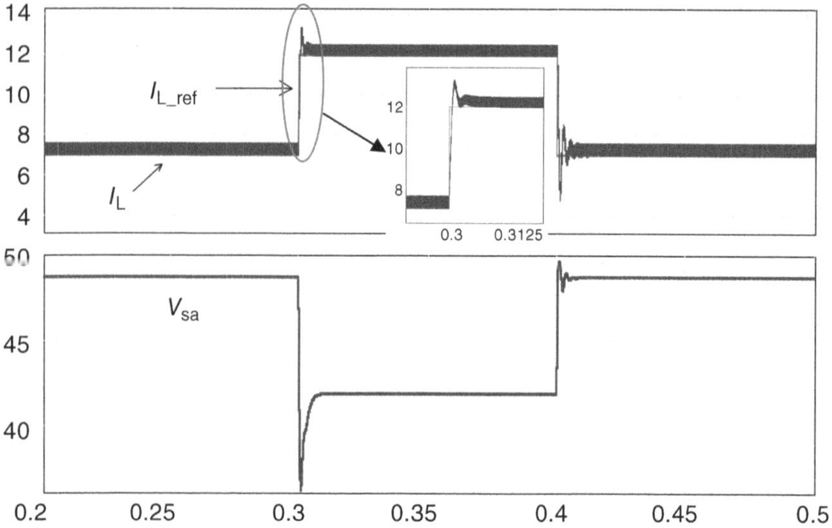

Figure 8.10 Simulation waveforms of the current controller.

to its steady-state value. Although the phase margin was designed to be 90 degrees, a slight overshoot is observed. This is attributed to the dominant poles of the closed-loop transfer function residing in the complex plane—a behavior that is also reflected in the MATLAB simulation results.

Since the MPPT system using the designed current controller shares the same structure as the Solar Boost converter, its detailed implementation is omitted and assigned as a term project to encourage further exploration by the reader.

9

Solar Three-Level Boost Converter

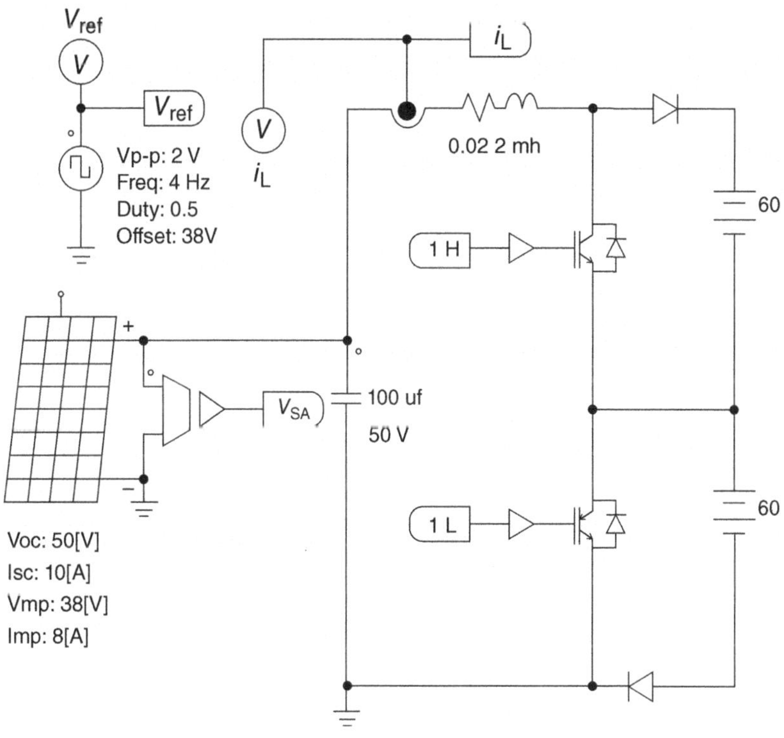

9.1 Controller Design

This chapter presents the control-system design for a photovoltaic configuration in which a solar panel is connected to a three-level boost converter. As demonstrated in Chapter 5, the transfer function of the three-level boost converter is identical to that of a conventional boost converter. Consequently, the dynamic characteristics of the solar boost converter and the three-level solar boost converter are equivalent.

Based on this equivalence, the emphasis of this chapter is not on re-deriving the state-space model, but rather on designing a voltage controller for the three-level solar boost converter using the previously derived transfer function of the conventional solar boost converter.

The MATLAB m-file used for controller design is provided in Table 9.1. Since either a voltage controller or a current controller may be employed, the results corresponding to the voltage controller are presented first.

Table 9.1 MATLAB m-file for three-level solar boost controller.

```
%  Solar 3-level Boost %
Voc=50;
Vmp=40;
Isc=10;
Imp=8;
alpha3  = Imp/Vmp;
beta3  = 2*Imp;

s=tf('s');
rL  = 0.02;      % rL=20 mohm
L  = 2.0e-3;     % L=2.0mh
C  = 100e-6;     % C=100uf
VB=60;

alpha  = alpha3;
beta  =  beta3;

G1  =  (2*VB/L)*(s+alpha/C)  /
(s^2+s*(rL/L+alpha/C)+(rL*alpha+1)/(L*C)  );
G2  =  (2*VB)/(L*C)  /
(s^2+s*(rL/L+alpha/C)+(rL*alpha+1)/(L*C)  );
G3  =  (-1/(2*C))  /  (s + alpha/C);
```

```
Kpv  =  1;  Kiv=1;
Gcv  =  Kpv*(s+Kiv)/s;

Kpi  =1;  Kii  =1;
Gci  =  Kpi*(s+Kii)/s;

Vsa_ref  =  38;
IL_ref   =  8;

%sisotool Voltage Single loop
T=sisoinit(1);
T.F.Value  =  Vsa_ref;
T.G.Value  =  G2;
T.C.Value  =  Gcv;
sisotool(T);

%sisotool Current Single loop
%  T=sisoinit(1);
%  T.F.Value  =  IL_ref;
%  T.G.Value  =  G1;
%  T.C.Value  =  Gci;
%  sisotool(T);
```

The controller that satisfies the design specifications of a bandwidth of 50 rad/sec and a phase margin of 90° is given by the following PI controller:

$$G_{CV}(s) = 0.0001675\frac{(s+2500)}{s}$$

Figure 9.1 illustrates the Bode plot and step response corresponding to this design. These plots confirm that the system meets the desired frequency-domain and time-domain performance requirements, achieving the specified bandwidth and phase margin.

To verify the controller's performance in simulation, the PSIM schematic is shown in Figure 9.2. This schematic includes the designed PI controller and gate-driving signals, demonstrating how the controller interacts with the system to regulate the output voltage.

The simulation results are presented in Figure 9.3. The waveforms align closely with the MATLAB analysis, validating the effectiveness of the designed controller. These results suggest that the controller will also perform reliably in practical implementation.

The MPPT controller and capacitor voltage balancing controller used in this system are identical to those implemented in the conventional solar boost converter. As such, the detailed implementation of these components is omitted and will be assigned as a term project for further exploration.

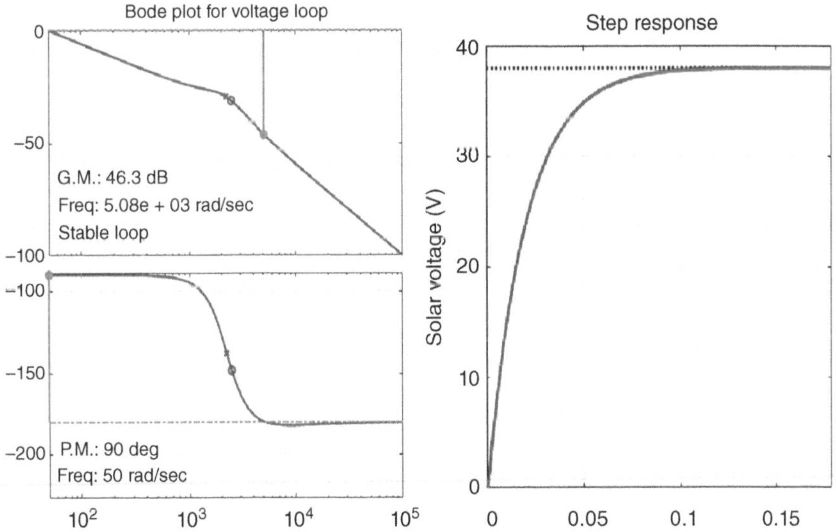

Figure 9.1 Bode plot and step response of the voltage controller.

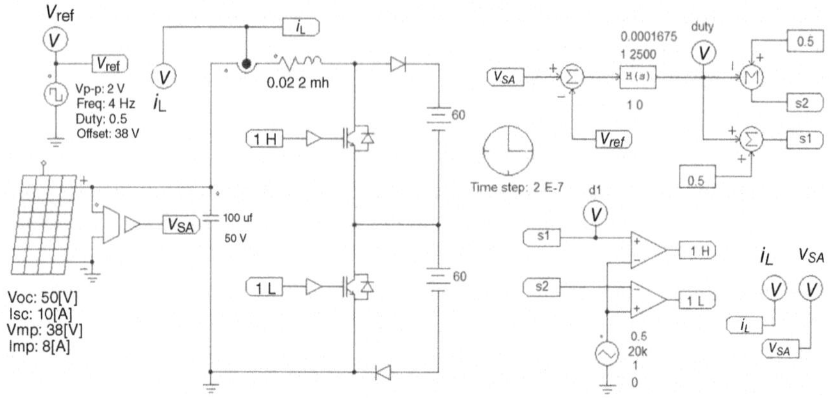

Figure 9.2 PSIM schematic of the solar three-level boost converter.

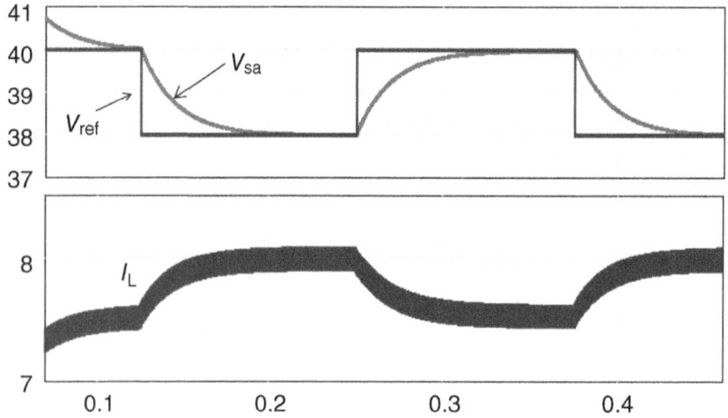

Figure 9.3 Simulation waveforms of the solar three-level boost converter.

Part I-C

Motor

The types and circuits of DC MOTOR are as follows.

1. UNI-POLAR DRIVING
2. BI-POLAR DRIVING

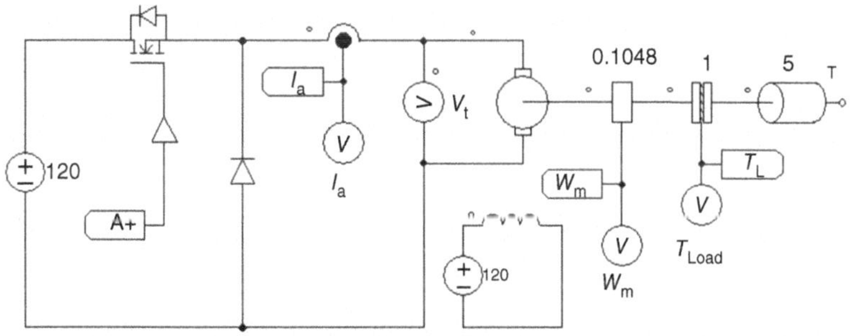

(a) Uni-polar driving circuit

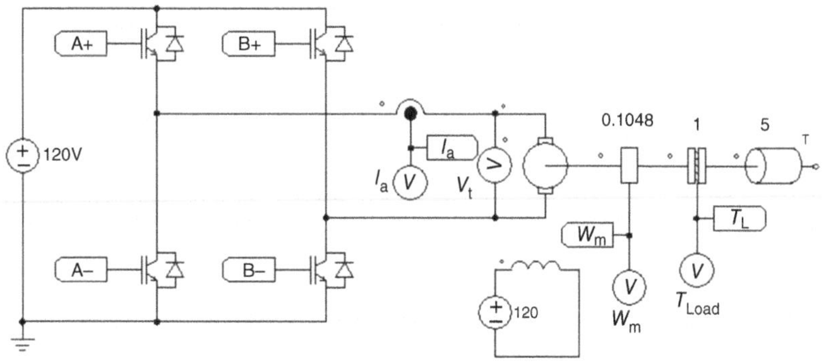

(b) Bi-polar driving circuit

10

DC Motor

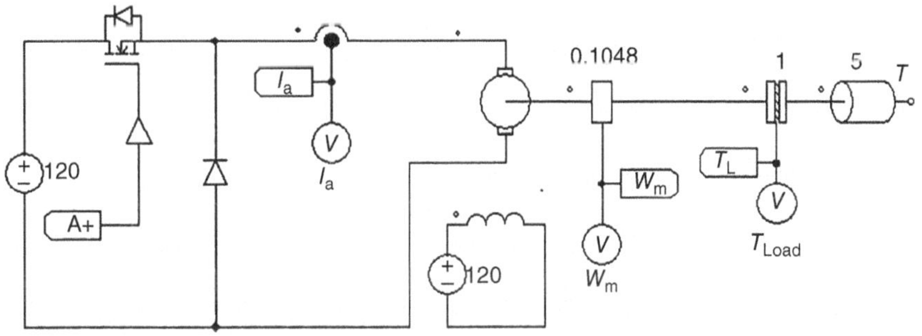

10.1 Motor Parameters

A direct current (DC) motor is an electromechanical device that converts electrical energy into mechanical motion through the interaction of magnetic fields. Operating on the principle of Lorentz force, the motor utilizes a direct current supply to produce rotational torque. DC motors are widely employed in applications that demand adjustable speed, rapid start-up, and precise torque control, such as robotics, electric vehicles, industrial automation, and compact household appliances.

Figure 10.1 presents the equivalent circuit of a separately excited DC motor. In this configuration, the armature winding and the field winding are supplied independently. The field winding is energized to generate the magnetic flux necessary for motor operation. Meanwhile, current in the armature circuit is supplied to the rotor through brushes and commutator segments, enabling mechanical rotation. The resulting electromagnetic torque is produced by the interaction between the armature current and the magnetic flux generated by the field winding.

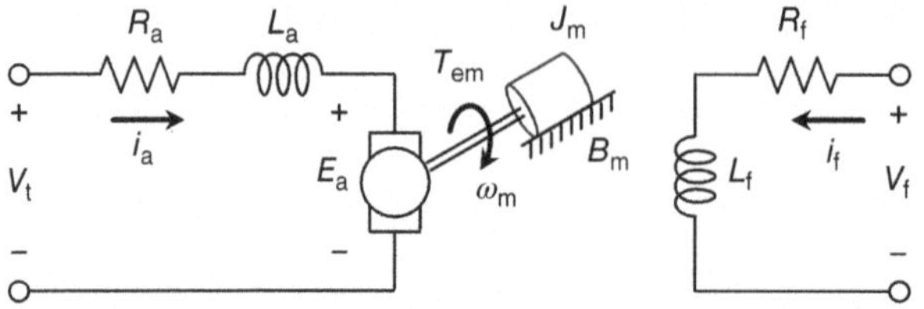

Figure 10.1 Equivalent circuit of a separately excited DC motor.

When a separately excited DC motor is powered with a field current i_f and an armature current i_a, it develops a back electromotive force (EMF) and a torque that counteracts the load torque at a specific operating speed. Since the armature and field windings are energized separately, the field current i_f remains independent of the armature current i_a. Changes in armature current do not affect the field current. Generally, i_f is significantly smaller than i_a.

The fundamental parameters associated with rotational motion in electric machines are defined as follows.

- Mechanical angular position (θ_m)
  The mechanical angular position θ_m represents the orientation of a rotating object with respect to a reference position. It is typically measured in radians or degrees and corresponds to displacement in linear motion.
- Mechanical angular velocity (ω_m)
  The mechanical angular velocity ω_m is defined as the time derivative of the angular position. When the mechanical angular position is expressed in radians, the mechanical angular velocity is given in radians per second (rad/sec):

  $$\omega_m = d\theta_m/dt$$

  In simulation environments such as the PSIM motor library, angular velocity is often measured using a speed sensor and expressed in revolutions per minute (rpm). The conversion from rpm to rad/sec is given by

  $$\omega_m\,[\text{rad/sec}] = \frac{2\pi}{60} \times [\text{rpm}] = 0.1047 \times [\text{rpm}]$$

- Electrical angle (θ_e) and mechanical angle (θ_m)
  The mechanical angle θ_m denotes the physical angular position of the motor shaft. In contrast, the electrical angle θ_e represents the angular position of the rotating magnetic field within the machine. The relationship between the mechanical and electrical angles depends on the number of poles P of the motor and is expressed as

  $$\theta_e = \frac{P}{2}\theta_m$$

Example:

- 2-pole motor (1 pole pair): $\theta_e = 1 \cdot \theta_m$
- 4-pole motor (2 pole pairs): $\theta_e = 2 \cdot \theta_m$.
- 6-pole motor (3 pole pairs): $\theta_e = 3 \cdot \theta_m$

- Torque (Γ)
  Torque is the rotational analogue of force and represents the tendency of a force to produce rotational motion about an axis. Just as force causes linear acceleration, torque causes angular acceleration. Torque is defined as

$$\Gamma = r \times F = rF \sin \theta$$

  where
  r is the distance from the axis of rotation to the point of force application,
  F is the applied force, and
  θ is the angle between the force vector and the lever arm.
  The SI unit of torque is Newton-meter ($\mathrm{N \cdot m}$). In some industrial contexts, the kilogram-force meter ($\mathrm{kgf \cdot m}$) is also used, with the conversion

$$1\ \mathrm{kgf \cdot m} = 9.8\ \mathrm{N \cdot m}$$

- Power (P)
  Mechanical power in a motor is defined as the product of torque and angular velocity

$$P = \Gamma \times \omega_m$$

  This relationship forms the basis for evaluating the mechanical output power of rotating electric machines.

DC Motor Mathematical Modeling The mathematical model of a DC machine, consistent with the PSIM Motor Library representation, is described by the following set of differential equations

$$V_t = E_a + L_a \frac{di_a}{dt} + R_a i_a$$

$$V_f = L_f \frac{di_f}{dt} + R_f i_f$$

$$E_a = L_{af} \cdot i_f \cdot \omega_m = K_\varphi \cdot \omega_m$$

$$T_{em} = L_{af} \cdot i_f \cdot i_a = K_\varphi \cdot i_a$$

$$J_m \frac{d\omega_m}{dt} = T_{em} - T_{load} \tag{10.1}$$

The variables and parameters are defined as follows:

V_t: Armature terminal voltage

V_f: Field winding voltage

E_a: Back EMF

i_a: Armature current

i_f: Field current

ω_m Mechanical angular velocity [rad/sec]

L_{af} Mutual inductance between field and armature windings

In most literature, the product $L_{af} \cdot i_f$ is referred to as the back EMF constant, denoted by K_φ. The value of L_{af} is typically determined under rated steady-state conditions. Given the rated values V_t, i_a, i_f, ω_m, the mutual inductance can be calculated as

$$L_{af} = \frac{V_{t_{rated}} - R_a \cdot i_{a_{rated}}}{i_{f_{rated}} \cdot \omega_{m_{rated}}} \tag{10.2}$$

Using rated parameters provided in the PSIM DC Machine Library shown below

$$L_{af} = \frac{120 - 10 \times 0.5}{1.6 \times 1200 \times \frac{2\pi}{60}} = 0.572[\text{H}]$$

Thus, the back EMF constant is

$$K_\varphi = L_{af} \cdot i_f = 0.9152 \tag{10.3}$$

The DC machine library provided by PSIM is illustrated in Figure 10.2.

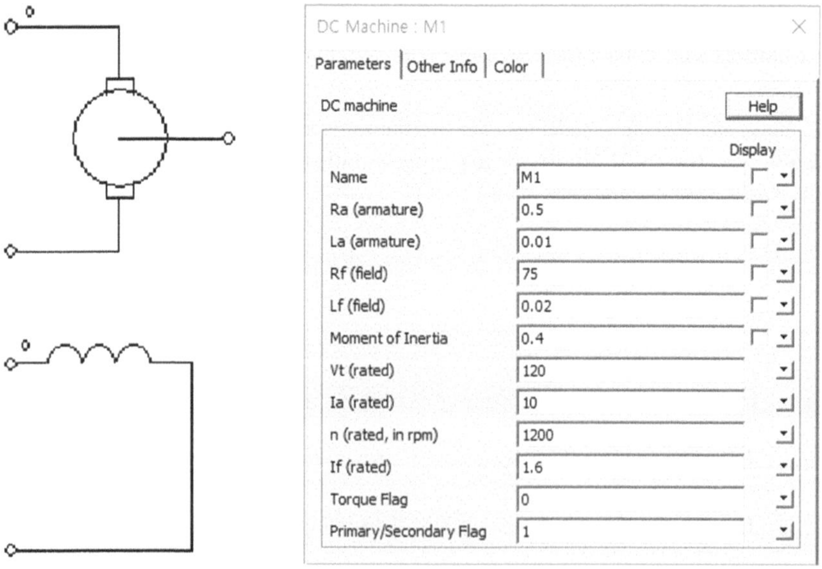

Figure 10.2 DC machine library provided by PSIM.

Steady-State Analysis The drive circuit under consideration, illustrated in Figure 10.3, operates under a constant load torque of $T_L = 50$ Nm. This configuration enables a direct comparison between analytically predicted steady-state values and the corresponding simulation results obtained using PSIM.

Steady-State Calculations Using (10.1) and the parameters given in Figure 10.2, the steady-state values are calculated as follows.

1) Field Current I_f: At steady state, the derivative term becomes zero. Thus,

$$I_f = \frac{V_f}{R_f} = \frac{120}{75} = 1.6 \text{ A}$$

2) Armature Current I_a: Assuming torque balance $T_{em} = T_L$,

$$I_a = \frac{T_{em}}{L_{af} \cdot I_f} = \frac{T_L}{L_{af} \cdot I_f} = \frac{50}{0.572 \times 1.6} = 54.6 \text{ A}$$

3) Back EMF E_a: $E_a = V_t - R_a i_a = 120 - 0.5 \times 54.6 = 92.7$ V

4) Angular Velocity ω_m: $\omega_m = \frac{E_a}{L_{af} \cdot I_f} = \frac{92.7}{0.572 \times 1.6} = 101.29$ rad/sec

Converting to rpm: $\omega_m \times \frac{60}{2\pi} = 101.29 \times \frac{60}{2\pi} = 967$ rpm

Simulation Analysis Using PSIM The PSIM simulation result, shown in Figure 10.3(b), confirms the accuracy of the above calculations.

- Armature current i_a peaks above 200 A during transients but stabilizes at 54.6 A.
- Field current i_f remains steady at 1.6 A.
- Motor speed settles at 967 rpm, equivalent to 101.2 rad/sec.
- Electromagnetic torque T_{em} initially exceeds 200 N · m during startup, then stabilizes to the load torque of 50 N · m.

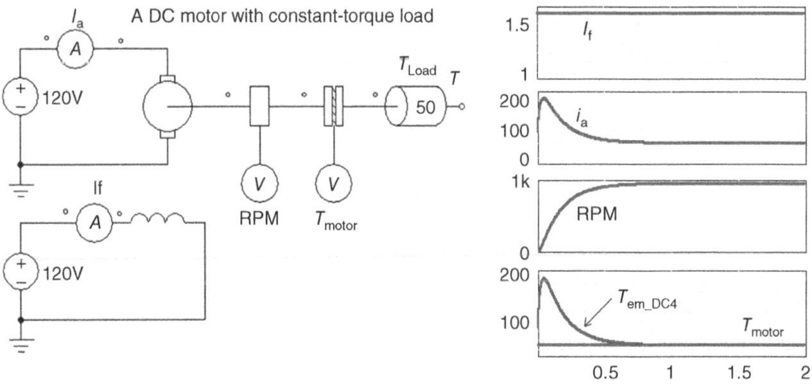

Figure 10.3 Driving circuit and waveforms of the DC machine.

Table 10.1 Steady-state values of the motor parameters.

```
%% DC Machine Steady-State Vaule %%

% rated value from DC Machine Library
Vt_rated = 120;
Ia_rated = 10;
If_rated = 1.6;
rpm_rated = 1200;

% Motor and Circuit Parameter
Ra = 0.5;
La = 0.01;
Rf = 75;
Lf = 0.02;
Jm = 0.3;

TL = 50;   % Load torque TL=50[Nm]
Vt=120;
Vf=120;
Te = Kpi*Ia;

% Step 1 : Mutual Inductance Laf
Laf =(Vt_rated-Ra*Ia_rated)
    /(If_rated*rpm_rated*0.1047)

% Step 2 : steady-state If
If = Vf/Rf

% Step 3 : steady-state Ia
Ia = TL/(Laf*If)

% Step 4 :  steady-state bemf Ea
Ea = Vt - Ra*Ia

% Step 5 : steady-state Wm
Wm = Ea/(Laf*If)
rpm = Wm*60/(2*3.1415)

%Step 6 : Kpi, Tem
Kphi = Laf*If
```

Overall, the simulation results are in excellent agreement with the analytical steady-state values. The MATLAB m-file for computing the steady-state motor parameters is presented in Table 10.1.

Parameter Variation and Verification To further validate the model, you may adjust the terminal voltage and load torque and observe whether the updated MATLAB steady-state values remain consistent with those obtained from PSIM simulations.

10.2 Uni-Polar Driving (Uni-Direction Drive)

Motor Drive Circuit When the DC motor modeling equations are expressed as in Eq. (10.1), the motor terminal voltage is composed of the voltage drop across the armature inductance and the back EMF. Based on the driving method, the equivalent circuits of the DC motor are illustrated in Figure 10.4.

- Figure (a) shows the equivalent circuit for uni-polar driving (uni-directional drive), which is commonly used for modeling and analysis.
- Figure (b) shows the equivalent circuit for bi-polar driving (bi-directional drive).

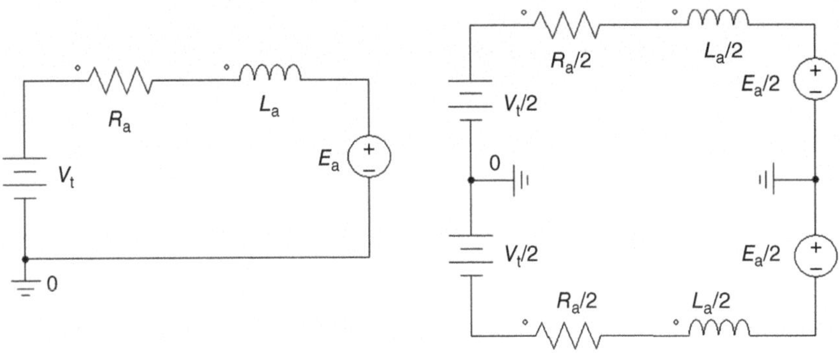

Figure 10.4 DC motor equivalent circuit for uni-polar and bi-polar driving methods. (a) Uni-polar driving and (b) Bi-polar driving.

In uni-polar driving, a buck converter is typically employed to control the motor in a single rotational direction. This configuration is well-suited for applications requiring speed control without reversal of rotation. In contrast, bi-polar driving utilizes a single-phase full-bridge circuit, enabling bi-directional motor operation by allowing both forward and reverse torque generation.

Because a single-phase inverter consists of two identical half-bridge circuits operating in opposite phases, the same control methodology introduced in the next chapter for single-phase inverter control can be directly applied to bi-polar drive systems. This structural similarity allows the inverter control strategy to be reused without modification, providing significant advantages in terms of controller design simplicity and implementation efficiency.

A more detailed discussion of bi-polar driving is presented in Chapter 11, where the operating principles and control design of single-phase inverters are examined in detail.

10.2.1 State-Averaged Equations and Transfer Function

We begin by considering a uni-directional drive circuit for a DC motor. In this configuration, the inductor L of the buck converter can be incorporated as part of the motor's armature circuit. As a result, the uni-directional drive can be implemented using only a switching device and a freewheeling diode. The circuit diagram for driving a DC motor in a single direction is shown in Figure 10.5.

The state equation of the motor is expressed as follows:

$$V_t = E_a + L_a \frac{di_a}{dt} + R_a i_a = K_\varphi \omega_m + L_a \frac{di_a}{dt} + R_a i_a$$

$$J_m \frac{d\omega_m}{dt} + B_m \omega_m = T_{em} - T_{load} = K_\varphi i_a - T_{load} \tag{10.4}$$

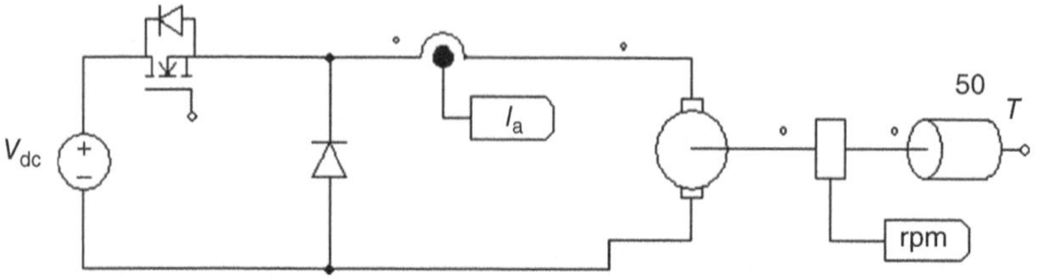

Figure 10.5 Uni-polar driving circuit diagram.

Depending on the ON/OFF state of the switch, the motor terminal voltage V_t alternates between V_{dc} and 0. By applying state averaging over the duty cycle d(ON-time) and $1 - d$ (OFF-time), the following state-averaged equations are obtained

$$\frac{di_a}{dt} = -\frac{R_a}{L_a}i_a - \frac{K_\varphi}{L_a}\omega_m + \frac{V_t}{L_a}d$$

$$\frac{d\omega_m}{dt} = \frac{K_\varphi}{J_m}i_a - \frac{B_m}{J_m}\omega_m - \frac{1}{J_m}T_{load}$$

These equations can be written in state-space matrix form as

$$\begin{pmatrix} \frac{di_a}{dt} \\ \frac{d\omega_m}{dt} \end{pmatrix} = \begin{pmatrix} -\frac{R_a}{L_a} & -\frac{K_\varphi}{L_a} \\ \frac{K_\varphi}{J_m} & -\frac{B_m}{J_m} \end{pmatrix} \begin{pmatrix} i_a \\ \omega_m \end{pmatrix} + \begin{pmatrix} \frac{V_t}{L_a} \\ 0 \end{pmatrix} d + \begin{pmatrix} 0 \\ -\frac{T_{load}}{J_m} \end{pmatrix} \tag{10.5}$$

This matrix equation is generally nonlinear due to the presence of the disturbance term related to the load torque T_{load}. To linearize the system for small-signal analysis, the disturbance term can be excluded by expressing the load torque in terms of the armature current

$$T_{load} = K_\varphi i_a \tag{10.6}$$

The armature current i_a can be decomposed into two components

$$i_a = i_{ff} + i_{fb}$$

where

- i_{ff} is the feed-forward current,
- i_{fb} is the feedback current controlled by the current controller.

The feedforward current can be calculated from the known load torque as

$$i_{ff} = \frac{T_{load}}{K_\varphi} \tag{10.7}$$

The load torque T_{load} can either be measured using a torque sensor or estimated through a load torque observer. After this compensation, the simplified linear state-space model is expressed as:

$$\dot{X} = AX + Bu$$

where:

$$X = \begin{pmatrix} \frac{di_a}{dt} \\ \frac{d\omega_m}{dt} \end{pmatrix}, \quad A = \begin{pmatrix} -\frac{R_a}{L_a} & -\frac{K_\varphi}{L_a} \\ \frac{K_\varphi}{J_m} & -\frac{B_m}{J_m} \end{pmatrix}, \quad B = \begin{pmatrix} \frac{V_t}{L_a} \\ 0 \end{pmatrix}, \quad U = d \tag{10.8}$$

Transfer Function Derivation The transfer function can be derived by taking the Laplace transform and computing the inverse of $(sI - A)$. Thus, the following transfer functions are obtained

1) Current to Duty Ratio Transfer Function

$$G_1(s) = \frac{I_a(s)}{d(s)} = \frac{\frac{V_t}{L_a}\left(s + \frac{B_m}{J_m}\right)}{s^2 + \left(\frac{R_a}{L_a} + \frac{B_m}{J_m}\right)s + \left(\frac{K_\varphi^2}{L_aJ_m} + \frac{R_aB_m}{L_aJ_m}\right)} \tag{10.9}$$

2) Speed to Duty Ratio Transfer Function

$$G_2(s) = \frac{\omega_m(s)}{d(s)} = \frac{\frac{V_tK_\varphi}{L_aJ_m}}{s^2 + \left(\frac{R_a}{L_a} + \frac{B_m}{J_m}\right)s + \left(\frac{K_\varphi^2}{L_aJ_m} + \frac{R_aB_m}{L_aJ_m}\right)} \tag{10.10}$$

3) Speed to Current Transfer Function

$$G_3(s) = \frac{\omega_m(s)}{I_a(s)} = \frac{\frac{K_\varphi}{J_m}}{s + \frac{B_m}{J_m}} \tag{10.11}$$

These transfer functions serve as the basis for control system design and stability analysis of the DC motor under uni-polar drive. The linearized state-space model and derived transfer functions allow for precise analysis of system dynamics and controller implementation.

10.2.2 Controller Design

Single-Loop Controller Structure The control system begins with the design of a single-loop controller, which consists of two main control loops

- Speed Controller
- Current Controller

Both controllers are designed using the same methodology applied to converter control systems, allowing for consistency and ease of implementation.

a) **Speed Controller Design**

The speed controller, responsible for regulating the motor's rotational speed ω_m, is configured as shown in Figure 10.6.

The frequency-domain design specifications for the speed controller are defined as follows:

- Bandwidth: 1 rad/sec
- Phase Margin: 90 deg

The MATLAB m-file for the speed controller is presented in Table 10.2. The speed controller design is fully implemented and ready for immediate execution, whereas the current controller design is masked for future use and is therefore not shown here.

By performing auto-tuning in MATLAB, the resulting speed controller is obtained as

$$C_1 = G_{cw}(s) = \frac{0.002185(s + 4.17)}{s}$$

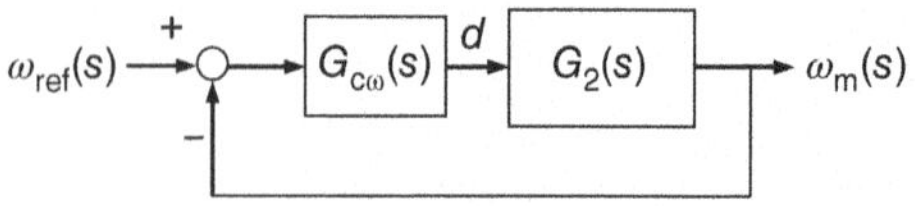

Figure 10.6 Speed controller configuration (single loop).

Table 10.2 MATLAB m-file for speed controller

```
Vt=100;
Ra = 0.5;
La = 0.01;
Jm = 0.4;
Bm=0;
TL = 50;    % Load torque
Laf=0.572;
If = 1.6;
Kphi = Laf*If;

s=tf('s');
Delta=s^2+(Ra/La+Bm/Jm)*s +
(Kphi^2+Ra*Bm)/(La*Jm);
```

```
G1=(Vt/La)*(s+Bm/Jm)/Delta;
G2=(Vt*Kphi/(La*Jm))/Delta;
G3=(Kphi/Jm)/(s+Bm/Jm) ;

%% Single loop %%
Kpw=1; Kiw=1;
Gcw=Kpw*(s+Kiw)/s;
%%%%% Speed %%%%%
Wref = 200;
T=sisoinit(1);
T.F.Value = Wref;
T.G.Value = G2;
T.C.Value = Gcw
sisotool(T);
```

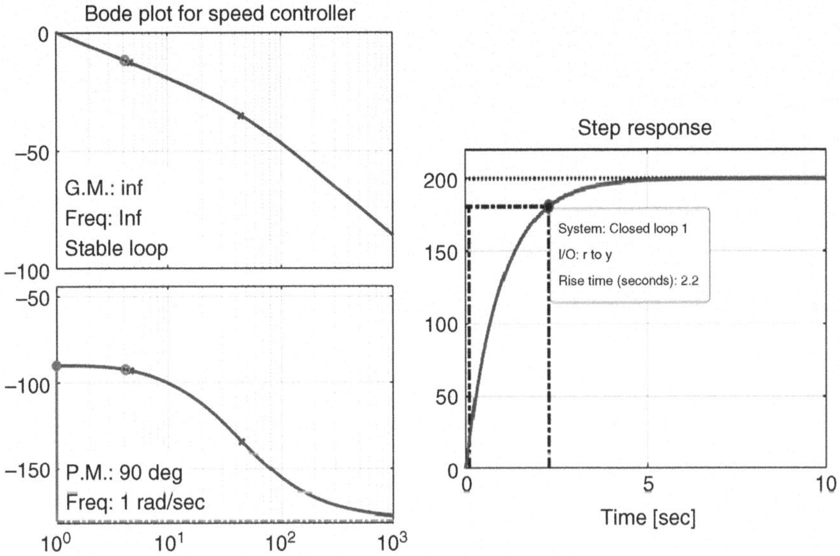

Figure 10.7 Bode plot and step response of the speed controller.

The Bode plot and step response of the designed controller are presented in Figure 10.7, confirming that the specified design conditions are successfully achieved.

PSIM Simulation of the Uni-Polar Drive The PSIM simulation circuit for uni-polar driving is depicted in Figure 10.8. The main features of the simulation setup are summarized as follows:

- The DC motor is driven by a buck converter.
- A speed sensor measures the motor speed in rpm.
- The conversion from rpm to angular velocity ω_m is performed using the factor 0.1047.
- A rotational load of $5\,N \cdot m$ is connected, and T_{load} is measured via a torque sensor.

The corresponding PSIM speed controller configuration is presented in Figure 10.9. A PI controller is implemented using the controller gains obtained from MATLAB. The control structure is simulated in PSIM, and the resulting waveforms are observed to evaluate the dynamic response.

The simulation results, shown in Figure 10.10, demonstrate that the motor speed accurately tracks the reference speed without overshoot, thereby confirming the effectiveness of the controller design and tuning.

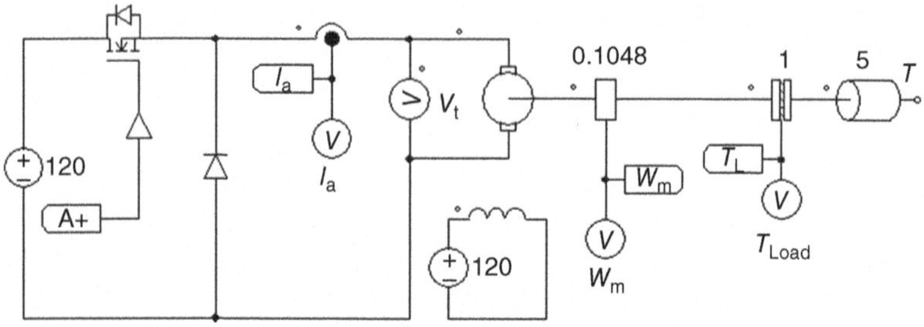

Figure 10.8 Circuit diagram of the uni-polar driving system.

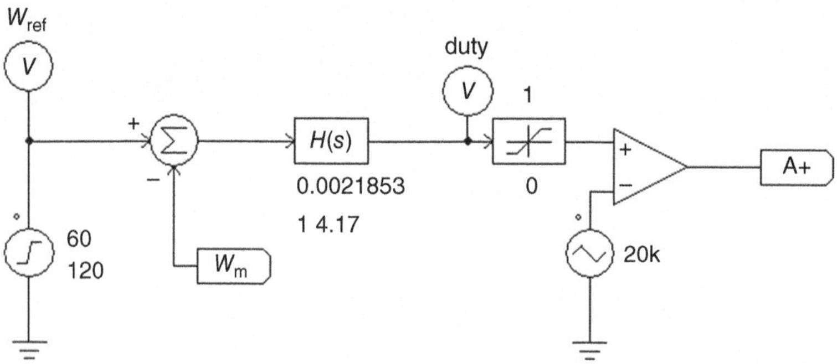

Figure 10.9 Speed controller configuration.

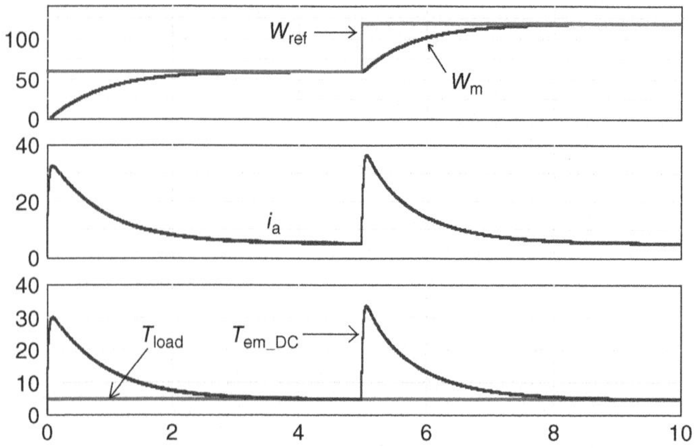

Figure 10.10 Waveforms of the speed controller response.

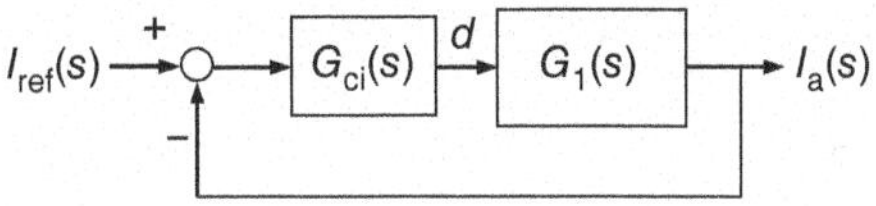

Figure 10.11 Current controller configuration (Single loop).

b) **Current Controller Design**

The current controller, responsible for regulating the armature current i_a, is configured as shown in Figure 10.11.

The frequency design specifications for the current controller are defined as follows:

- Bandwidth = 2000 rad/sec
- Phase margin = 60 deg

The MATLAB m-file for current controller is presented in Table 10.3. The motor parameters and transfer function definitions are identical to those used in the speed controller design:

By performing auto-tuning in MATLAB, the designed current controller is obtained as

$$G_{ci}(s) = \frac{0.71(s + 1220)}{s}$$

The Bode plot and step response of the designed controller are presented in Figure 10.12, confirming that the target design specifications are successfully achieved.

The PSIM circuit diagram for the current controller is shown in Figure 10.13. Since the motor and power circuit components remain unchanged, only the updated control section is highlighted.

The corresponding simulation waveforms are displayed in Figure 10.14. The results demonstrate that the current waveform follows the reference accurately, confirming consistency between the PSIM simulation and the MATLAB-based controller design.

Table 10.3 MATLAB m-file for current controller.

```
%% Single loop : Current %%
Iref= TL/Kphi;
T=sisoinit(1);
T.F.Value  = Iref;
T.G.Value  = G1;
T.C.Value  = Gci;
sisotool(T);
```

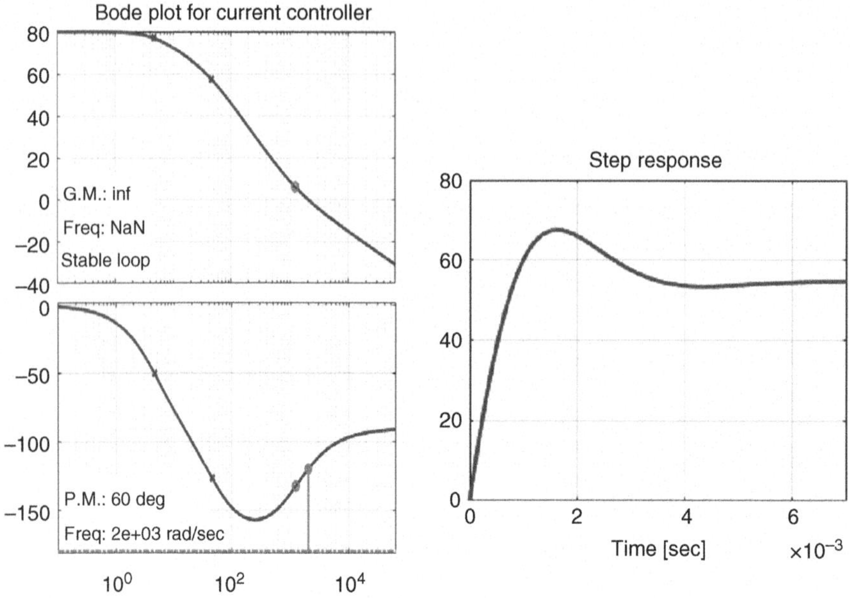

Figure 10.12 Bode plot and step response of the current controller.

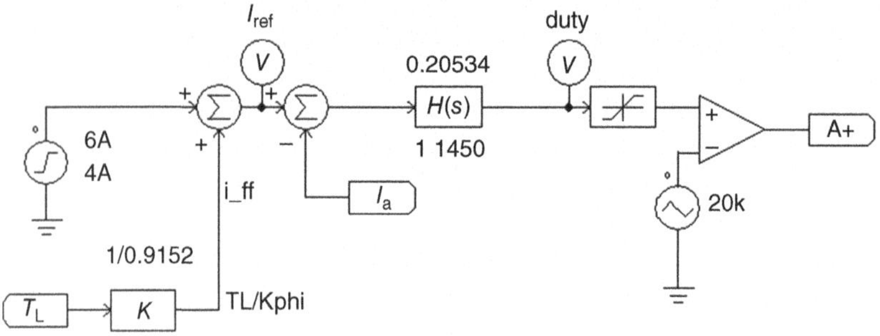

Figure 10.13 Current controller block diagram.

c) Double-Loop Controller Design

The design of a double-loop controller is presented in this section. The overall control structure is illustrated in Figure 10.15.

An m-file for the double-loop controller design is provided in Table 10.4. The motor parameters and transfer function definitions remain the same as those used in the previous controller designs.

The frequency design specifications for both the speed controller and current controller are presented in Table 10.5.

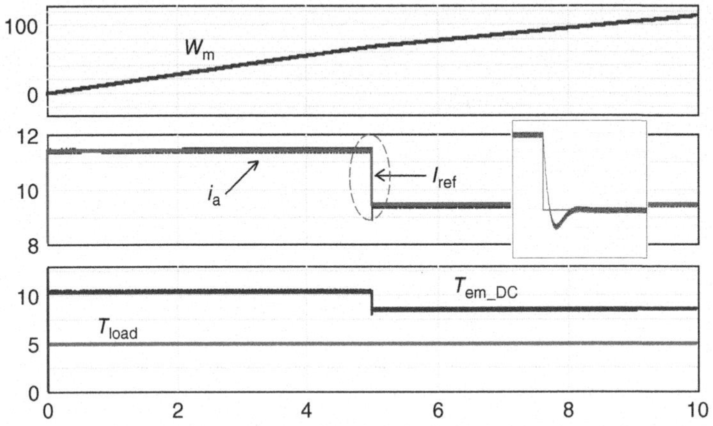

Figure 10.14 Waveforms of the current controller response.

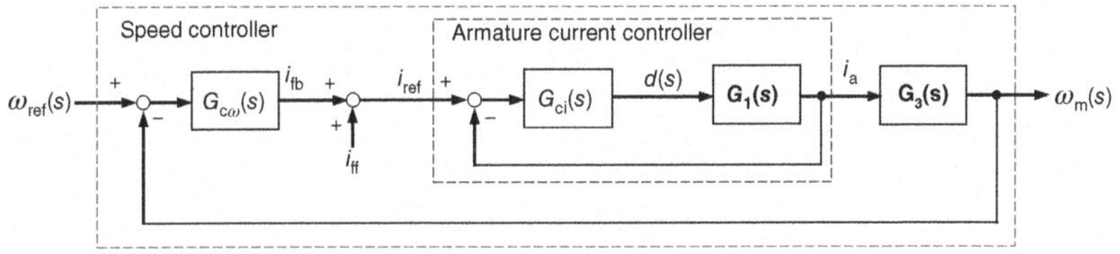

Figure 10.15 Double-loop control architecture.

Table 10.4 MATLAB m-file for double-loop controller.

```
Kpw=1; Kiw=1;
Gcw=Kpw*(s+Kiw)/s;
Kpi=1; Kii=1;
Gci=Kpi*(s+Kii)/s;

Wref = 200;
T=sisoinit(6);
T.F.Value = Wref;
T.G1.Value = G1;
T.C2.Value = Gci;

T.G2.Value = G3;
T.C1.Value = Gcw;
sisotool(T);
```

Table 10.5 Design specification for double loop controller.

	Bandwidth	Phase margin
Speed controller (C_1)	1 rad/sec	90 deg
Current controller (C_2)	2400 rad/sec	60 deg

By performing auto-tuning in MATLAB, the following controller expressions are obtained:

- Speed Controller:

$$C_1 = G_{cw}(s) = \frac{0.437(s + 0.0175)}{s}$$

- Current Controller:

$$C_2 = G_{ci}(s) = \frac{0.2(s + 1450)}{s}$$

The Bode plots for both the speed and current controllers are shown in Figure 10.16, confirming that the design specifications have been successfully achieved.

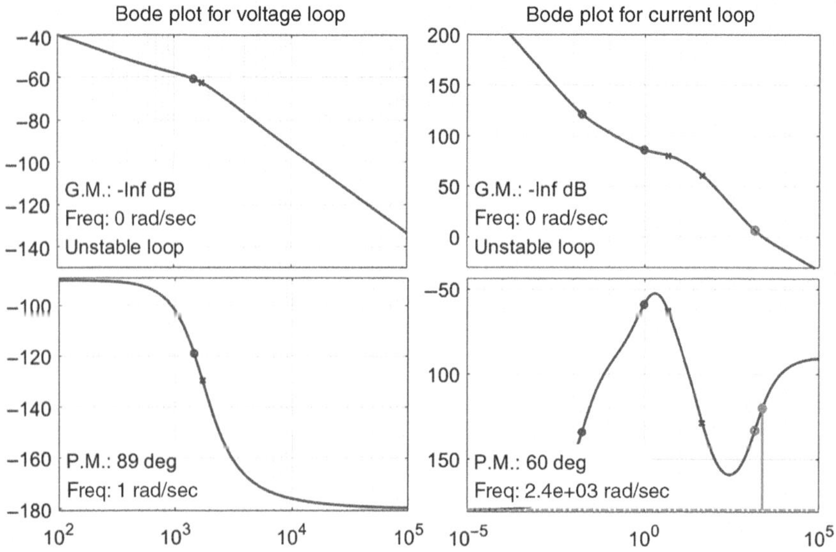

Figure 10.16 Bode plots of the speed controller and current controller. (a) Speed controller and (b) Current controller.

Figure 10.17 Step response of the double-loop controller.

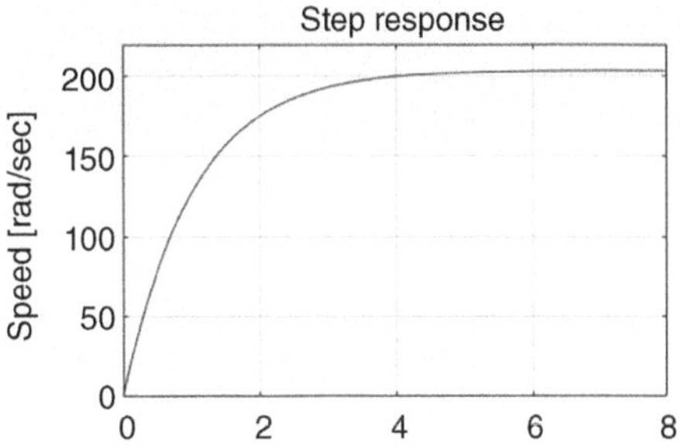

The step response of the double-loop controller is shown in Figure 10.17, further validating the system's dynamic performance.

The corresponding PSIM circuit diagram for the double loop controller is presented in Figure 10.18. In this implementation:

- The output of the speed controller i_{fb} is summed with the feedforward current ($i_{\mathrm{ff}} = \frac{T_{\mathrm{load}}}{K_{\varphi}}$) to form the reference input for the current controller.
- The controller gains are implemented based on the values obtained from MATLAB

The simulation results, shown in Figure 10.19, demonstrate the effectiveness of the controller:

- The motor speed accurately reaches the reference value without overshoot, indicating excellent speed control performance.
- The current controller operates as expected, ensuring stable and precise current regulation.

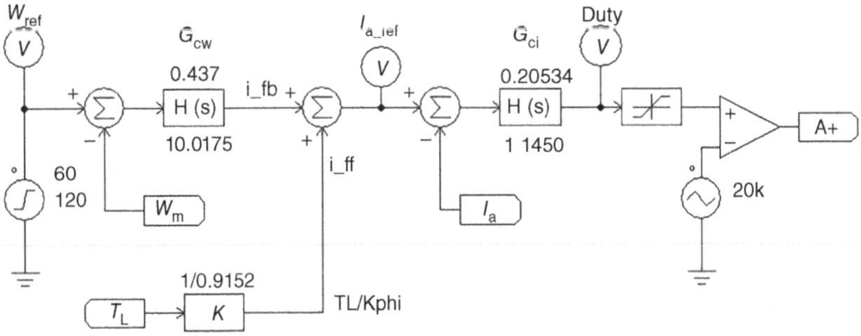

Figure 10.18 Double-loop controller block diagram.

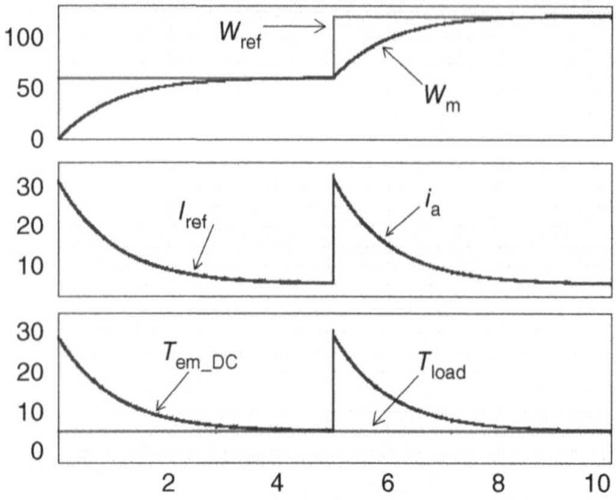

Figure 10.19 Simulation waveforms of the double-loop controller response.

10.3 Bi-Polar Driving (Bi-Directional Drive)

Figure 10.20(a) illustrates a bi-directional DC motor drive circuit implemented using an H-bridge composed of four switching devices. This configuration allows the armature current to flow in both directions, thereby enabling forward and reverse motor operation. For comparison, Figure 10.20(b) presents a single-phase grid-connected power conditioning system (PCS). By examining these two circuits, the following key equivalences can be identified.

- The motor's armature resistance and inductance, R_a and L_a are analogous to the filter resistance and inductance, r_L and L in the single-phase PCS.
- The motor's back EMF, E_a is analogous to the grid voltage, V_{grid}.

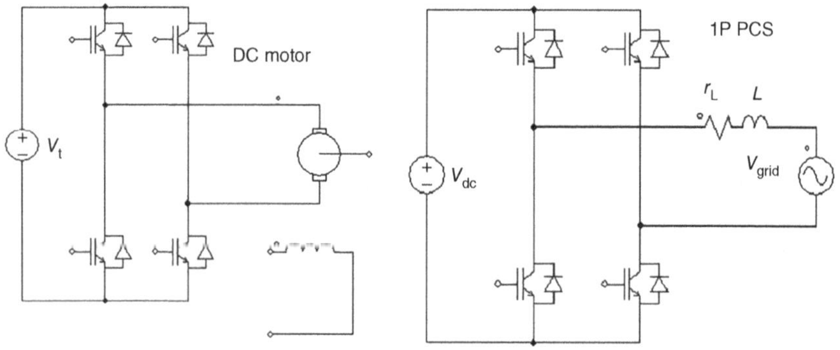

Figure 10.20 Demonstration of the duality between DC Motor and PCS circuits. (a) DC Motor Bi-directional drive circuit and (b) Single-phase PCS.

Despite these similarities, a fundamental difference exists between the two systems.

- The back EMF E_a is a sinusoidal waveform whose magnitude and frequency vary proportionally with the motor's angular velocity ω_m of the motor.
- In contrast, the grid voltage V_{grid} is a fixed-amplitude sinusoid with constant frequency.

Because the single-phase inverter circuit introduced in Chapter 11 consists of two complementary half-bridge legs, the same control methodology developed for inverter operation can be directly applied to the bi-polar DC motor drive. This structural equivalence significantly simplifies the design and implementation of the motor control system, as illustrated in Figure 10.21.

In the H-bridge configuration, Phase A and Phase B operate with a 180° phase difference, satisfying the relationships

$$A+ = B- \quad \text{and} \quad A- = B+$$

As a result

- When the gate signal is high, the motor terminal voltage is $+V_t$.
- When the gate signal is low, the motor terminal voltage is $-V_t$.

Accordingly, the application of $+V_t$ corresponds to the duty cycle d, while the application of $-V_t$ corresponds to the complementary interval $(1 - d)$.

It should be emphasized that the mechanical dynamics of the motor are independent of the duty cycle itself and are governed solely by the electromagnetic torque and load conditions.

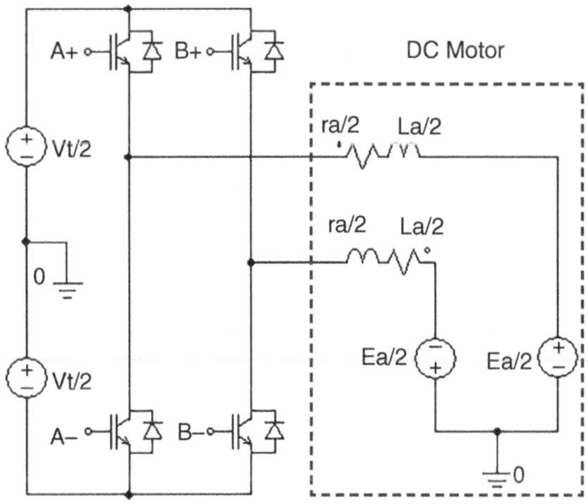

Figure 10.21 Bi-polar driving circuit analysis.

10.3.1 State-Averaged Equations and Transfer Function

The state equations of the DC motor can be written as

$$V_t = E_a + L_a\frac{di_a}{dt} + R_a i_a = K_\varphi \omega_m + L_a\frac{di_a}{dt} + R_a i_a$$

$$J_m\frac{d\omega_m}{dt} + B_m\omega_m = T_{em} - T_{load} = K_\varphi i_a - T_{load} \tag{10.12}$$

During operation, the motor terminal voltage V_t alternates between $+V_{dc}$ and $-V_{dc}$, depending on the switching states of the H-bridge. Applying state averaging over the ON interval d and the OFF interval $(1 - d)$ yields the following averaged dynamics.

- Electrical dynamics

  ON time:

  $$\frac{di_a}{dt} = -\frac{R_a}{L_a}i_a - \frac{K_\varphi}{L_a}\omega_m + \frac{V_t}{L_a}$$

  OFF time:

  $$\frac{di_a}{dt} = -\frac{R_a}{L_a}i_a - \frac{K_\varphi}{L_a}\omega_m - \frac{V_t}{L_a}$$

- Mechanical dynamics (identical for both states):

  $$\frac{d\omega_m}{dt} = \frac{K_\varphi}{J_m}i_a - \frac{B_m}{J_m}\omega_m - \frac{1}{J_m}T_{load}$$

  Averaging the electrical dynamics over one switching period gives

  $$\frac{di_a}{dt} = -\frac{R_a}{L_a}i_a - \frac{K_\varphi}{L_a}\omega_m + \frac{V_t}{L_a}(2d - 1)$$

  which can be rewritten as

  $$\frac{di_a}{dt} = -\frac{R_a}{L_a}i_a - \frac{K_\varphi}{L_a}\omega_m + \frac{2V_t}{L_a}d - \frac{V_t}{L_a}$$

  The averaged system can be expressed in matrix form as

  $$\dot{X} = AX + Bu + F$$

where the affine term F introduces nonlinearity. To eliminate this term and obtain a linear state-space model, the duty cycle d is decomposed as

$$d = d_{ff} + d_{fb}$$

with

$$d_{ff} = 0.5\ (\text{feed} - \text{forward term}),\quad d_{fb}\ (\text{feedback control input})$$

Substituting this decomposition removes the constant term, yielding the linearized model

$$\dot{X} = AX + Bu$$

where

$$X = \begin{pmatrix} i_a \\ \omega_m \end{pmatrix}, \quad A = \begin{pmatrix} -\dfrac{R_a}{L_a} & -\dfrac{K_\varphi}{L_a} \\ \dfrac{K_\varphi}{J_m} & -\dfrac{B_m}{J_m} \end{pmatrix}, \quad B = \begin{pmatrix} \dfrac{2V_t}{L_a} \\ 0 \end{pmatrix}, \quad u = d_{fb} \tag{10.13}$$

Compared with the uni-polar drive case, the system matrix A remains unchanged, while the input matrix B is doubled due to the bi-polar voltage swing. To maintain consistency in control design and allow reuse of the same controller gains as in the uni-directional case, the feedback duty input d_{fb} must be scaled by a factor of 0.5.

10.3.2 Transfer Function Derivation

Based on the state-averaged model derived in the previous section, the following transfer functions are obtained.

1) Current-to-duty ratio transfer function

$$G_1(s) = \frac{I_a(s)}{d(s)} = \frac{2 \cdot \frac{V_t}{L_a}\left(s + \frac{B_m}{J_m}\right)}{s^2 + \left(\frac{R_a}{L_a} + \frac{B_m}{J_m}\right)s + \left(\frac{K_\varphi^2}{L_aJ_m} + \frac{R_aB_m}{L_aJ_m}\right)} \tag{10.14}$$

2) Speed-to-duty ratio transfer function

$$G_2(s) = \frac{\omega_m(s)}{d(s)} = \frac{2 \cdot \frac{V_tK_\varphi}{L_aJ_m}}{s^2 + \left(\frac{R_a}{L_a} + \frac{B_m}{J_m}\right)s + \left(\frac{K_\varphi^2}{L_aJ_m} + \frac{R_aB_m}{L_aJ_m}\right)} \tag{10.15}$$

3) Speed-to-current transfer function

$$G_3(s) = \frac{\omega_m(s)}{I_a(s)} = \frac{\frac{K_\varphi}{J_m}}{s + \frac{B_m}{J_m}} \tag{10.16}$$

Compared with the uni-directional drive case, the bi-polar drive results in a doubling of the magnitude of the transfer functions due to the symmetric $\pm V_t$ voltage excitation. Consequently, the controller gains must be reduced by a factor of two to preserve equivalent closed-loop dynamics. Notably, the pole and zero locations of the transfer functions remain unchanged.

10.3.3 PSIM Simulation of Bi-Directional Driving

The PSIM simulation schematic for the bi-directional drive is shown in Figure 10.22. The system is configured using a single-phase H-bridge, allowing the motor terminals to receive voltages of +120 V and −120 V. A rotational load of 5 N · m is applied, and both speed and torque sensors are included, maintaining consistency with the uni-directional drive configuration presented earlier.

a) **Speed Controller Design**

The speed controller for the bi-directional drive is designed by reducing the proportional gain of the uni-directional controller to 50% of its original value. This adjustment is required because the plant transfer function of the bi-directional drive has twice the gain of the uni-directional case. To achieve equivalent closed-loop performance, the controller gain must therefore be scaled accordingly. The configuration of the speed controller is shown in Figure 10.23.

The output of the speed controller, denoted as d_{fb}, is combined with a feedforward offset of +0.5 to generate the duty cycle d_{a} for Phase A. Since Phase B operates with a 180° phase shift relative to Phase A, its duty cycle d_{b} is obtained by adding the same feedforward offset to the negative feedback term $-d_{\text{fb}}$.

The simulation results shown in Figure 10.24 demonstrate that the speed controller accurately maintains the desired steady-state speed without overshoot. However, when the reference speed is subjected to a step change from +50 to −50 rad/sec, noticeable distortion appears in the speed response due to regenerative current effects. To mitigate this distortion, it is recommended that the reference speed be gradually ramped or transitioned through intermediate values, for example $+50 \rightarrow 0 \rightarrow -50$, rather than applying an abrupt reversal.

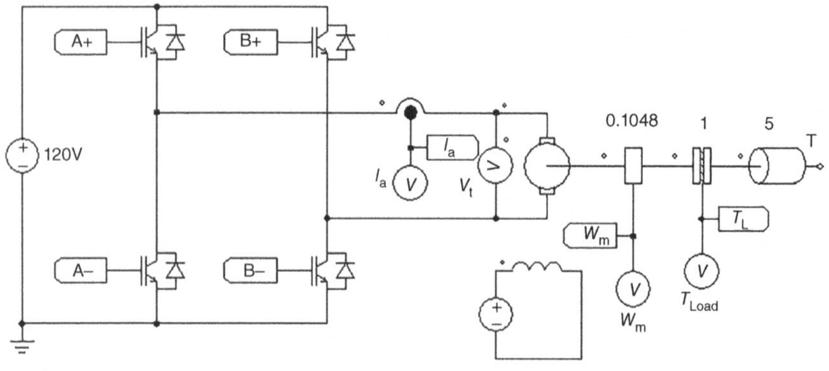

Figure 10.22 PSIM schematic for bi-directional driving configuration.

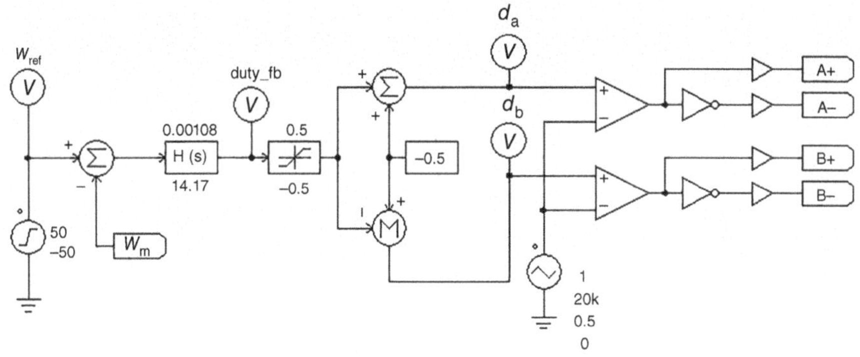

Figure 10.23 Speed controller circuit diagram.

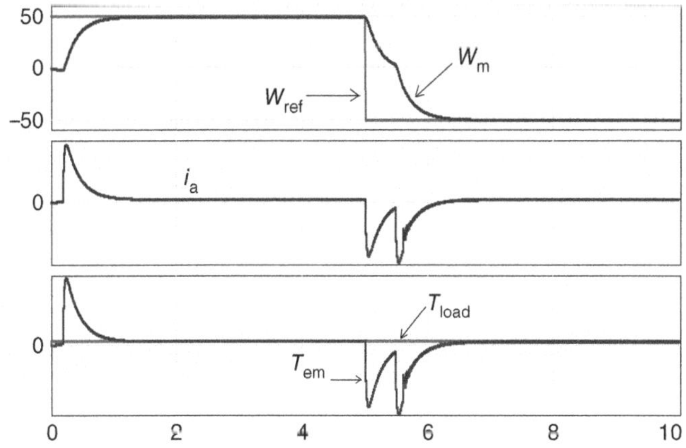

Figure 10.24 Waveforms of the speed controller response.

b) **Current Controller Design**
The current controller is also designed by scaling the P-gain from the unidirectional drive by 50% to account for the doubled transfer function gain in the bi-directional drive. The duty cycle generation circuit for each phase remains unchanged from the previous section. Figure 10.25 presents current controller circuit diagram.
The simulation results, shown in Figure 10.26, confirm that the system operates correctly and matches the expected design values, with precise regulation of the armature current.

c) **Double-Loop Controller**
For the double-loop controller, the only modification relative to the uni-directional control scheme is a reduction of the proportional gain of

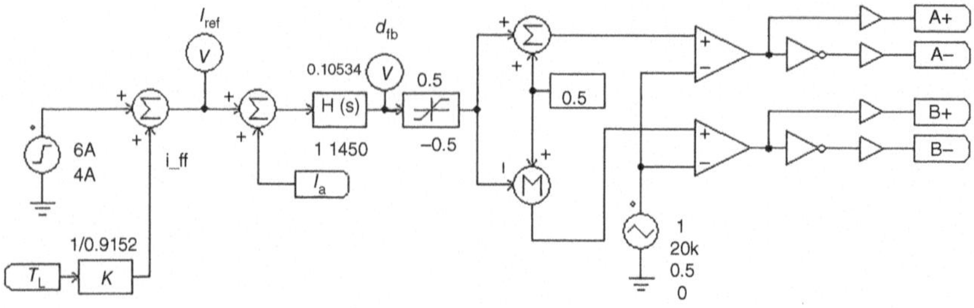

Figure 10.25 Current controller circuit diagram.

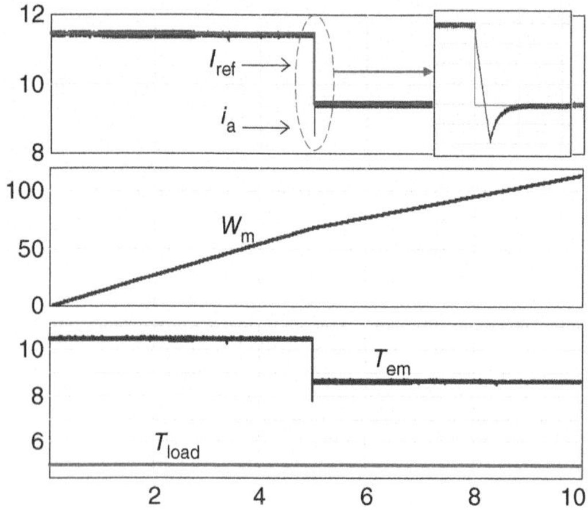

Figure 10.26 Waveforms of the current controller response.

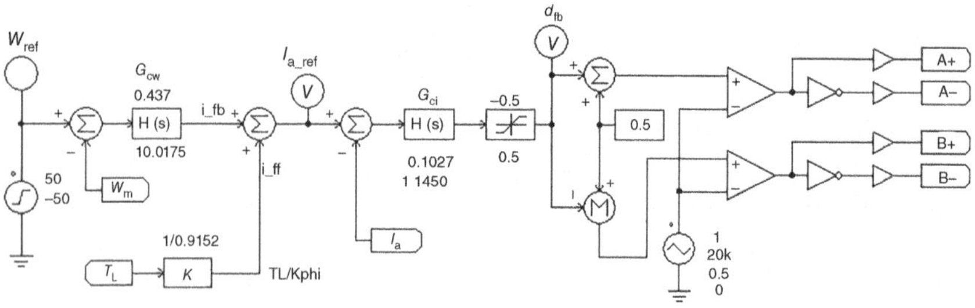

Figure 10.27 Double-loop controller circuit diagram.

the current controller to 50% of its original value. This adjustment compensates for the doubled plant gain in the bi-directional drive. All other aspects of the control architecture remain unchanged. Figure 10.27 shows the circuit diagram of the double-loop controller.

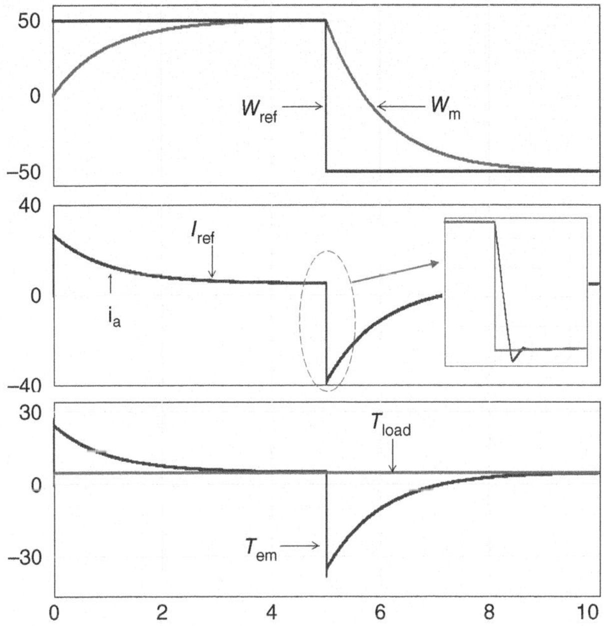

Figure 10.28 Waveforms of the double-loop controller response.

The simulation waveforms shown in Figure 10.28 confirm that the system operates as designed. In particular, when the reference speed transitions from a positive to a negative value, no distortion is observed in the speed response. This improved behavior is attributed to the effective action of the inner current controller, which regulates the motor current based on speed feedback, thereby ensuring smooth and stable operation during speed reversals.

d) **C-Block Digital Conversion**

To implement the controllers in a digital control environment, sampling time of $T_s = 50\ us$ is employed. Using this sampling period, the continuous-time controllers are converted into discrete-time form. The resulting digital controller expressions for the speed controller and the current controller are given as follows.

- Speed Controller:

$$C_1 = G_{cw}(z) = \frac{0.437(z-1)}{z-1}$$

- Current Controller:

$$C_2 = G_{ci}(z) = \frac{0.1095(z-0.939)}{z-1}$$

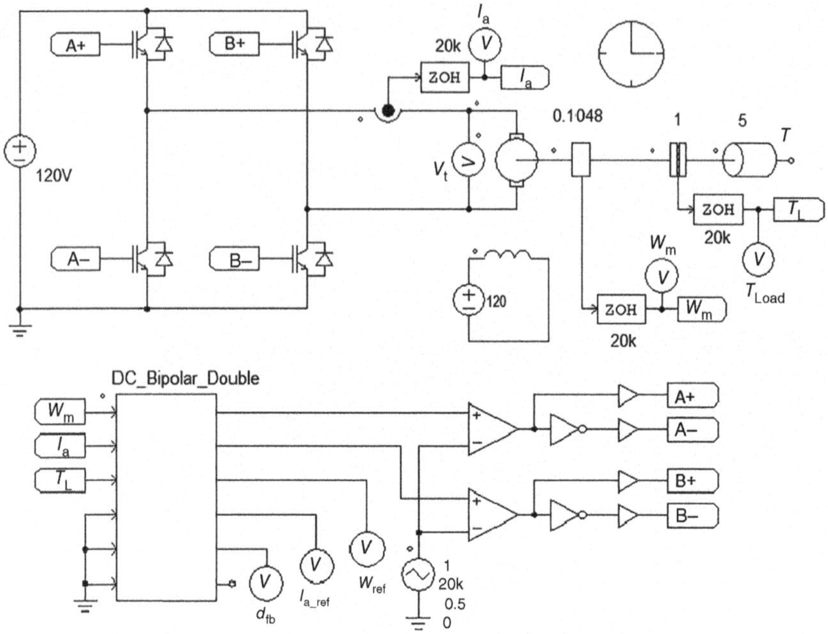

Figure 10.29 Double-loop controller (C block) circuit diagram.

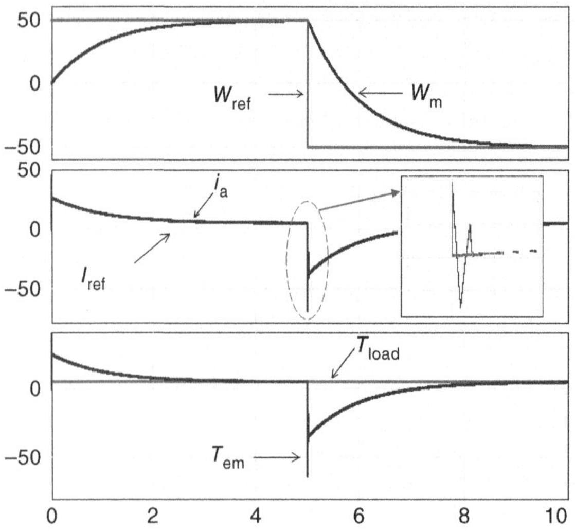

Figure 10.30 Simulation waveforms of the double-loop controller response.

Table 10.6 C-block code for the double loop controller (DC_Bipolar_Double).

```
#define Kphi   0.9152  // bemf coeff.

#define KWp    0.437   // speed: p gain
#define KWi    1       // speed: I gain

#define Kip    0.1095 //current: p gain
#define Kii    0.939  //current: I gain

double Wm, Ia, TL;
double Wm_ref, Ia_ref, Wm_err, Ia_err;
double I_fb, I_ff, d_fb, duty_a, duty_b;

static double Wm_err_old=0.0, I_fb_old=0.0;
static double d_old=0.0, Ia_err_old=0.0;

/**** Input Variable    ****/
Wm      = in[0];
Ia      = in[1];
TL      = in[2];

if (t < 5.0) Wm_ref =  50;
else         Wm_ref = -50;

/****** Speed Controller *******/
Wm_err = Wm_ref - Wm;
I_fb = I_fb_old+KWp*Wm_err-
          KWp*KWi*Wm_err_old;
I_fb_old=I_fb;
Wm_err_old=Wm_err;

I_ff =  TL/Kphi;
Ia_ref = I_fb + I_ff;

/**** Current controller ****/
Ia_err = Ia_ref - Ia;
d_fb = d_old +Kip*Ia_err -
          Kip*Kii*Ia_err_old;
d_old=d_fb;
Ia_err_old=Ia_err;

if      (d_fb >  0.5) d_fb= 0.5;
else if (d_fb < -0.5) d_fb=-0.5;
duty_a =  0.5 + d_fb;
duty_b =  0.5 - d_fb;

out[0] = duty_a;
out[1] = duty_b;
out[2] = Wm_ref;
out[3] = Ia_ref;
out[4] = d_fb;
```

The corresponding PSIM simulation circuit implementing these digital controllers using C-blocks is shown in Figure 10.29.

The implementation of this double-loop controller C-block is shown in Table 10.6.

The simulation results in Figure 10.30 confirm that the digital control system reproduces the same performance as the analog controller.

Part II

AC Power Conversion Systems

11

Single-Phase Inverter

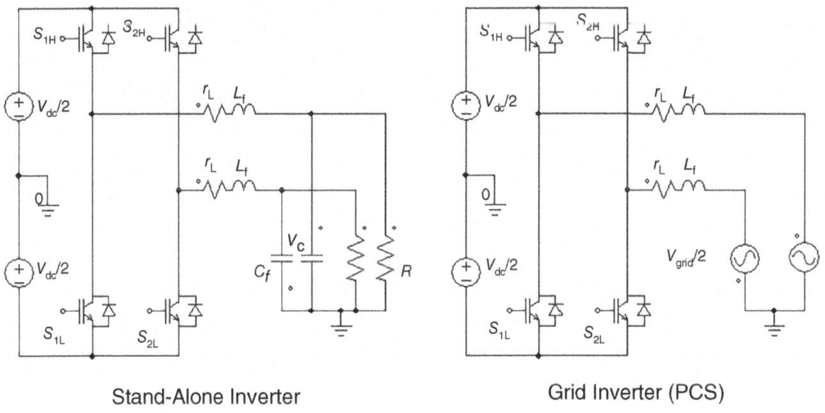

11.1 Circuit Representation Methods for Inverter Systems

In DC–DC converter systems, current flows unidirectionally from the input to the output, with the input and output grounds typically connected, forming a common ground configuration. Consequently, the input ground is conventionally used as the output ground reference.

However, applying this same grounding approach to inverter circuits—such as single-phase or three-phase inverters—introduces certain challenges due to the nature of AC power generation

- Inverters produce both positive and negative voltages and currents, which complicates the generation of true bipolar (i.e., negative) outputs when referenced to a single-sided ground.
- A common mode voltage (V_{n0}) arises between the input and output grounds due to the asymmetrical switching states, which can degrade output signal quality.

Figure 11.1 illustrates a three-phase inverter circuit. In the *R*-phase leg, the state-space equations based on the switching conditions in Figure 11.1(a) are as follows:

- When the upper switch (SW) is ON

$$V_{dc} = r_L i_L + L\frac{di_L}{dt} + V_{rn} + V_{n0}$$

- When the lower switch (SW) is ON

$$0 = r_L i_L + L\frac{di_L}{dt} + V_{rn} + V_{n0}$$

In this case, the value of V_{n0} varies between 0 and V_{dc}, resulting in a DC offset of $+V_{dc}/2$.

To mitigate this issue, as illustrated in Figure 11.1(b), the input power source is split into two equal parts, and the midpoint voltage is designated as the reference ground. This midpoint is then shared with the output, creating a symmetric configuration.

Under this arrangement, the switching states alternately apply $+V_{dc}/2$ and $-V_{dc}/2$, enabling the generation of a true AC voltage or current waveform with zero offset. This book adopts the circuit configuration shown in Figure 11.1(b) for modeling and controller design, ensuring balanced bipolar output and eliminating common-mode voltage concerns.

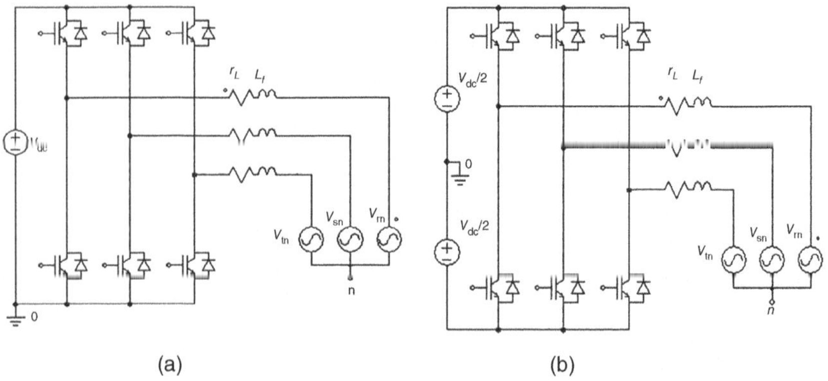

Figure 11.1 Three-phase inverter circuit topologies. (a) Inverter circuit with ground offset and (b) Inverter circuit with symmetric grounding.

To implement the input voltage division shown in Figure 11.1(b), the DC power supply V_{dc} is split into two equal voltages using series-connected electrolytic capacitors. Electrolytic capacitors are widely employed for this purpose because of their high capacitance-to-volume ratio. In practice, capacitors with voltage ratings in the range of 450 to 500 V are commonly used. However, for applications delivering a 380 VAC output, the corresponding rectified DC-link voltage exceeds the rated insulation voltage of a single electrolytic capacitor. Consequently, the use of a single capacitor becomes inadequate from a voltage-rating standpoint, necessitating a series capacitor configuration for safe operation.

To meet these requirements, capacitors are connected in series, effectively increasing the total withstand voltage to approximately 900–1000 V. Since the series connection reduces the equivalent capacitance to half that of a single capacitor, additional capacitors are added in parallel to restore the required total capacitance. This configuration ensures both sufficient voltage insulation and appropriate capacitance values. The complete schematic of the input voltage-splitting circuit is illustrated in Figure 11.2.

Single-Phase Inverter Circuit Representation Figure 11.3(a) shows a conventional single-phase inverter circuit, commonly employed in practical applications. The dynamic behavior of this circuit can be described by the following state-space equations, which depend on the switching states:

- Switches S_{1H} and S_{2L} ON:

$$V_{\mathrm{dc}} = 2\left(r_L i_L + L\frac{di_L}{dt}\right) + V_{\mathrm{grid}}$$

- Switches S_{2H} and S_{1L} ON:

$$-V_{\mathrm{dc}} = 2\left(r_L i_L + L\frac{di_L}{dt}\right) + V_{\mathrm{grid}}$$

- Only one switch ON:

$$0 = r_L i_L + L\frac{di_L}{dt} + V_{\mathrm{grid}}$$

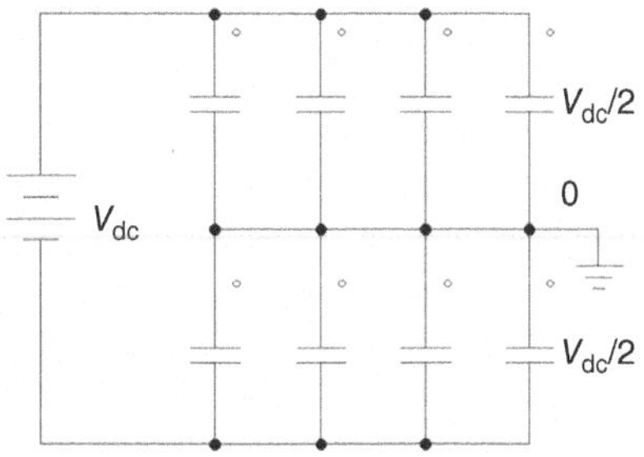

Figure 11.2 Schematic of the power supply division circuit.

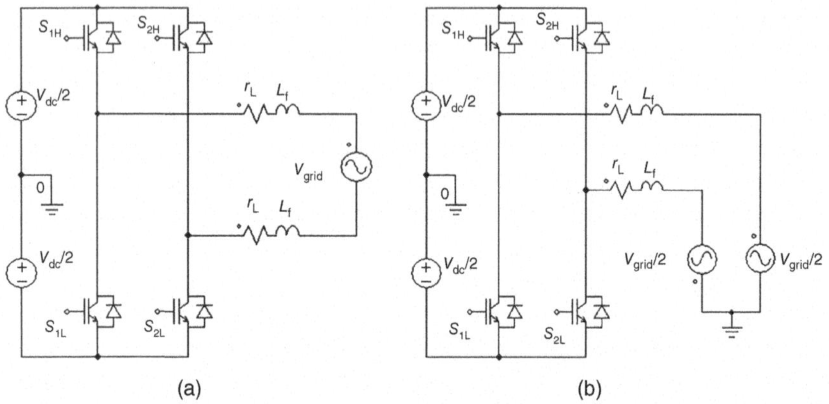

Figure 11.3 Circuit representation of a single-phase inverter. (a) Conventional representation and (b) Unified representation.

These equations indicate that the inverter bridge can apply one of three discrete voltage levels to the load: $+V_{dc}$, $-V_{dc}$, or 0 V, depending on the switch configuration.

Figure 11.3(b) represents the unified representation method that will be used throughout this textbook. While the conventional representation shown in Figure 11.3(a) is standard in many applications, the structure in Figure 11.3(b) is more suitable for modeling and controller design purposes.

Inverter circuits are broadly categorized into single-phase and three-phase configurations. A single-phase inverter typically consists of two bridge arms, while a three-phase inverter comprises three. To establish a unified modeling framework applicable to both configurations, the single-phase inverter can be restructured by conceptually removing one bridge arm from a three-phase inverter. This modified structure is shown in Figure 11.3(b), in which the single-phase inverter is represented as a two-leg system derived from a three-phase topology.

This approach makes it possible to use the same analysis and controller design methods for both single-phase and three-phase systems, resulting in a more consistent and easier-to-implement design process.

In the unified configuration, the grid voltage V_{grid} is symmetrically divided into two components, $\pm V_{grid}/2$, with the midpoint serving as the circuit ground. This arrangement effectively emulates a three-phase inverter with one phase removed.

Careful attention must be paid to the polarity of the grid voltage after it is split into two sources. Since the two legs of a single-phase inverter operate with a 180° phase shift, the separated voltage sources must also be connected with opposite polarities to maintain this phase relationship.

Inverter Classifications Inverters are broadly classified into two main types based on their output characteristics:

- Constant voltage constant frequency (CVCF) inverters
- Variable voltage variable frequency (VVVF) inverters

1) **CVCF Inverters**

CVCF inverters are designed to deliver a fixed output voltage and frequency. These inverters are commonly employed in applications requiring stable and regulated AC power, such as power transmission systems and grid-connected systems. Depending on their interaction with the utility grid, CVCF inverters can be further divided into the following types:

a) **Grid-Tied Inverter (PCS: Power Conditioning System)**
A power conditioning system (PCS) operates in parallel with the utility grid and is capable of both injecting power into the grid and drawing power from it, as illustrated in Figure 11.4. Since the grid behaves as an infinite bus, injecting power does not affect the grid voltage amplitude—only the current magnitude can be controlled.
   - Control target: The inductor current is the primary control variable.
   - Power factor control: To maintain unity power factor ($PF = 1$), the inverter current is regulated to remain in phase with the grid voltage.
   - Frequency synchronization: The inverter's output frequency is synchronized with the grid using a phase-locked loop (PLL), which precisely tracks the grid frequency.

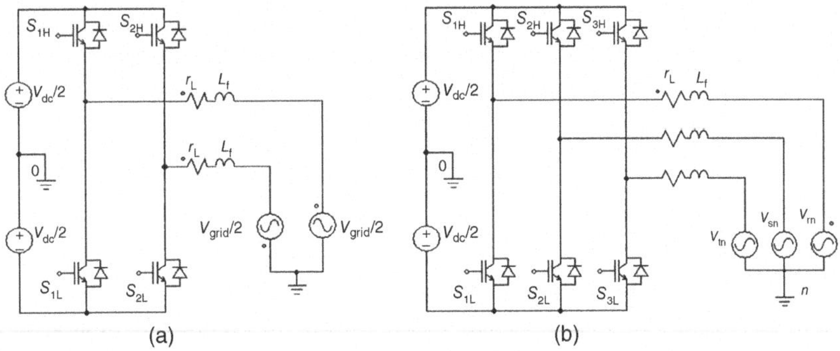

Figure 11.4 Circuit diagrams of grid-tied PCS systems. (a) Single-phase PCS. and (b) Three-phase PCS.

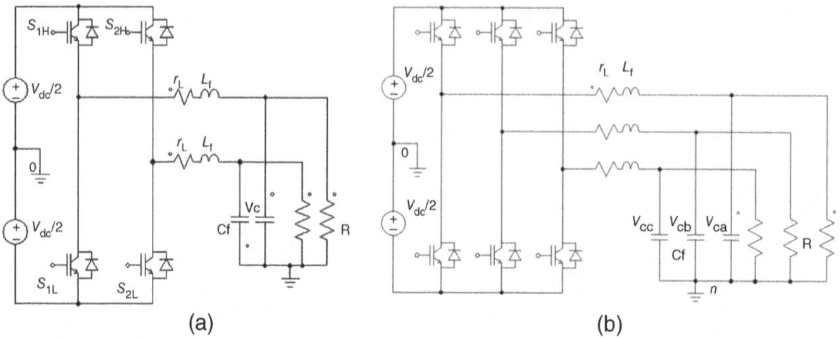

Figure 11.5 Circuit diagrams of stand-alone inverter systems. (a) Single-phase stand-alone system and (b) Three-phase stand-alone system.

b) **Stand-Alone System**

Unlike a PCS, a stand-alone inverter system operates independently of the utility grid and must autonomously generate both the output voltage amplitude and frequency. This configuration is illustrated in Figure 11.5.

- Control target: The output voltage is controlled by regulating the voltage across the output capacitor, which directly sets the AC output voltage level.
- Load dependency: The output current is determined by the load characteristics rather than being actively controlled.
- Fixed frequency: The system operates at a standard frequency (typically 50 or 60 Hz) to ensure compatibility with conventional AC loads.

2) **VVVF Inverters**

VVVF inverters are designed to dynamically adjust both the output voltage and frequency. These inverters are widely used in motor drive systems for precise speed and torque control of AC motors in applications such as industrial automation, HVAC systems, and electric vehicles.

- Output characteristics: The output voltage and frequency are continuously modulated in real time according to the control strategy and motor operating conditions.

Figure 11.6(a) illustrates a typical VVVF inverter-based motor drive system, and Figure 11.6(b) shows the equivalent circuit model of a three-phase AC motor. A structural comparison between the VVVF inverter in Figure 11.6(a) and the three-phase PCS system in Figure 11.4(b) reveals that both circuits share the same fundamental topology. This is because each phase of an AC motor can be modeled as a series R–L circuit with an internal back electromotive force (EMF), denoted as V_{b_emf}.

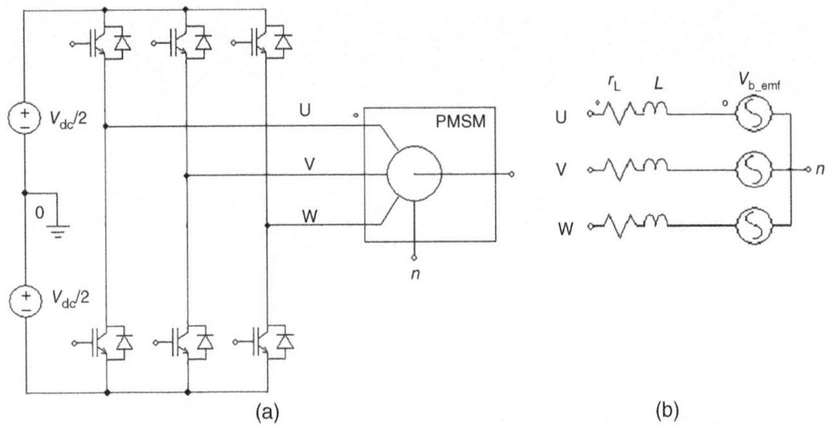

Figure 11.6 VVVF inverter circuit and ac motor model. (a) AC motor drive circuit and (b) Equivalent circuit of AC motor.

- In PCS systems, the grid voltage V_{grid} is a fixed-voltage, fixed-frequency source.
- In motor drive systems, the back-EMF V_{b_emf} is a variable-voltage, variable-frequency source that varies with motor speed.

Despite the differences in their source characteristics, both systems are governed by the same underlying voltage relationship:

$$V_{grid} \equiv V_{b\_emf}$$

This commonality allows both systems to be analyzed using a unified mathematical model, simplifying both control system design and performance analysis.

11.2 State-Space Averaged Equations and Transfer Functions

The circuit configuration of a Stand-alone single-phase inverter system is illustrated in Figure 11.7. The key distinction between Figure 11.7(a) and (b) lies in how the output voltage V_c is defined and measured.

- Figure 11.7(a): The output voltage V_c is measured across a single capacitor phase.
- Figure 11.7(b): The output voltage V_c is defined as the sum of the two-capacitor voltage.

For modeling and simulation, the configuration shown in Figure 11.7(a) is adopted throughout this textbook. This choice preserves consistency with

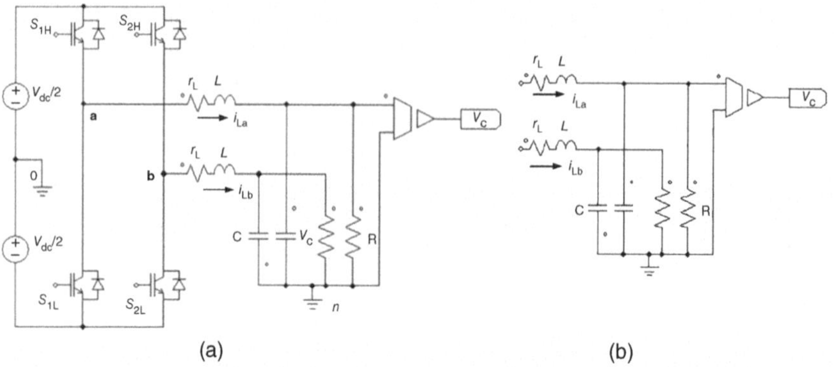

Figure 11.7 Stand-alone single-phase inverter and output voltage configuration. (a) Single capacitor voltage and (b) Double capacitor voltage.

three-phase circuit analysis and simplifies the formulation of the state-space model.

In practical hardware implementation, however, directly measuring the split single-phase AC voltage, as depicted in Figure 11.7(a), can be difficult. For this reason, the configuration illustrated in Figure 11.7(b) is commonly preferred. In this arrangement, the voltage across the inverter bridge is divided by two and treated as the output voltage V_c, which is more convenient for sensing and control purposes.

Single-Phase Operation In the circuit shown in Figure 11.7(a), the a-phase and b-phase inverter bridges operate independently with a phase difference of 180°. The key operating characteristics are as follows:

- The inductor currents of the two legs satisfy the condition:

  $$i_{La} = -i_{Lb}$$

  where i_{La} and i_{Lb} are the inductor currents of the a-phase and b-phase, respectively
- The switching states are limited to combinations that avoid both high-side switches being ON simultaneously. Thus, the allowed switching states are: 00, 01, and 10. The 11 state is prohibited.

For modeling purposes, the a-phase inductor current i_{La} is denoted simply as i_L. The steady-state equations for the a-phase operation are given below:

- Switch ON state:

$$\frac{V_{dc}}{2} = r_L i_L + L\frac{di_L}{dt} + v_c, \quad i_L = C\frac{dv_c}{dt} + \frac{v_c}{R} \tag{11.1}$$

- Switch OFF state:

$$-\frac{V_{\mathrm{dc}}}{2} = r_L i_L + L\frac{di_L}{dt} + v_c, \quad i_L = C\frac{dv_c}{dt} + \frac{v_c}{R} \tag{11.2}$$

Averaged Model Derivation The averaged model, using state-space averaging techniques, is formulated as follows:

$$L\frac{di_L}{dt} = \left(\frac{V_{\mathrm{dc}}}{2} - v_c - r_L i_L\right)d + \left(-\frac{V_{\mathrm{dc}}}{2} - v_c - r_L i_L\right)(1-d) \tag{11.3}$$

Simplifying

$$L\frac{di_L}{dt} = V_{\mathrm{dc}}d - \frac{V_{\mathrm{dc}}}{2} - v_c - r_L i_L$$

$$C\frac{dv_c}{dt} = i_L - \frac{v_c}{R}$$

The resulting state-space averaged model for the stand-alone inverter is written as

$$\begin{pmatrix} \frac{di_L}{dt} \\ \frac{dv_c}{dt} \end{pmatrix} = \begin{pmatrix} -\frac{r_L}{L} & -\frac{1}{L} \\ \frac{1}{C} & -\frac{1}{RC} \end{pmatrix} \begin{pmatrix} i_L \\ v_c \end{pmatrix} + \begin{pmatrix} \frac{V_{\mathrm{dc}}}{L} \\ 0 \end{pmatrix} d + \begin{pmatrix} -\frac{V_{\mathrm{dc}}}{2L} \\ 0 \end{pmatrix} \tag{11.4}$$

This can be written in matrix form as

$$\dot{X} = AX + BU + F.$$

where

$$\dot{X} = \begin{pmatrix} \frac{di_L}{dt} \\ \frac{dv_c}{dt} \end{pmatrix}, \quad \mathrm{A} = \begin{pmatrix} -\frac{r_L}{L} & -\frac{1}{L} \\ \frac{1}{C} & -\frac{1}{RC} \end{pmatrix}, \quad \mathrm{B} = \begin{pmatrix} \frac{V_{\mathrm{dc}}}{L} \\ 0 \end{pmatrix}, \quad \mathrm{U} = d, \quad F = \begin{pmatrix} -\frac{V_{\mathrm{dc}}}{2L} \\ 0 \end{pmatrix} \tag{11.5}$$

This model forms the basis for control system analysis and PI controller design in stand-alone inverter systems.

Linearization of the State-Space Model In the state-space model derived earlier, the term F introduces nonlinearity into the system equations. To linearize the model and eliminate this offset term, the duty cycle d is decomposed into feed-forward (d_{ff}) and feedback (d_{fb}) components.

$$d = d_{ff} + d_{fb} \tag{11.6}$$

Substituting $d = d_{ff} + d_{fb}$ into the original state equation and solving for d_{ff}, we find

$$d_{ff} = 0.5 \tag{11.7}$$

This eliminates the offset term (F), resulting in a linear form

$$\dot{X} = AX + Bd_{fb} \tag{11.8}$$

In this equation, d_{ff} represents the feed-forward duty, and d_{fb} represents the feedback duty. The feed-forward duty (d_{ff}) is a pre-calculated value added to the plant input, but it is not used in controller design. Only d_{fb} is used as the input to the plant. Figure 11.8 illustrates the configuration of the feedback control system. In Figure 11.8(a), which shows a commonly used feedback system, the controller output ($d = d_{fb}$) is directly input to the plant transfer function (G). In Figure 11.8(b), a feed-forward system structure is shown, where the feedback controller output (d_{fb}) is combined with the feed-forward compensation ($d_{ff} = 0.5$), and their sum ($d = d_{ff} + d_{fb}$) is used as the input to the plant.

Equivalence with Buck Converter Model The final state equation $\dot{X} = AX + Bd_{fb}$ retains the same matrix values A and B values as those of the buck converter system, as expressed in Eq. (2.4). Consequently, the transfer function derived from this system is identical to that of the buck converter. Expressing this equivalently, the system can be represented as follows.

- Inductor Current to Duty Ratio

$$G_1(s) = \frac{I_L(s)}{d(s)} = \frac{\frac{V_{in}}{L}\left(s + \frac{1}{RC}\right)}{s^2 + \left(\frac{1}{RC} + \frac{r_L}{L}\right)s + \frac{1}{LC}} \tag{11.9}$$

- Capacitor Voltage to Duty Ratio

$$G_2(s) = \frac{v_c(s)}{d(s)} = \frac{\frac{V_{in}}{LC}}{s^2 + \left(\frac{1}{RC} + \frac{r_L}{L}\right)s + \frac{1}{LC}} \tag{11.10}$$

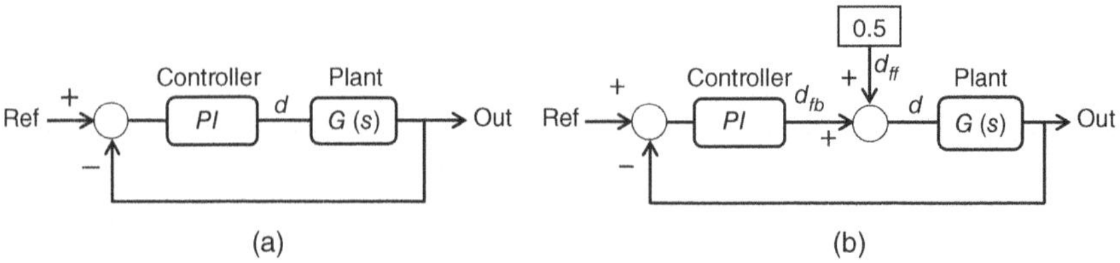

Figure 11.8 Block diagram of the feedback control system. (a) General feedback configuration and (b) Feed-forward compensation configuration.

- Capacitor Voltage to Inductor Current

$$G_3(s) = \frac{v_c(s)}{I_L(s)} = \frac{1/C}{s + (1/RC)} \tag{11.11}$$

11.3 Controller Design

In single-phase inverter systems, the output current is an AC sinusoidal waveform. Consequently, the reference signal is also sinusoidal, differing from the constant (DC) reference typically encountered in other control applications. Because the reference signal varies continuously, the notion of steady-state error differs from that of DC tracking; specifically, it does not accumulate in the same manner. In such cases, a proportional (*P*) controller is often sufficient to track the sinusoidal reference within an acceptable error margin, particularly when the control bandwidth is high. The use of a proportional–integral (*PI*) controller is less common in these applications, as the integrator introduces phase lag, which can degrade system stability.

Steady-State Error Analysis To better understand the implications of controller selection, we consider the steady-state error behavior for both *P* and *PI* controllers. This concept is fundamental in control engineering and is typically introduced in undergraduate control systems courses.

Let the open-loop transfer function of the unity feedback system be defined as

$$G(s) = G_c(s)G_p(s)$$

Where $G_c(s)$ represents the controller transfer function (e.g., $G_{ci}(s)$ or $G_{cv}(s)$), and $G_p(s)$ is the plant transfer function (e.g., $G_1(s)$ or $G_2(s)$).

In control theory, system type is determined by the number of integrators, or poles at the origin, in the open-loop transfer function: Type 0 systems have no integrators and Type 1 systems have one integrator. The system type determines the steady-state error performance for various input types (e.g., step, ramp, sinusoidal). Both $G_1(s)$ and $G_2(s)$ are classified as Type 0 systems. When a *PI* controller is introduced, adding an integrator (a pole at the origin), the system becomes Type 1. A Type 1 system yields zero steady-state error for a step input. However, for more complex input types, such as ramp or sinusoidal signals, a finite steady-state error remains.

Mathematical Expression of Steady-State Error The steady-state error equation is calculated using the final value theorem, as shown in Table 11.1.

Table 11.1 Equation for steady-state error.

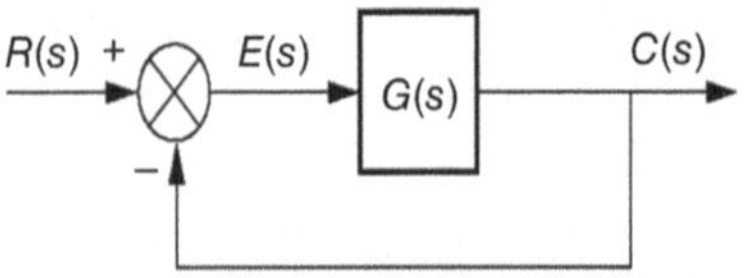

$$E(s) = R(s) - C(s)$$
$$C(s) = E(s)G(s) \Rightarrow$$

$$E(s) = \frac{R(s)}{1 + G(s)}$$

Applying the final value theorem:

$$e(\infty) = \lim_{t \to \infty} e(t) = \lim_{s \to 0} sE(s) = \lim_{s \to 0} \frac{sR(s)}{1 + G(s)}$$

Let us consider two common input types.

1) Step input $R(s) = \frac{A}{s}$

$$e(\infty) = \frac{sR(s)}{1 + \lim\limits_{s \to 0} G(s)} = \frac{s \cdot \frac{A}{s}}{1 + \lim\limits_{s \to 0} G(s)} = 0 \text{ (for Type 1 system)}$$

2) Ramp input $R(s) = \frac{A}{s^2}$

$$e(\infty) = \frac{sR(s)}{1 + \lim\limits_{s \to 0} G(s)} = \frac{s \cdot \frac{A}{s^2}}{1 + \lim\limits_{s \to 0} G(s)} = \frac{s \cdot \frac{A}{s^2}}{1 + \lim\limits_{s \to 0} G(s)} \text{ (finite for Type 1 system).}$$

The steady-state error in this case can be further expressed as

- Current control error $e(\infty) = \frac{1}{V_{dc} K_i K_p R C^2}$
- Voltage control error $e(\infty) = \frac{1}{V_{dc} K_i K_p}$

where V_{dc} is the DC link voltage and K_p, K_i are proportional and integral gain.

In feedback control systems with a DC reference, zero steady-state error can be achieved with a *PI* controller. However, in the case of AC sinusoidal reference signals, a residual steady-state error typically persists. For this reason, effective sinusoidal tracking in single phase inverters can be achieved using either

- *P* controller with sufficiently high bandwidth, or
- *PI* controller implemented in the DQ reference frame via the single-phase DQ transformation method.

Overview of Control Strategies This chapter introduces the AC instantaneous value control method, which is applicable to a wide range of single-phase systems, including stand-alone inverters, grid-tied inverters, and pulse width modulation (PWM) rectifiers.

Chapter 14 presents a detailed discussion of the single-phase DQ control method. This technique employs an all-pass filter (APF) to transform sinusoidal signals into DC-like quantities, thereby enabling the use of PI controllers. To fully understand the single-phase DQ control approach, it is recommended that readers first become familiar with the three-phase DQ transformation discussed in Chapter 12.

For this reason, a comprehensive treatment of single-phase DQ controllers is deferred to Chapter 14.

11.3.1 Current Controller in Stand-Alone Inverter

The *AC instantaneous value control method* is based on deriving current commands from the inverter's equivalent circuit model using known circuit parameters. Since the reference signal is an AC sinusoidal waveform, a simple proportional (*P*) controller is typically employed. Although this control scheme introduces a steady-state error, it offers advantages such as structural simplicity and stable operation, making it suitable for stand-alone inverter applications.

Figure 11.9 illustrates the equivalent circuit for the inverter's a-phase at point "a" in Figure 11.7(a). The equivalent circuit for point "b" is identical and therefore omitted.

Let i_L denote the inductor current in the a-phase, which corresponds to i_{La}. Applying Kirchhoff's voltage law (KVL), the voltage equation is given by

$$V_{\text{an}} = r_L i_L + L\frac{di_L}{dt} + v_c \tag{11.12}$$

Discretizing the equation using a finite-difference approximation yield

$$V_{\text{an}}(k+1) = r_L i_L(k) + L\frac{i_L(k+1) - i_L(k)}{T_s} + v_c(k) \tag{11.13}$$

Assuming that the current controller perfectly compensates the current error in the next control cycle, we set the reference current as $i_{\text{ref}}(k) = i_L(k+1)$. With the

Figure 11.9 Equivalent circuit of the inverter a-phase.

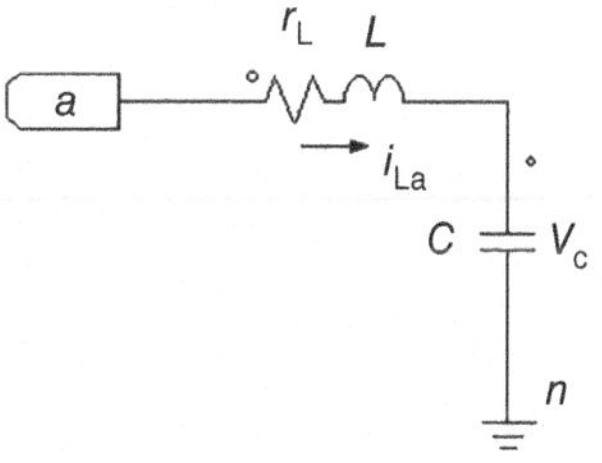

sinusoidal reference given by $i_{\text{ref}} = I_p \cos\theta$, the output of the current controller becomes

$$V_{\text{an}}^*(k+1) = \frac{L}{T_s}[i_{\text{ref}}(k) - i_L(k)] + r_L i_L(k) + v_c(k) \tag{11.14}$$

Here, T_s denotes the switching period. For example, if $L = 1$ mH and the switching frequency is 20 kHz, the gain $K = L/T_s = 20$.

It is important to note that $i_L(k)$, being the inverter's output current, is an AC signal. Consequently, the reference current $i_{\text{ref}}(k)$ is also AC. The output command $V_{\text{an}}^*(k+1)$ in Eq. (11.14) is thus calculated by amplifying the current error through a proportional gain K and adding the instantaneous capacitor voltage $v_c(k)$. This formulation characterizes a P controller.

To determine the relationship between the controller output V_{an}^* and the duty cycle d_{fb}, the continuous-time state-space model previously introduced in Eq. (11.4) can be rewritten as

$$L\frac{di_L}{dt} + v_c + r_L i_L = V_{\text{dc}} d_{fb}$$

Recognizing the left-hand side as V_{an}^*, we obtain:

$$V_{\text{an}}^* = V_{\text{dc}} d_{fb} \implies d_{fb} = \frac{V_{\text{an}}^*}{V_{\text{dc}}} = \frac{K[i_{\text{ref}} - i_L] + r_L i_L + v_c}{V_{\text{dc}}} \tag{11.15}$$

The total duty input to the inverter is:

$$d = d_{ff} + d_{fb} = 0.5 + d_{fb} \tag{11.16}$$

Figure 11.10 illustrates the control structure for generating the a-phase duty signal d_a.

To generate the b-phase duty d_b, the same procedure is followed. In single-phase systems, the b-phase current is 180° out of phase with the a-phase current, i.e., $i_{Lb} = -i_{La}$. Therefore, the b-phase reference current is

$$i_{\text{ref},b} = -I_p \cos\theta$$

Likewise, the capacitor voltage polarity for the b-phase is inverted. Applying the equivalent circuit model to the b-phase gives

$$V_{\text{bn}}^* = -r_L i_L - L\frac{di_L}{dt} - v_c = -V_{\text{an}}^*.$$

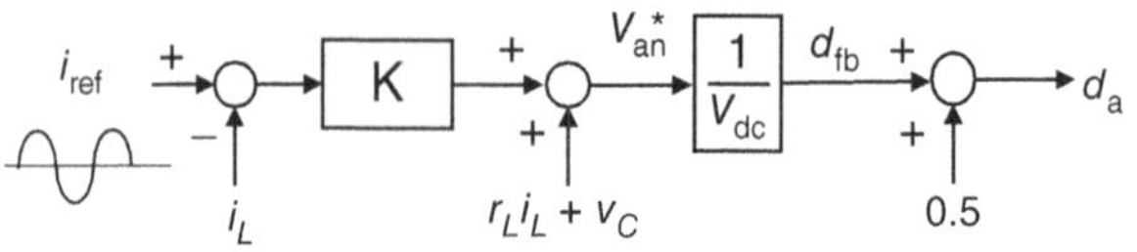

Figure 11.10 Current controller configuration for a-phase.

This results in the feedback duty for the b-phase being the negative of that for the a-phase

$$d_b = 0.5 - d_{fb} \tag{11.17}$$

The final duty expressions for each phase are

$$d_{fb} = \frac{K[i_{\text{ref}} - i_L] + r_L i_L + v_c}{V_{\text{dc}}}$$

$$d_a = 0.5 + d_{fb}, \quad d_b = 0.5 - d_{fb} \tag{11.18}$$

Figure 11.11 shows the PWM signal generation based on these duty values. The control signals d_a and d_b are compared with a carrier triangular waveform to generate the PWM signals. Assuming the carrier has an amplitude of 1 V (i.e., ranges from 0 to 1 V), both d_a and d_b must satisfy

$$0 < d_a,\ d_b < 1 \Longrightarrow -0.5 < d_{fb} < +0.5$$

This constraint reflects the behavior of the single-phase inverter operating as a voltage source inverter (VSI), which inherently performs voltage step-down operations. Consequently, the output voltage v_c cannot exceed the input voltage V_{dc}. Since the reference for each leg is biased around $V_{\text{dc}}/2$, the maximum possible magnitude of the output is $V_{\text{dc}}/2$. Assuming ideal current tracking (i.e., zero current error), the condition $-0.5 < d_{fb} < 0.5$ must always hold.

PSIM Simulation Using Continuous Time and Discrete-Time Controllers A single-phase inverter and its control circuit were configured and simulated using PSIM. Two types of controllers were implemented: a continuous-time analog controller and a discrete-time digital controller. The analog controller, composed of standard

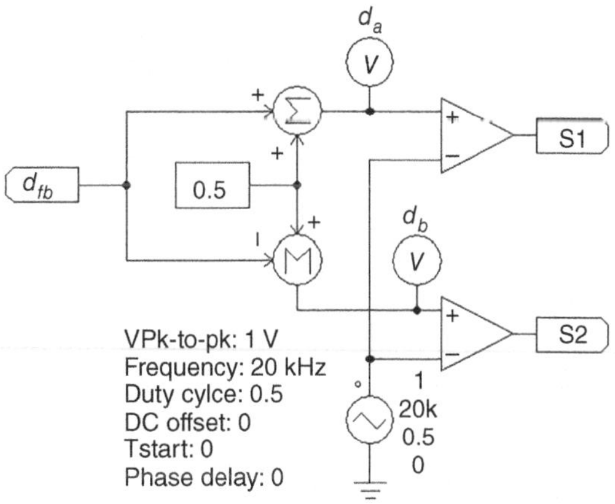

Figure 11.11 PWM configuration for generating switching signals.

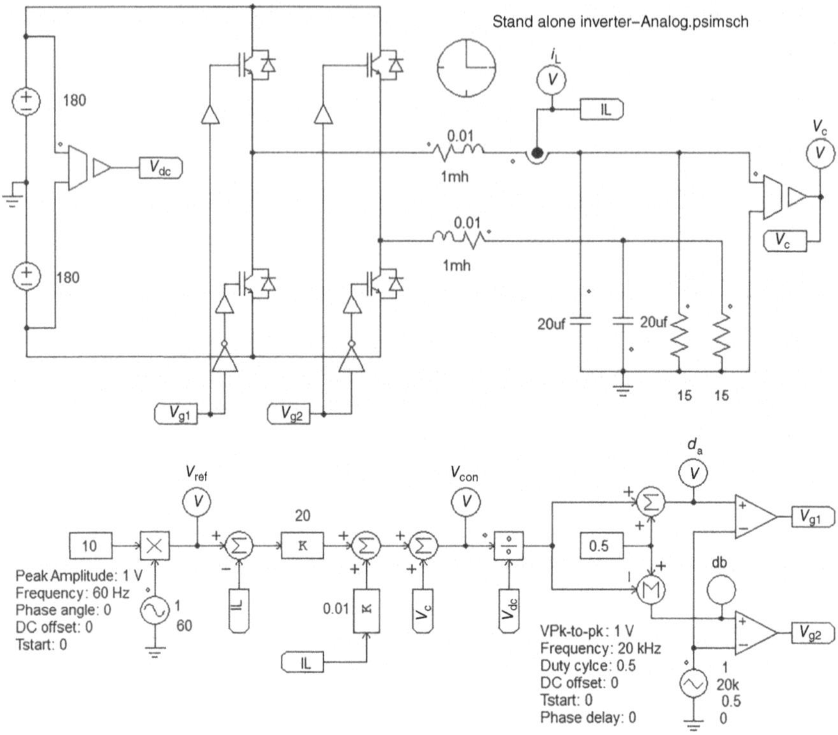

Figure 11.12 Analog controller simulation for stand-alone inverter.

circuit components, and its simulation results are shown in Figures 11.12 and 11.13. The discrete-time controller, implemented using a C-code block in conjunction with a zero-order holder (ZOH), is presented in Figures 11.14 and 11.15.

In the analog controller circuit shown in Figure 11.12, the control algorithm described in the previous section is implemented using functional blocks and arithmetic circuits. Key control variables—including the AC current reference I_{ref}, a-phase voltage command V_{an}^{*}, feedback duty d_{fb}, and phase duties d_a and d_b–are implemented according to the derived design equations.

The system incorporates a PWM generation circuit operating at a switching frequency of 20 kHz using a symmetrical triangular carrier waveform. As discussed in Chapter 1, a sawtooth carrier is typically employed for unidirectional converters, whereas a symmetrical triangular waveform is preferred for inverter PWM schemes to accommodate bipolar voltage and current characteristics.

The corresponding simulation results are shown in Figure 11.13. In subfigure (a), the full-cycle waveform is displayed, while subfigure (b) presents a magnified view over half a cycle. The inductor current i_L closely tracks the sinusoidal

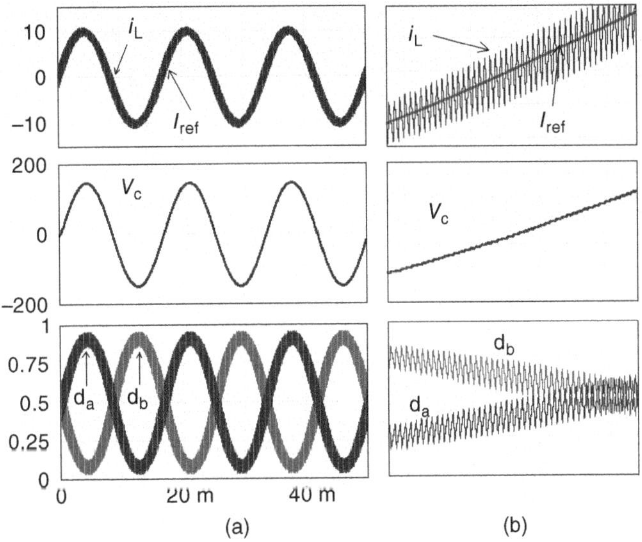

Figure 11.13 Simulation waveforms of current, voltage, and duty ratios. (a) Full time-scale. (b) Magnified view.

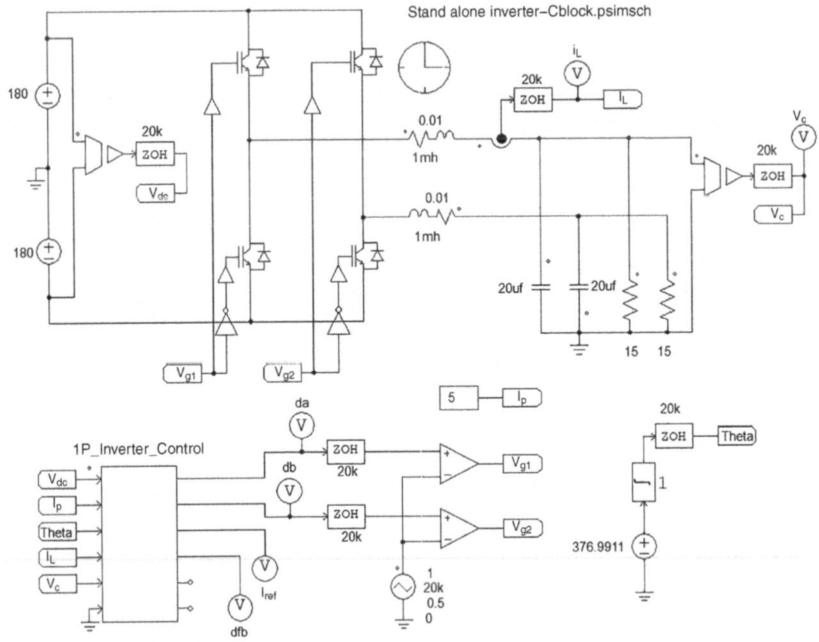

Figure 11.14 C-block schematic for single-phase stand-alone inverter.

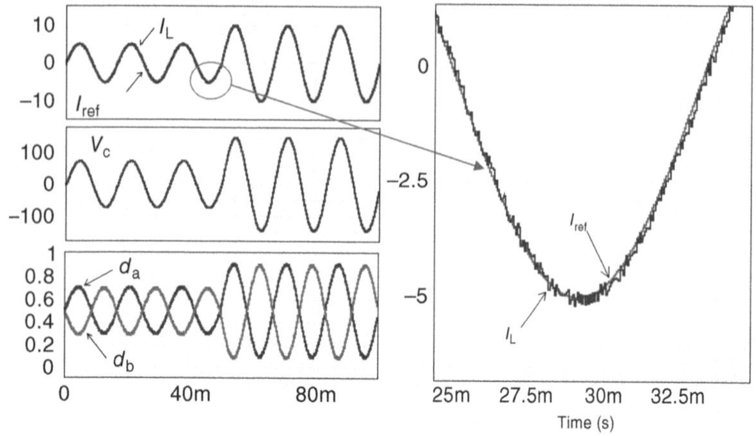

Figure 11.15 Stand-alone inverter simulation using C-block controller. (a) Full-cycle waveform and (b) Magnified waveform.

reference I_{ref}, demonstrating good current regulation. The duty signals d_a and d_b exhibit symmetrical behavior centered around 0.5. The output voltage v_c, developed across a 15 Ω load resistor, reaches an AC amplitude of 150 V. Since the DC bus voltage is 180 V, the peak duty ratio is calculated as 150 V/180 V = 0.83, which matches the observed simulation results.

When implementing a controller in hardware, it is typically realized in a microcontroller (e.g., DSP) via C-code. To emulate this structure, the controller was also implemented using PSIM's C-block environment with discrete-time execution. Figure 11.14 illustrates the C-block schematic, while the ZOH block ensures that analog input signals—such as voltage, current, and angle—are sampled at fixed intervals. Typically, the sampling period matches the switching period of the triangular PWM carrier, ensuring synchronization among the sampling, switching, and C-code execution cycles. This approach effectively replicates the behavior of an interrupt service routine (ISR) in a real-time embedded system.

The C-block receives inputs such as the DC voltage, current magnitude I_p, angular frequency ($\theta = \omega t = 2\pi f t = 377t$), inductor current i_L, and output capacitor voltage v_c. The computed outputs are the phase duties d_a and d_b. Additional outputs, such as I_{ref} and d_{fb}, are included for monitoring intermediate variables and control behavior. In practical implementations, d_a and d_b are not directly used to generate PWM signals. Instead, they are written to the ePWM (enhanced PWM) module registers inside the microcontroller after being scaled by the PWM period, thereby generating hardware PWM signals identical to those in the simulation. The angular frequency θ, while shown here as a simulation circuit using a ZOH, is typically generated within a PLL routine in embedded applications.

Table 11.2 C-block code: 1P_Inverter_Control.

```
double Vdc, Ip, Theta, IL, Vc;
double duty_a,duty_b, dfb;
double  Iref, Ierr, Vcon;

/****** start here ******/
Vdc   = in[0];
Ip    = in[1];
Theta = in[2];
IL    = in[3];
Vc    = in[4];

if (t>0.05) Ip=2*Ip;
Iref  = Ip*sin(Theta);
Ierr  = Iref - IL;

Vcon =20*Ierr + 0.01*IL +Vc;
dfb = Vcon/Vdc;

if (dfb > 0.5) dfb = 0.5;
if (dfb <-0.5) dfb =-0.5;

duty_a = 0.5 + dfb;
duty_b = 0.5 - dfb;

out[0]  = duty_a;
out[1]  = duty_b;
out[2]  = Iref;
out[3]  = dfb;
```

The corresponding C-code for the digital controller is shown in Table 11.2. It implements the same control logic used in the analog simulation but with an increased current reference amplitude (from 5 to 10 A) to better demonstrate the controller's tracking capability.

Figure 11.15 presents the simulation results from the C-block-based controller. The first waveform (a) shows the inductor current i_L and the reference current I_{ref}, the second shows the output voltage v_c, and the third shows the PWM duties d_a and d_b. Subfigure (b) provides a magnified view. As demonstrated, the dynamic behavior of the system with the C-block controller matches that of the analog implementation, confirming the correctness of the digital realization.

11.3.2 Current Controller in Grid-Tied Inverter (PCS)

A PCS refers to a configuration in which the inverter output is directly connected to the utility grid. In this system, the output capacitor voltage v_c used in a stand-alone inverter is effectively replaced by the grid voltage v_{grid}. Since the inverter operates in a CVCF environment dictated by the grid, voltage regulation via feedback control is not feasible. As a result, only current control is applicable.

In typical PCS operation, the output current of the inverter is regulated to remain in phase with the grid voltage, thereby achieving unity power factor operation (i.e., P.F. = 1). The primary role of the current controller is to ensure that the output inductor current closely tracks the desired reference current.

Figure 11.16 illustrates the single-phase PCS configuration. In this topology, the grid voltage v_{grid} is modeled as being evenly split across the inverter's a-phase

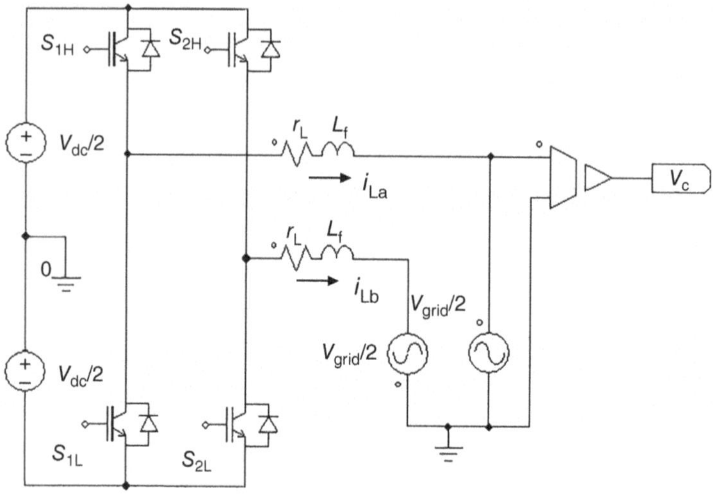

Figure 11.16 Single-phase PCS circuit configuration.

and b-phase legs. The representation is analogous to that of the stand-alone inverter discussed in the previous chapter. For example, if the DC input voltage is $V_{dc} = 360\ V$ and the grid voltage is $v_{grid} = 220\ V_{ac}$, then the split input and output voltages are $V_{dc}/2 = 180\ V$ and $v_{grid}/2 = 110\ V_{ac}$. The peak value of $110\ V_{ac}$ is approximately 155 V, meaning that the inverter in PCS mode functions as a step-down VSI, converting 180 V_{DC} to an AC output of 155 V peak.

It is important to emphasize that in this modeling approach, the stand-alone inverter output voltage v_c corresponds to half the grid voltage. Although the line voltage is 220 V_{ac} in reality, it is represented as 110 V_{ac} in the simulation to match the midpoint-based reference used in single-phase full-bridge inverters.

Figure 11.17 shows the same PCS circuit implemented with a C-block in PSIM. The only structural difference compared to the stand-alone inverter is that the output capacitor is replaced with a voltage source representing the grid. The control algorithm implemented in the C block remains unchanged.

The simulation waveforms are shown in Figure 11.18. The inductor current closely follows the reference signal, confirming effective current regulation. An important observation concerns the duty ratios d_a and $d_b$when the reference current is increased from 5 to 10 A. In the PCS simulation, these duty ratios change only slightly. In contrast, in the stand-alone inverter case presented earlier in Figure 11.15, the duty ratios increase by approximately a factor of two.

This difference stems from the computation of the current command voltage V^*_{an}, which includes the sum of the proportional term and the output voltage. In the stand-alone inverter configuration, the output voltage v_c is dynamically controlled

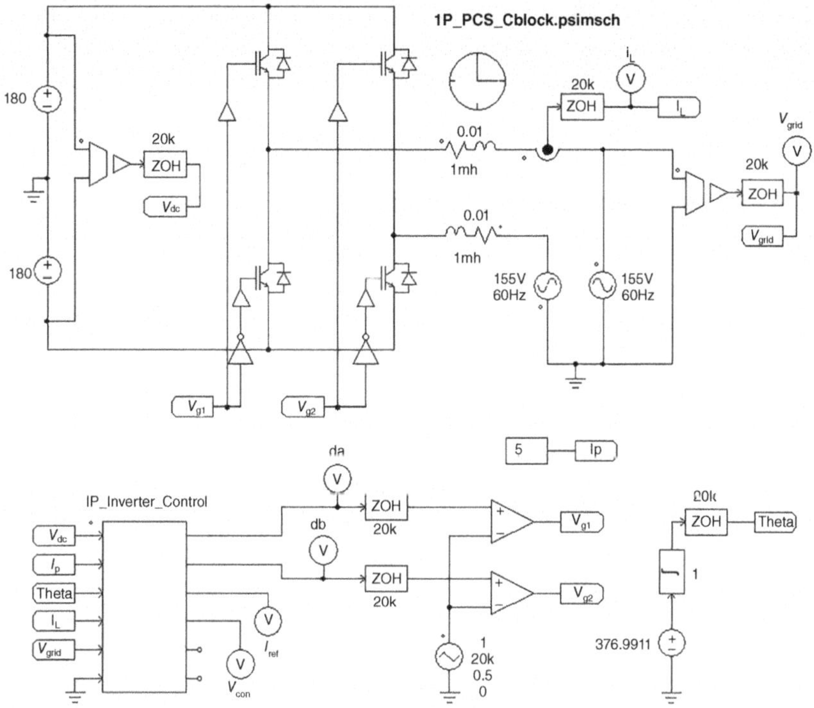

Figure 11.17 Single-phase PCS C-block schematic.

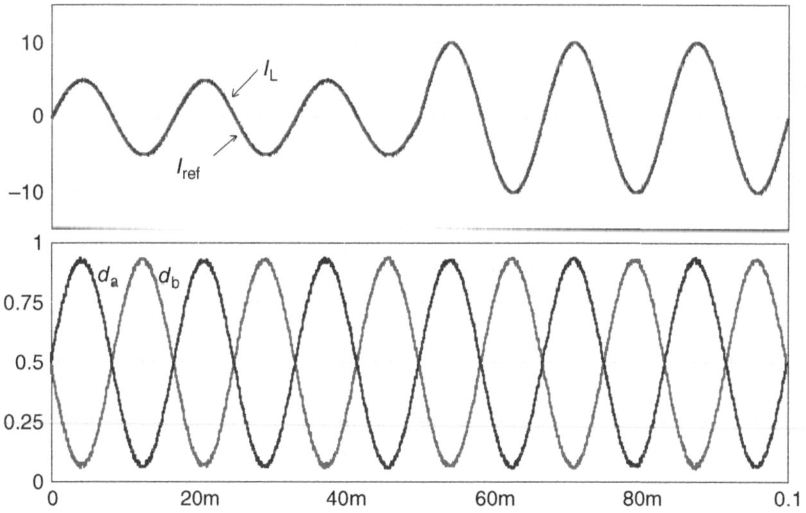

Figure 11.18 PCS simulation (C-block): waveforms of inductor current and duty ratio.

and varies with load conditions. By contrast, in the PCS configuration, the output voltage is fixed by the grid and remains constant. Consequently, changes in the reference current have a much smaller effect on the resulting duty cycle in PCS operation.

11.3.3 PWM Converter (Rectifier) Design

A system that converts a DC input into an AC output is referred to as an inverter. If the output is connected to the grid, the system is classified as a PCS. If the inverter operates independently from the grid, it is categorized as a stand-alone inverter. Conversely, when the system receives an AC input from the grid and outputs DC power—typically to a capacitor or battery—it is known as an AC/DC converter, more specifically, a PWM converter or rectifier.

In a typical PCS configuration, a DC power source such as a battery is connected at the input, while the AC grid is connected at the output. This system has power sources on both sides—DC at the input and AC at the output—allowing bidirectional energy flow. The roles of input and output can be reversed depending on the control strategy applied.

To understand how power flows in such systems, consider the two cases shown in Figure 11.19.

In both cases, power flows from the side with a higher voltage (V_1) to the side with a lower voltage (V_2). The current entering the power source terminals (V_1, V_2) is defined as positive, a convention adopted to emphasize the perspective of power being delivered to a load. In power electronics, the load-side (the controlled side) is typically where voltage or power is regulated. Accordingly, power received by the load-side is considered positive, and power supplied by the source-side is considered negative.

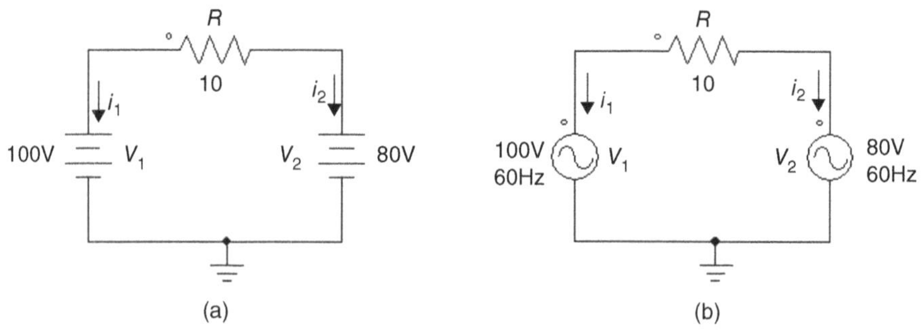

Figure 11.19 Energy flow circuit between two power sources. (a) DC case and (b) AC cases.

For the DC case, assume $V_1 = 100\,\text{V}$ and $i_1 = -2\,\text{A}$. Then,

- Power at the input: $P_1 = V_1 \cdot i_1 = 100 \cdot (-2) = -200\,\text{W}$
- Power at the output: $P_2 = V_2 \cdot i_2 = 80 \cdot (2) = 160\,\text{W}$
- Power loss in the resistor: $P_R = i^2 R = 40[\text{W}]$

This satisfies the power balance equation: $P_1 = P_2 + P_R$. In Figure 11.20(a), the constant DC current results in a constant DC power flow, making the direction of energy transfer easy to interpret.

In contrast, in the AC case, both voltage and current are sinusoidal. The alternating nature of AC makes it difficult to determine the direction of power flow based solely on instantaneous current polarity. However, the phase relationship between voltage and current provides meaningful insight. As shown in Figure 11.20(b):

- On the source side (V_1 and i_1), voltage and current are out of phase, resulting in a power waveform oscillating from −200 to 0 W.
- On the load side (V_2 and i_2), voltage and current are in phase, producing a power waveform ranging from 0 to 160 W.

The phase information allows us to distinguish between power source and load, despite the sinusoidal nature of AC signals. In summary, PWM converters enable bidirectional power control in AC/DC systems, and analyzing power flow—especially in AC cases—requires careful examination of phase relationships between voltage and current.

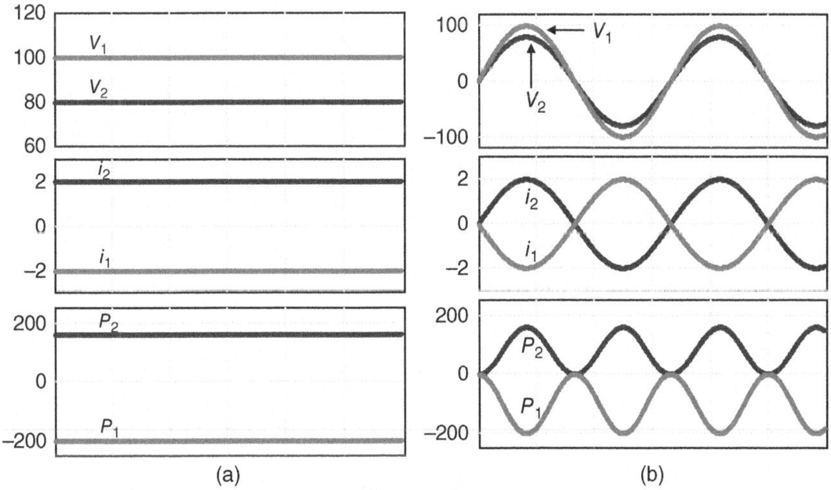

Figure 11.20 Power flow waveform. (a) DC power source waveform and (b) AC power source waveform.

PCS and Rectifier Operation Using a Single-Phase Inverter A single-phase inverter can operate either as a PCS or as a rectifier, depending solely on the control method—without any hardware modification. In PCS (inverter) mode, the input is a DC source V_{dc}, and the output is the AC grid voltage V_{grid}; thus, power flows from DC to AC. Conversely, in rectifier mode, the input is V_{grid} and the output is V_{dc}, meaning that power flows from AC to DC.

As shown in Figure 11.20(b), in PCS mode, the inductor current flows *into* the grid and is in phase with the grid voltage. In contrast, in rectifier mode, the inductor current flows *out of* the grid and is out of phase with the grid voltage.

Figure 11.21 shows the schematic diagram of a bidirectional PCS/rectifier circuit. By analyzing the a-phase leg, it can be observed that:

- In PCS mode, the circuit operates fundamentally as a buck converter.
- In rectifier mode, it behaves as a boost converter.

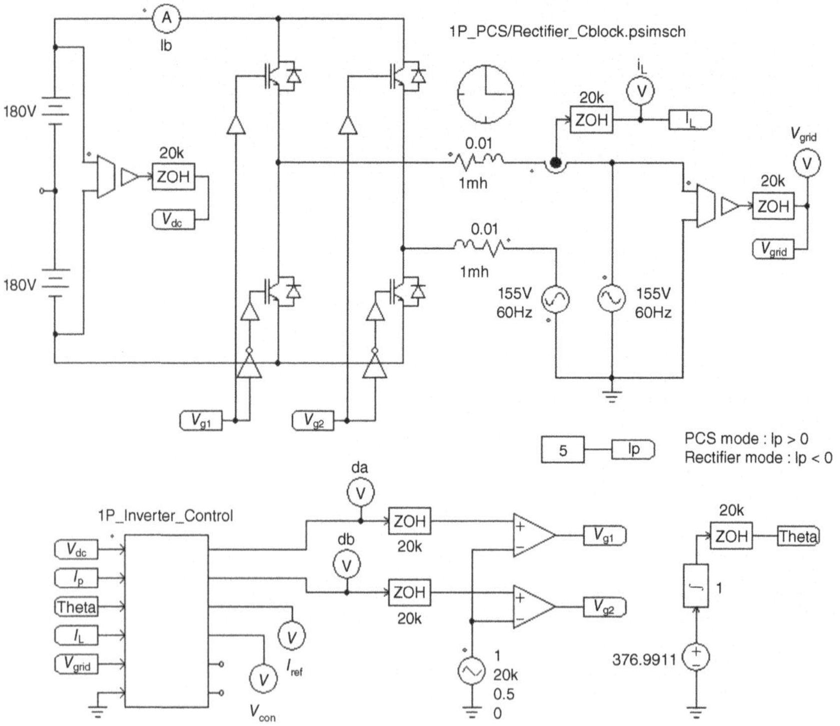

Figure 11.21 Schematic of PCS (Rectifier) circuit.

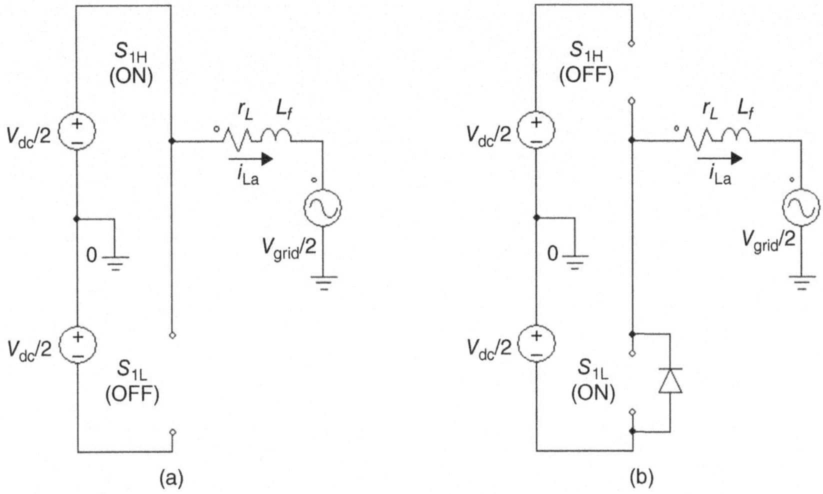

Figure 11.22 Buck converter behavior in PCS mode. (a) Switch ON and (b) Switch OFF.

PCS Mode: Buck Converter Operation In PCS mode, the converter steps down the voltage; thus, the input voltage V_{dc} must be greater than the peak of V_{grid}. The inductor current flows into the grid, maintaining an in-phase condition with the grid voltage (Figure 11.22).

Since it operates as a buck converter, the voltage is stepped down, and the input voltage (V_{dc}) is greater than the peak value of the grid voltage (V_{grid}). In PCS mode, the inductor current flows into the grid voltage (V_{grid}), maintaining an in-phase relationship.

Rectifier Mode: Boost Converter Operation In rectifier mode, the converter boosts the voltage, with V_{grid} as the input and V_{dc} as the output. Consequently, V_{dc} must be higher than the peak value of V_{grid}. The inductor current flows out of the grid, leading to an out-of-phase condition between voltage and current (Figure 11.23).

Although the current waveforms differ in phase, the underlying converter topology and operation remain identical. Therefore, mode switching can be accomplished by adjusting the sign of the current amplitude in the reference signal.

The reference current is typically defined as:

$$i_{ref} = I_p \sin\theta, \quad I_P > 0 \ \text{(PCS mode)}, \quad I_P < 0 \ \text{(Rectifier mode)}$$

In PCS mode, since the inverter output is tied to the grid, voltage control is not feasible, and only current control is implemented. Similarly, in rectifier mode, if the output is connected to a stiff DC source such as a battery or regulated voltage V_{dc}, voltage regulation is again not possible, and only the current is controlled.

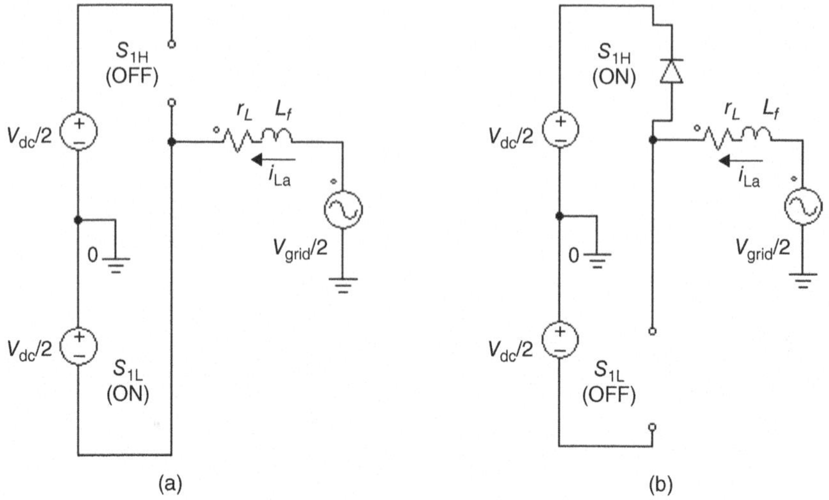

Figure 11.23 Boost converter behavior in Rectifier mode. (a) Switch ON and (b) Switch OFF.

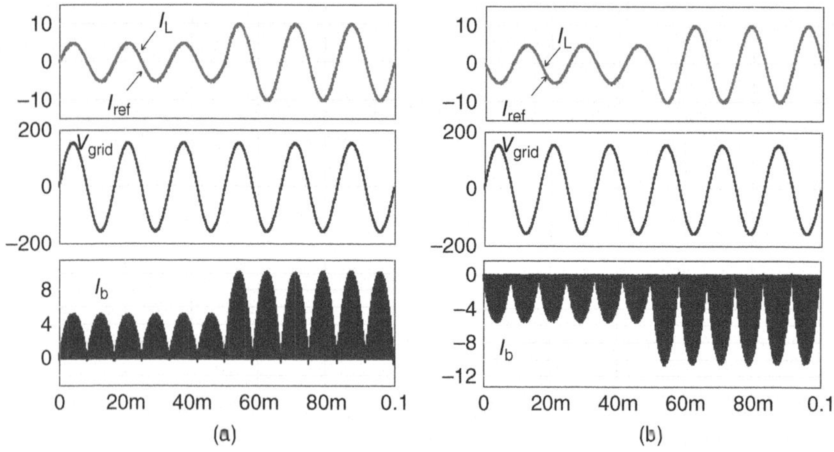

Figure 11.24 Inductor current control waveforms for PCS and Rectifier modes. (a) PCS mode ($I_P = 5$ A) and (b) Rectifier mode ($I_P = -5$ A).

Thus, the operating mode is entirely determined by the phase of the reference current, while the hardware and controller structure remain unchanged.

This control strategy simplifies the system design, allowing a unified controller structure for both PCS and rectifier operation, with mode switching implemented solely by altering the sign of I_P. The simulated inductor current waveforms for each mode are shown in Figure 11.24.

Voltage Control in Rectifier Mode When the output of a rectifier is not maintained by a stiff voltage source but instead consists of a capacitor and a resistive load, active *output voltage regulation* becomes feasible. Figure 11.25 illustrates the schematic of the single-phase rectifier circuit. The objective of the controller is twofold: (i) to regulate the DC output voltage according to the design specification, and (ii) to maintain grid current in synchronism with the grid voltage, thereby achieving unity power factor. Please note that each capacitor in the series connection should have a capacitance of $2C$ so that the resulting equivalent capacitance becomes C.

In voltage-control operation (Figure 11.26), a double-loop control structure is employed. The inner loop implements the same current control strategy as

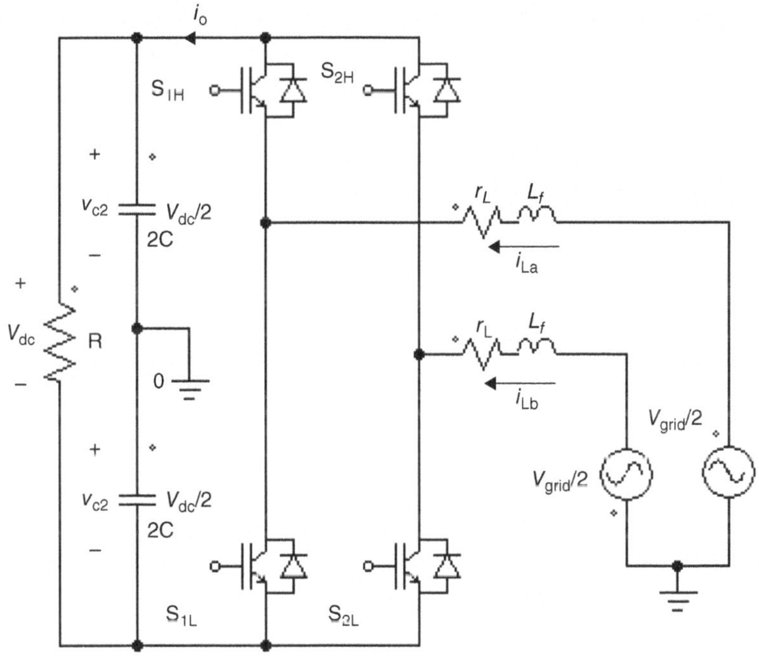

Figure 11.25 Schematic diagram of the single-phase rectifier circuit.

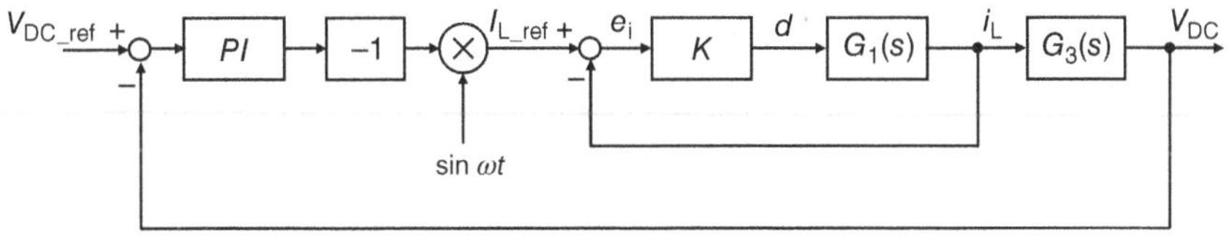

Figure 11.26 Double-loop controller configuration for rectifier mode.

used in PCS operation, while the outer loop is a *PI* controller that regulates the output DC voltage. The *PI* controller generates the current amplitude command, denoted as I_P.

This current amplitude reference is multiplied by −1 before entering the current controller. The negative sign compensates for the direction of current flow, ensuring consistency with PCS operation. The resulting current amplitude is then modulated by the sinusoidal function of the grid angle, producing the time-varying current reference:

$$I_{\mathrm{L\_ref}} = I_P \sin \omega t$$

This reference current serves as the input to the inner-loop current controller. In a single-phase rectifier, the DC output voltage inherently contains a 120-Hz ripple—twice the grid frequency—which necessitates careful tuning of the PI controller. A practical and effective method is to employ the DQ transformation, which converts sinusoidal variables into DC quantities in the synchronous reference frame, thereby simplifying the selection of PI gains. The detailed procedure and design considerations will be presented in Chapter 14.

In this chapter, since the operation of the rectifier is analogous to that of a boost converter, the same design procedure and MATLAB m-files used for PI tuning of the boost converter can be applied directly. This approach enables precise determination of the PI gains and ensures that, despite the presence of the 120-Hz ripple, the average steady-state DC output voltage converges accurately to the reference value.

Small-Signal Modeling of the Rectifier

Step 1. Current Equation

For the a-phase current, the state equation depends on the switching state of the upper device S_{1H} and the lower device S_{1L}.

- High ON(S_{1H}): with duty ratio d_a

$$\frac{V_{\mathrm{grid}}}{2} = r_L i_{La} + L\frac{di_{La}}{dt} + \frac{v_{\mathrm{dc}}}{2}$$

- Low ON(S_{1L}): with duty ratio (1-d_a)

$$\frac{V_{\mathrm{grid}}}{2} = r_L i_{La} + L\frac{di_{La}}{dt} - \frac{v_{\mathrm{dc}}}{2}$$

Averaging these two conditions yields

$$\frac{di_{La}}{dt} = -\frac{r_L}{L} i_{La} - \frac{v_{\mathrm{dc}}}{2L}(2d_a - 1) + \frac{V_{\mathrm{grid}}}{2L}$$

Step 2. Voltage Equation

The DC-link voltage dynamics are expressed as

$$\frac{dv_{\mathrm{dc}}}{dt} = \frac{1}{C}\left(i_o - \frac{v_{\mathrm{dc}}}{R}\right)$$

where the output current is

$$i_o = (d_a i_{La} + d_b i_{Lb})$$

For the single-phase rectifier, this expression simplifies to

$$i_o = d_a i_{La} - (1 - d_a) i_{La} = (2d_a - 1) i_{La}$$

Thus, the voltage equation becomes

$$\frac{dv_{\mathrm{dc}}}{dt} = \frac{1}{C}(2d_a - 1) i_{La} - \frac{v_{\mathrm{dc}}}{RC}$$

Step 3. State-Averaged Equations

The resulting equations form a nonlinear system. To simplify, the duty ratio is decomposed into

$$d_a = d_{fb} + d_{ff} \quad \Rightarrow \quad d_{ff} = 0.5$$

Substituting yields the state-averaged model

$$\frac{d}{dt} i_{La} = -\frac{r_L}{L} i_{La} - \frac{d_{fb}}{L}(v_{\mathrm{dc}}) + \frac{V_{\mathrm{grid}}}{2L}$$

$$\frac{d}{dt} v_{\mathrm{dc}} = \frac{2}{C} \mathrm{d}_{fb} \cdot i_{La} - \frac{v_{\mathrm{dc}}}{RC}$$

Step 4. Small-Signal Linearization

To derive the small-signal model, perturbations are applied:

$$i_{La} = I_L + \hat{i}_{La}, \quad d_{d,fb} = D_a + \hat{d}_{fb}, \quad v_{\mathrm{dc}} = V_{\mathrm{DC}} + \hat{v}_{\mathrm{dc}}$$

Current equation:

$$\frac{d}{dt}(I_L + \hat{i}_{La}) = -\frac{r_L}{L}(I_L + \hat{i}_{La}) - \frac{(D_a + \hat{d}_{fb})}{L}(V_{\mathrm{DC}} + \hat{v}_{\mathrm{dc}}) + \left(\frac{V_{\mathrm{grid}}}{2L}\right)$$

$$\mathrm{DC} \Rightarrow 0 = -\frac{r_L}{L} I_L + \frac{1}{L} D_a V_{\mathrm{DC}} - \frac{V_{\mathrm{grid}}}{2L} \rightarrow D_a = \frac{V_{\mathrm{grid}}}{2V_{\mathrm{DC}}}$$

$$\mathrm{AC} \Rightarrow \frac{d}{dt}(\hat{i}_{La}) = -\frac{r_L}{L}(\hat{i}_{La}) - \frac{D_a}{L}(\hat{v}_{dc}) - \frac{V_{DC}}{L}(\hat{d}_{fb})$$

Voltage equation:

$$\frac{d}{dt}(V_{\mathrm{DC}} + \hat{v}_{\mathrm{dc}}) = \frac{2}{C}(D_a + \hat{d}_{fb})(I_L + \hat{i}_{La}) - \frac{1}{\mathrm{RC}}(V_{\mathrm{DC}} + \hat{v}_{\mathrm{dc}})$$

$$\text{DC} \Rightarrow 0 = \frac{2}{C} D_a I_L - \frac{V_{\text{DC}}}{\text{RC}} \quad \rightarrow I_L = \frac{V_{\text{DC}}}{2RD_a}$$

$$\text{AC} \Rightarrow \frac{\text{d}}{dt}(\widehat{v}_{\text{dc}}) = \frac{2}{C}(D_a \cdot \widehat{i}_{La} + I_L \cdot \widehat{d}_{fb}) - \frac{\widehat{v}_{\text{dc}}}{\text{RC}}$$

Step 5. Final Small-Signal Model

In compact matrix form, the small-signal model is:

$$\begin{pmatrix} \dfrac{\text{d}(\widehat{i}_{La})}{dt} \\ \dfrac{\text{d}(\widehat{v}_{\text{dc}})}{dt} \end{pmatrix} = \begin{pmatrix} -\dfrac{r_L}{L} & -\dfrac{D_a}{L} \\ \dfrac{2D_a}{C} & -\dfrac{1}{\text{RC}} \end{pmatrix} \begin{pmatrix} \widehat{i}_{La} \\ \widehat{v}_{\text{dc}} \end{pmatrix} + \begin{pmatrix} -\dfrac{V_{\text{DC}}}{L} \\ \dfrac{2I_L}{C} \end{pmatrix} \widehat{d}_{fb} \tag{11.19}$$

with steady-state values:

$$V_{\text{DC}} = \frac{V_{\text{grid}}}{2D_a}, \quad I_L = \frac{V_{\text{DC}}}{2RD_a}$$

The transfer function is derived using the MATLAB built-in function `ss2tf`, which converts the state-space model into its equivalent transfer function representation. The MATLAB m-file developed for this design is presented in Table 11.3.

Table 11.3 MATLAB m-file for Rectifier voltage controller.

```
%-- 1-Phase PFC Rectifier  Design--%
%% Current direction : Grid -> DC
s = tf('s');
Vgrid = 310;
L = 1.0e-3;
C = 2000e-6;
R = 50;
rL = 0.01;

Da = 0.4;
VDC = Vgrid/(2*Da);
IL =  VDC/(2*R*Da);

A = [-rL/L,    -Da/L;
     2*Da/C,-1/(R*C)];
B = [-VDC/L;
     2*IL/C];
C1 = [1,0]; C2 = [0,1]; Dx = 0;
[num1,den1] = ss2tf(A,B,C1,Dx);
[num2,den2] = ss2tf(A,B,C2,Dx);
G1 = tf(num1, den1);
G2 = tf(num2, den2);
G3 = tf(num2,num1);

%% double loop controller
Vref = 450;
Kpv = 1; Kiv = 1;
Gcv = Kpv*(s+Kiv)/s;
Kpi = 1; Kii = 1;
Gci = Kpi*(s+Kii)/s;

T=sisoinit(6);
T.F.Value = Vref;
T.G1.Value = G1;
T.C2.Value = Gci;
T.G2.Value = G3;
T.C1.Value = Gcv;
sisotool(T);
```

Controller Design and Verification The design specifications for the double controllers are presented in Table 11.4.

Figure 11.27 illustrates the SISOTOOL auto-tuning results. Subfigure (a) presents the voltage loop Bode plot, while subfigure (b) shows the current loop response. The design successfully satisfies the target specifications—100 rad/s bandwidth with a 90° phase margin for the voltage loop, and 2400 rad/s bandwidth with a 60° phase margin for the current loop.

The controller gains obtained from auto-tuning, both in continuous- and discrete-time forms, are listed in Table 11.5. The discrete-time designs are based on $f_{sw} = 20$ kHz.

Figure 11.28 confirms that the closed-loop response meets the design specifications, with zero steady-state error and a rise time of approximately 22 ms.

PSIM Simulation Figure 11.29 presents the PSIM schematic of the rectifier voltage control circuit. The output of the voltage controller corresponds to the current

Table 11.4 Design specification for the double-loop controller of the rectifier.

Design specification	Voltage controller (C_1)	Current controller (C_2)
Bandwidth [rad/sec]	100	2400
Phase margin [deg]	90	60

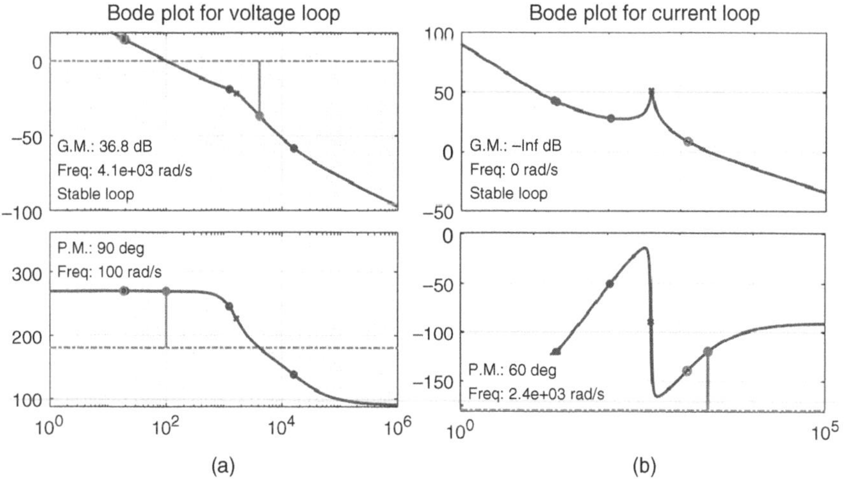

Figure 11.27 Bode plots of the designed controllers. (a) Voltage controller and (b) Current controller.

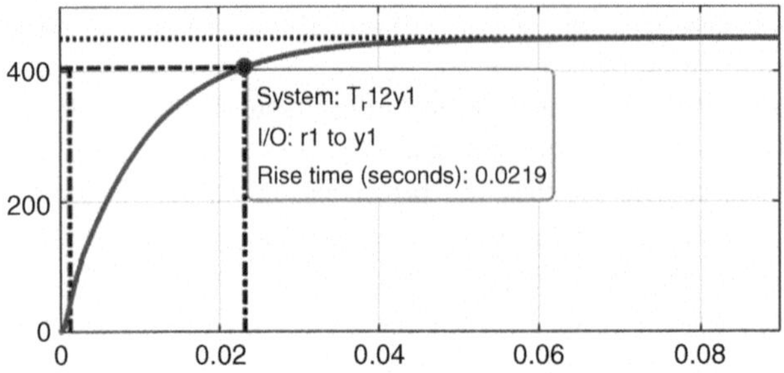

Figure 11.28 Step response of the designed controllers.

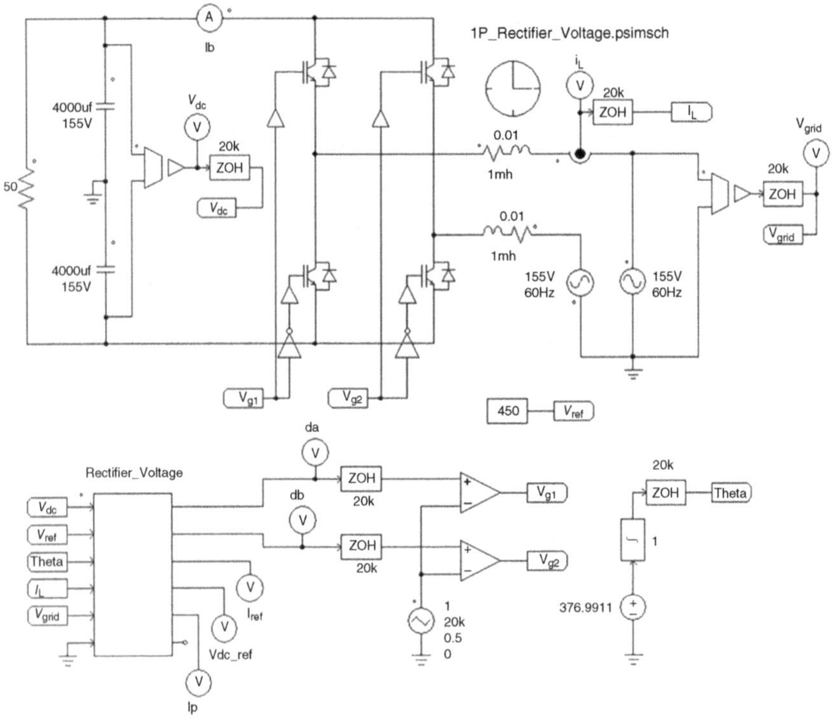

Figure 11.29 Schematic diagram of the rectifier voltage control circuit.

Table 11.5 Controller gains in continuous-time and discrete-time formats.

	Voltage controller (C_1)	Current controller (C_2)
Continuous time	$\frac{0.2648(s+18.6)}{s}$	$-\frac{0.00537(s+1248)}{s}$
Discrete time	$\frac{0.2669(z-0.999)}{z-1}$	$-\frac{0.00569(z-0.948)}{z-1}$

Table 11.6 C-block code: Rectifier_Voltage.

```
#define Kp 0.2669
#define Ki 0.999
double Vdc,  Theta, IL, Vc;
double duty_a,duty_b,Ip,Iref;
double Ierr,Vcon,Vdc_ref,Verr;
static double Verr_old =0.0;
static double Ip_old=0.0;
/****** start here ******/
Vdc     = in[0];
Vdc_ref = in[1];
Theta   = in[2];
IL      = in[3];
Vc      = in[4];

if   (t>0.4)          Vdc_ref=500;
else  if (t>0.2) Vdc_ref=550;
if (Vdc < 0.01)  Vdc = 0.01;
/**** Voltage Controller  ***/
Verr =  (Vdc_ref - Vdc);
Ip=Ip_old+Kp*Verr-Kp*Ki*Verr_old;
Ip_old=Ip;
Verr_old=Verr;

/*Same as Current Controller */
Iref  = -Ip*sin(Theta);
Ierr  = Iref  - IL;
Vcon =20*Ierr + 0.01*IL +Vc;

if (Vcon >  Vdc/2) Vcon= Vdc/2;
if (Vcon < -Vdc/2) Vcon=-Vdc/2;

duty_a = 0.5 + Vcon/Vdc;
duty_b = 0.5 - Vcon/Vdc;

out[0] = duty_a;
out[1] = duty_b;
out[2] = Iref;
out[3] = Vdc_ref;
```

amplitude I_P, which is used to generate the sinusoidal reference current expressed as $I_{\text{ref}} = I_P \sin \omega t$.

The C-code implementation of the voltage controller is given in Table 11.6. It employs a standard PI structure, consistent with the PCS current controller design.

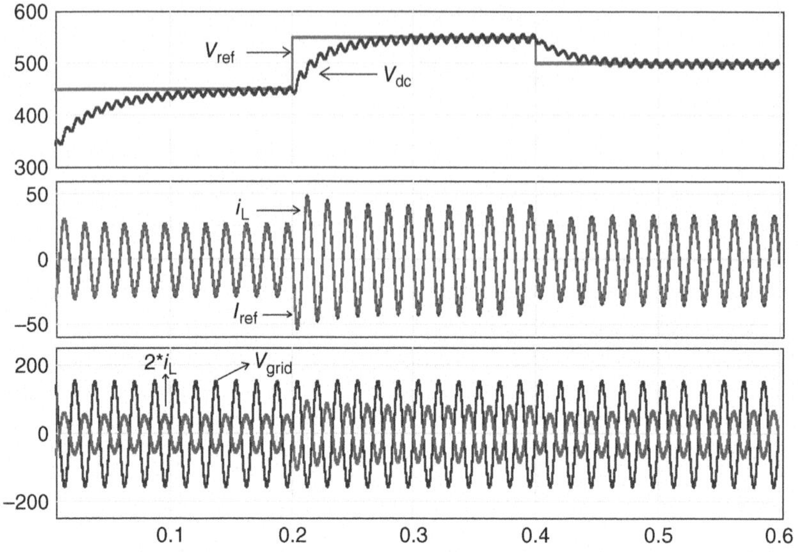

Figure 11.30 Simulation waveforms of the rectifier voltage control.

As shown in Figure 11.30, the DC-link voltage, despite containing a 120 Hz ripple component, is effectively regulated to the reference value. This result validates the proposed controller and confirms its ability to satisfy the performance requirements under sinusoidal grid conditions. In addition, due to the reversed orientation of the current sensor, the grid current is observed to flow in the opposite direction, that is, away from the grid voltage source.

12

Three-Phase Inverter

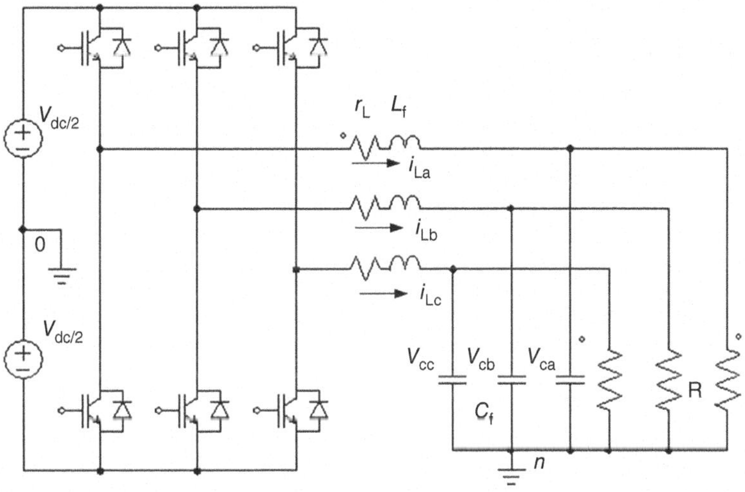

Three-Phase Stand-Alone Inverter.

12.1 State-Space Averaged Equations for a Stand-Alone System

Figure 12.1 shows the circuit configuration of a three-phase inverter, modeled as a three-phase four-wire system that includes the neutral point. For systems using line-to-line voltages (e.g., in a three-wire configuration), it becomes necessary to compute the neutral point voltage in order to derive the corresponding phase voltages.

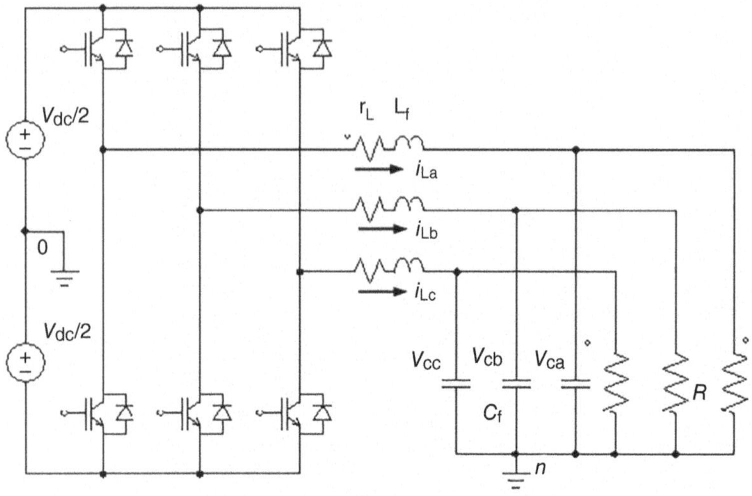

Figure 12.1 Configuration of a three-phase stand-alone inverter circuit.

For a balanced system, the transformation relationships between line-to-line voltages v_{ab}, v_{bc}, v_{ca} and phase voltages v_{an}, v_{bn}, v_{cn} are given as:

$$
\begin{aligned}
v_{bn} &= (v_{bc} - v_{ab})/3 \\
v_{an} &= v_{bn} + v_{ab} \\
v_{ca} &= v_{ab} - 3v_{an} \\
v_{cn} &= (v_{ca} - v_{bc})/3
\end{aligned}
\tag{12.1}
$$

Each inverter leg (a-phase, b-phase, and c-phase) operates independently, with a 120° phase shift between them. The output currents i_{La}, i_{Lb}, i_{Lc} have identical magnitudes but differ in phase, reflecting the balanced nature of the three-phase system. In this analysis, we derive the state-space averaged equations for a balanced three-phase inverter. In the circuit, the capacitor voltages v_{ca}, v_{cb}, v_{cc} are equivalent to the phase voltages v_{an}, v_{bn}, v_{cn}, respectively.

State Equations for a-Phase The state equation of the a-phase circuit depends on the ON/OFF status of the switching device:

- Switch ON:

$$\frac{V_{dc}}{2} = r_L i_a + L\frac{di_a}{dt} + v_{ca}, \qquad i_a = C\frac{dv_{ca}}{dt} + \frac{v_{ca}}{R}$$

- Switch OFF:

$$-\frac{V_{dc}}{2} = r_L i_a + L\frac{di_a}{dt} + v_{ca}, \qquad i_a = C\frac{dv_{ca}}{dt} + \frac{v_{ca}}{R}$$

To derive the state-averaged equations, the switching duty ratio d is introduced. Averaging the ON and OFF states over one switching period yields

$$L\frac{di_a}{dt} = \left(\frac{V_{dc}}{2} - v_{ca} - r_L i_a\right) d + \left(-\frac{V_{dc}}{2} - v_{ca} - r_L i_a\right)(1-d)$$
$$= V_{dc} d - \frac{V_{dc}}{2} - v_{ca} - r_L i_a$$

$$C\frac{dv_{ca}}{dt} = i_a - \frac{v_{ca}}{R}$$

The state-averaged equation for the a-phase of a three-phase inverter is expressed as follows.

$$\begin{pmatrix} \frac{di_a}{dt} \\ \frac{dv_{ca}}{dt} \end{pmatrix} = \begin{pmatrix} -\frac{r_L}{L} & -\frac{1}{L} \\ \frac{1}{C} & -\frac{1}{RC} \end{pmatrix} \begin{pmatrix} i_a \\ v_{Ca} \end{pmatrix} + \begin{pmatrix} \frac{V_{dc}}{L} \\ 0 \end{pmatrix} d + \begin{pmatrix} -\frac{1}{2L} V_{dc} \\ 0 \end{pmatrix} \tag{12.2}$$

In compact matrix form, it becomes $\dot{X} = AX + BU + F$.

$$\dot{X} = \begin{pmatrix} \frac{di_a}{dt} \\ \frac{dv_{ca}}{dt} \end{pmatrix}, \quad \mathrm{A} = \begin{pmatrix} -\frac{r_L}{L} & -\frac{1}{L} \\ \frac{1}{C} & -\frac{1}{RC} \end{pmatrix}, \quad \mathrm{B} = \begin{pmatrix} \frac{V_{dc}}{L} \\ 0 \end{pmatrix}, \quad F = -\frac{1}{2}\begin{pmatrix} \frac{V_{dc}}{L} \\ 0 \end{pmatrix}, \quad \mathrm{U} = d$$

The complete state-averaged model for the three-phase inverter (a, b, c phases) is expressed as

$$\begin{pmatrix} \frac{di_{abc}}{dt} \\ \frac{dv_{abc}}{dt} \end{pmatrix} = \begin{pmatrix} -\frac{r_L}{L} & -\frac{1}{L} \\ \frac{1}{C} & -\frac{1}{RC} \end{pmatrix} \begin{pmatrix} i_{abc} \\ v_{abc} \end{pmatrix} + \begin{pmatrix} \frac{V_{dc}}{L} \\ 0 \end{pmatrix} d_{abc} + \begin{pmatrix} -\frac{1}{2L} V_{dc} \\ 0 \end{pmatrix} \tag{12.3}$$

Here, $i_{abc} = [i_a\ i_b\ i_c]^T$, $v_{abc} = [v_{an}\ v_{bn}\ v_{cn}]^T$, $U = d_{abc} = [d_a\ d_b\ d_c]^T$ are defined, and when expressed in matrix form, it becomes $\dot{X} = AX + BU + F$.

As in the case of a single-phase inverter, the three phase state averaged equation is a nonlinear system and must be linearized. To facilitate this, the duty cycle vector d_{abc} is decomposed into feedback and feed-forward components.

$$d_{abc} = d_{fb\_abc} + d_{ff\_abc}$$

Substituting $U = d_{abc}$ into the state-averaged equation:

$$\dot{X} = AX + B(d_{fb\_abc} + d_{ff\_abc}) + F$$

Rearranging terms yields the linearized control model:

$$\dot{X} = AX + Bd_{fb\_abc}$$

Here, the feed-forward duty component is set as:

$$d_{ff\_abc} = 0.5$$

This offset centers the switching signal around the midpoint of the duty cycle, aligning with typical PWM modulation strategies.

Instantaneous AC Current Control Law The instantaneous alternating current (AC) current control method used for single-phase inverters can be directly extended to the three-phase system. Applying the same principles independently to the *a*-, *b*-, and *c*-phases yields the following control laws:

- *a*-phase:

$$V_{a\_con} = \frac{L_a}{T_s}(I_{a,\text{ref}} - I_a) + r_L i_a + v_{an} \implies d_a = \frac{V_{a\_con}}{V_{dc}} + 0.5$$

- *b*-phase:

$$V_{b\_con} = \frac{L_b}{T_s}(I_{b,\text{ref}} - I_b) + r_L i_b + v_{bn} \implies d_b = \frac{V_{b\_con}}{V_{dc}} + 0.5$$

- *c*-phase:

$$V_{c\_con} = \frac{L_c}{T_s}(I_{c,\text{ref}} - I_c) + r_L i_c + v_{cn} \implies d_c = \frac{V_{c_{con}}}{V_{dc}} + 0.5 \tag{12.4}$$

These equations calculate the required control voltages and corresponding PWM duty cycles for each phase to regulate the inductor current based on the instantaneous error between the reference and measured currents.

The reference current values for each a, b, and c phase are defined as sine waves with a phase difference of 120°, and the mathematical expressions are as follows:

$$I_{a,\text{ref}} = \left(\frac{2}{3}\right) I_P \sin \omega t, \omega = 2\pi f$$

$$I_{b,\text{ref}} = \left(\frac{2}{3}\right) I_P \sin(\omega t - 2\pi/3)$$

$$I_{c,\text{ref}} = \left(\frac{2}{3}\right) I_P \sin(\omega t + 2\pi/3) \tag{12.5}$$

where:

- I_P is the peak amplitude of the reference current
- $\omega = 2\pi f$ is the angular frequency of the desired output

The factor (2/3) is a scaling coefficient used to ensure power balance across the three phases in the inverter system. The gain used in the control term,

$$K = \frac{L_{\text{abc}}}{T_s} \tag{12.6}$$

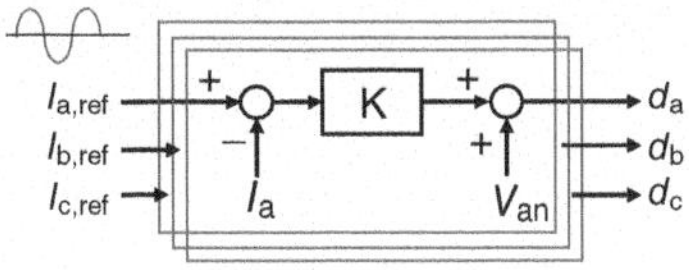

Figure 12.2 Three-phase instantaneous AC current controller configuration.

is determined by the inductance of each phase and the sampling period T_s, and it serves as the proportional coefficient in the current controller.

Figure 12.2 illustrates the block diagram of the three-phase current controller. Each phase operates independently under a feed-forward plus feed-back control structure, generating duty cycles for PWM modulation that achieve fast and accurate current regulation.

PSIM Simulation The simulation schematic for the three-phase stand-alone inverter system is shown in Figure 12.3. The control logic is implemented in C based on the structure shown in Figure 12.2. Each block in the diagram is directly translated into equivalent *C* statements.

When implementing controllers in *C*, numerical representation must be handled carefully. For example, the reference current scaling factor (2/3) must be written as 2.0/3.0 or 2./3. in *C*-code. Writing it as 2/3 results in integer division, which yields zero and causes erroneous waveforms. Alternatively, the constant

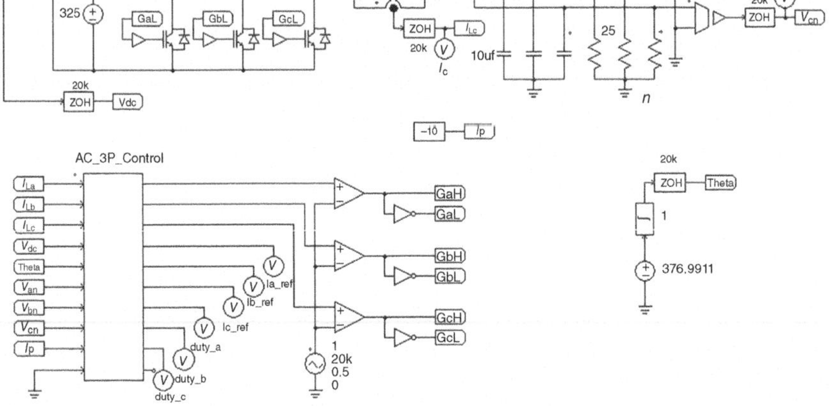

Figure 12.3 Three-phase stand-alone inverter circuit schematic.

Table 12.1 C-block code: AC_3P_Control.

```
#define PI      3.1415
#define Kp      20
#define rL      0.01

double Ia_ref, Ib_ref, Ic_ref
double Ia_err, Ib_err, Ic_err;
double Va_con, Vb_con, Vc_con;
double duty_a, duty_b, duty_c;
double ILa, ILb, ILc,Vdc,theta;
double Van, Vbn, Vcn, Ip;

// Input Variable
ILa    = in[0];
ILb    = in[1];
ILc    = in[2];
Vdc    = in[3];
Theta  = in[4];
Van    = in[5];
Vbn    = in[6];
Vcn    = in[7];
Ip     = in[8];

if (t >0.04) Ip = Ip*1.5;

Ia_ref=(2./3.)*Ip*sin(theta);
Ib_ref=(2./3.)*Ip*sin(theta-2*PI/3);
Ic_ref=(2./3.)*Ip*sin(theta-4*PI/3);

Ia_err = Ia_ref - ILa;
Ib_err = Ib_ref - ILb;
Ic_err = Ic_ref  - ILc;

Va_con = Kp*Ia_err+rL*ILa+Van;
Vb_con = Kp*Ib_err+rL*ILb+Vbn;
Vc_con = Kp*Ic_err+rL*ILc+Vcn;

duty_a = Va_con/(Vdc) + 0.5;
duty_b = Vb_con/(Vdc) + 0.5;
duty_c = Vc_con/(Vdc) + 0.5;

out[0] = duty_a;     // R
out[1] = duty_b;     // S
out[2] = duty_c;     // T
out[3] = Ia_ref;
out[4] = Ib_ref;
out[5] = Ic_ref;
out[6] = duty_a;
out[7] = duty_b;
out[8] = duty_c;
```

value 0.66667 can be used directly. In addition, attention must be paid to variable declarations and type consistency, especially in embedded systems where numerical precision and real-time performance are critical.

The implementation of the three-phase stand-alone inverter C-block is presented in Table 12.1.

PSIM Simulation Results The simulation results are presented in Figure 12.4. Figure 12.4(a) shows the phase currents, phase voltages, and corresponding PWM duty cycles. The waveforms demonstrate sinusoidal behavior, indicating correct tracking of the reference signals. Figure 12.4(b) displays the current tracking error and the phase difference between voltage and current. A small phase shift is observed due to the reactive power component introduced by the output filter capacitor C_f.

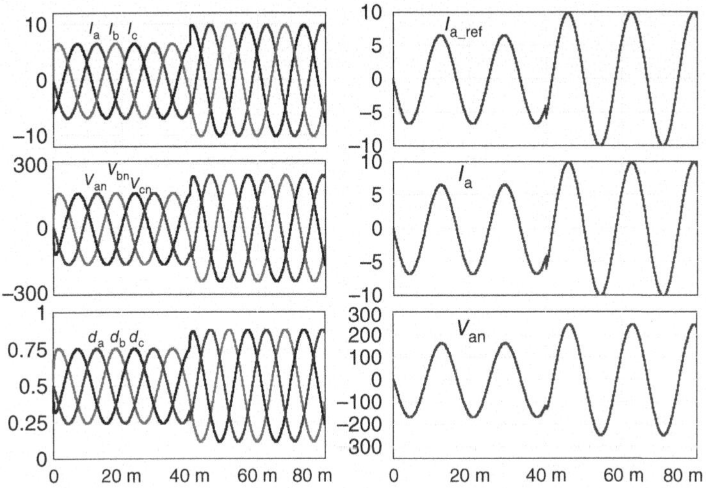

Figure 12.4 Simulation waveforms of the three-phase stand-alone inverter. (a) Three-phase current, voltage, and duty cycle and (b) a-phase current and voltage.

This phase difference can be minimized by reducing the output capacitance. For example, decreasing C_f from 10 to 1 μF significantly reduces the phase lag between current and voltage, thereby improving the power factor.

Operation of a Three-Phase PCS By replacing the output capacitor with a three-phase grid, the inverter operates as a three-phase power conditioning system (PCS). Figure 12.5 illustrates the PCS circuit configuration in which the capacitor is substituted by a three-phase grid. Assuming a standard 380 V three-phase, four-wire system, each phase provides 220 V RMS (with a peak voltage of 310 V) at a frequency of 60 Hz. The *b*-phase and *c*-phase voltages are phase-shifted by −120° and + 120°, respectively, relative to the a-phase.

The simulation results are presented in Figure 12.6. In part (a), waveforms under PCS mode ($I_P > 0$)) are shown. The grid voltage and current are in phase, indicating that power flows from the direct current (DC) input (650 V_{dc}) to the AC output (380 V_{ac}). The input current (I_{in}), including a 180 Hz ripple, flows at approximately +10 A.

Similar to the single-phase rectifier operation, reversing the polarity of I_P (i.e., $I_P < 0$) causes the system to operate as a three-phase rectifier. The waveforms in part (b) show operation under $I_P = -10\,A$. In this case, the voltage and current waveforms are out of phase, and the input current (I_{in}) flows at −10 A, indicating reverse power flow from the AC grid (380 V_{ac}) to the DC side (650 V_{dc}). Thus, during PCS mode, the system behaves like a buck converter, while in rectifier mode, it behaves as a boost converter.

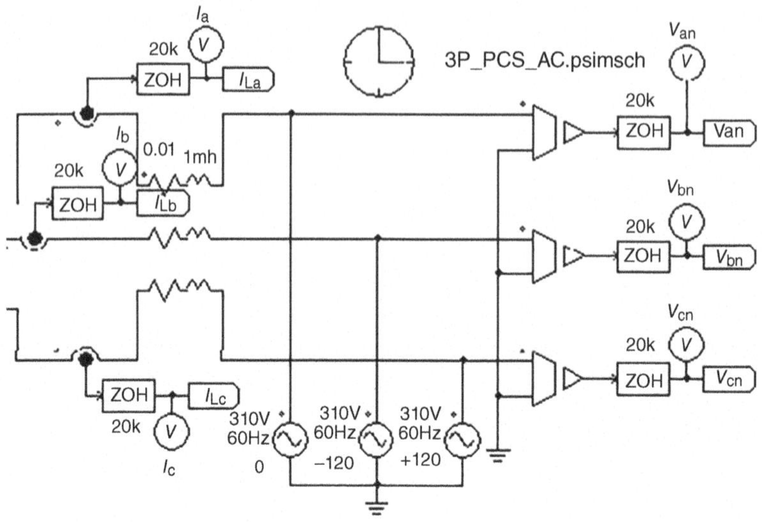

Figure 12.5 Configuration of the three-phase PCS circuit (only a portion of the circuit).

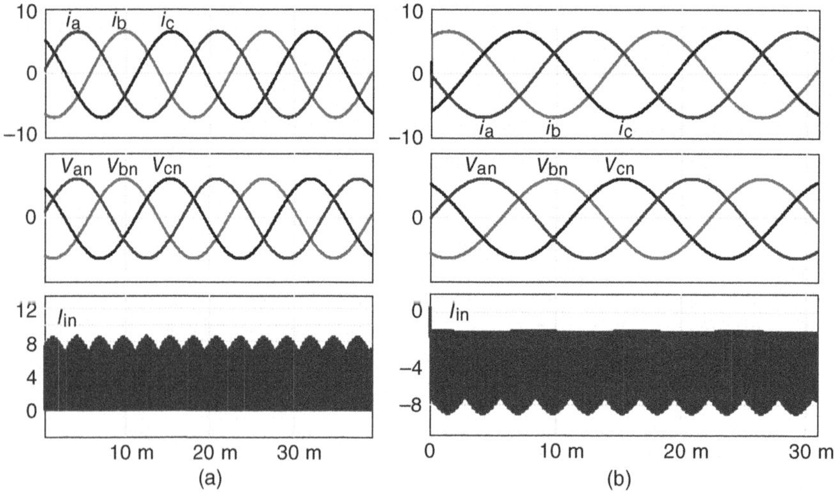

Figure 12.6 Simulation waveforms of the three-phase PCS/rectifier. (a) PCS mode ($I_p > 0$) and (b) Rectifier mode ($I_p < 0$).

This approach is known as instantaneous AC control method. In this method, the reference currents for each phase are sinusoidal waveforms, and the controller typically consists of a proportional (P) controller with gain K. The main advantage of this method lies in its simple structure and ease of implementation, even with analog circuitry such as operational amplifiers. However, it suffers from notable drawbacks, including difficulty in controlling power factor and reactive power, which limit its use in modern systems. Furthermore, since the reference current is time-varying, steady-state error can occur in tracking the desired current.

To address the limitations of the instantaneous AC current control method (commonly referred to as AC control), the DQ control method is widely adopted in modern power electronics applications. This method converts time-varying AC signals into constant DC values using a coordinate transformation, typically through a Park transformation. Although this transformation imposes a significant computational burden, it is made feasible using high-speed digital signal processors (DSPs) and other advanced microcontrollers.

The DQ control method enables precise power factor control and is now the industry-standard approach for grid-tied and stand-alone inverters. In this method, the reference current components for each phase are expressed as DC quantities:

$$I_{d,\mathrm{ref}} = \left(\frac{2}{3}\right) I_P, \quad I_{q,\mathrm{ref}} = \left(\frac{2}{3}\right) I_q \tag{12.7}$$

Compared to the instantaneous AC control method, the DQ approach eliminates the time-varying sinusoidal component (e.g., sin ωt), yielding steady DC values of the same magnitude. This not only simplifies the control process but also reduces the number of variables from three (phases a, b, c) to two (d and q components), thereby enhancing control efficiency and system stability.

Let us analyze the steady-state error in the instantaneous AC control method. When using a proportional (P) controller, as in the case of single-phase inverter control, a steady-state error persists for sinusoidal reference inputs. However, for step inputs, the steady-state error is zero. To eliminate the error in sinusoidal tracking, a proportional–integral (PI) controller must be used. However, PI control is only effective for DC reference signals.

To accommodate this requirement, the sinusoidal reference current $I_{\mathrm{ref}} = I_P \sin\omega t$ is decomposed into two distinct components: the magnitude I_P and the sinusoidal carrier $\sin\omega t$, as illustrated in Figure 12.7.

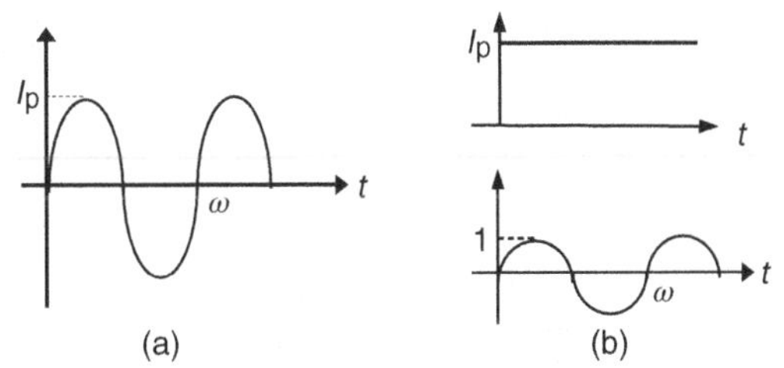

Figure 12.7 Decomposition of the sinusoidal reference current. (a) Sinusoidal reference current and (b) Separation into magnitude and frequency.

If the sinusoidal carrier component $\sin\omega t$ can be measured or generated accurately, the magnitude I_P can be used as a DC reference input for the controller. This enables the use of a *PI* controller to eliminate steady-state error.

The transformation of a three-phase AC signal into DC form typically involves two steps:

- $\alpha\beta$ transformation (Clarke transform): Convert the three-phase system (a, b, c), which has 120° phase differences, into two orthogonal components (α, β) with a 90° phase difference.
- amplitude and frequency separation: Extract the magnitude (amplitude) and the carrier frequency (e.g., ωt) from the $\alpha\beta$ components. The magnitude is then used as the control variable in the PI controller.

This transformation-based control approach not only improves dynamic performance and steady-state accuracy but also simplifies the implementation of advanced control strategies, such as active and reactive power regulation.

12.2 DQ Transformation

In this chapter, the three-phase AC system under discussion is a balanced system, defined as one in which the three voltage (or current) waveforms are of equal magnitude and spaced 120° apart in phase. The mathematical representation of such a system is given by:

$$v_a(t) = M\cos(\omega t + \varphi)$$

$$v_b(t) = M\cos\left(\omega t - \frac{2\pi}{3} + \varphi\right)$$

$$v_c(t) = M\cos\left(\omega t + \frac{2\pi}{3} + \varphi\right) \tag{12.8}$$

where M represents the magnitude, ωt is the angular frequency, and φ is the initial phase.

For example, a three-phase AC system with $M = 100$ and $\omega = 2\,\pi * 60 = 377$ rad/s is illustrated as waveforms in Figure 12.8.

In systems with a single frequency—such as CVCF (constant voltage, constant frequency) or VVVF (variable voltage, variable frequency)—the phase variables can be uniformly represented using phasors. A phasor describes each sinusoidal signal by its magnitude and initial phase. Because the system is balanced, a single amplitude M and phase φ are sufficient to describe all three phases. For example, the a-phase can be written as:

$$v_a(t) = M\angle\varphi$$

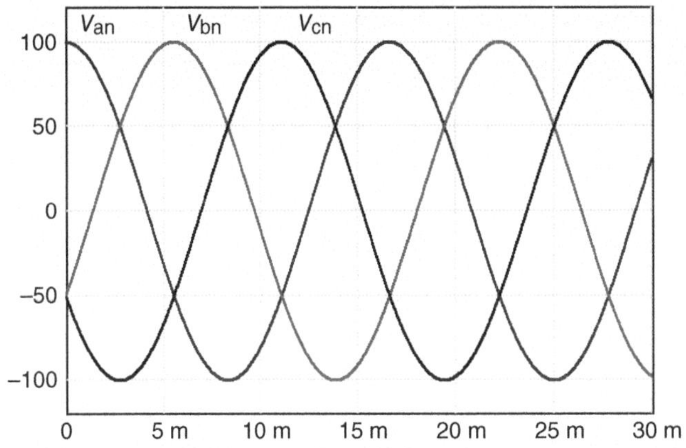

Figure 12.8 Balanced three-phase AC system.

The *b*- and *c*-phases are then understood to be shifted versions of the a-phase by ±120°, respectively.

a) **Clarke Transformation ($abc-to-\alpha\beta$ Transformation)**
The Clarke transformation converts a three-phase (*abc*) AC system into a two-phase orthogonal (αβ) AC system. This is useful for control purposes, as it reduces complexity while preserving all essential information. Let us denote a three-phase system with a magnitude *M* and an initial phase of 0 as f_a, f_b, f_c. When f_a, f_b, f_c are projected onto the X-axis (α-axis), the following relationship is obtained. Using the balanced system condition, $f_a+f_b+f_c=0$, the α–axis transformation is expressed as

$$f_\alpha = f_a - f_b\cos(\pi/3) - f_c\cos(\pi/3) = f_a - \frac{1}{2}(f_b+f_c) = \frac{3}{2}f_a$$

To normalize and preserve the magnitude (i.e., to make $f_\alpha = f_a$), the result is scaled by $\frac{2}{3}$, then

$$f_\alpha = f_a$$

When f_a, f_b, f_c are projected onto the Y-axis (β-axis), the β-component becomes:

$$f_\beta = 0 + f_b\sin(2\pi/3) - f_c\sin(2\pi/3) = \frac{\sqrt{3}}{2}(f_b - f_c) = \frac{\sqrt{3}}{2}(f_a + 2f_b)$$

To maintain consistent magnitude, this too is scaled by $\frac{2}{3}$, resulting in:

$$f_\beta = \frac{1}{\sqrt{3}}(f_b - f_c) = \frac{1}{\sqrt{3}}(f_a + 2f_b)$$

The Clarke's transformation expressed in matrix form is as follows.

$$\begin{bmatrix}\alpha\\ \beta\end{bmatrix}=\begin{bmatrix}1 & 0 & 0\\ 0 & \dfrac{1}{\sqrt{3}} & \dfrac{-1}{\sqrt{3}}\end{bmatrix}\begin{bmatrix}a\\ b\\ c\end{bmatrix},\quad \begin{bmatrix}a\\ b\\ c\end{bmatrix}=\begin{bmatrix}1 & 0\\ -\dfrac{1}{2} & \dfrac{\sqrt{3}}{2}\\ -\dfrac{1}{2} & -\dfrac{\sqrt{3}}{2}\end{bmatrix}\begin{bmatrix}\alpha\\ \beta\end{bmatrix} \tag{12.9}$$

This transformation reduces the three-phase (abc) variables—each 120° apart—into two orthogonal components (α, β) that are 90° apart in phase as illustrated in Figure 12.9. Importantly, both f_α and f_β remain AC signals.

Figure 12.10 shows the waveform resulting from the Clarke transformation. It is notable that V_α corresponds to the original phase voltage V_{an}, and that the transformation does not alter the magnitude of the signals. One important observation from the waveforms is that V_α leads V_β, consistent with the 90° phase difference between them.

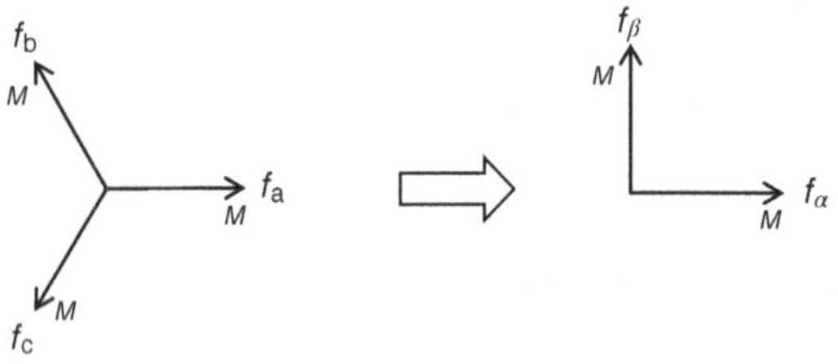

Figure 12.9 Clarke transformation.

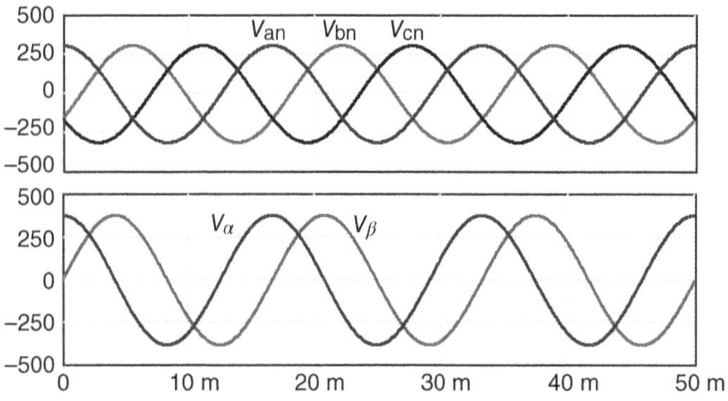

Figure 12.10 Waveforms of the Clarke transformation.

b) **Park Transformation ($\alpha\beta - to - dq$ Transformation)**

The $\alpha\beta$ components resulting from the Clarke transformation are AC quantities and are commonly expressed as:

$$\alpha = M\cos(\omega t), \quad \beta = M\cos\left(\omega t - \frac{\pi}{2}\right).$$

If the angular frequency $\cos(\omega t)$ is known, α can be interpreted as DC quantities in a rotating reference frame—where α leads β by 90°. This transformation from the stationary $\alpha\beta$ frame to the rotating dq frame is known as the Park transformation.

When projecting f_α and f_β onto the rotating d-axis (aligned with the rotating vector), the transformation yields:

$$f_d = f_\alpha \cos\emptyset + f_\beta \sin\emptyset$$

Similarly, projecting onto the orthogonal q-axis results in:

$$f_q = f_\alpha \cos(90+\emptyset) + f_\beta \cos\emptyset = -f_\alpha \sin(\emptyset) + f_\beta \cos\emptyset$$

In matrix form, the Park transformation ($\alpha\beta$ to dq) is expressed as:

$$\begin{bmatrix} d \\ q \end{bmatrix} = \begin{bmatrix} \cos\emptyset & \sin\emptyset \\ -\sin\emptyset & \cos\emptyset \end{bmatrix} \begin{bmatrix} \alpha \\ \beta \end{bmatrix}, \quad \begin{bmatrix} \alpha \\ \beta \end{bmatrix} = \begin{bmatrix} \cos\emptyset & -\sin\emptyset \\ \sin\emptyset & \cos\emptyset \end{bmatrix} \begin{bmatrix} d \\ q \end{bmatrix} \tag{12.10}$$

A key point to understand is that the $\alpha\beta$ components are AC signals with a 90° phase difference, while the dq components are DC signals (in steady state) within the rotating reference frame. This transformation effectively converts sinusoidal signals into constant values, greatly simplifying control algorithms, especially when using *PI* controllers. The graphical representation is shown in Figure 12.11.

Figure 12.12 summarizes the complete sequence of coordinate transformations from the three-phase *abc* frame to the *dq* frame via the intermediate $\alpha\beta$ frame.

An important note: In this textbook, three-phase AC waveforms are defined using *cosine functions*. That is, the phase voltages are expressed in terms of $\cos\omega t$, and this convention is used throughout the dq transformation process.

What happens if the three-phase AC system is expressed using *sine* functions instead of *cosine* functions? This can be verified through simulation. Although the

Figure 12.11 Park transformation ($\alpha\beta$ to dq).

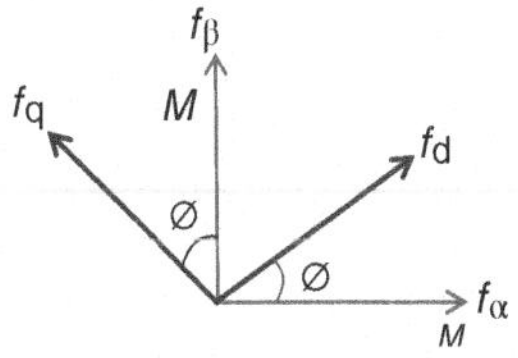

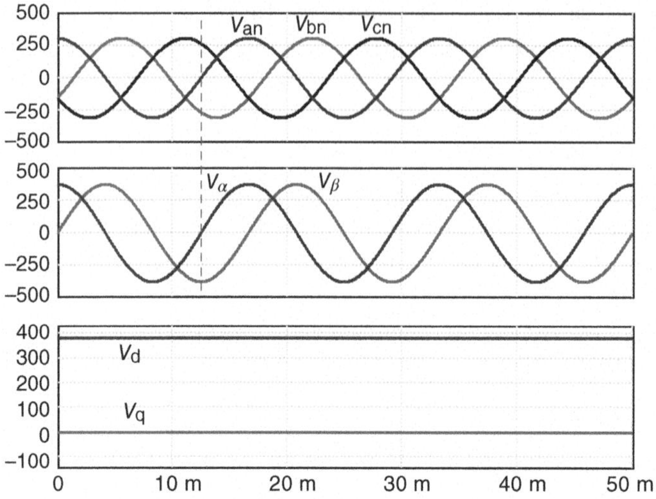

Figure 12.12 Transformation waveforms (abc → αβ → dq).

$\alpha\beta$ waveforms remain unchanged, the resulting *dq* values differ because of the difference in phase reference.

For example, if the system is defined using *sine* functions, the simulation may yield:

$$V_d = 0\,V, \quad V_q = -310\,V$$

This discrepancy arises because the phase angle θ differs between *cosine*- and *sine*-based definitions. In this textbook, the phase angle is referenced as 0° for *cosine* input. For *sine*-based inputs, the equivalent initial phase angle should be set to +90°.

This consideration is especially important when using *dq* transformation blocks in simulation tools such as PSIM. Be sure to verify the reference phase angle used in your model to ensure consistency and correct interpretation of the *dq* components.

In conclusion, when expressing a three-phase AC system, if *cosine* functions are used, the initial phase angle should be set to 0°. If *sine* functions are used instead, the initial phase angle must be set to **+90°** to ensure that the resulting *dq* components align correctly with the standard transformation equations.

Summary of* dq *Transformation (AC → DC) The process of transforming a three-phase AC signal into a DC representation via the *dq* frame involves the following steps:

1) $abc \rightarrow \alpha\beta$ transform (Clarke transformation)

$$\begin{bmatrix} \alpha \\ \beta \end{bmatrix} = \begin{bmatrix} 1 & 0 & 0 \\ 0 & \dfrac{1}{\sqrt{3}} & \dfrac{-1}{\sqrt{3}} \end{bmatrix} \begin{bmatrix} a \\ b \\ c \end{bmatrix}$$

2) $\alpha\beta \rightarrow dq$ transform (Park transformation)

$$\begin{bmatrix} d \\ q \end{bmatrix} = \begin{bmatrix} \cos Ø & \sin Ø \\ -\sin Ø & \cos Ø \end{bmatrix} \begin{bmatrix} \alpha \\ \beta \end{bmatrix}$$

3) $abc \rightarrow dq$ (Combined transformation)

$$\begin{bmatrix} d \\ q \end{bmatrix} = \begin{bmatrix} \cos Ø & \sin Ø \\ -\sin Ø & \cos Ø \end{bmatrix} \begin{bmatrix} 1 & 0 & 0 \\ 0 & \dfrac{1}{\sqrt{3}} & \dfrac{-1}{\sqrt{3}} \end{bmatrix} \begin{bmatrix} a \\ b \\ c \end{bmatrix} = \begin{bmatrix} \cos Ø & \dfrac{\sin Ø}{\sqrt{3}} & -\dfrac{\sin Ø}{\sqrt{3}} \\ -\sin Ø & \dfrac{\cos Ø}{\sqrt{3}} & -\dfrac{\cos Ø}{\sqrt{3}} \end{bmatrix} \begin{bmatrix} a \\ b \\ c \end{bmatrix}$$

Summary of Inverse DQ Transformation (DC → AC) To reconstruct the original three-phase AC signals from the DQ frame, the following inverse transformations are applied:

1) $dq \rightarrow \alpha\beta$ transform (Inverse Park transformation)

$$\begin{bmatrix} \alpha \\ \beta \end{bmatrix} = \begin{bmatrix} \cos Ø & -\sin Ø \\ \sin Ø & \cos Ø \end{bmatrix} \begin{bmatrix} d \\ q \end{bmatrix}$$

2) $\alpha\beta \rightarrow abc$ transform (Inverse Clarke transformation)

$$\begin{bmatrix} a \\ b \\ c \end{bmatrix} = \begin{bmatrix} 1 & 0 \\ -\dfrac{1}{2} & \dfrac{\sqrt{3}}{2} \\ -\dfrac{1}{2} & -\dfrac{\sqrt{3}}{2} \end{bmatrix} \begin{bmatrix} \alpha \\ \beta \end{bmatrix}$$

3) $dq \rightarrow abc$ (Combined Inverse transformation)

$$\begin{bmatrix} a \\ b \\ c \end{bmatrix} = \begin{bmatrix} 1 & 0 \\ -\dfrac{1}{2} & \dfrac{\sqrt{3}}{2} \\ -\dfrac{1}{2} & -\dfrac{\sqrt{3}}{2} \end{bmatrix} \begin{bmatrix} \cos Ø & -\sin Ø \\ \sin Ø & \cos Ø \end{bmatrix} \begin{bmatrix} d \\ q \end{bmatrix}$$

$$= \begin{bmatrix} \cos Ø & -\sin Ø \\ \dfrac{-\cos Ø + \sqrt{3}\sin Ø}{2} & \dfrac{\sin Ø + \sqrt{3}\cos Ø}{2} \\ \dfrac{-\cos Ø - \sqrt{3}\sin Ø}{2} & \dfrac{\sin Ø - \sqrt{3}\cos Ø}{2} \end{bmatrix} \begin{bmatrix} d \\ q \end{bmatrix}$$

This compact matrix formulation allows for efficient implementation of dq transformations in digital controllers and simulation environments. It is especially important to maintain consistency in the reference angle θ, particularly when switching between *sine*- and *cosine*-based definitions.

Implementation of Coordinate Transformation Using PSIM Function Blocks PSIM provides built-in function blocks for coordinate transformations, including *abc-to-dq*, *abc-to-*$\alpha\beta$, and their respective inverse transformations. These blocks can be conveniently integrated into simulation circuits. When using these PSIM library functions, it is essential to consider several implementation details to ensure correct interpretation of results.

In this textbook, active power is aligned with the d-axis, and reactive power with the q-axis. This convention is commonly adopted in power electronics applications. However, in motor control systems, the q-axis is typically associated with torque, while the d-axis corresponds to the flux component. To maximize torque (or power), field-oriented control (FOC) is used by setting the d-axis component to zero. As different software tools may adopt different axis conventions, special attention must be paid when interpreting or configuring coordinate transformations.

To illustrate this, consider the waveform results obtained using PSIM's transformation blocks. Figure 12.13 shows the implementation of $abc \rightarrow \alpha\beta \rightarrow dq$ transformations. When a three-phase voltage source is applied as shown in Figure 12.14, the resulting waveforms can be used to validate proper axis alignment.

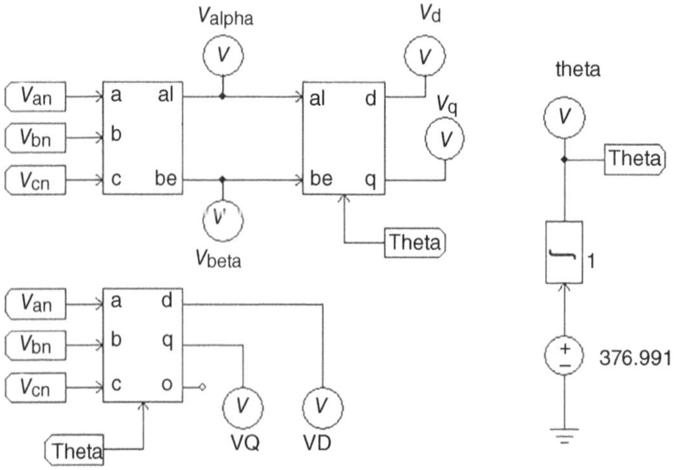

Figure 12.13 Transformation using PSIM function blocks ($abc \rightarrow \alpha\beta \rightarrow dq$).

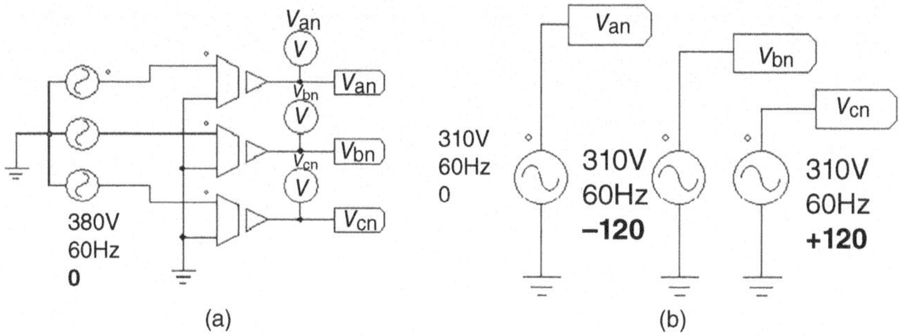

Figure 12.14 Initial phase angle = 0° (incorrect configuration). (a) Three-phase sinusoidal voltage source and (b) Three single-phase sine sources.

Figure 12.14 presents two input circuit configurations. (a) uses PSIM's built-in three-phase sinusoidal voltage source set to 380 V line-to-line RMS, which corresponds to 220 V phase RMS. (b) constructs a three-phase source using three individual sine wave voltage sources, each with a peak amplitude of 310 V, frequency of 60 Hz, and phase angles of 0°, −120°, and + 120°, respectively. Both configurations generate an equivalent balanced three-phase AC system with an initial phase angle of 0°. As expected, the simulation results for both configurations are identical.

The resulting waveforms are shown in Figure 12.15. The simulation yields:

$$V_q = 310\,V, \quad V_d = 0\,V$$

This indicates that active power is aligned with the *q*-axis and reactive power with the *d*-axis which contradicts the earlier assumption of active power on the *d*-axis. To correct this, the input voltage waveform must be shifted by 90°, effectively changing the input from a *sine* wave to a *cosine* waveform.

To achieve the desired *dq* alignment where $V_d = 310\ V$, $V_q = 0\ V$, the initial phase angle of the voltage source should be set to *90*°. The modified circuits and

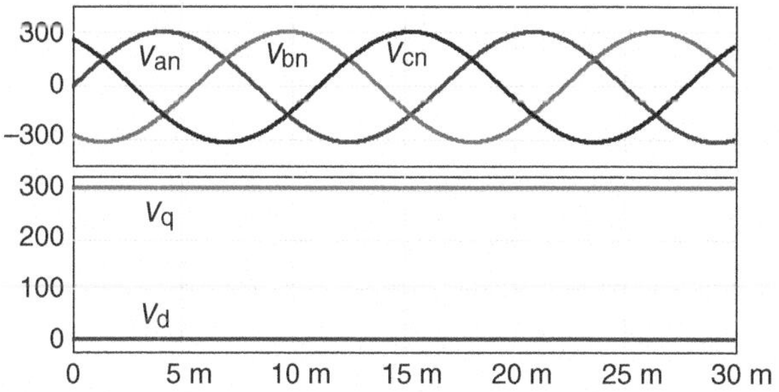

Figure 12.15 Waveform output with initial phase = 0° (incorrect result).

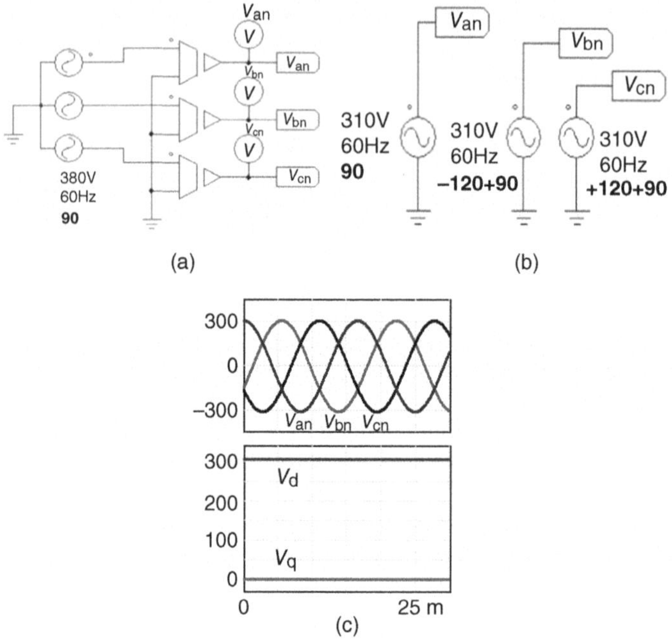

Figure 12.16 Initial phase = 90° (correct configuration). (a) Three-phase sine, (b) Three single-phase sines and (c) Correct dq result.

their corresponding simulation results are shown in Figure 12.16. These results now correctly reflect the expected outcome described earlier and match the theoretical *dq* transformation waveforms introduced in Figure 12.12.

In summary, when using PSIM's *dq* transformation blocks, it is crucial to align the input waveform's phase angle with the transformation convention used in the control strategy. In this chapter, a *cosine*-based input is assumed, requiring the initial phase angle to be set to 90° in simulations.

Phase Alignment and Implementation of DQ Transformation Using C Code in PSIM Let us now consider the reason why the initial phase angle must be set to 90° in the simulation. At the beginning of this chapter, the three-phase AC system was mathematically described using *cosine* functions (e.g., $cos\omega t$). However, the default AC voltage sources in PSIM generate signals based on *sine* functions. To align these with the *cosine*-based formulation used in the *dq* transformation equations, the initial phase of each source should be set to 90°, effectively converting $sin(\omega t + 90°) = cos\omega t$. Therefore, all three-phase AC sources in PSIM should be configured with an initial phase of 90° to ensure consistency with the theoretical models.

Since the built-in PSIM function blocks cannot be deployed directly to DSP hardware, it is advisable to implement all control functions, including coordinate transformations, using C language. This ensures portability and allows for real-time execution on embedded platforms.

Figure 12.17 shows the schematic implementation of the *dq* transformation using a C-block in PSIM, along with the corresponding *C*-code. The simulation results are shown in Figure 12.18, where the input signals are in the *abc*-frame, Output 1 provides the $\alpha\beta$ components, and Output 2 provides the *dq* components.

To perform the inverse *dq* transformation, both the schematic and *C*-code are shown in Figure 12.19, with corresponding waveforms in Figure 12.20. In this case, the input is in the *dq* frame, Output 1 gives the $\alpha\beta$ components, and Output 2 provides the reconstructed *abc* signals.

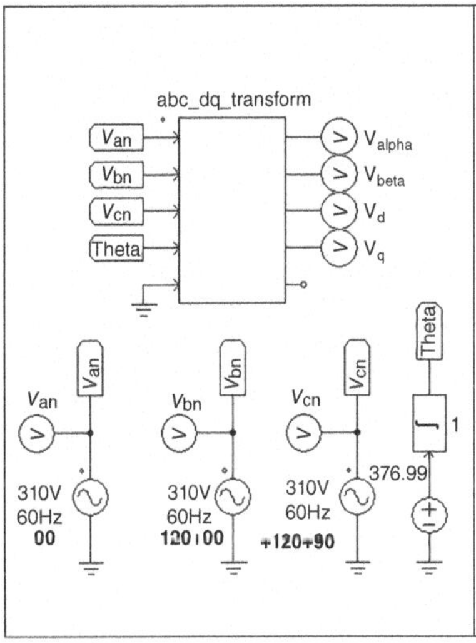


```
double Vd, Vq, theta;
double Valpha, Vbeta, Van, Vbn, Vcn;
/**** Input Variable ****/
Van =in[0];
Vbn =in[1];
Vcn =in[2];
Theta = in[3];

Valpha = Van;
Vbeta = (Vbn-Vcn) / sqrt(3);
Vd = Valpha*cos(theta) +Vbeta*sin(theta);
Vq = -Valpha*sin(theta) +Vbeta*cos(theta);

out[0] = Valpha;
out[1] = Vbeta;
out[2] = Vd;
out[3] = Vq;
```

Figure 12.17 *dq* conversion using C-block.

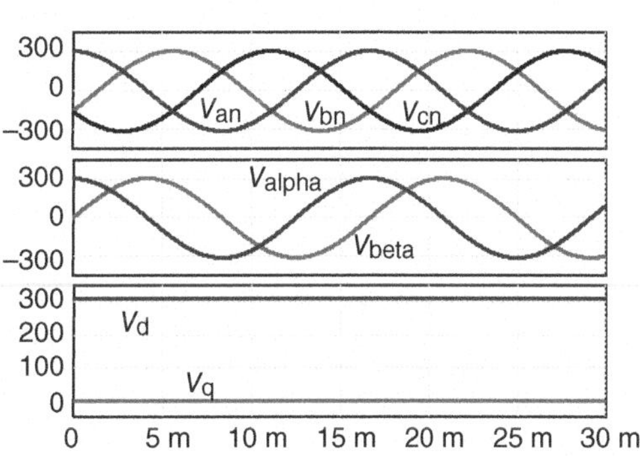

Figure 12.18 Simulation waveforms for dq transformation ($abc \rightarrow \alpha\beta \rightarrow dq$).

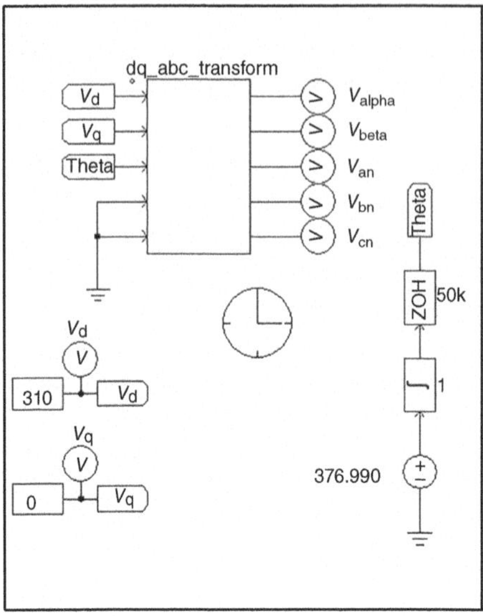


```
double Vd, Vq, theta;
double Valpha, Vbeta,Van,Vbn,Vcn;
/**** Input Variable    ****/
Vd = in[0];
Vq = in[1];
theta = in[2];

Valpha = Vd*cos(theta)-Vq*sin(theta);
Vbeta = Vd*sin(theta)+Vq*cos(theta);

Van = Valpha;
Vbn = -0.5*Valpha+sqrt(3)*Vbeta/2;
Vcn = -0.5*Valpha-sqrt(3)*Vbeta/2;

out[0] = Valpha;
out[1] = Vbeta;
out[2] = Van;
out[3] = Vbn;
out[4] = Vcn;
```

Figure 12.19 Inverse DQ conversion using C-block.

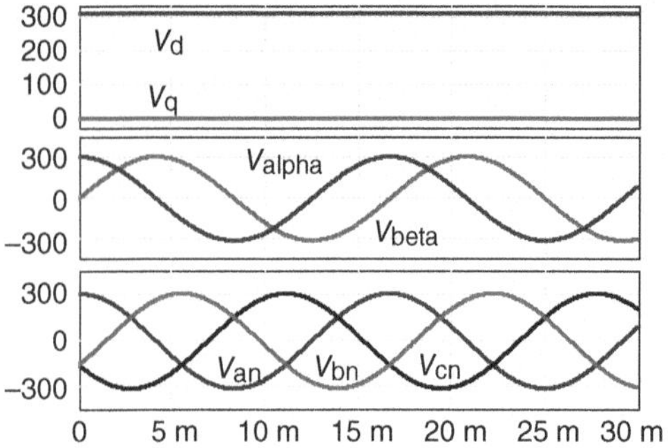

Figure 12.20 Simulation waveforms for inverse *dq* transformation ($dq \rightarrow \alpha\beta \rightarrow abc$).

In practical applications, coordinate-axis conventions may vary depending on the system type.

- In grid-connected PCS, the *d*-axis typically corresponds to active power, and the *q*-axis to reactive power.
- In motor control systems, the *q*-axis is generally aligned with torque, and the *d*-axis with the flux component.

Therefore, it is essential to clearly define the *d*- and *q*-axis roles when designing controllers. Throughout this chapter, we adopt the convention that the *d*-axis represents active power, and all controller designs and analyses will follow this convention accordingly.

Effect of V_d and V_q on Output Waveforms In both PCS and motor control applications, it is common practice to set either the V_d or V_q (or equivalently I_d or I_q) component to zero as part of the control strategy:

- In PCS, to achieve maximum power transfer and a unity power factor, the active power component V_d is maximized, while the reactive power component V_q is set to zero.
- In motor control, the torque-producing (active power) component is maximized on the q-axis, and the flux (reactive power) component on the d-axis is set to zero through FOC.

What Happens When $V_q \neq 0$? To explore this, consider three cases:

1) Case 1: $V_d = 100, V_q = 0$
2) Case 2: $V_d = 50, V_q = 50$
3) Case 3: $V_d = 50, V_q = -50$

From these values, we can derive the resultant magnitude M and phase angle ϕ using the following equations:

$$M = \sqrt{V_d^2 + V_q^2}, \quad \phi = \tan^{-1}\frac{V_q}{V_d}$$

The calculated values for each case are summarized in Table 12.2.

Figure 12.21 shows the simulation results for these three cases. The waveform peak values (M) match the magnitudes calculated above. The phase shifts are also evident:

- In *Case 1*, the zero-crossing point is at 4.17 ms.
- In *Case 2*, it occurs at 2.09 ms, which is 2.08 ms earlier, indicating a leading phase.
- In *Case 3*, it occurs at 6.25 ms, which is 2.08 ms later, indicating a lagging phase.

Given that the period of a 60 Hz waveform is 16.67 ms, a time shift of 2.08 ms corresponds to a phase angle of 45°. Thus, *Case 2* leads *Case 1* by 45°, and *Case 3* lags *Case 1* by 45°, confirming the calculated phase angles.

Table 12.2 Magnitude and phase angle for various V_d and V_q.

V_d	V_q	M	ϕ
100	0	100	0°
50	50	70.7	+45°
50	−50	70.7	−45°

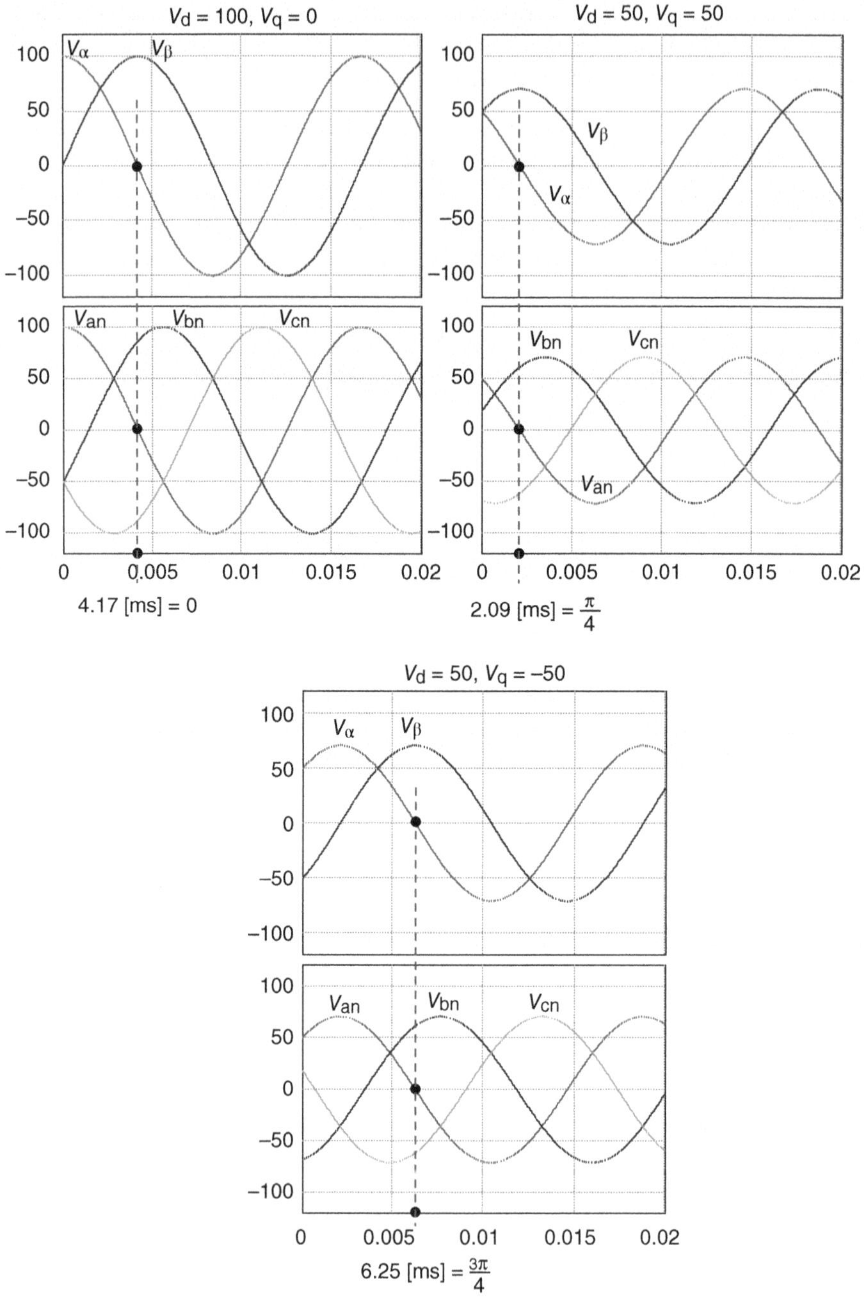

Figure 12.21 Waveform variations for different V_d, V_q conditions.

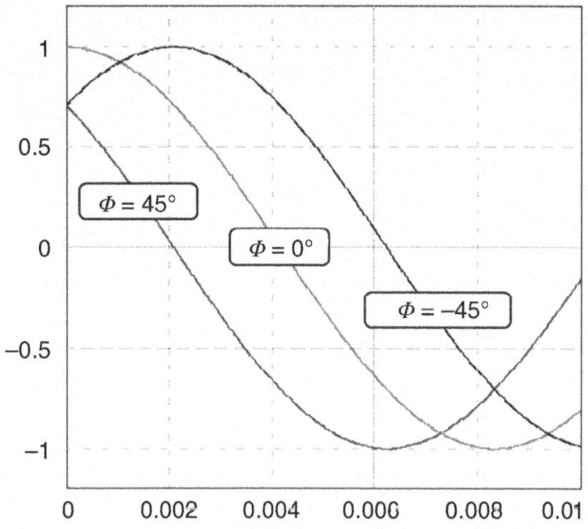

Figure 12.22 Phase shift analysis for various V_d, V_q combinations.

Figure 12.22 further illustrates the phase shifts corresponding to each of the three conditions. The figure clearly demonstrates that the phase difference between the waveforms is 45°, consistent with the analytical results.

Transformation of State Equations from* abc *to* dq *Coordinates To convert the state-space equations expressed in the *abc* frame into the *dq* frame, we use the previously derived $dq - to - abc$ transformation, which can be expressed in matrix form as:

$$\begin{bmatrix} a \\ b \\ c \end{bmatrix} = P \begin{bmatrix} d \\ q \end{bmatrix}$$

where the transformation matrix P is defined as:

$$P = \begin{bmatrix} \cos \emptyset & -\sin \emptyset \\ \dfrac{-\cos \emptyset + \sqrt{3}\sin \emptyset}{2} & \dfrac{\sin \emptyset + \sqrt{3}\cos \emptyset}{2} \\ \dfrac{-\cos \emptyset - \sqrt{3}\sin \emptyset}{2} & \dfrac{\sin \emptyset - \sqrt{3}\cos \emptyset}{2} \end{bmatrix}, \quad \emptyset = \omega t$$

To compute the time derivative of the abc-frame variables, we apply the product rule:

$$\frac{d}{dt}(abc) = \frac{d}{dt}\left(P \cdot \begin{bmatrix} d \\ q \end{bmatrix}\right) = \left(\frac{d}{dt}P\right) \cdot \begin{bmatrix} d \\ q \end{bmatrix} + P \cdot \frac{d}{dt}(dq)$$

Differentiating P with respect to time gives:

$$\left(\frac{d}{dt}P\right)\cdot\begin{bmatrix} d \\ q \end{bmatrix} = \omega\begin{bmatrix} -\sin\emptyset & -\cos\emptyset \\ \dfrac{\sin\emptyset+\sqrt{3}\cos\emptyset}{2} & \dfrac{\cos\emptyset-\sqrt{3}\sin\emptyset}{2} \\ \dfrac{\sin\emptyset-\sqrt{3}\cos\emptyset}{2} & \dfrac{\cos\emptyset+\sqrt{3}\sin\emptyset}{2} \end{bmatrix}\begin{bmatrix} d \\ q \end{bmatrix} = \omega P\begin{bmatrix} -q \\ d \end{bmatrix}$$

Substituting this result into the total derivative yields:

$$\frac{d}{dt}(abc) = \omega P\begin{bmatrix} -q \\ d \end{bmatrix} + P\cdot\frac{d}{dt}(dq) \tag{12.11}$$

Current Equation in* dq *Frame Reorganizing the linearized state-averaged equations for the abc frame, as previously developed in Eq. (12.3).

$$\begin{pmatrix} \dfrac{di_{abc}}{dt} \\ \dfrac{dv_{abc}}{dt} \end{pmatrix} = \begin{pmatrix} -\dfrac{r_L}{L} & -\dfrac{1}{L} \\ \dfrac{1}{C} & -\dfrac{1}{RC} \end{pmatrix}\begin{pmatrix} i_{abc} \\ v_{abc} \end{pmatrix} + \begin{pmatrix} \dfrac{V_{\mathrm{dc}}}{L} \\ 0 \end{pmatrix} d_{fb\_abc} \tag{12.12}$$

Here, $i_{abc} = [i_a\ i_b\ i_c]^T$, $V_{abc} = [V_a\ V_b\ V_c]^T$, $d_{fb\_abc} = [d_a\ d_b\ d_c]^T$ are defined, and in matrix form, it is expressed as $\dot{X} = AX + BU$. The duty cycle consists of feedback components (d_{fb_abc}) and feed-forward components (d_{ff_abc}), with d_{fb_abc} being applied in the modeling. Here, $d_{ff\_abc} = 0.5$.

Focusing on the current dynamics:

$$\frac{di_{abc}}{dt} = -\frac{r_L}{L}i_{abc} - \frac{1}{L}v_{abc} + \frac{V_{dc}}{L}d_{fb\_abc}$$

Transforming all variables i_{abc}, v_{abc}, d_{fb_abc} to their d, q counterparts, we apply the previous derivative identity:

$$\omega P\begin{bmatrix} -i_q \\ i_d \end{bmatrix} + P\cdot\frac{d}{dt}\left(\begin{bmatrix} i_d \\ i_q \end{bmatrix}\right) = -\frac{r_L}{L}P\cdot\begin{bmatrix} i_d \\ i_q \end{bmatrix} - \frac{1}{L}P\cdot\begin{bmatrix} v_d \\ v_q \end{bmatrix} + \frac{V_{\mathrm{dc}}}{L}P\cdot\begin{bmatrix} d_{fb\_d} \\ d_{fb\_q} \end{bmatrix}$$

Multiplying both sides by P^{-1}, and simplifying using the orthogonality of the transformation, we isolate the dq-frame current equation:

$$\frac{d}{dt}\left(\begin{bmatrix} i_d \\ i_q \end{bmatrix}\right) = \omega\begin{bmatrix} i_q \\ -i_d \end{bmatrix} - \frac{r_L}{L}\cdot\begin{bmatrix} i_d \\ i_q \end{bmatrix} - \frac{1}{L}\cdot\begin{bmatrix} v_d \\ v_q \end{bmatrix} + \frac{V_{\mathrm{dc}}}{L}\cdot\begin{bmatrix} d_{fb\_d} \\ d_{fb\_q} \end{bmatrix} \tag{12.13}$$

Voltage Equation in* dq *Frame From the abc voltage dynamics in Eq. (12.12):

$$\frac{dv_{abc}}{dt} = \frac{1}{C}i_{abc} - \frac{1}{RC}v_{abc}$$

Transforming into the dq frame:

$$\omega P \begin{bmatrix} -v_q \\ v_d \end{bmatrix} + P \cdot \frac{d}{dt}\left(\begin{bmatrix} v_d \\ v_q \end{bmatrix}\right) = \frac{1}{C} P \cdot \begin{bmatrix} i_d \\ i_q \end{bmatrix} - \frac{1}{RC} P \cdot \begin{bmatrix} v_d \\ v_q \end{bmatrix}$$

Solving for the dq-frame voltage derivatives:

$$\frac{d}{dt}\left(\begin{bmatrix} v_d \\ v_q \end{bmatrix}\right) = \omega \begin{bmatrix} v_q \\ -v_d \end{bmatrix} - \frac{1}{RC}\begin{bmatrix} v_d \\ v_q \end{bmatrix} + \frac{1}{C}\begin{bmatrix} i_d \\ i_q \end{bmatrix} \tag{12.14}$$

These dq-domain equations provide a convenient form for implementing control algorithms such as *PI* current control, decoupling control, or FOC, as they eliminate the time-varying nature of the original *abc*-frame variables under balanced conditions.

Verification of* dq*-Axis Transformation Using an Open-Loop Control System Figure 12.23 illustrates the block diagram of a control system based on state equations expressed in the dq-coordinate frame. In this configuration, the reference and feedback signals—comprising i_d, i_q, v_d, v_q—are used to generate error signals, which serve as inputs to the controller. The controller outputs, d_{fb_d} and d_{fb_q}, are then transformed into *abc*-phase duty commands, to which a feedforward duty term $d_{ff} = 0.5$ is added. The resulting signals are used as the input duty cycles for the inverter.

To validate the correctness of the coordinate transformation from the *abc* frame to the dq frame, an open-loop simulation is employed by excluding the controller and feedback components. These omitted blocks are indicated as dashed lines in the figure. This simplification is necessary because the design of the controller requires the dq-axis transfer functions $G_1(s)$ and $G_2(s)$, and verifying the transformation process independently ensures accuracy before progressing to controller design.

The simplest and most effective method for verification involves computing steady-state values directly from the dq-domain state equations and comparing

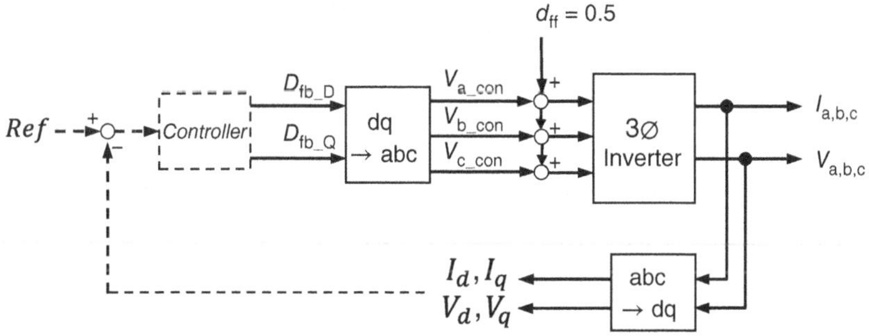

Figure 12.23 Block diagram of dq-axis control system (Open loop).

them with the results obtained from PSIM simulations. Under steady-state conditions, the derivative terms are set to zero.

Steady-State Equations The steady-state current components can be derived from the voltage equations:

$$I_D = \frac{V_D}{R} - \omega C V_Q, \quad I_Q = \frac{V_Q}{R} + \omega C V_D \tag{12.15}$$

The corresponding steady-state duty cycle components can be obtained from the current equations:

$$D_{fb\_D} = \frac{V_D + r_L I_D - \omega L I_Q}{V_{dc}}, \quad D_{fb\_Q} = \frac{V_Q + r_L I_Q + \omega L I_D}{V_{dc}} \tag{12.16}$$

If the transformation from *abc* to *dq* coordinates is correctly implemented, the reference voltages V_{ref_d} and V_{ref_q} should match the steady-state values of the simulated V_D and V_Q waveforms. Similarly, the analytically calculated I_D and I_Q values should agree with the steady-state values of i_d and i_q observed in simulation.

Therefore, the following duty commands are applied in the open-loop simulation:

$$d_{fb\_d} = \frac{V_{ref\_d} + r_L i_d - \omega L i_q}{V_{dc}}, \quad d_{fb\_q} = \frac{V_{ref\_q} + r_L i_q + \omega L i_d}{V_{dc}} \tag{12.17}$$

The transformation to *d*, *q* coordinates is considered valid if the simulation results satisfy:

$$V_{ref\_d} = V_D, \quad V_{ref\_q} = V_Q, \quad I_D = i_d, \quad I_Q = i_q$$

In these expressions, V_D, V_Q represent the steady-state values of the simulated V_d, V_q waveforms, and I_D, I_Q are corresponding steady-state values of the i_d, i_q.

Figure 12.24 illustrates the simulation schematic configured for open-loop verification of the dq-axis transformation.

The implementation of this open-loop controller C-block is shown in Table 12.3.

Figure 12.25 presents the simulation results for the open loop system. In the first waveform, the calculated steady-state value of I_D matches the inductor current I_d in the steady state, although discrepancies are observed during the transient response. This is expected since the steady-state value is derived under the assumption that the derivatives are zero. The second waveform shows the behavior of I_Q, which also deviates during the transient period but converges to the steady-state value. The third and fourth waveforms illustrate the voltages, confirming that the actual V_d follows the reference voltage V_{ref_d}, while V_q remains at zero as intended. These results verify the correctness of the *dq*-axis transformation and its implementation in the open-loop simulation.

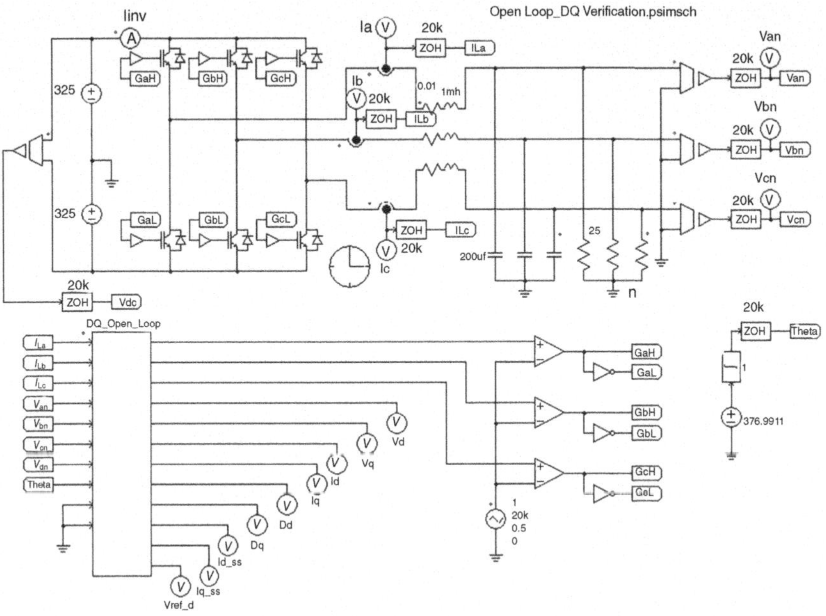

Figure 12.24 Schematic for open-loop *dq* verification.

Three-Phase Stand-Alone Inverter Decoupling Model Thus far, we have developed the state-averaged model of the three-phase inverter in the *abc*-coordinate system and subsequently transformed it into the *dq*-coordinate system as illustrated in Eqs. (12.13 and 12.14). The next step is to design the controller, which is typically performed using transfer functions. Since transfer functions apply to linear systems, a challenge arises: the state equations in the *dq*-coordinate system exhibits nonlinear characteristics.

The nonlinear nature stems from the presence of terms involving ωi_q, ωi_d, ωV_q, and V_d, which couples the dynamics of different state variables. Rewriting the state equations in the *dq*-frame:

$$\frac{d}{dt}\left(\begin{bmatrix} i_d \\ i_q \end{bmatrix}\right) = \omega \begin{bmatrix} i_q \\ -i_d \end{bmatrix} - \frac{r_L}{L} \cdot \begin{bmatrix} i_d \\ i_q \end{bmatrix} - \frac{1}{L} \cdot \begin{bmatrix} v_d \\ v_q \end{bmatrix} + \frac{V_{\text{dc}}}{L} \cdot \begin{bmatrix} d_{fb\_d} \\ d_{fb\_q} \end{bmatrix}$$

$$\frac{d}{dt}\left(\begin{bmatrix} v_d \\ v_q \end{bmatrix}\right) = \omega \begin{bmatrix} v_q \\ -v_d \end{bmatrix} - \frac{1}{RC} \begin{bmatrix} v_d \\ v_q \end{bmatrix} + \frac{1}{C} \begin{bmatrix} i_d \\ i_q \end{bmatrix}$$

The state variables to be controlled are i_d, i_q, V_d, V_q. For a system to be linear, its state equations must consist solely of linear combinations of state variables, their derivatives, and inputs. However, in the above equations, the presence of cross-coupling terms like ωi_q, ωV_q renders the system nonlinear.

Table 12.3 C-block code: DQ_Open_Loop.

```
#define Lf       0.001  //  1 [mh]
#define rL       0.01   // 10 [mohm]
#define Cf       0.0002 // 200 [uf]
#define R        25     // 25 [ohm]
#define Omega    377    // 60 [Hz]

double ILa,ILb,ILc,Van,Vbn,Vcn,Vdc;
double Valpha,Vbeta,Ialpha,Ibeta;
double Vd_ref, Vq_ref, Va_con, Vb_con,
Vc_con, theta, Vd, Vq;
double duty_a, duty_b, duty_c;
double ILd, ILq, Dd,Dq,Idss,Iqss;

/**** Input Variable    ****/
ILa    = in[0];
ILb    = in[1];
ILc    = in[2];
Van    = in[3];
Vbn    = in[4];
Vcn    = in[5];
Vdc    = in[6];
theta  = in[7];

if (Vdc == 0 ) Vdc =0.1;

Valpha=(2./3.)*Van-(1./3.)*Vbn-(1./.3)*Vcn;
Vbeta = (Vbn-Vcn)/sqrt(3);
Vd= Valpha*cos(theta) + Vbeta*sin(theta);
Vq=-Valpha*sin(theta) + Vbeta*cos(theta);

Ialpha=(2./3.)*ILa-(1./3.)*ILb-(1./3.)*ILc;
Ibeta = (ILb-ILc)/sqrt(3);
ILd = Ialpha*cos(theta)+Ibeta*sin(theta);
ILq =-Ialpha*sin(theta)+Ibeta*cos(theta);

if (t < 0.3) Vd_ref= 110*sqrt(2.0);
else         Vd_ref= 220*sqrt(2.0);
Vq_ref=0;

//steady-state duty, Open loop
Dd=(Vd_ref+rL*ILd-Lf*Omega*ILq)/Vdc;
Dq=(Vq_ref+rL*ILq+Lf*Omega*ILd)/Vdc;

Valpha=Dd*cos(theta) - Dq*sin(theta);
Vbeta =Dd*sin(theta) + Dq*cos(theta);

Va_con=Valpha;
Vb_con=-0.5*Valpha + sqrt(3)*Vbeta/2;
Vc_con=-0.5*Valpha - sqrt(3)*Vbeta/2;

// duty = dff + dfb = 0.5 + dfb
duty_a =  0.5 + Va_con;
duty_b =  0.5 + Vb_con;
duty_c =  0.5 + Vc_con;

Idss = Vd/R - Omega*Cf*Vq;
Iqss = Vq/R + Omega*Cf*Vd;

out[0]  = duty_a;  // Inverter R
out[1]  = duty_b;  // Inverter S
out[2]  = duty_c;  // Inverter T
out[3]  = Vd;
out[4]  = Vq;
out[5]  = ILd;
out[6]  = ILq;
out[7]  = Dd;
out[8]  = Dq;
out[9]  = Idss;
out[10] = Iqss;
out[11] = Vd_ref;
```

For instance, the differential equation for i_d contains ωi_q making the dynamics of i_d dependent on i_q. Consequently, the control inputs d_{fb_d} and d_{fb_q} cannot act independently. To overcome this, feedforward compensation can be applied to eliminate the coupling terms, thereby linearizing the system.

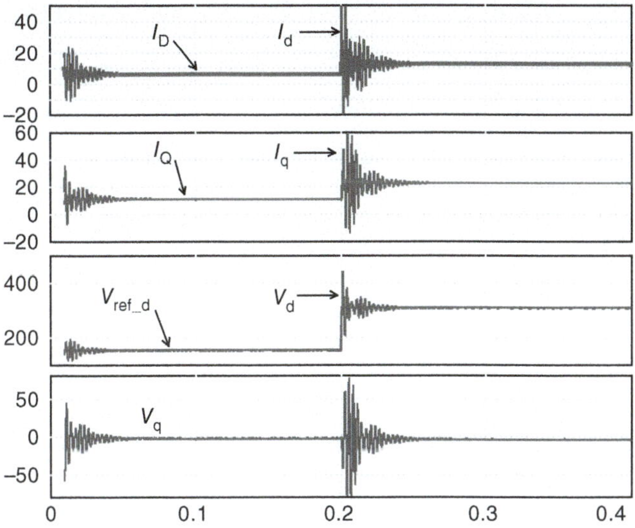

Figure 12.25 Simulation results of *dq*-axis open-loop verification.

Let the control inputs be decomposed as:

$$d_{fb\_d} = d_d + d_{fd}, \quad d_{fb\_q} = d_q + d_{fq}$$

The feedforward components that cancel the cross-coupling terms are:

$$d_{fd} = -\frac{\omega L i_q}{V_{dc}}, \quad d_{fq} = \frac{\omega L i_d}{V_{dc}}$$

Substituting this compensation into the current equations yields a linearized system:

$$\frac{d}{dt}\left(\begin{bmatrix} i_d \\ i_q \end{bmatrix}\right) = -\frac{r_L}{L} \cdot \begin{bmatrix} i_d \\ i_q \end{bmatrix} - \frac{1}{L} \cdot \begin{bmatrix} v_d \\ v_q \end{bmatrix} + \frac{V_{dc}}{L} \cdot \begin{bmatrix} d_d \\ d_q \end{bmatrix} \tag{12.18}$$

One might argue that the current equations remain nonlinear due to the presence of V_d and V_q as state variables. However, since current dynamics are significantly faster than voltage dynamics, it is a common and practical assumption to treat V_d and V_q as quasi-static during the current control time step *dt*. In the case of a grid-tied inverter (PCS), these voltages can be considered constant, and in a stand-alone inverter, they may be approximated as slowly varying measured values.

The voltage equations are expressed as:

$$\frac{d}{dt}\begin{bmatrix} v_d \\ v_q \end{bmatrix} = \omega \begin{bmatrix} v_q \\ -v_d \end{bmatrix} - \frac{1}{RC}\begin{bmatrix} v_d \\ v_q \end{bmatrix} + \frac{1}{C}\begin{bmatrix} i_d \\ i_q \end{bmatrix}$$

By decomposing the current i_d and i_q into feedback and feedforward components:

$$i_d = i_{d,ff} + i_{d,fb}, \quad i_q = i_{q,ff} + i_{q,fb}$$

and applying feedforward compensation to cancel the nonlinear terms:

$$i_{d,ff} = -\omega C v_q, \quad i_{q,ff} = \omega C v_d$$

If $i_{d,fb} = i_d$ is controlled by the current controller, the final linearized dq-axis voltage equation is expressed as follows.

$$\frac{d}{dt}\begin{bmatrix} v_d \\ v_q \end{bmatrix} = -\frac{1}{RC}\begin{bmatrix} v_d \\ v_q \end{bmatrix} + \frac{1}{C}\begin{bmatrix} i_d \\ i_q \end{bmatrix} \tag{12.19}$$

The linearized model is also referred to as the *decoupling model*. When organized in matrix form, it can be represented as follows.

$$\begin{pmatrix} \frac{d}{dt} i_{dq} \\ \frac{d}{dt} v_{dq} \end{pmatrix} = \begin{pmatrix} -\frac{r_L}{L} & -\frac{1}{L} \\ \frac{1}{C} & -\frac{1}{RC} \end{pmatrix} \begin{pmatrix} i_{dq} \\ v_{dq} \end{pmatrix} + \begin{pmatrix} \frac{V_{dc}}{L} \\ 0 \end{pmatrix} d_{dq} \tag{12.20}$$

Let $i_{dq} = [i_d \; i_q]^T$, $v_{dq} = [v_d \; v_q]^T$ and $U = d_{dq} = [d_d \; d_q]^T$, then the state-space representation becomes

$$\dot{X} = AX + BU$$

A key observation is that the matrices A and B in this model are identical to those of the *buck converter*. The state-averaged equation of a buck converter is expressed in Eq. (2.3).

$$\begin{pmatrix} \frac{d}{dt} i_L \\ \frac{d}{dt} V_C \end{pmatrix} = \begin{pmatrix} -\frac{r_L}{L} & -\frac{1}{L} \\ \frac{1}{C} & -\frac{1}{RC} \end{pmatrix} \begin{pmatrix} i_L \\ V_C \end{pmatrix} + \begin{pmatrix} \frac{V_{in}}{L} \\ 0 \end{pmatrix} d \tag{2.3}$$

This reveals that the decoupled dq-axis model and the buck converter model share identical dynamics. In earlier discussions, it was noted that a single-phase inverter can be modeled as two buck converters operating in parallel with a 180° phase shift. Extending this concept, a three-phase inverter can be represented as three buck converters with 120° phase shifts. Through dq transformation, these are converted into two orthogonal axes with a 90° phase difference, ultimately yielding a system structure equivalent to that of the buck converter.

The transfer functions of the decoupling model, analogous to those of a buck converter, are derived as:

- Current transfer function:

$$G_1(s) = \frac{I_{dq}(s)}{d_{dq}(s)} = \frac{\frac{V_{dc}}{L}\left(s + \frac{1}{RC}\right)}{s^2 + \left(\frac{r_L}{L} + \frac{1}{RC}\right)s + \frac{1}{LC}\left(1 + \frac{r_L}{R}\right)} \tag{12.21}$$

- Voltage transfer function:

$$G_2(s) = \frac{V_{dq}(s)}{d_{dq}(s)} = \frac{\frac{V_{dc}}{LC}}{s^2 + \left(\frac{r_L}{L} + \frac{1}{RC}\right)s + \frac{1}{LC}\left(1 + \frac{r_L}{R}\right)} \tag{12.22}$$

- Impedance transfer function:

$$G_3(s) = \frac{V_{dq}(s)}{I_{dq}(s)} = \frac{\frac{1}{C}}{s + \frac{1}{RC}} \tag{12.23}$$

These linearized transfer functions form the basis for designing *PI* or advanced controllers in three-phase stand-alone inverter.

12.3 Controller Design

This section presents the controller design methodology for a three-phase stand-alone inverter and a grid-connected system (PCS/Rectifier). The design begins with the three-phase stand-alone inverter, followed by the PCS.

12.3.1 Three-Phase Stand-Alone Inverter

a) Current Controller Design

Figure 12.26 illustrates the current control structure based on the decoupled model. The controller operates by computing the feedback duty cycle in the *dq*-current controller using the error between the reference and measured

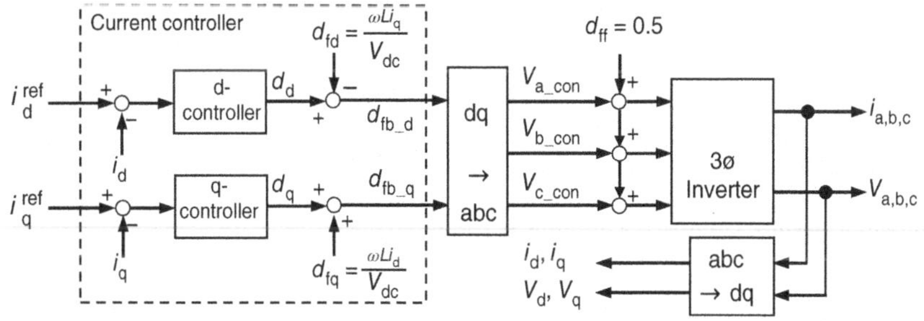

Figure 12.26 Current control configuration based on the decoupled model.

dq-axis currents. This feedback duty is then combined with the feedforward duty components (d_{fd}, d_{fq}) to form the total duty input (d_{fb_d}, d_{fb_q}), which is applied to the inverter.

Since the control objective is the regulation of dq-axis currents, the system adopts a single-loop configuration. This section outlines the structure and design process of the dq-current controller.

The current controller is designed using the decoupled model transfer functions $G_1(s)$. The current controller design follows the same methodology used in the buck converter. Using MATLAB's SISOTOOL, optimal *PI* gains are selected via auto-tuning.

1) Design Specifications:
   - Bandwidth: 4000 rad/sec
   - Phase margin: 60°
2) Controller Gains (Continuous-Time) by Auto-tuning:

$$G_{ci}(s) = \frac{K_{pi}(s + K_{ii})}{s} = \frac{0.001833(s + 3310)}{s}$$

3) Discrete time conversion ($f_{sw} = 20$ kHz, $T_{sw} = 50\ us$)

$$G_{ci}(z) = \frac{0.0021(z - 0.874)}{z - 1}$$

Note: The PI controller gains are identical for both the d-axis and q-axis.

The MATLAB m-file for current controller is presented in Table 12.4.

Figure 12.27 shows the SISOTOOL auto-tuning results. Subfigure (a) presents the Bode plot, and subfigure (b) displays the root locus and step response.

The design meets the target specifications – 4000 rad/sec bandwidth, 60° phase margin, zero steady-state error, and a rising time of 17 ms.

To validate the controller, time-domain simulations were performed using PSIM. The *PI* gains obtained from MATLAB were implemented in the PSIM C-block, and the decoupled model was encoded accordingly. The simulation circuit is shown in Figure 12.28. The designed *PI* gain was applied, and the method proposed in the decoupled model was coded into the C-block.

The implementation of this current controller C-block is presented in Table 12.5.

Simulation Results Figure 12.29 shows the step response of the dq-axis current controller. Subfigure (a) displays the reference and measured dq-axis inductor

Table 12.4 MATLAB m-file for current controller of the decoupled model.

```
%%%  3-Phase Inverter Current Controller  Design by Decoupled model %%
s=tf('s');
Vin = 650;
Omega = 377;
L = 1e-3;    % L=1mh
rL = 0.01;   % rL = 10mohm
C = 100e-6;  % C=100uf
R = 25;

G1=(Vin/L)*(s+1/(R*C))/(s^2+s*(rL/L+1/(R*C))+(1+rL/R)/(L*C));% Idq/ddq
G2=Vin/(L*C)/(s^2+s*(rL/L + 1/(R*C))+(1+rL/R)/(L*C));% Vdq/ddq
G3=(1/C) / (s + 1/(R*C) );   % Vdq(s)/Idq(s)

% Bandwidth=4000[rad/sec], Phase margin = 60 [deg], fsw=20[kHz]
Kpi =1; Kii =1;   % Initial gain
Gci = Kpi*(s+Kii)/s;

Vd_ref = 220*sqrt(2);
Id_ref = Vd_ref/R;
Iq_ref = Omega*C*Vd_ref;

%sisotool Current Single loop
T=sisoinit(1);
T.F.Value = Id_ref;
T.G.Value = G1;
T.C.Value = Gci;
sisotool(T);
```

currents. Subfigure (b) is a zoomed in view of the response. The measured response perfectly matches the MATLAB step response, validating the controller design.

Figure 12.30 shows the phase voltage and current waveforms: (a) plots the instantaneous phase voltage and current, (b) highlights the phase shift between the a-phase voltage and current. This phase difference is due to the reactive component introduced by the 100 μF output capacitor. Reducing the capacitor to 10 μF can significantly reduce the phase shift.

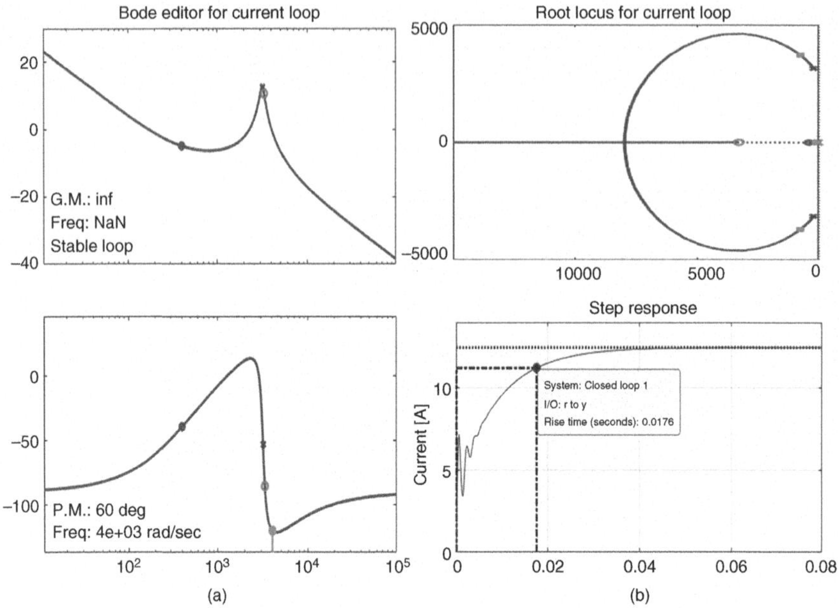

Figure 12.27 Current controller design using SISOTOOL. (a) Bode plot (b) Root-locus and step response plot.

b) **Voltage Controller Design**

Figure 12.31 illustrates the voltage regulation structure based on the decoupled model. The controller computes the feedback duty cycle d_{dq} in the dq-voltage controller by processing the error between the reference voltage and the measured dq-axis voltage. This feedback duty is then combined with the feedforward duty components d_{fd}, d_{fq} to generate the total duty command (d_{fb_d}, d_{fb_q}), which is applied to the inverter.

Since the control target is the dq-axis voltage, the system operates under a *single-loop control* configuration. This section details the structure and design process of the dq-voltage controller.

The voltage controller design uses the transfer function $G_2(s)$. The voltage controller design follows the same methodology used in the buck converter. Using MATLAB's SISOTOOL, optimal *PI* gains are selected via auto-tuning.

1) Design Specifications:
   - Bandwidth: 50 [rad/sec]
   - Phase margin: 90 [deg]
2) Controller Gains (continuous-time) by Auto-tuning:

$$G_{cv}(s) = \frac{K_{pv}(s + K_{iv})}{s} = \frac{0.000007(s + 10990)}{s}$$

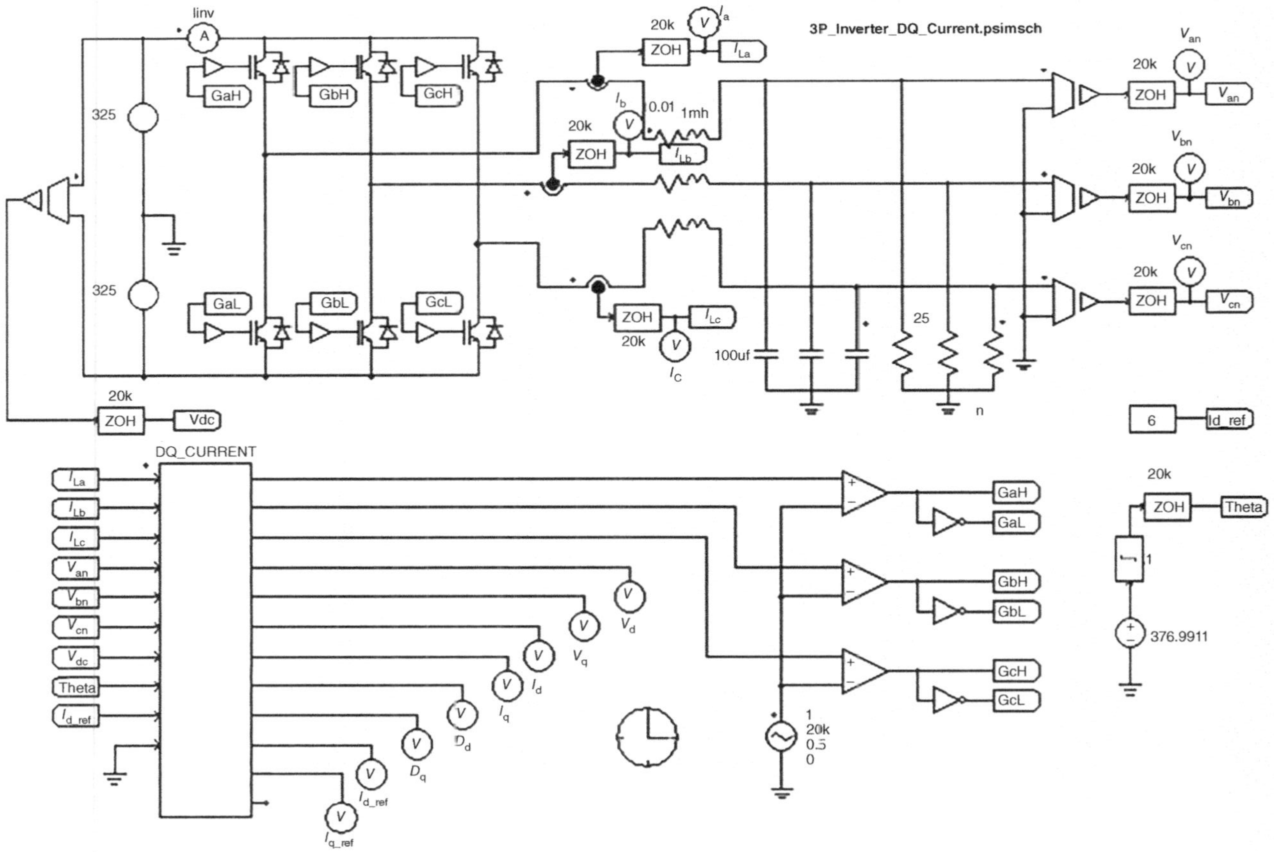

Figure 12.28 Simulation schematic for current control using the decoupled model.

Table 12.5 C-block code: DQ_CURRENT.

```
#define Lf          0.001 // 1 [mh]
#define rL          0.01  // 10  [mohm]
#define Cf          0.0001 //  100 [uf]
#define R           25     //  25 [ohm]
#define Omega       377    // 60 [Hz]
#define Kip         0.0021  // Kp
#define Kii         0.874   //Kiz

double ILa,ILb,ILc,Van,Vbn,Vcn,Vdc,theta;
double Valpha,Vbeta,Ialpha,Ibeta,Vd,Vq;
double ILd,ILq, Vd_ref, Vq_ref;
double Va_con, Vb_con, Vc_con, dd,dq;
double Id_ref, duty_a, duty_b, duty_c;
double dfb_d,dfb_q,Iq_ref,Id_err,Iq_err;
static double dd_old=0.0, dq_old=0.0,
       Id_err_old=0.0, Iq_err_old=0.0;

/**** Input Variable   ****/
ILa       = in[0];
ILb       = in[1];
ILc       = in[2];
Van       = in[3];
Vbn       = in[4];
Vcn       = in[5];
Vdc       = in[6];
theta     = in[7];
Id_ref    = in[8];

if (Vdc ==0 ) Vdc =0.1;
if (t >0.1) Id_ref = 12.4;

Valpha=(2./3.)*Van-(1/3.)*Vbn-(1/3.)*Vcn;
Vbeta = (Vbn-Vcn)/sqrt(3);
Vd =  Valpha*cos(theta)+Vbeta*sin(theta);
Vq = -Valpha*sin(theta)+Vbeta*cos(theta);

Ialpha=(2./3.)*ILa-(1./3.)*ILb-(1./3.)*ILc;
Ibeta = (ILb-ILc)/sqrt(3);
ILd =  Ialpha*cos(theta)+Ibeta*sin(theta);
ILq = -Ialpha*sin(theta)+Ibeta*cos(theta);

Id_err = Id_ref - ILd;
dd=dd_old+Kip*Id_err-
Kip*Kii*Id_err_old;
dd_old=dd;
Id_err_old=Id_err;

Iq_ref = Omega*Cf*Vd;
Iq_err = Iq_ref - ILq;
dq=dq_old+Kip*Iq_err-
Kip*Kii*Iq_err_old;
dq_old=dq;
Iq_err_old=Iq_err;

dfb_d = dd -  Lf*Omega*ILq/(Vdc);
dfb_q = dq +  Lf*Omega*ILd/(Vdc);

Valpha=dfb_d*cos(theta) -
df b_q*sin(theta);
Vbeta =dfb_d*sin(theta) +
dfb_q*cos(theta);

Va_con = Valpha;
Vb_con = -0.5*Valpha +
sqrt(3)*Vbeta/2;
Vc_con = -0.5*Valpha -
sqrt(3)*Vbeta/2;

// duty = dff + dfb = 0.5 + dfb
duty_a =  0.5 + Va_con;
duty_b =  0.5 + Vb_con;
duty_c =  0.5 + Vc_con;

out[0] = duty_a; // Inverter R
out[1] = duty_b; // Inverter S
out[2] = duty_c; // Inverter T

out[3] = Vd;
out[4] = Vq;
out[5] = ILd;
out[6] = ILq;
out[7] = dd;
out[8] = dq;
out[9] = Id_ref;
out[10]= Iq_ref;
```

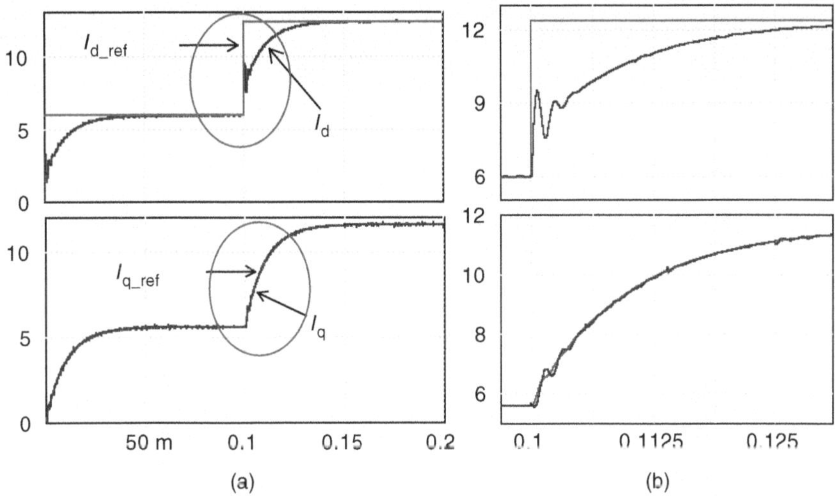

Figure 12.29 Current control response for step input. (a) dq -axis current response and (b) Zoomed-in waveform.

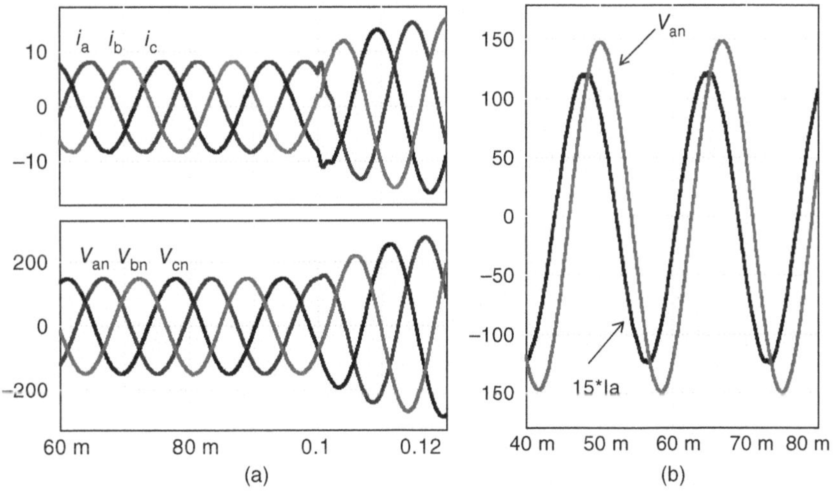

Figure 12.30 Phase current and voltage. (a) Phase current and voltage and (b) a-phase current and voltage (with phase difference).

3) Discrete time conversion ($f_{sw} = 20\ kHz$, $T_{sw} = 50\ us$)

$$G_{cv}(z) = \frac{0.000007(z - 0.451)}{z - 1}$$

Note: The *PI* controller gains are identical for both the *d*-axis and *q*-axis.

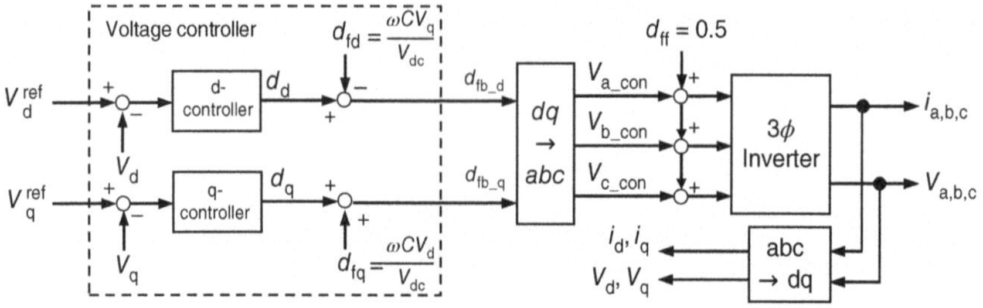

Figure 12.31 Voltage control configuration based on the decoupled model.

Table 12.6 MATLAB m-file for voltage controller.

```
%%%  sisotool Voltage Single loop  %%%
% Bandwidth= 50[rad/sec],
% Phase margin = 90 [deg], fsw=20[kHz]
Kpv =1; Kiv =1; % Initial gain
Gcv = Kpv*(s+Kiv)/s;
Vd_ref = 220*sqrt(2);

T=sisoinit(1);
T.F.Value = Vd_ref;
T.G.Value = G2;
T.C.Value = Gcv;
sisotool(T);
```

The MATLAB m-file used for voltage controller is presented in Table 12.6. The circuit model setup is the same as that of the current controller; only the controller design is changed.

Figure 12.32 shows the results of SISOTOOL auto-tuning: Subfigure (a) displays Bode plot and Subfigure (b) shows the Root locus and step response. The design achieves the target bandwidth of 50 rad/s and phase margin of 90°, with zero steady-state error. The rising time is approximately 44 ms.

PSIM Implementation and C-Block Code The PSIM schematic is the same as the current control system, with the only change being the C-block code. The implementation of this voltage controller C-block is shown in Table 12.7.

Figure 12.33 presents the PSIM simulation waveforms for the voltage controller. Figure 12.33 (a) shows the *dq*-axis reference voltage and measured capacitor voltage, while Figure 12.33 (b) illustrates the inverter's phase voltages and currents.

The step response to a change in $V_{d,ref}$ aligns closely with the MATLAB simulation results, validating the controller's accuracy and effectiveness.

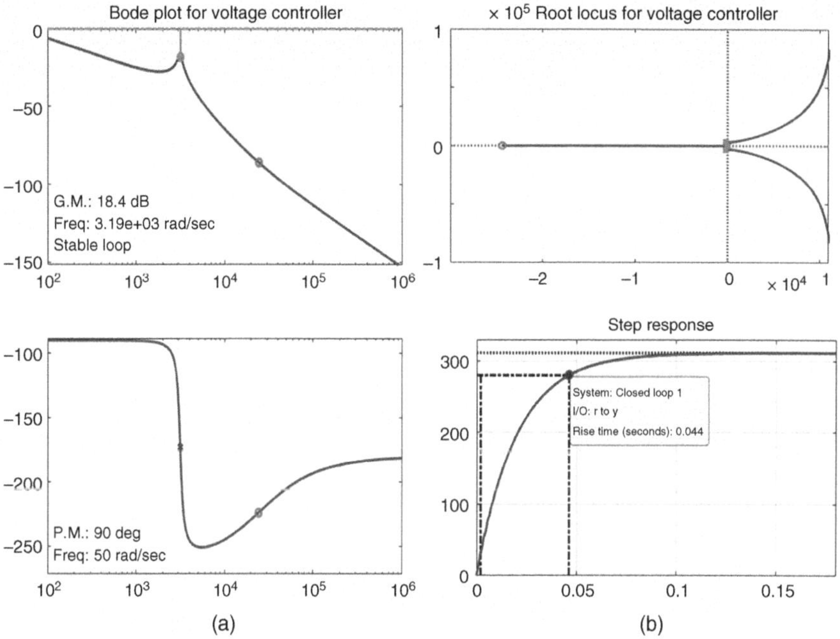

Figure 12.32 Voltage controller design using SISOTOOL. (a) Bode plot and (b) Root-locus and step response plot.

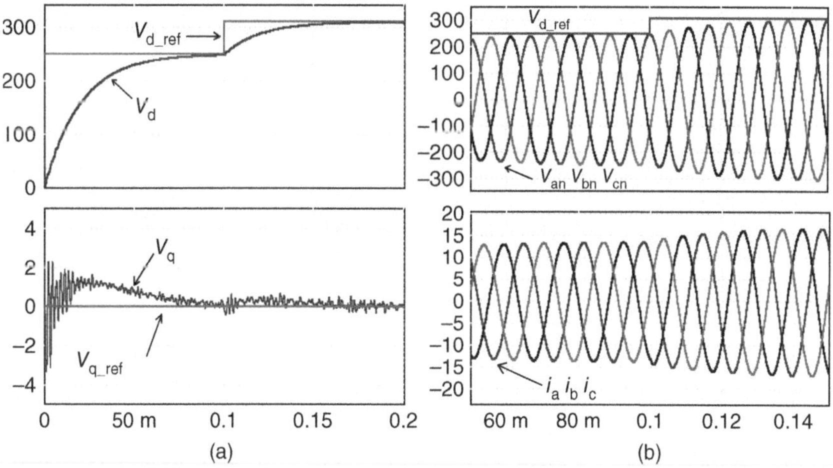

Figure 12.33 Simulation waveforms of the voltage controller. (a) dq-axis voltage waveform and (b) Phase voltage and current waveforms.

Table 12.7 C-block code: DQ_Voltage.

```
#define Lf        0.001    // 1 [mh]
#define rL        0.01     // 10  [mohm]
#define Cf        0.0001   // 100 [uf]
#define R         25        // 25 [ohm]
#define Omega     377      // 60 [Hz]

#define Kvp 0.000007  // P-gain  Kvp
#define Kvi 0.451     // I -gain Kvz

Double ILa,ILb,ILc,Van,Vbn,Vcn,Vdc,theta;
doubleValpha,Vbeta,Ialpha,Ibeta,Id_ref;
double Vd_ref,Vq_ref,Va_con,Vb_con, Vc_con;
double ILd,ILq,duty_a,duty_b,duty_c,dd,dq,
dfb_d, dfb_q,Vd,Vq;
double Iq_ref,Id_err,Iq_err,Vd_err,Vq_err;
static double dd_old=0.0, dq_old=0.0,
 Id_err_old=0.0, Iq_err_old=0.0;
static double Vd_err_old=0.0, Vq_err_old=0.0;

/**** Input Variable   ****/
ILa          = in[0];
ILb          = in[1];
ILc          = in[2];
Van         = in[3];
Vbn         = in[4];
Vcn         = in[5];
Vdc         = in[6];
theta      = in[7];
Id_ref     = in[8];

if (Vdc ==0 ) Vdc =0.1;
if (t < 0.1) Vd_ref = 250;
else Vd_ref = 310;
Vq_ref = 0;

Valpha=(2./3.)*Van-(1./3.)*Vbn-(1./3.)*Vcn;
Vbeta = (Vbn-Vcn)/sqrt(3);
Vd =  Valpha*cos(theta) + Vbeta*sin(theta);
Vq = -Valpha*sin(theta) + Vbeta*cos(theta);

Ialpha=(2./3.)*ILa-(1./3.)*ILb-(1./3.)*ILc;
Ibeta = (ILb-ILc)/sqrt(3);
ILd =  Ialpha*cos(theta) + Ibeta*sin(theta);
ILq = -Ialpha*sin(theta) + Ibeta*cos(theta);

Vd_err = Vd_ref - Vd;
dd = dd_old + Kvp*Vd_err
 -Kvp*Kvi*Vd_err_old;
dd_old=dd;
Vd_err_old=Vd_err;

Vq_err = Vq_ref - Vq;
dq = dq_old +Kvp*Vq_err -
Kvp*Kvi*Vq_err_old;
dq_old=dq;
Vq_err_old=Vq_err;

dfb_d = dd -  Cf*Omega*Vq/Vdc;
dfb_q = dq +  Cf*Omega*Vd/Vdc;
Valpha = dfb_d*cos(theta) -
dfb_q*sin(theta);
Vbeta  = dfb_d*sin(theta) +
dfb_q*cos(theta);

Va_con = Valpha;
Vb_con = -0.5*Valpha +
sqrt(3)*Vbeta/2;
Vc_con = -0.5*Valpha -
sqrt(3)*Vbeta/2;

// duty = dff + dfb = 0.5 + dfb
duty_a =  0.5 + Va_con;
duty_b =  0.5 + Vb_con;
duty_c =  0.5 + Vc_con;

out[0] = duty_a;  // Inverter R
out[1] = duty_b;  // Inverter S
out[2] = duty_c;  // Inverter T

out[3] = Vd;
out[4] = Vq;
out[5] = ILd;
out[6] = ILq;
out[7] = dd;
out[8] = dq;
out[9] = Vd_ref;
out[10]= Vq_ref;
```

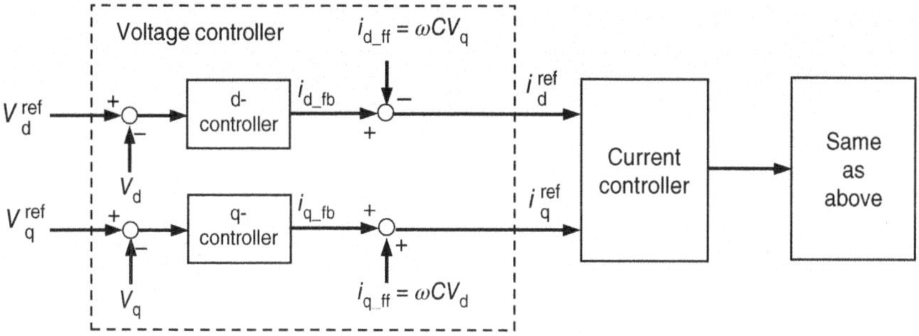

Figure 12.34 Double-loop control configuration based on the decoupled model.

c) Double-Loop Controller Design

Figure 12.34 presents the double-loop regulation structure based on the decoupled model. In this configuration, the outer voltage controller computes the feedback current components (i_{d_fb}, i_{q_fb}) by processing the error between the reference voltage and the measured *dq*-axis voltage. These feedback currents are then added to the feedforward currents (i_{d_ff}, i_{q_ff}) to form the reference currents (i_d^{ref}, i_q^{ref}), which are used as inputs to the inner current controller.

The current control structure remains the same as previously described. Since both voltage and current in the *dq*-domain are controlled, the system adopts a double-loop control architecture.

The double-loop controller is designed using the decoupled model transfer functions $G_1(s)$ and $G_3(s)$. As shown in Figure 12.35, the overall control structure consists of two cascaded single-loop controllers: the inner loop is the current controller

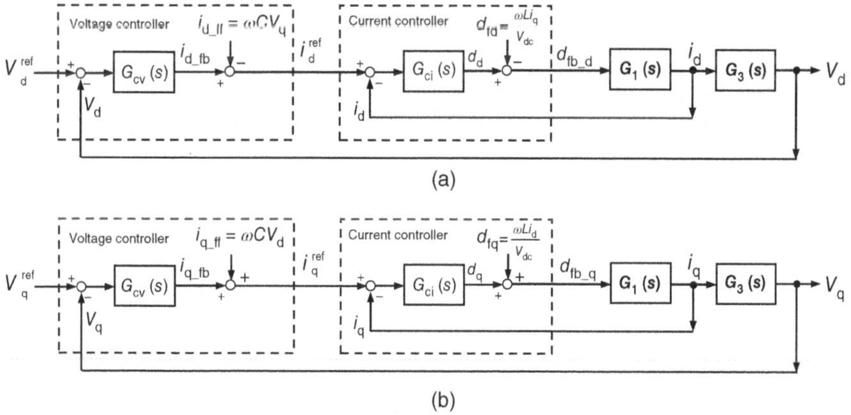

Figure 12.35 Double-loop controller configuration. (a) d-axis controller configuration and (b) q-axis controller configuration.

Table 12.8 Design specification for double-loop controller.

Design specification	Voltage controller (C_1)	Current controller (C_2)
Bandwidth [rad/sec]	50	4000
Phase margin [deg]	90	60

Table 12.9 Controller gains in continuous-time and discrete-time formats.

	Voltage controller (C_1)	Current controller (C_2)
Continuous time	$\frac{0.0177(s+114)}{s}$	$\frac{0.00189(s+3023)}{s}$
Discrete time	$\frac{0.0208(z-0.995)}{z-1}$	$\frac{0.00217(z-0.887)}{z-1}$

Note: The *PI* controller gains are identical for both the *d*-axis and *q*-axis.

$G_{ci}(s)$, which operates with fast dynamics, and the outer loop is the voltage controller $G_{cv}(s)$, designed for slower regulation.

Both controllers are implemented using *PI* structures, and their gains must be determined in both continuous-time and discrete-time domains.

1) Design criteria are selected to ensure stable and fast regulation while preventing interaction between the inner and outer loops. The frequency-domain design specifications for the double loop controllers are summarized in Table 12.8.
2) Controller Gains (Continuous and Discrete-Time) by Auto-Tuning (f_{sw} = 20 *kHz*) are presented in Table 12.9.
   The MATLAB m-file for double-loop controller is presented in Table 12.10.
   The frequency response (Bode plot) of both controllers is shown in Figure 12.36:
   - The left plot represents the voltage controller, which satisfies the design targets of 50 rad/sec bandwidth and 90° phase margin.
   - The right plot represents the current controller, with a bandwidth of 4000 rad/sec and 60° phase margin.

The closed-loop step response from V_d^{ref} to V_d is shown in Figure 12.37. As expected, the waveform exhibits no overshoot. The settling time is 77 [ms], and the rising time is 44 [ms]. These results confirm the proper tuning and effectiveness of the double-loop controller.

Table 12.10 MATLAB m-file for double-loop controller.

```
% 3-Phase Double Loop Controller Design

Vin = 650;
Omega = 377;
L = 1e-3;     % L=1mh
rL = 0.01;   % rL = 10mohm
C = 100e-6; % C=100uf
R = 25;

Den=(s^2+s*(rL/L+1/(R*C))+(1+rL/R)/(L*C));
G1=(Vin/L)*(s+1/(R*C)) / Den;   % Idq/d
G2=Vin/(L*C) / Den;               % Vdq/d
G3=(1/C)/(s+1/(R*C));      % Vdq(s)/Idq(s)

Vd_ref = 220*sqrt(2);

Kpi =1; Kii =1;
Gci = Kpi*(s+Kii)/s;
Kpv =1; Kiv =1;
Gcv = Kpv*(s+Kiv)/s;

%%%  sisotool Double Loop  %%%
T=sisoinit(6);
T.F.Value = Vd_ref;
T.G1.Value = G1;
T.C2.Value = Gci;

T.G2.Value = G3;
T.C1.Value = Gcv;
sisotool(T);
```

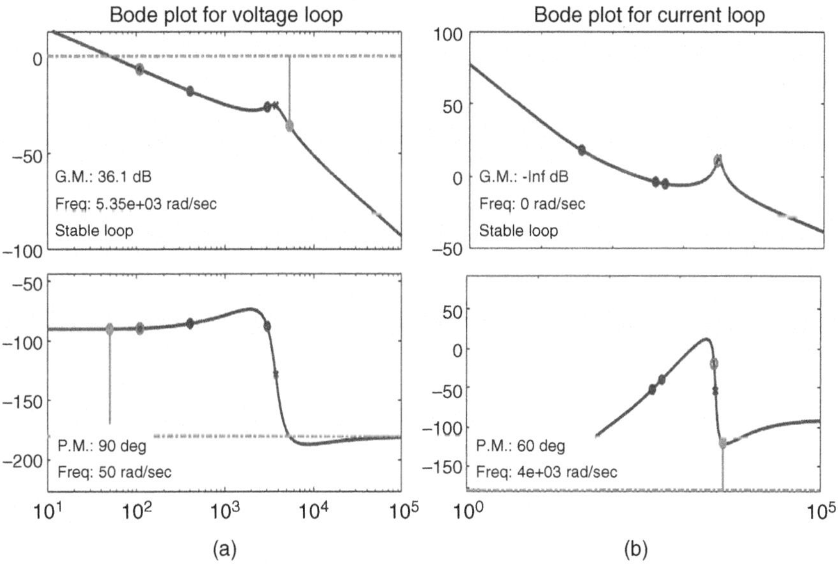

Figure 12.36 Bode plots of the voltage and current controllers.

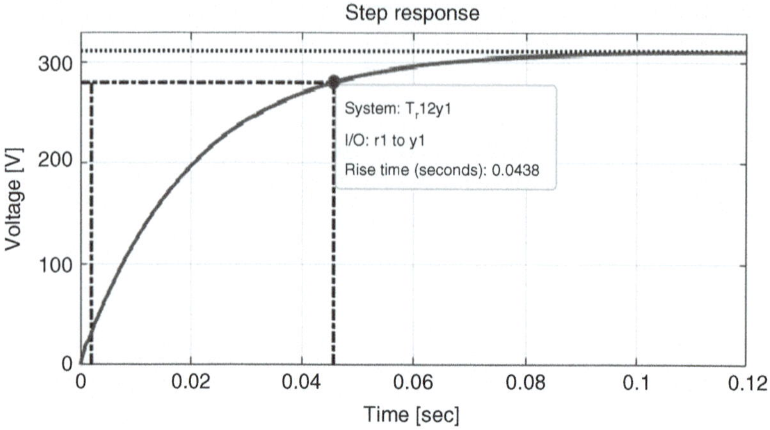

Figure 12.37 Step response of the closed-loop voltage control system.

PSIM Simulation of the Double-Loop Controller To verify the performance of the designed *PI* gains, a PSIM simulation was conducted using the double-loop control architecture. The circuit configuration in the simulation schematic remains identical to the current and voltage controllers in previous sections. The only modification lies in the C-block code, which implements the double-loop control logic. The implementation of this double-loop controller C-block is shown in Table 12.11.

Figure 12.38 presents the simulation waveforms of the double-loop controller. In this test, the d-axis reference voltage V_d^{ref} was step-changed from 155 V to 310 V, while the q-axis reference voltage V_q^{ref} was fixed at 0 V.

- The d-axis voltage V_d accurately tracks the reference with zero overshoot, as designed.
- The q-axis voltage V_q is regulated to zero with high precision.
- The d-axis current i_d attempts to follow the reference i_{d_ref}. Initially, a tracking error appears due to the rapidly changing reference current during the transient period. However, once the reference stabilizes, the steady-state error reduces to zero. This is expected behavior, as the current lags while trying to follow the continuously updating reference caused by the changing V_d.

To further investigate performance, the bandwidth of the voltage controller was reduced to 10 rad/sec. As shown in Figure 12.39, this results in a slower voltage rise time, which in turn reduces the rate of change in the reference current. This gentler current transition improves overall control stability and dynamic response.

Table 12.11 C-block code: DQ_Double.

```
#define Lf   0.001   // 1 [mh]
#define rL   0.01    // 10 [mohm]
#define Cf   0.0001 // 100 [uf]
#define R    25      // 25 [ohm]
#define Omega 377   // 60 [Hz]

#define Kip   0.0021  // Current P- Kip
#define Kii   0.887   // Currentr I-Kiz

#define Kvp   0.021   // Voltage P-  Kvp
#define Kvi   0.995   // Voltage I   Kvz

double ILa, ILb, ILc, Van, Vbn, Vcn, Vdc,
theta, Vd_ref, Valpha, Vbeta;
double Ialpha, Ibeta, Vd, Vq, ILd, ILq;
double Vq_ref, Id_ref, Iq_ref, Va_con,
Vb_con, Vc_con, duty_a, duty_b, duty_c;
double Id_fb, Iq_fb, Id_ff, Iq_ff;
double dd, dq, dfb_d, dfb_q, Id_err, Iq_err,
Vd_err, Vq_err;
static double Id_fb_old=0, Vd_err_old=0,
Iq_fb_old=0, Vq_err_old=0;
static double dd_old=0.0, dq_old=0.0,
 Id_err_old=0.0, Iq_err_old=0.0;
```

```
Id_fb=Id_fb_old+Kvp*Vd_err-Kvp*Kvi*Vd_err_old;
Id_fb_old=Id_fb;
Vd_err_old=Vd_err;

/****** Current controller dq *******/
Id_ff = Cf*Omega*Vq;
Id_ref= Id_fb-Id_ff;  // id reference

Vq_err = Vq_ref - Vq;
Iq_fb=Iq_fb_old+Kvp*Vq_err-Kvp*Kvi*Vq_err_old;
Iq_fb_old=Iq_fb;
Vq_err_old=Vq_err;

Iq_ff =  Cf*Omega*Vd;
Iq_ref= Iq_fb+Iq_ff;//iq reference
Id_err = Id_ref - ILd;
dd = dd_old + Kip*Id_err -Kip*Kii*Id_err_old;
dd_old=dd;
Id_err_old=Id_err;

Iq_err = Iq_ref - ILq;
dq = dq_old +Kip*Iq_err-Kip*Kii*Iq_err_old;
dq_old=dq;
Iq_err_old=Iq_err;
```

```
/**** Input Variable    ****/
ILa        = in[0];
ILb        = in[1];
ILc        = in[2];
Van        = in[3];
Vbn        = in[4];
Vcn        = in[5];
Vdc        = in[6];
theta      = in[7];

if (Vdc == 0)Vdc =0.01; divide by 0

Valpha=(2./3.)*Van-(1/3.)*Vbn-(1/3.)*Vcn;
Vbeta = (Vbn-Vcn)/sqrt(3);
Vd =  Valpha*cos(theta) + Vbeta*sin(theta);
Vq = -Valpha*sin(theta) + Vbeta*cos(theta);

Ialpha=(2./3.)*ILa-(1/3.)*ILb-(1/3.)*ILc;
Ibeta =(ILb-ILc)/sqrt(3);
ILd =  Ialpha*cos(theta)+Ibeta*sin(theta);
ILq = -Ialpha*sin(theta)+Ibeta*cos(theta);

if (t < 0.1) Vd_ref = 155;
else Vd_ref = 310;
Vq_ref = 0;

/****** Voltage controller dq *******/
Vd_err = Vd_ref - Vd;

dfb_d = dd  - Lf*Omega*ILq/(Vdc);
dfb_q = dq +  Lf*Omega*ILd/(Vdc);
/****** End of controller ***********/

Valpha=dfb_d*cos(theta)-dfb_q*sin(theta);
Vbeta= dfb_d*sin(theta)+dfb_q*cos(theta);

Va_con = Valpha;
Vb_con = -0.5*Valpha + sqrt(3)*Vbeta/2;
Vc_con = -0.5*Valpha - sqrt(3)*Vbeta/2;

// duty = dff + dfb = 0.5 + dfb
duty_a =  0.5 + Va_con;
duty_b =  0.5 + Vb_con;
duty_c =  0.5 + Vc_con;

out[0] = duty_a;    // Inverter R
out[1] = duty_b;    // Inverter S
out[2] = duty_c;    // Inverter T
out[3] = Vd;
out[4] = Vq;
out[5] = ILd;
out[6] = ILq;
out[7] = dd;
out[8] = dq;
out[9] = Id_ref;
out[10]= Iq_ref;
```

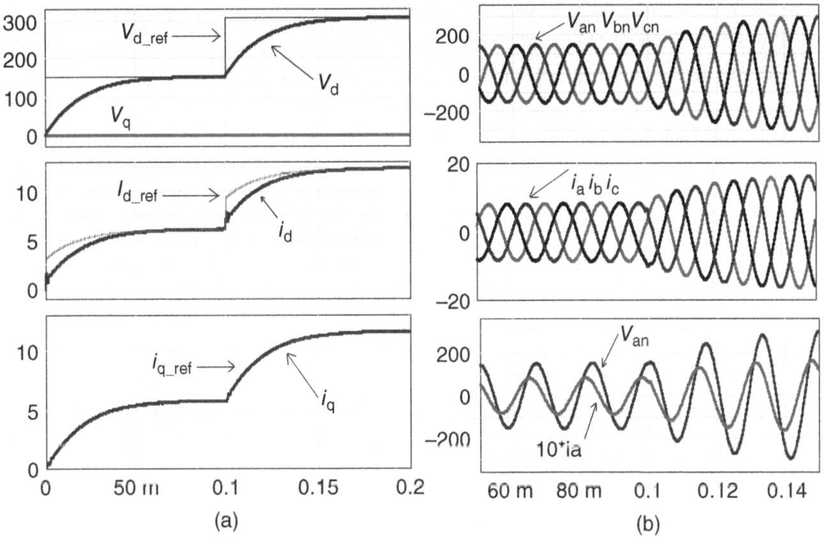

Figure 12.38 Simulation waveforms of the double-loop controller. (a) dq-axis voltages and currents and (b) abc-phase voltages and currents.

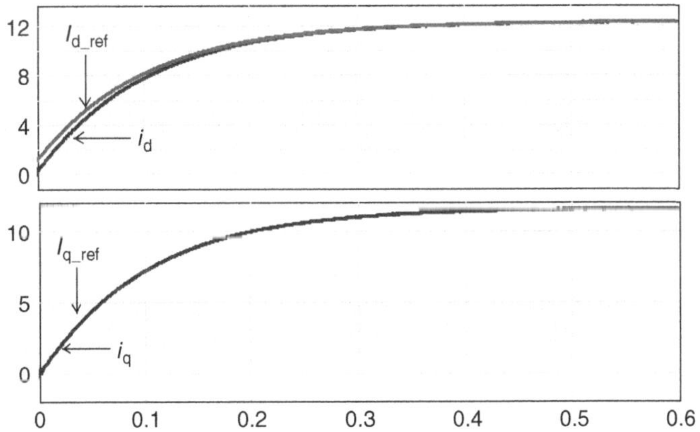

Figure 12.39 Simulation results for voltage controller bandwidth of 10 rad/sec.

12.3.2 Three-Phase Grid-Tied Inverter (PCS)

A three-phase power conversion system (PCS) can be obtained from the stand-alone three-phase inverter by replacing the output capacitor with the three-phase grid. Because the terminal voltage is now imposed by the grid, direct voltage regulation is not feasible; only the inductor current can be controlled.

Consequently, the dq-current controller developed in the previous chapter can be reused without structural modification for PCS operation.

Rewriting current equations of the PCS in the decoupling model:

$$\frac{d}{dt}\left(\begin{bmatrix} i_d \\ i_q \end{bmatrix}\right) = \omega \begin{bmatrix} i_q \\ -i_d \end{bmatrix} - \frac{r_L}{L} \cdot \begin{bmatrix} i_d \\ i_q \end{bmatrix} - \frac{1}{L} \cdot \begin{bmatrix} V_d \\ V_q \end{bmatrix} + \frac{V_{\text{dc}}}{L} \cdot \begin{bmatrix} d_{fb\_d} \\ d_{fb\_q} \end{bmatrix}$$

Let the control inputs be decomposed as:

$$d_{fb\_d} = d_d + d_{fd}, \quad d_{fb\_q} = d_q + d_{fq}$$

Since grid voltage V_d and V_q are calculated using dq-transformation, choose the decoupling feedforward terms as

$$d_{fd} = \frac{V_d - \omega L i_q}{V_{\text{dc}}}, \quad d_{fq} = \frac{V_q + \omega L i_d}{V_{\text{dc}}}$$

which cancels the grid-voltage and cross-coupling terms. Substituting yields the linearized current subsystem

$$\frac{d}{dt}\left(\begin{bmatrix} i_d \\ i_q \end{bmatrix}\right) = -\frac{r_L}{L} \cdot \begin{bmatrix} i_d \\ i_q \end{bmatrix} + \frac{V_{\text{dc}}}{L} \cdot \begin{bmatrix} d_d \\ d_q \end{bmatrix}$$

Hence, each axis reduces to an identical first-order plant with transfer function

$$G_{1,\text{pcs}}(s) = \frac{I_{dq}(s)}{d_{dq}(s)} = \frac{\frac{V_{\text{dc}}}{L}}{\left(s + \frac{r_L}{L}\right)}$$

Therefore, the same design specifications and PI tuning used for the stand-alone inverter's current controller apply directly to the PCS, provided the decoupling/feedforward is enabled.

Figure 12.40 shows the PSIM schematic for the 3-phase PCS. The circuit is nearly identical to the dq-current control system presented earlier, except the output capacitor has been replaced with the grid, and the controller sets $i_{q\_ref} = 0$. A step change is applied to i_{d_ref}, increasing from 6 A to 12 A at $t = 0.1$ seconds.

Figure 12.41 presents the simulation results for both PCS mode and Rectifier mode operations.

- Figure 12.41(a): PCS mode

  In this mode, i_{d_ref} is positive, indicating power is delivered from the DC side to the grid. The d-axis current is controlled from 6 A to 12 A with zero steady-state error, and i_q is regulated to 0 A, indicating zero reactive power. The a-phase current i_a is in phase with the grid voltage V_{an}, showing unity power factor operation. The inverter current I_{inv} exhibits a 180 Hz ripple, and its average value follows the inductor current precisely, increasing from 6 A to 12 A.

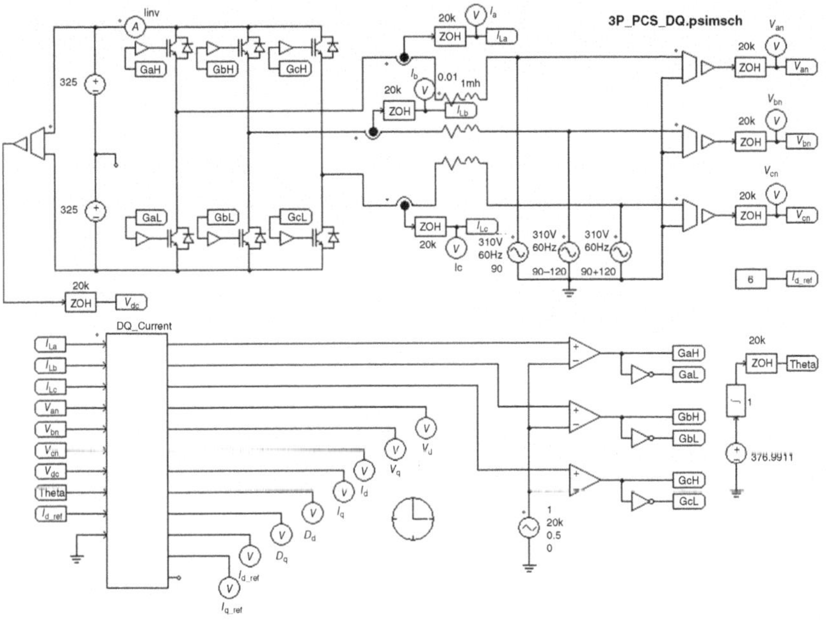

Figure 12.40 Schematic of the three-phase PCS.

- Figure 12.41(b): Rectifier mode

  In this case, i_{d_ref} is negative, indicating that power flows from the grid to the DC side. The d-axis current is regulated from −6 A to −12 A with zero steady-state error, and again i_q remains at 0 A. The a-phase current i_a is 180° out of phase with the grid voltage V_{an}, demonstrating that power is now being drawn from the grid. As in PCS mode, the inverter current shows a 180 Hz ripple, with an average value matching the inductor current.

12.3.3 PWM Rectifier (AC/DC PFC Converter)

When the output of a rectifier is not a pure voltage source but instead consists of a capacitor and a resistor, closed-loop control of the DC output voltage becomes possible. Such a configuration is referred to as a three-phase AC/DC converter or PWM rectifier, and its control methodology is analogous to that of a single-phase rectifier. The key distinction lies in the application of the dq-transformation, which yields a DC reference current for control purposes.

Figure 12.42 illustrates the schematic diagram of the three-phase rectifier circuit. The control objectives are twofold:

1) DC voltage regulation—to maintain the DC output voltage at its reference value according to the design specification.

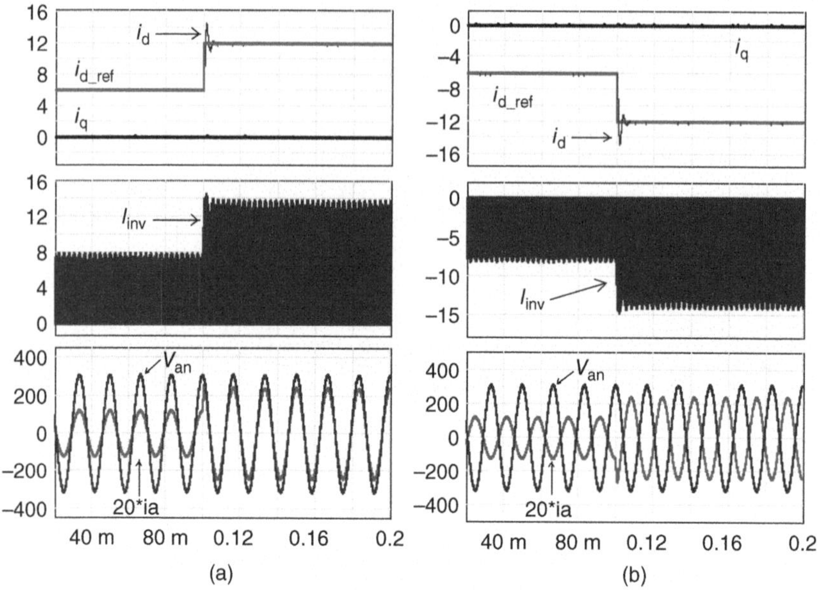

Figure 12.41 Bi-directional current flow in PCS and Rectifier modes. (a) PCS mode ($i_{d\_ref} = 6A$) and (b) Rectifier mode ($i_{d\_ref} = -6A$).

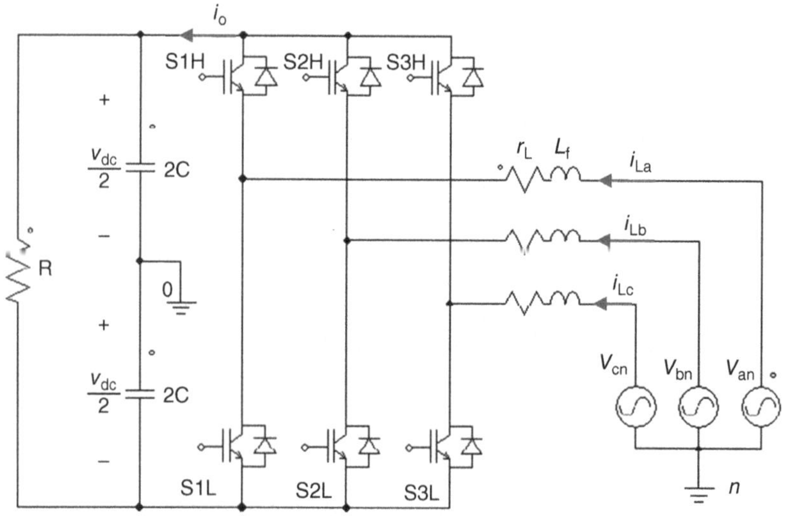

Figure 12.42 Schematic diagram of the three-phase rectifier circuit.

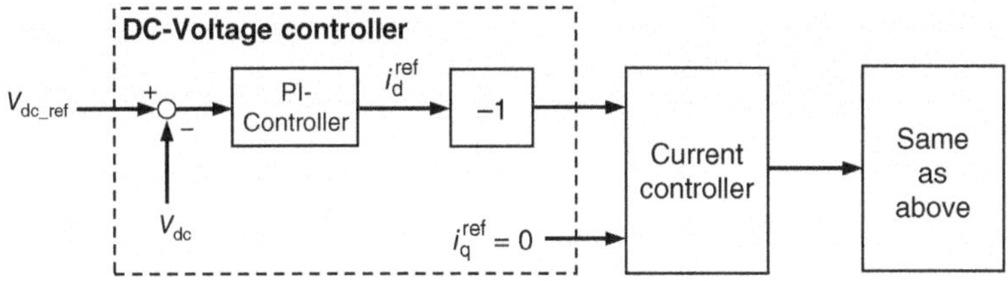

Figure 12.43 PI voltage controller configuration for the three-phase AC/DC system.

2) Power quality enhancement—to synchronize the grid current with the grid voltage, thereby ensuring unity power factor operation.

Please note that each capacitor in the series connection should have a capacitance of **2*C*** so that the resulting equivalent capacitance becomes ***C***.

The controller structure is presented in Figure 12.43. The measured DC output voltage is compared against its reference, and the resulting error is processed by a *PI* controller. The *PI* controller generates the reference direct-axis current, i_d^{ref}. Before being applied to the current controller, this value is multiplied by −1 to compensate for the direction of current flow, thereby maintaining consistency with the PCS operation. Meanwhile, the quadrature-axis current reference, i_q^{ref}, is set to zero since no reactive power is required in the DC output stage.

Small-Signal Modeling of the Rectifier The small-signal modeling procedure of the three-phase rectifier is carried out systematically in six steps.

Step 1. Current Equation

For the a-phase current, the state equation depends on the switching state of the upper device S_{1H} and the lower device S_{1L}.

- High ON(S_{1H}): with duty ratio d_a

$$V_{an} = r_L i_{La} + L\frac{di_{La}}{dt} + \frac{v_{dc}}{2}$$

- Low ON(S_{1L}): with duty ratio (1-d_a)

$$V_{an} = r_L i_{La} + L\frac{di_{La}}{dt} - \frac{v_{dc}}{2}$$

Averaging these two conditions yields:

$$\frac{di_{La}}{dt} = -\frac{r_L}{L}i_{La} - \frac{v_{dc}}{2L}(2d_a - 1) + \frac{V_{an}}{L}$$

Expanding this expression to three phases gives:

$$\frac{d}{dt}i_{L,abc} = -\frac{r_L}{L}i_{L,abc} - \frac{v_{dc}}{2L}(2d_{abc} - 1) + \frac{V_{abc,n}}{L} \tag{12.24}$$

Step 2. Voltage Equation

The DC-link voltage dynamics are expressed as:

$$\frac{dv_{dc}}{dt} = \frac{1}{C}\left(i_o - \frac{v_{dc}}{R}\right)$$

where the output current is:

$$\begin{aligned} i_o &= (d_a i_{La} + d_b i_{Lb} + d_c i_{Lc}) - ((1 - d_a)i_{La} + (1 - d_b)i_{Lb} + (1 - d_c)i_{Lc}) \\ &= (2d_a - 1)i_{La} + (2d_b - 1)i_{Lb} + (2d_c - 1)i_{Lc} \end{aligned}$$

Thus, the voltage equation becomes:

$$\frac{dv_{dc}}{dt} = \frac{1}{C}((2d_a - 1)i_{La} + (2d_b - 1)i_{Lb} + (2d_c - 1)i_{Lc}) - \frac{v_{dc}}{RC} \tag{12.25}$$

Step 3. State-Averaged Equations

The nonlinear system represented by Eqs. (12.24) and (12.25) is simplified by decomposing the duty ratio:

$$d_{abc} = d_{abc,fb} + d_{abc,ff} \quad \rightarrow d_{abc,ff} = 0.5$$

Substituting, the state-averaged model becomes:

$$\frac{d}{dt}i_{L,abc} = -\frac{r_L}{L}i_{L,abc} - \frac{v_{dc}}{L}d_{abc,fb} + \frac{V_{abc,n}}{L} \tag{12.26}$$

$$\frac{dv_{dc}}{dt} = \frac{1}{C}(2d_{a,fb} \cdot i_{La} + 2d_{b,fb} \cdot i_{Lb} + 2d_{c,fb} \cdot i_{Lc}) - \frac{v_{dc}}{RC} \tag{12.27}$$

Step 4. DQ Transform

Applying the Park transformation:

$$\begin{bmatrix} a \\ b \\ c \end{bmatrix} = P \cdot \begin{bmatrix} d \\ q \end{bmatrix}, \quad \begin{bmatrix} d \\ q \end{bmatrix} = P^{-1} \cdot \begin{bmatrix} a \\ b \\ c \end{bmatrix}$$

where transformation matrix P and its inverse P^{-1} are defined as:

$$P = \begin{bmatrix} \cos\emptyset & -\sin\emptyset \\ \dfrac{-\cos\emptyset + \sqrt{3}\sin\emptyset}{2} & \dfrac{\sin\emptyset + \sqrt{3}\cos\emptyset}{2} \\ \dfrac{-\cos\emptyset - \sqrt{3}\sin\emptyset}{2} & \dfrac{\sin\emptyset - \sqrt{3}\cos\emptyset}{2} \end{bmatrix},$$

$$P^{-1} = \begin{bmatrix} \cos\emptyset & \dfrac{\sin\emptyset}{\sqrt{3}} & -\dfrac{\sin\emptyset}{\sqrt{3}} \\ -\sin\emptyset & \dfrac{\cos\emptyset}{\sqrt{3}} & -\dfrac{\cos\emptyset}{\sqrt{3}} \end{bmatrix}, \quad \emptyset = \omega t$$

The time derivative of the three-phase variables can be expressed as:

$$\frac{d}{dt}(abc) = \omega P \begin{bmatrix} -q \\ d \end{bmatrix} + P \cdot \frac{d}{dt}(dq)$$

Applying this transformation to (12.26) and (12.27) yields the following dq-frame model:

$$\frac{d}{dt}\left(\begin{bmatrix} i_d \\ i_q \end{bmatrix}\right) = \omega \begin{bmatrix} i_q \\ -i_d \end{bmatrix} - \frac{r_L}{L} \cdot \begin{bmatrix} i_d \\ i_q \end{bmatrix} - \frac{v_{\text{dc}}}{L} \cdot \begin{bmatrix} d_d \\ d_q \end{bmatrix} + \frac{1}{L} \begin{bmatrix} V_d \\ V_q \end{bmatrix} \tag{12.28}$$

$$\frac{dv_{\text{dc}}}{dt} = \frac{3}{C}(d_d \cdot i_d + d_q \cdot i_q) - \frac{v_{\text{dc}}}{RC} \tag{12.29}$$

Additionally, the following relationship holds between the duty ratios and the inductor currents in both frames:

$$\begin{bmatrix} \mathrm{d}_a & \mathrm{d}_b & \mathrm{d}_c \end{bmatrix} \begin{bmatrix} \mathrm{i}_{La} \\ \mathrm{i}_{Lb} \\ \mathrm{i}_{Lc} \end{bmatrix} = \begin{bmatrix} \mathrm{d}_d & \mathrm{d}_q \end{bmatrix} P^T P \begin{bmatrix} \mathrm{i}_d \\ \mathrm{i}_q \end{bmatrix}$$

Since

$$P^T P = \frac{3}{2} \begin{bmatrix} 1 & 0 \\ 0 & 1 \end{bmatrix}$$

the equivalence between the three-phase and dq reference frames is preserved.

Step 5. Small-Signal Linearization

Using the decoupling model:

$$d_d = d_{d,fd} + d_{d,ff}, \quad d_q = d_{q,fd} + d_{q,ff} \Rightarrow \quad d_{d,ff} = \frac{\omega L i_q}{v_{dc}}, \quad d_{q,ff} = -\frac{\omega L i_d}{v_{dc}}$$

The linearized equations become:

$$\frac{d}{dt}\left(\begin{bmatrix} i_d \\ i_q \end{bmatrix}\right) = -\frac{r_L}{L} \cdot \begin{bmatrix} i_d \\ i_q \end{bmatrix} - \frac{v_{dc}}{L} \cdot \begin{bmatrix} d_{d,fb} \\ d_{q,fb} \end{bmatrix} + \frac{1}{L} \begin{bmatrix} V_d \\ V_q \end{bmatrix} \tag{12.30}$$

$$\frac{dv_{\text{dc}}}{dt} = \frac{3}{C}(d_d \cdot i_d + d_q \cdot i_q) - \frac{v_{\text{dc}}}{RC} \tag{12.31}$$

Applying perturbations:

$$i_{dq} = I_{DQ} + \hat{i}_{dq}, \quad d_{d,fb} = D_d + \hat{d}_{d,fb}, \qquad d_{q,fb} = D_q + \hat{d}_{q,fb}, \quad v_{\text{dc}} = V_{\text{DC}} + \hat{v}_{\text{dc}}$$

From the current equation:

$$\frac{d}{dt}(I_{DQ} + \hat{i}_{dq}) = -\frac{r_L}{L}(I_{DQ} + \hat{i}_{dq}) - \frac{(V_{\text{DC}} + \hat{v}_{\text{dc}})}{L}(D_{DQ} + \hat{d}_{dq,fb}) + \frac{1}{L}(V_{DQ} + \hat{v}_{dq})$$

$$\text{DC} \Rightarrow 0 = -\frac{r_L}{L} I_{DQ} - \frac{1}{L} D_{DQ} \cdot V_{DC} + \frac{1}{L} V_{DQ} \Rightarrow D_{DQ} = \frac{V_{DQ}}{V_{\text{DC}}}$$

$$AC \Rightarrow \frac{d}{dt}(\hat{i}_{dq}) = -\frac{r_L}{L}(\hat{i}_{dq}) - \frac{D_{DQ}}{L}(\hat{v}_{dc}) - \frac{V_{DC}}{L}(\hat{d}_{dq,fb})$$

From the voltage equation:

$$\frac{d}{dt}(V_{DC} + \hat{v}_{dc}) = \frac{3}{C} \cdot ((D_D + \hat{d}_{d,fb}) \cdot (I_D + \hat{i}_d) + (D_Q + \hat{d}_{q,fb}) \cdot (I_Q + \hat{i}_q)) - \frac{V_{DC} + \hat{v}_{dc}}{RC}$$

$$DC \Rightarrow 0 = \frac{3}{C} \cdot ((D_D \cdot I_D) + (D_Q \cdot I_Q)) - \frac{V_{DC}}{RC} \Rightarrow I_D = \frac{V_{DC}}{3RD_D}$$

$$AC \Rightarrow \frac{d}{dt}(\hat{v}_{dc}) = \frac{3}{C} \cdot ((D_D \cdot \hat{i}_d + D_Q \cdot \hat{i}_q) + (I_D \cdot \hat{d}_{d,fb} + I_Q \cdot \hat{d}_{q,fb})) - \frac{\hat{v}_{dc}}{RC}$$

Step 6. Final Small-Signal Model

In compact matrix form, the small-signal model is expressed as:

$$\begin{pmatrix} \dfrac{d\hat{i}_{dq}}{dt} \\ \dfrac{d(\hat{v}_{dc})}{dt} \end{pmatrix} = \begin{pmatrix} -\dfrac{r_L}{L} & -\dfrac{D_{DQ}}{L} \\ \dfrac{3D_{DQ}}{C} & -\dfrac{1}{RC} \end{pmatrix} \begin{pmatrix} \hat{i}_{dq} \\ \hat{v}_{dc} \end{pmatrix} + \begin{pmatrix} -\dfrac{V_{DC}}{L} \\ \dfrac{3I_{DQ}}{C} \end{pmatrix} \hat{d}_{dq,fb} \tag{12.32}$$

with steady-state values:

$$V_{DC} = \frac{V_{DQ}}{D_{DQ}}, \quad I_L = \frac{V_{DC}}{3RD_D}$$

The transfer function is derived using the MATLAB built-in function `ss2tf`, which converts the state-space model into its equivalent transfer function representation. The MATLAB m-file for rectifier voltage controller is presented in Table 12.12.

Controller Design and Verification The design specifications for the double loop controllers are presented in Table 12.13.

Figure 12.44 illustrates the SISOTOOL auto-tuning results. Subfigure (a) presents the voltage loop Bode plot, while Subfigure (b) shows the current loop response. The design successfully satisfies the target specifications––100 rad/sec bandwidth with a 90° phase margin for the voltage loop, and 2400 rad/s bandwidth with a 60° phase margin for the current loop.

Using the auto-tuning feature in MATLAB's SISOTOOL, the double loop controller was designed for a switching frequency of 20 kHz. The resulting continuous and discrete-time gains are summarized in Table 12.14.

The closed-loop step response of the designed controllers is shown in Figure 12.45. The output voltage accurately tracks the reference, achieving the specified performance. No steady-state error is observed, and the rise time is approximately 22 ms.

Table 12.12 MATLAB m-file for Rectifier voltage controller.

```
%-- 3-Phase PFC Controller Design --%
%% Current direction : Grid -> Dc

%% Power Circuit Parameters %%
s = tf('s');
VD = 310;
L = 1.0e-3;
C = 2000e-6;
R = 50;
rL = 0.01;

Dq = 0.0;
Dd = 0.4;

Vdc = VD/Dd;
ID = Vdc/(3*R*Dd);
IQ = 0;

%% Small signal model
A = [ -rL/L -Dd/L;
      3*Dd/(C) -1/(R*C)];
B = [-Vdc/L;
      3*ID/(C)];
C= [1 0]; D=0;

[num,den] = ss2tf(A,B,C,D);
G1 = tf(num,den);
C2 = [0 1];
[num2, den2] = ss2tf(A,B,C2, D);
G2 = tf(num2, den2);
G3 = tf(num2, num);

%% double loop controller
Vref = 700.0;
Kpv = 1; Kiv = 1;
Gcv = Kpv*(s+Kiv)/s;

Kpi = 1; Kii = 1;
Gci = Kpi*(s+Kii)/s;

T=sisoinit(6);
T.F.Value = Vref;
T.G1.Value = G1;
T.C2.Value = Gci;

T.G2.Value = G3;
T.C1.Value = Gcv;

sisotool(T);
```

Table 12.13 Design specification for double-loop controller.

Design specification	Voltage controller (C_1)	Current controller (C_2)
Bandwidth [rad/sec]	100	2400
Phase margin [deg]	90	60

PSIM Simulation The schematic of the three-phase PWM rectifier used for DC voltage control is illustrated in Figure 12.46.

The C-block code for the *DC_Voltage_Rectifier*, provided in Table 12.15, is implemented to evaluate the dynamic response of the controller. In this test, the reference DC voltage V_{dc_ref} is programmed to step from 670 to 700 V and then back to

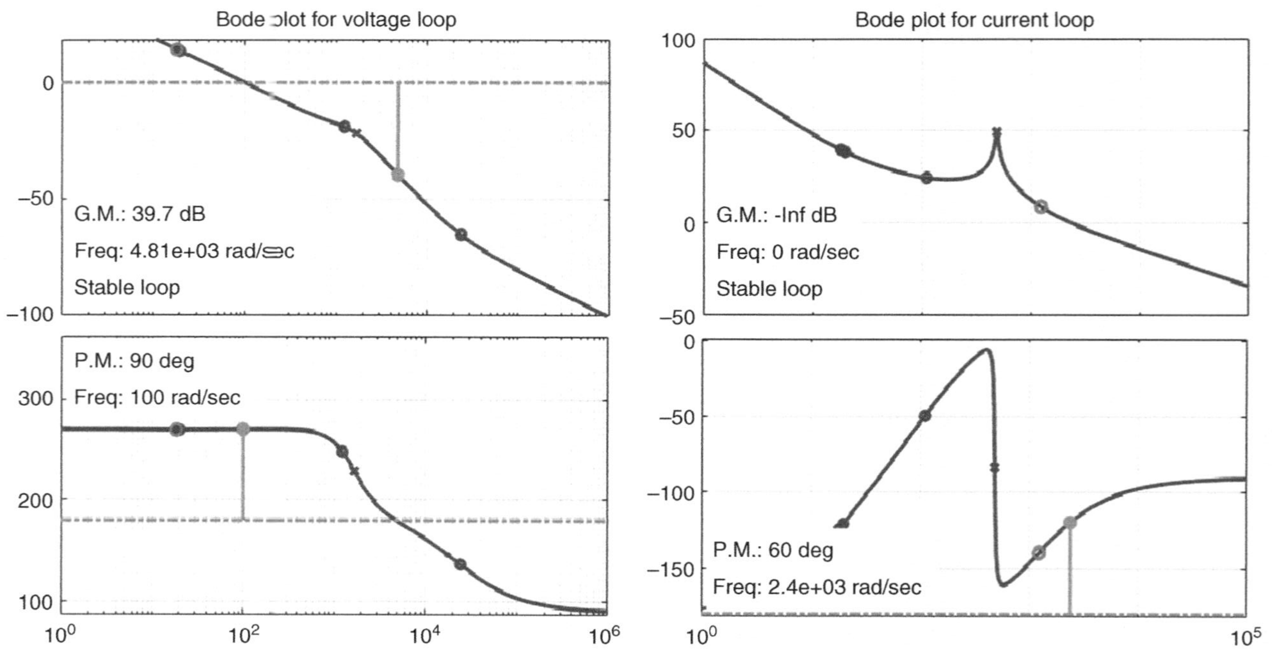

Figure 12.44 Bode plots of the designed controllers. (a) Voltage controller and (b) Current controller.

Table 12.14 Controller gains in continuous-time and discrete-time formats.

	Voltage controller (C_1)	Current controller (C_2)
Continuous time	$\frac{0.1819(s + 18.5)}{s}$	$-\frac{0.00264(s + 1240)}{s}$
Discrete time	$\frac{0.1839(z - 0.999)}{z - 1}$	$-\frac{0.0028(z - 0.948)}{z - 1}$

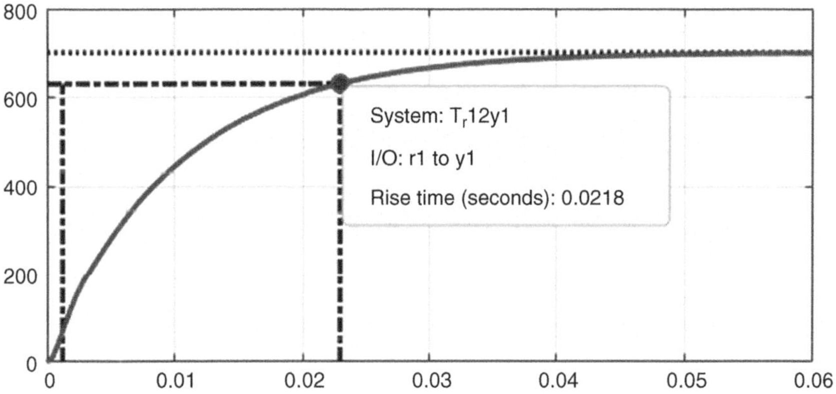

Figure 12.45 Step response of the designed controllers.

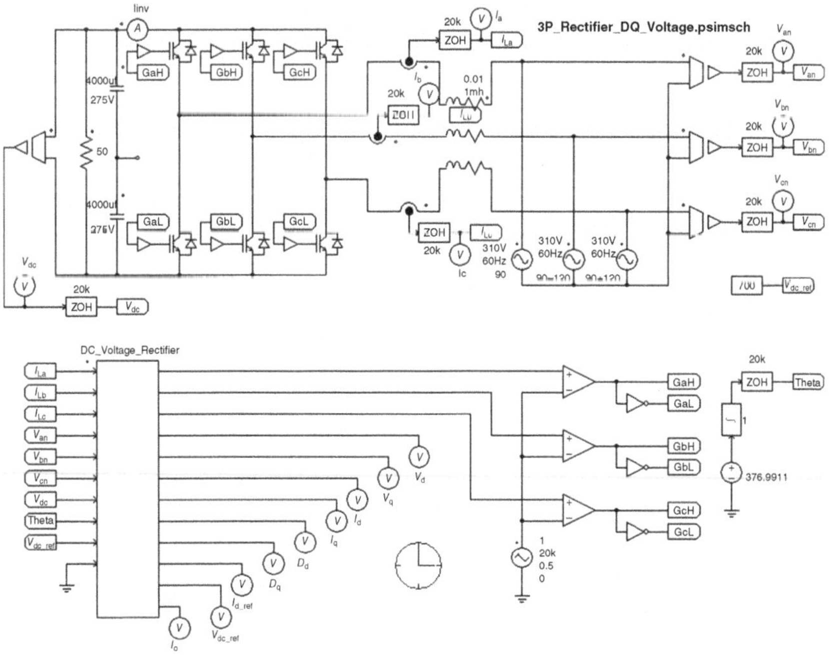

Figure 12.46 Schematic of the three-phase PWM Rectifier for DC voltage control.

Table 12.15 C-block code: DC_Voltage_Rectifier.

```
#define Lf      0.001     // 1  [mh]
#define rL      0.01      // 10   [mohm]
#define R       50        // 50 [ohm]
#define Omega 377         // 60 [Hz]

#define Kpv     0.1839    // voltage
#define Kiv     0.999
#define Kpi     0.0028    // current
#define Kii     0.948

double ILa, ILb, ILc, Van, Vbn, Vcn, Vdc,
theta, Id_ref;  // Input Variable
double Valpha, Vbeta,  Ialpha, Ibeta, Vd, Vq,
ILd, ILq, dd, dq; // DQ variable
double Vd_ref, Vq_ref, Va_con, Vb_con, Vc_con;
double duty_a, duty_b, duty_c, Io;
double dfb_d, dfb_q, Iq_ref, Id_err, Iq_err,
Vdc_err, Vdc_ref;
static double dd_old=0.0, dq_old=0.0,
Id_err_old=0.0,Iq_err_old=0.0,Vdc_err_sum=0.0;
static double Id_ref_old=0.0, Vdc_err_old=0.0;
/**** Input Variable   ****/
ILa     = in[0];
ILb     = in[1];
ILc     = in[2];
Van     = in[3];
Vbn     = in[4];
Vcn     = in[5];

Id_err =   -(Id_ref - ILd);
dd = dd_old+Kpi*Id_err-Kpi*Kii*Id_err_old;
Id_err_old=Id_err;
dd_old=dd;

Iq_ref = 0;
//Iq_ref = Omega*Cf*Vdc;
Iq_err = -(Iq_ref - ILq);
dq = dq_old+Kpi*Iq_err -Kpi*Kii*Iq_err_old;
dq_old=dq;
Iq_err_old=Iq_err;

dfb_d = dd + Lf*Omega*ILq/(Vdc);
dfb_q = dq - Lf*Omega*ILd/(Vdc);

Valpha = dfb_d*cos(theta)-dfb_q*sin(theta);
Vbeta  = dfb_d*sin(theta)+dfb_q*cos(theta);

Va_con = Valpha;
Vb_con = -0.5*Valpha + sqrt(3)*Vbeta/2;
Vc_con = -0.5*Valpha - sqrt(3)*Vbeta/2;

// duty = dff + dfb = 0.5 + dfb
duty_a =  0.5 + Va_con;
duty_b =  0.5 + Vb_con;
duty_c =  0.5 + Vc_con;
```

```
Vdc     = in[6];
theta   = in[7];
Vdc_ref= in[8];

if (Vdc <0.1 ) Vdc =0.1;
if (t  < 0.2)      Vdc_ref = 670.;
else if ( t< 0.3) Vdc_ref = 700.;
else               Vdc_ref = 670.;

Valpha =(2./3.)*Van - (1/3.)*Vbn -(1/3.)*Vcn;
Vbeta = (Vbn-Vcn)/sqrt(3);
Vd =  Valpha*cos(theta) + Vbeta*sin(theta);
Vq = -Valpha*sin(theta) + Vbeta*cos(theta);

Ialpha =(2./3.)*ILa - (1/3.)*ILb -(1/3.)*ILc;
Ibeta = (ILb-ILc)/sqrt(3);
ILd =  Ialpha*cos(theta) + Ibeta*sin(theta);
ILq = -Ialpha*sin(theta) + Ibeta*cos(theta);

Vdc_err = (Vdc_ref-Vdc);
Id_ref =Id_ref_old + Kpv*Vdc_err-
Kpv*Kiv*Vdc_err_old;
Id_ref_old= Id_ref;
Vdc_err_old=Vdc_err;

if ( duty_a > 1.0 )          duty_a=1.0;
else if ( duty_a < 0.0 )     duty_a=0.0;
if ( duty_b > 1.0 )          duty_b=1.0;
else if ( duty_b < 0.0 )     duty_b=0.0;
if ( duty_c > 1.0 )          duty_c=1.0;
else if ( duty_c < 0.0 )     duty_c=0.0;

Io = duty_a*ILa +duty_b*ILb +duty_c*ILc;

out[0] = duty_a;    // Inverter R
out[1] = duty_b;    // Inverter S
out[2] = duty_c;    // Inverter T

out[3] = Vd;
out[4] = Vq;
out[5] = ILd;
out[6] = ILq;
out[7] = Id;
out[8] = Iq;
out[9]=Id_ref;
out[10]=Vdc_ref;
out[11] = Io;
```

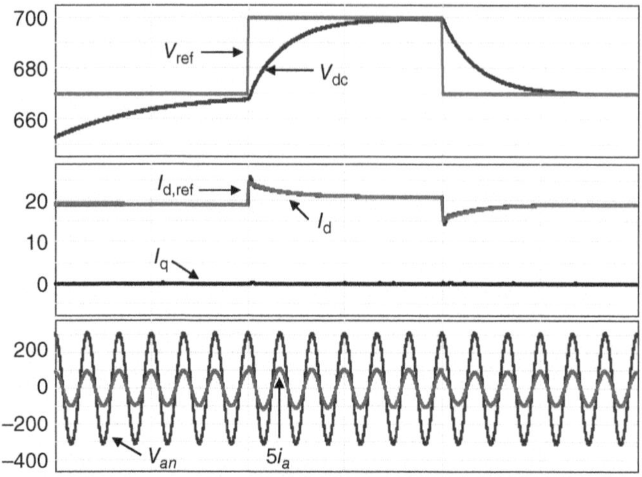

Figure 12.47 Simulation waveforms of the three-phase AC/DC rectifier.

670 V. The current controller structure remains identical to that used in the PCS design.

The corresponding simulation waveforms are presented in Figure 12.47.

a) DC Voltage Response – The output voltage precisely reaches the reference value without overshoot, and the steady-state error is zero.
b) Grid Current Response – The direction of current flow is opposite to that of PCS operation, as expected in rectifier mode where power is drawn from the grid and delivered to the DC side. The a-phase current i_a remains synchronized with the grid voltage V_{an}, thereby confirming unity power factor operation.

12.4 Space Vector Pulse Width Modulation

Pulse width modulation (PWM) is a fundamental technique for controlling the output voltage of voltage-source inverters (VSIs). Among various PWM strategies, sinusoidal PWM (SPWM) and space vector PWM (SVPWM) are the most widely adopted in industrial and academic applications. Although both methods aim to produce a three-phase voltage system with minimal harmonic distortion, their underlying principles and performance characteristics differ significantly.

In SPWM, three sinusoidal reference signals, each phase-shifted by 120°, are compared with a common high-frequency triangular carrier wave. In contrast, SVPWM represents the three-phase system as a single rotating reference vector in the $\alpha\beta$-plane and synthesizes the inverter switching states to reproduce this vector. Compared with conventional SPWM, SVPWM provides higher DC-bus voltage utilization and lower harmonic distortion.

The voltage utilization ratio indicates the effectiveness with which the available DC-bus voltage is utilized to generate the fundamental AC output voltage. The maximum line-to-line voltage for SPWM is $\frac{1}{2}V_{dc}$, while that for SVPWM is $\frac{\sqrt{3}}{2}V_{dc}$. Thus, the voltage utilization ratios are 78.5% for SPWM and 90.7% for SVPWM, respectively. Consequently, SVPWM produces a higher output-voltage amplitude without increasing the DC-bus voltage, resulting in greater power capability for inverter drives.

SPWM Min–Max Compensation for SVPWM Throughout this book, the controller designs have primarily employed the SPWM method. In certain cases, however, it is desirable to apply SVPWM to achieve higher output voltage. In such cases, the min–max compensation technique offers a convenient means to realize SVPWM behavior while retaining the simplicity of the SPWM framework.

This technique generates SVPWM-equivalent signals directly from the existing SPWM structure—without requiring coordinate transformation or explicit sector identification. The method modifies the instantaneous SPWM reference signals by compensating for their minimum and maximum values, thereby producing a modulation pattern equivalent to that of SVPWM. The detailed control block diagram is presented in Figure 12.48.

Figure 12.49 illustrates the phase-a waveform obtained from a simulation in which the min–max compensation method was applied to the SPWM to realize SVPWM. Here, V_{a_con} represents the conventional SPWM waveform, d_a denotes the SVPWM waveform, and V_{sn} indicates the offset value introduced by the min–max compensation.

To implement the current controller of the stand-alone inverter shown in Figure 12.28 using SVPWM, the C-block DQ_CURRENT of Table 12.5 was modified as shown in Table 12.16.

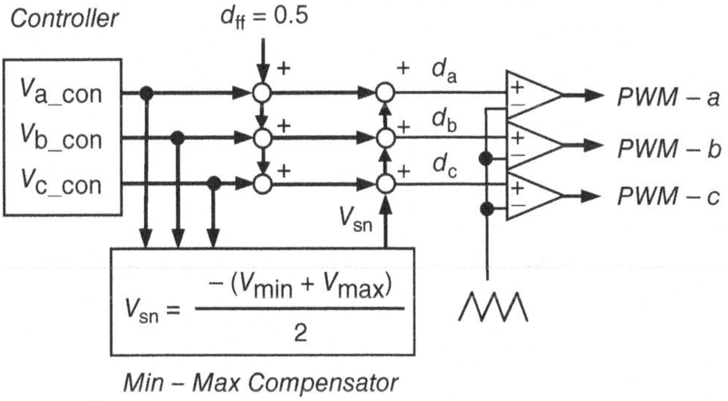

Figure 12.48 Implementation of SVPWM using min–max compensation of SPWM.

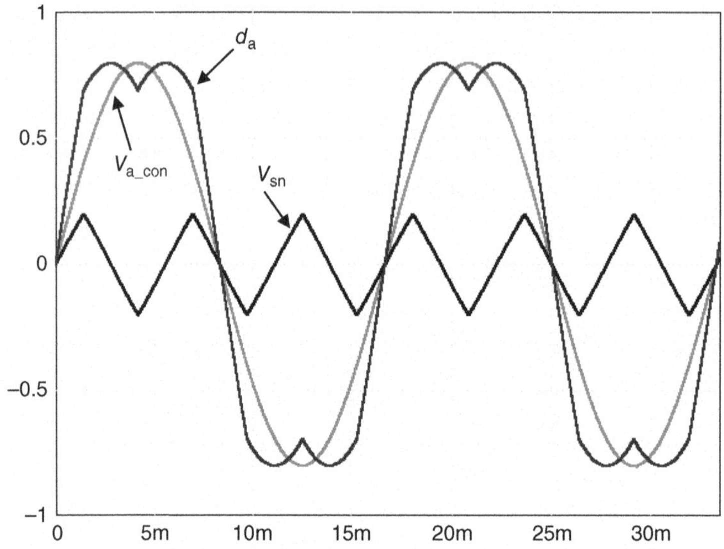

Figure 12.49 SPWM and SVPWM waveforms for *a*-phase.

Table 12.16 C-block:DQ_CURRENT_Space_Vector.

Same as Table 12.5 except for the following SVPWM section:

```
// Current Controller  (same as Table 12.5)
Va_con = Valpha;
Vb_con = -0.5*Valpha + sqrt(3)*Vbeta/2;
Vc_con = -0.5*Valpha - sqrt(3)*Vbeta/2;

/*********Insert Space Vector ***********/
double Vmax = fmax(fmax(Va_con, Vb_con), Vc_con);
double Vmin = fmin(fmin(Va_con, Vb_con), Vc_con);
double Vsn = -(Vmin+Vmax)/2;
double m = 1.154; // Modulation scaling factor

// duty = dff + dfb = 0.5 + dfb
duty_a =  0.5 + (Va_con/m + Vsn);
duty_b =  0.5 + (Vb_con/m + Vsn);
duty_c =  0.5 + (Vc_con/m + Vsn);
/*********End of Space Vector ***********/

out[0] = duty_a; // Inverter R
out[1] = duty_b; // Inverter S
out[2] = duty_c; // Inverter T
```

(Remaining portions of the code are identical to Table 12.5.*)*

13

Solar PCS

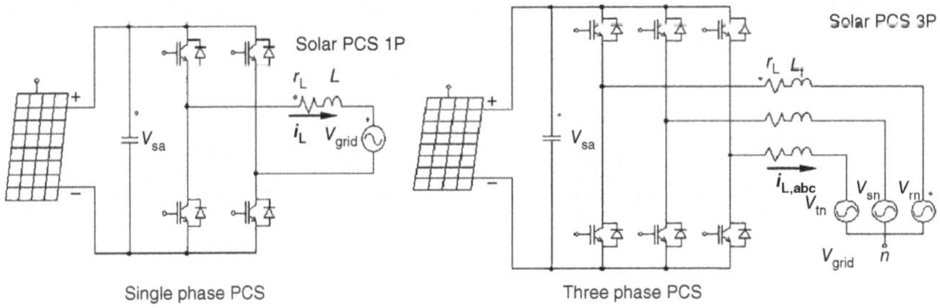

13.1 Single-Phase Solar PCS

Power conversion systems (PCS) refer to inverter-based configurations in which the output is directly interfaced with the electrical grid. A solar PCS specifically utilizes photovoltaic (PV) solar cells as the primary energy source, converting solar energy into grid-compatible electrical power.

The principal functions of a solar PCS can be summarized as follows:

- Maximum Power Extraction – To maximize the harvested energy, the voltage across the solar cells must be regulated to operate at the maximum power point (MPP), where the product of voltage and current yields the maximum power output.
- Grid Synchronization – To ensure efficient energy transfer and minimize reactive power exchange, the inverter's output current must remain synchronized in both frequency and phase with the grid voltage.

Figure 13.1 illustrates the configurations of solar PCS systems for single-phase and three-phase applications.

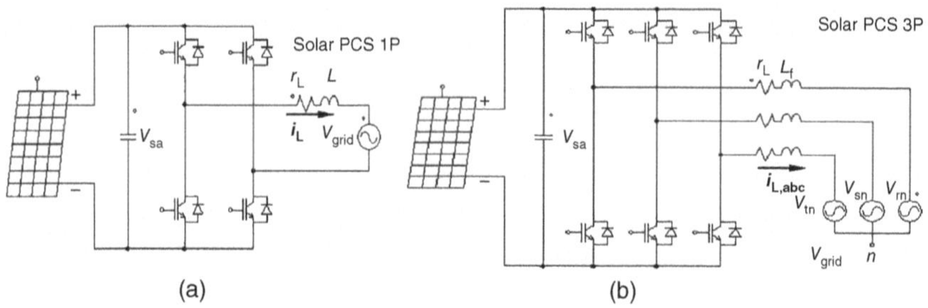

Figure 13.1 Structure of solar PCS. (a) Single-phase configuration and (b) Three-phase configuration.

Since the solar cells are directly connected to the input of either single-phase or three-phase PCS systems, the same current control strategy can be applied to both topologies. Similarly, the voltage controller responsible for maximum power point tracking (MPPT) can employ control methodologies widely adopted in conventional solar converter designs.

Figure 13.2 presents the schematic of a single-phase solar PCS, where the PV array supplies power to the input of the PCS. The key state variables in this system are defined as follows:

- The inductor current at the output stage, denoted as i_L.
- The capacitor voltage at the input stage connected to the PV array, denoted as v_{sa}.

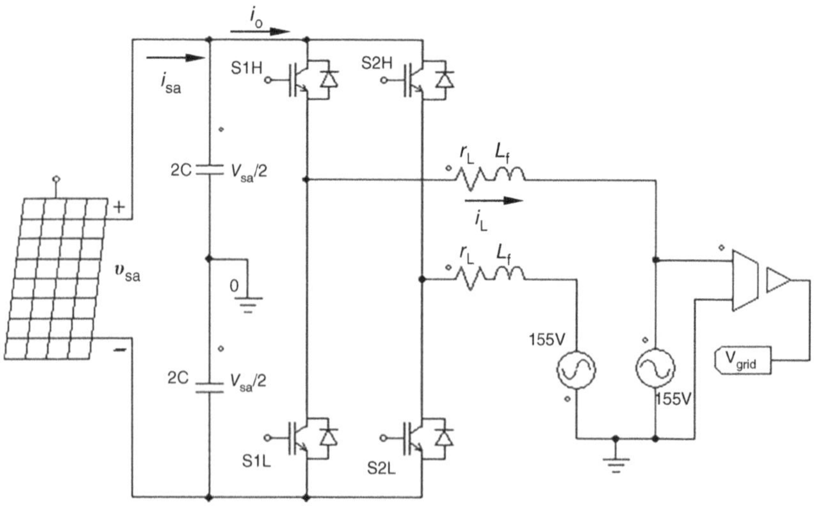

Figure 13.2 Single-phase solar PCS schematic.

13.1.1 State-Averaged Equations and Transfer Functions

Step 1. Current Equation

For the a-phase inductor current, the state equation depends on the switching state of the upper device S_{1H} and the lower device S_{1L}.

- Upper device ON(S_{1H}) with duty ratio d_a:

$$\frac{v_{sa}}{2} = r_L i_{La} + L\frac{di_{La}}{dt} + V_{grid}$$

- Lower device ON(S_{1L}) with duty ratio (1−d_a):

$$-\frac{v_{sa}}{2} = r_L i_{La} + L\frac{di_{La}}{dt} + V_{grid}$$

By averaging these two switching conditions, the dynamic equation is obtained:

$$\frac{di_{La}}{dt} = -\frac{r_L}{L} i_{La} + \frac{v_{sa}}{2L}(2d_a - 1) - \frac{V_{grid}}{L} \quad (13.1)$$

Step 2. Voltage Equation

The capacitor voltage across the solar cells is expressed as:

$$i_{sa} = C\frac{dv_{sa}}{dt} + i_o$$

where the inverter output current is:

$$i_o = (d_a i_{La} + d_b i_{Lb})$$

For the single-phase PCS, this reduces to:

$$i_o = d_a i_{La} - (1 - d_a) i_{La} = (2d_a - 1) i_{La}$$

Thus, the voltage equation becomes:

$$\frac{dv_{sa}}{dt} = -\frac{1}{C}(2d_a - 1) i_{La} + \frac{i_{sa}}{C} \quad (13.2)$$

Step 3. State-Averaged Equations

The above current and voltage equations form a nonlinear system. To simplify, the duty ratio is decomposed into a feedback and feedforward component:

$$d_a = d_{fb} + d_{ff} \quad \Rightarrow \quad d_{ff} = 0.5$$

Substituting yields the state-averaged model:

$$\frac{d}{dt} i_{La} = -\frac{r_L}{L} i_{La} + \frac{d_{fb}}{L}(v_{sa}) - \frac{V_{grid}}{L} \quad (13.3)$$

$$\frac{d}{dt} v_{sa} = -\frac{2}{C} d_{fb} \cdot i_{La} + \frac{i_{sa}}{C} \quad (13.4)$$

Step 4. Small-Signal Linearization

Small-signal perturbations are introduced as:

$$i_{La} = I_L + \hat{i}_{La},\ d_{d,fb} = D_a + \hat{d}_{fb},\qquad v_{sa} = V_{SA} + \hat{v}_{sa}$$

- Current equation:

$$\frac{d}{dt}(I_L + \hat{i}_{La}) = -\frac{r_L}{L}(I_L + \hat{i}_{La}) + \frac{(D_a + \hat{d}_{fb})}{L}(V_{SA} + \hat{v}_{sa}) - \left(\frac{V_{grid}}{L}\right)$$

$$\text{DC} \Rightarrow 0 = -\frac{r_L}{L}I_L + \frac{1}{L}D_a V_{SA} - \frac{V_{grid}}{L} \rightarrow D_a = \frac{V_{grid}}{V_{SA}}$$

$$\text{AC} \Rightarrow \frac{d}{dt}(\hat{i}_{La}) = -\frac{r_L}{L}(\hat{i}_{La}) + \frac{D_a}{L}(\hat{v}_{sa}) + \frac{V_{SA}}{L}(\hat{d}_{fb})$$

- Voltage equation:

The solar current i_{sa} is expressed as a piecewise linear relation:

$$i_{sa} = -\alpha v_{sa} + \beta.$$

Thus,

$$\frac{d}{dt}(V_{SA} + \hat{v}_{sa}) = -\frac{2}{C}(D_a + \hat{d}_{fb})(I_L + \hat{i}_{La}) - \frac{\alpha}{C}(V_{SA} + \hat{v}_{sa}) + \frac{\beta}{C}$$

$$\text{DC} \Rightarrow 0 = -\frac{2}{C}D_a I_L - \frac{\alpha}{C}V_{SA} + \frac{\beta}{C} \quad \rightarrow I_L = \frac{-\alpha V_{SA} + \beta}{2D_a}$$

$$\text{AC} \Rightarrow \frac{d}{dt}(\hat{v}_{sa}) = -\frac{2}{C}(D_a \cdot \hat{i}_{La} + I_L \cdot \hat{d}_{fb}) - \frac{\alpha}{C}\hat{v}_{sa}$$

Step 5. Final Small-Signal Model

The small-signal dynamics can be expressed in compact matrix form:

$$\begin{pmatrix} \dfrac{d(\hat{i}_{La})}{dt} \\ \dfrac{d(\hat{v}_{sa})}{dt} \end{pmatrix} = \begin{pmatrix} -\dfrac{r_L}{L} & \dfrac{D_a}{L} \\ -\dfrac{2D_a}{C} & -\dfrac{\alpha}{C} \end{pmatrix} \begin{pmatrix} \hat{i}_{La} \\ \hat{v}_{sa} \end{pmatrix} + \begin{pmatrix} \dfrac{V_{SA}}{L} \\ -\dfrac{2I_L}{C} \end{pmatrix} \hat{d}_{fb} \tag{13.5}$$

with steady-state operating points:

$$V_{SA} = \frac{V_{grid}}{D_a},\quad I_L = \frac{-\alpha V_{SA} + \beta}{2D_a}$$

Derivation of Transfer Functions for Solar PCS The frequency-domain transfer function is derived from the state-space representation.

1) Duty to Inductor Current Transfer Function

$$G_1(s) = \frac{\hat{i}_{La}(s)}{\hat{d}_{fb}(s)} = \frac{\frac{V_{SA}}{L}\left(s + \frac{\alpha}{C}\right) - \frac{2D_a I_L}{LC}}{s^2 + \left(\frac{r_L}{L} + \frac{\alpha}{C}\right)s + \frac{(r_L \cdot \alpha + 2D_a^2)}{LC}} \tag{13.6}$$

2) Duty to Solar Voltage Transfer Function

$$G_2(s) = \frac{\hat{v}_{sa}(s)}{\hat{d}_{fb}(s)} = \frac{-\frac{2}{C}\left[I_L\left(s+\frac{r_L}{L}\right)+\frac{DV_{SA}}{L}\right]}{s^2+\left(\frac{r_L}{L}+\frac{\alpha}{C}\right)s+\frac{\left(r_L\cdot\alpha+2D_a^2\right)}{LC}} \tag{13.7}$$

3) Voltage–Current Transfer Function

$$G_3(s) = \frac{\hat{v}_{sa}(s)}{\hat{i}_{La}(s)} = \frac{-\frac{2}{C}\left[I_L\left(s+\frac{r_L}{L}\right)+\frac{DV_{SA}}{L}\right]}{\frac{V_{SA}}{L}\left(s+\frac{\alpha}{C}\right)-\frac{2D_aI_L}{LC}} \tag{13.8}$$

The transfer functions can also be obtained programmatically using MATLAB's `ss2tf`, which converts a state-space realization (`A, B, C, D`) into its equivalent rational transfer-function form. In this work, we use `ss2tf` to extract $G_1(s)$ and $G_2(s)$ by selecting the output rows corresponding to i_{La} and v_{sa} and the input column associated with the duty perturbation $d_{d,fb}$. The MATLAB m-file for solar PCS controller is presented in Table 13.1.

13.1.2 Controller Design

The control architecture integrates the voltage-loop approach commonly used in solar power converters with the current-loop strategy employed in single-phase power conditioning systems (PCS). The controller design is carried out under the assumption that the reference voltage V_{ref} is provided by the MPPT algorithm.

The MPPT block generates V_{sa_ref}, which is regulated by the outer voltage controller. The voltage-loop output sets the amplitude of the current reference, which is then multiplied by a synchronized sine wave to enforce phase alignment between the inductor (grid) current and the grid voltage. This arrangement simultaneously achieves maximum power extraction from the PV array and unity power factor operation. Figure 13.3 depicts the resulting AC *instantaneous control* configuration driven by the MPPT voltage reference.

The design specifications of the solar PCS are summarized in Table 13.2.

Remark on phase margin: the maximum achievable phase margin for the voltage loop, given the plant and constraints, is approximately **84°**. While larger phase

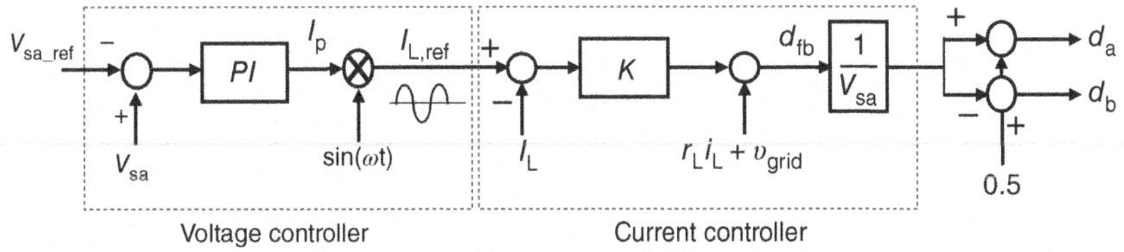

Figure 13.3 Configuration of the AC instantaneous controller.

Table 13.1 MATLAB m-file for single-phase solar PCS.

```
% Solar PCS  : Single phase %
Voc = 400;
Isc = 16;
Vmp = 360;
Imp = 12;
alpha = Imp/Vmp;
beta = 2*Imp;

s = tf('s');
rL= 0.01;          % 34m[ohm]
L = 1e-3;          % 2.4m[H]
C = 2000e-6;       % 2000 uf
Vgrid = 155;       % half voltage

% Steady state value %
Da  = 0.43;
VSA = Vgrid / Da;
IL = Imp / (2*Da);

%% State Averaging equation
A = [-rL/L,    Da/L;
     -2*Da/C,-alpha/(C)];
B = [VSA/L;
    -2*IL/C];

C1 = [1,0]; C2 = [0,1]; Dx = 0;

[num1,den1] = ss2tf(A,B,C1,Dx);
[num2,den2] = ss2tf(A,B,C2,Dx);

G1 = tf(num1, den1);
G2 = tf(num2, den2);
G3 = tf(num2,num1);

%% double loop controller
Vref = 450;
Kpv = 1; Kiv = 1;
Gcv = Kpv*(s+Kiv)/s;
Kpi = 1; Kii = 1;
Gci = Kpi*(s+Kii)/s;

T=sisoinit(6);
T.F.Value = Vref;
T.G1.Value = G1;
T.C2.Value = Gci;
T.G2.Value = -G3;
T.C1.Value = Gcv;
sisotool(T);
```

Table 13.2 Design specification for solar PCS double-loop controller.

Design specification	Voltage controller (C_1)	Current controller (C_2)
Bandwidth [rad/sec]	100	4800
Phase margin [deg]	**84**	60

margins generally correlate with reduced overshoot (with an idealized 90° implying near-zero overshoot for a second-order approximation), the attainable 84° setting provides a robust response in this application. As shown in the time-domain results below, the actual closed-loop response exhibits a small overshoot due to the higher-order and discrete-time characteristics of the system.

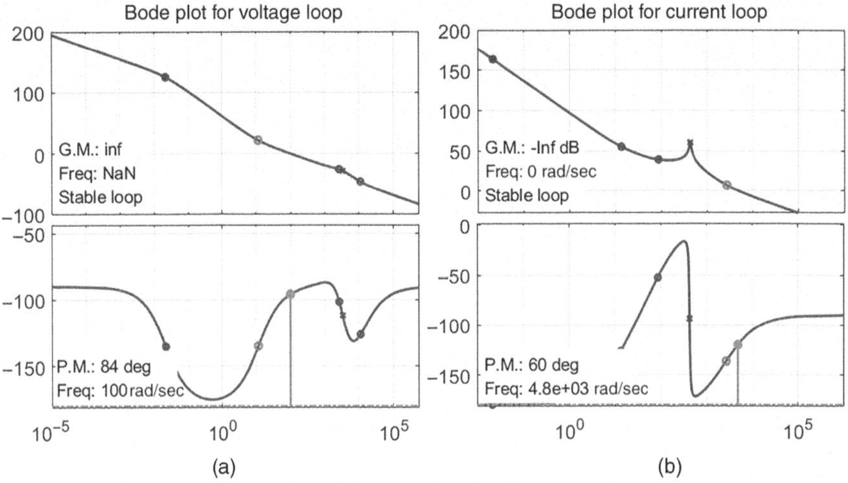

Figure 13.4 Bode plots of the double-loop controllers. (a) Voltage controller and (b) Current controller.

Figure 13.4 shows the SISOTOOL auto-tuning outcomes. Subfigure (a) presents the voltage-loop Bode plot, and subfigure (b) presents the current-loop response. The design meets the specifications: a 100 rad/sec bandwidth with **84°** phase margin for the voltage loop, and a 4800 rad/sec bandwidth with 60° phase margin for the current loop.

The controller gains obtained through auto-tuning are summarized in Table 13.3, presented in both continuous-time and discrete-time (difference-equation) forms. Discrete-time parameters are computed for a switching (and sampling) frequency of $f_{sw} = 20$ kHz

As shown in Figure 13.5, the closed-loop response satisfies the design objectives. The output tracks the reference without steady-state error, with approximately 7.7% overshoot and a settling time of 17 ms.

Table 13.3 Controller gains in continuous-time and discrete-time formats.

	Voltage controller (C_1)	Current controller (C_2)
Continuous time	$\frac{0.2345(s + 11.5)}{s}$	$\frac{0.0114(s + 2670)}{s}$
Discrete time	$\frac{0.23564(z - 0.999)}{z - 1}$	$\frac{0.0127(z - 0.906)}{z - 1}$

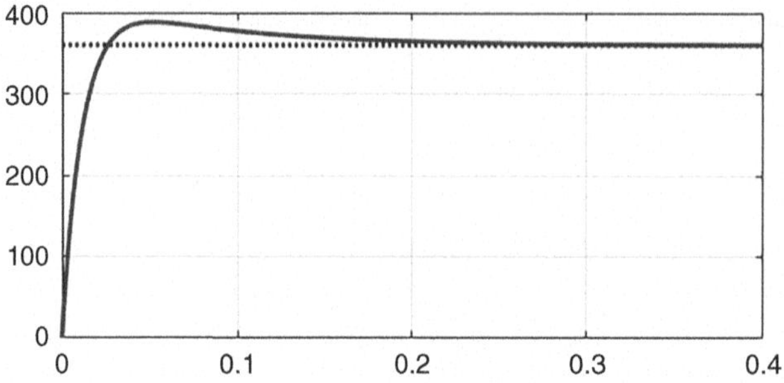

Figure 13.5 Step response of the double-loop controllers.

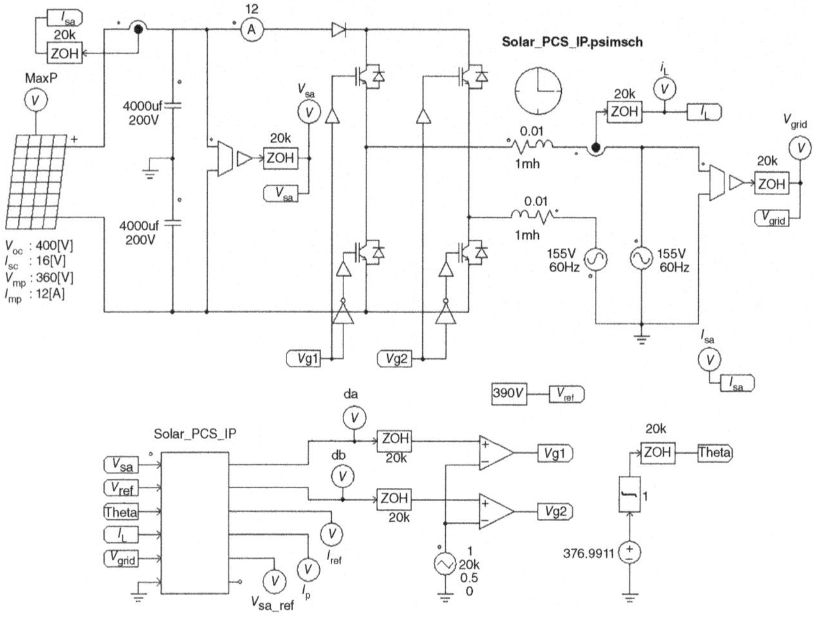

Figure 13.6 Schematic diagram of the single-phase solar PCS.

PSIM Simulation Figure 13.6 shows the PSIM schematic of the single-phase solar PCS. The output of the voltage loop determines the current amplitude I_P, which is then used to synthesize the sinusoidal current reference synchronized with the grid, expressed as $I_{ref} = I_P \sin \omega t$.

The voltage controller is implemented in C as the `Solar_PCS_1P` C-block, employing a standard *PI* structure consistent with the PCS current-loop design.

Table 13.4 C-block code: Solar_PCS_1P.

```
double Vsa, Vsa_ref,Theta,IL,Vg;
double duty_a,duty_b, Verr;
double  Iref, Ierr,Vcon,dfb,Ip;
static double Ip_old = 16.0,
Verr_old=0.0;

#define Kp 0.2356    // P-gain
#define Ki  0.999    // I-gain
/****** start here ******/
Vsa      = in[0];
Vsa_ref  = in[1];
Theta    = in[2];
IL       = in[3];
Vg       = in[4];

if (Vsa <1.0) Vsa = 1.0;
if   (t>0.9)       Vsa_ref= 360;
else if (t>0.6)    Vsa_ref= 340;
else if (t>0.4)    Vsa_ref= 360;

Verr = (Vsa - Vsa_ref);
```

```
Ip=Ip_old+Kp*Verr-Kp*Ki*Verr_old;
Ip_old= Ip;
Verr_old= Verr;

Iref = Ip*sin(Theta);
Ierr = Iref - IL;

Vcon = 20*Ierr + 0.01*IL +Vg;
Dfb = Vcon/Vsa;

if (dfb >  0.5)  dfb= 0.5;
if (dfb <- 0.5)  dfb=-0.5;

duty_a = 0.5 + dfb;
duty_b = 0.5 - dfb;

out[0] = duty_a;
out[1] = duty_b;
out[2] = Iref;
out[3] = dfb;
out[4] = Vsa_ref;
```

It regulates the PV-side voltage to track the MPPT-generated reference while simultaneously commanding the current amplitude to ensure unity power factor operation. The C-code for this single-phase PCS controller C-block is presented in Table 13.4.

The control strategy employed for the single-phase system can be readily extended to the three-phase PCS. The voltage controller remains unchanged, while the current controller is implemented using the three-phase instantaneous value control method. Since this method has already been explained in detail in Chapter 12, no further description is provided here.

Figure 13.7 presents representative simulation waveforms. As expected for single-phase systems, the PV (capacitor) voltage exhibits a 120 Hz ripple—a consequence of the twice-line-frequency power pulsation—yet it remains tightly regulated around the reference. The inductor (grid) current is phase-aligned with the grid voltage, enabling efficient power transfer while minimizing reactive power.

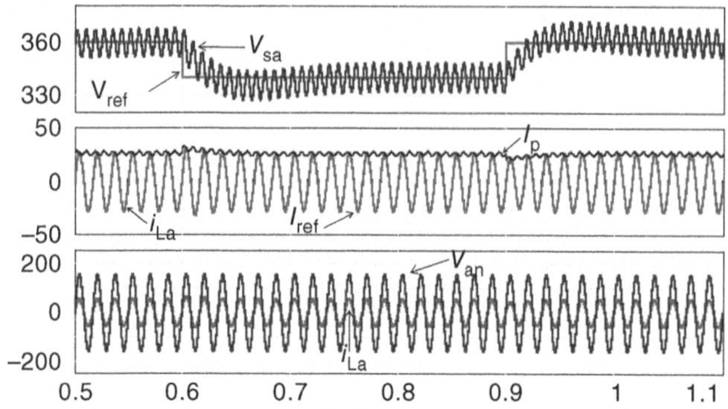

Figure 13.7 Simulation waveforms of the single-phase solar PCS.

13.2 Three-Phase Solar PCS Design

The controller design is now extended to a three-phase solar PCS. The schematic of the three-phase solar PCS is presented in Figure 13.8. In this configuration, the input stage is modified to interface with the PV array, while the control architecture follows the same fundamental structure as the single-phase case. Specifically, the system employs an outer voltage control loop—driven by MPPT—that determines the reference current amplitude, and an inner current control loop that ensures grid synchronization.

13.2.1 State-Averaging Equations and Transfer Functions

Let $i_{L,abc} \triangleq [i_{La}\ i_{Lb}\ i_{Lc}]^T$, $d_{abc} \triangleq [d_a\ d_b\ d_c]^T$, and $V_{abc,n} \triangleq [V_{an}\ V_{bn}\ V_{cn}]^T$ where v_{sa} represents the PV capacitor voltage. Note that each capacitor in the series connection should have a capacitance of $2C$ so that the resulting equivalent capacitance becomes C.

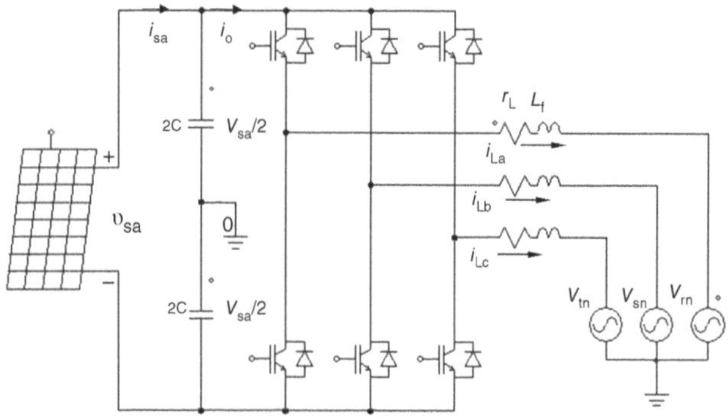

Figure 13.8 Schematic of the three-phase solar PCS.

Step 1. Current Averaging Equation

By applying switch-state averaging, in a manner analogous to the single-phase derivation, the inductor current dynamics can be expressed as

$$\frac{di_{L,abc}}{dt} = -\frac{r_L}{L} i_{L,abc} - \frac{(1-2d_{abc})}{2L} v_{sa} - \frac{V_{abc,n}}{L} \tag{13.9}$$

where $1 \triangleq [1\ 1\ 1]^T$ and the product $(1-2d_{abc})v_{sa}$ is interpreted element-wise; that is, the scalar voltage v_{sa} multiplies each component of the vector $(1-2d_{abc})$.

Step 2. Voltage Averaging Equation

The capacitor dynamics of the PV array are given by

$$\frac{dv_{sa}}{dt} = \frac{1}{C}(i_{sa} - i_o)$$

with i_{sa} linearized around the operating point as $i_{sa} = -\alpha v_{sa} + \beta$. Accounting for the switch ON/OFF states,

$$\frac{dv_{sa}}{dt} = \frac{1}{C}((1-2d_a)i_{La} + (1-2d_b)i_{Lb} + (1-2d_c)i_{Lc} - \alpha v_{sa} + \beta)$$

Or, in compact form,

$$\frac{dv_{sa}}{dt} = \frac{(1-2d_{abc})i_{L,abc}}{C} + \frac{(-\alpha v_{sa} + \beta)}{C} \tag{13.10}$$

Step 3. State-Averaged Equations

Decompose the duty ratio into feedback and feedforward parts,

$$d_{abc} = d_{abc,fb} + d_{abc,ff} \quad \Rightarrow \quad d_{abc,ff} = 0.5$$

yields the state-averaged model:

$$\frac{di_{L,abc}}{dt} = -\frac{r_L}{L} i_{L,abc} + \frac{(d_{abc,fb})}{L} v_{sa} - \frac{V_{abc,n}}{L} \tag{13.11}$$

$$\frac{dv_{sa}}{dt} = -\frac{(2d_{abc,fb})i_{L,abc}}{C} + \frac{(-\alpha v_{sa} + \beta)}{C} \tag{13.12}$$

Step 4. DQ Transformation

Employ the power-invariant Clarke–Park transformation. Let $i_{DQ} \triangleq [i_d\ i_q]^T$, $d_{DQ} \triangleq [d_d\ d_q]^T$ and $V_{DQ} \triangleq [V_d\ V_q]^T$. With electrical angular frequency ω,

$$\frac{d}{dt}\left(\begin{bmatrix} i_d \\ i_q \end{bmatrix}\right) = \omega \begin{bmatrix} i_q \\ -i_d \end{bmatrix} - \frac{r_L}{L} \cdot \begin{bmatrix} i_d \\ i_q \end{bmatrix} + \frac{v_{sa}}{L} \cdot \begin{bmatrix} d_d \\ d_q \end{bmatrix} - \frac{1}{L}\begin{bmatrix} V_d \\ V_q \end{bmatrix} \tag{13.13}$$

Using the identity $d_{abc} i_{L,abc} = d_{DQ} P^T P i_{DQ} = \frac{3}{2} d_{DQ} i_{DQ}$ (power-invariant Park, $P^T P = \frac{3}{2} I$)

$$\frac{dv_{sa}}{dt} = -\frac{3}{C}(d_d \cdot i_d + d_q \cdot i_q) + \frac{(-\alpha v_{sa} + \beta)}{C} \tag{13.14}$$

Step 5. Small-Signal Linearization with Decoupling

Adopt standard feed-forward decoupling

$$d_{\mathrm{d}} = d_{\mathrm{d,fd}} + d_{\mathrm{d,ff}},\ d_{\mathrm{q}} = d_{\mathrm{q,fd}} + d_{\mathrm{q,ff}} \Rightarrow d_{\mathrm{d,ff}} = \frac{\omega L i_{\mathrm{q}}}{v_{\mathrm{sa}}},\quad d_{\mathrm{q,ff}} = -\frac{\omega L i_{\mathrm{d}}}{v_{\mathrm{sa}}}$$

With this choice, the ω dependent-coupling cancels in the nominal model.

$$\frac{\mathrm{d}}{\mathrm{d}t}\left(\begin{bmatrix} i_{\mathrm{d}} \\ i_{\mathrm{q}} \end{bmatrix}\right) = -\frac{r_{\mathrm{L}}}{L}\cdot\begin{bmatrix} i_{\mathrm{d}} \\ i_{\mathrm{q}} \end{bmatrix} + \frac{v_{\mathrm{sa}}}{L}\cdot\begin{bmatrix} d_{\mathrm{d,fb}} \\ d_{\mathrm{q,fb}} \end{bmatrix} - \frac{1}{L}\begin{bmatrix} V_{\mathrm{d}} \\ V_{\mathrm{q}} \end{bmatrix} \tag{13.15}$$

$$\frac{\mathrm{d}v_{\mathrm{sa}}}{\mathrm{d}t} = -\frac{3}{C}(d_{\mathrm{d}}\cdot i_{\mathrm{d}} + d_{\mathrm{q}}\cdot i_{\mathrm{q}}) + \frac{(-\alpha v_{\mathrm{sa}} + \beta)}{C} \tag{13.16}$$

Applying perturbations:

$$i_{\mathrm{dq}} = I_{\mathrm{DQ}} + \hat{i}_{\mathrm{dq}},\quad d_{d,fb} = D_{\mathrm{d}} + \hat{d}_{\mathrm{d,fb}},\quad d_{\mathrm{q,fb}} = D_{\mathrm{q}} + \hat{d}_{\mathrm{q,fb}},\quad v_{\mathrm{sa}} = V_{\mathrm{SA}} + \hat{v}_{\mathrm{sa}}$$

From the current equation:

$$\frac{\mathrm{d}}{\mathrm{d}t}(I_{\mathrm{DQ}} + \hat{i}_{\mathrm{dq}}) = -\frac{r_{\mathrm{L}}}{L}(I_{\mathrm{DQ}} + \hat{i}_{\mathrm{dq}}) + \frac{(V_{\mathrm{SA}} + \hat{v}_{\mathrm{sa}})}{L}(D_{\mathrm{DQ}} + \hat{d}_{\mathrm{dq,fb}}) - \frac{1}{L}(V_{\mathrm{DQ}} + \hat{v}_{\mathrm{dq}})$$

$$\mathrm{DC} \Rightarrow 0 = -\frac{r_{\mathrm{L}}}{L}I_{\mathrm{DQ}} + \frac{1}{L}V_{\mathrm{SA}}\cdot D_{\mathrm{DQ}} - \frac{1}{L}V_{\mathrm{DQ}} \to D_{\mathrm{DQ}} = \frac{V_{\mathrm{DQ}}}{V_{\mathrm{SA}}}$$

$$\mathrm{AC} \Rightarrow \frac{\mathrm{d}}{\mathrm{d}t}(\hat{i}_{\mathrm{dq}}) = -\frac{r_{\mathrm{L}}}{L}(\hat{i}_{\mathrm{dq}}) + \frac{D_{\mathrm{DQ}}}{L}(\hat{v}_{\mathrm{sa}}) + \frac{V_{\mathrm{SA}}}{L}(\hat{d}_{\mathrm{dq,fb}})$$

From the voltage equation:

$$\frac{\mathrm{d}}{\mathrm{d}t}(V_{\mathrm{SA}} + \hat{v}_{\mathrm{sa}}) = -\frac{3}{C}\cdot((D_{\mathrm{D}} + \hat{d}_{\mathrm{d,fb}})\cdot(I_{\mathrm{D}} + \hat{i}_{\mathrm{d}}) + (D_{\mathrm{Q}} + \hat{d}_{\mathrm{q,fb}})\cdot(I_{\mathrm{Q}} + \hat{i}_{\mathrm{q}})) + \frac{(-\alpha(V_{\mathrm{SA}} + \hat{v}_{\mathrm{sa}}) + \beta)}{C}$$

$$\mathrm{DC} \Rightarrow 0 = -\frac{3}{C}\cdot((D_{\mathrm{D}}\cdot I_{\mathrm{D}}) + (D_{\mathrm{Q}}\cdot I_{\mathrm{Q}})) + \frac{(-\alpha V_{\mathrm{SA}} + \beta)}{C} \to I_{\mathrm{D}} = \frac{(-\alpha V_{\mathrm{SA}} + \beta)}{3D_{\mathrm{D}}} = \frac{I_{\mathrm{SA}}}{3D_{\mathrm{D}}}$$

$$\mathrm{AC} \Rightarrow \frac{\mathrm{d}}{\mathrm{d}t}(\hat{v}_{\mathrm{dc}}) = -\frac{3}{C}\cdot((D_{\mathrm{D}}\cdot\hat{i}_{\mathrm{d}} + D_{\mathrm{Q}}\cdot\hat{i}_{\mathrm{q}}) + (I_{\mathrm{D}}\cdot\hat{d}_{\mathrm{d,fb}} + I_{\mathrm{Q}}\cdot\hat{d}_{\mathrm{q,fb}})) + \frac{-\alpha}{C}(\hat{v}_{\mathrm{sa}})$$

Step 6. Final Small-Signal Model

In compact matrix form, the small-signal model is expressed as:

$$\begin{pmatrix} \frac{\mathrm{d}\hat{i}_{\mathrm{dq}}}{\mathrm{d}t} \\ \frac{\mathrm{d}(\hat{v}_{\mathrm{sa}})}{\mathrm{d}t} \end{pmatrix} = \begin{pmatrix} -\frac{r_L}{L} & \frac{D_{\mathrm{DQ}}}{L} \\ -\frac{3D_{\mathrm{DQ}}}{C} & \frac{-\alpha}{C} \end{pmatrix}\begin{pmatrix} \hat{i}_{\mathrm{dq}} \\ \hat{v}_{\mathrm{sa}} \end{pmatrix} + \begin{pmatrix} \frac{V_{\mathrm{SA}}}{L} \\ -\frac{3I_{\mathrm{DQ}}}{C} \end{pmatrix}\hat{d}_{\mathrm{dq,fb}} \tag{13.17}$$

Table 13.5 MATLAB m-file for three-phase solar PCS.

```
% Solar PCS 3-Phase PI Design %
Voc = 700;Isc = 20;Vmp = 650;Imp = 16;
alpha = Imp/Vmp; beta = 2*Imp;

s = tf('s');
rL= 0.01;   L = 1e-3; C = 2e-3;
Van = 310;

%% steady-state value
VD=310;   VQ=0;
VSA = Vmp; ISA = Imp;
DD  = VD/VSA; ID = ISA/(3*DD);

%% State-Averaging Equation
A = [-rL/L,     DD/L;
      -3*DD/C,-alpha/C];
B = [VSA/L;
      -3*ID/C];
C1 = [1,0]; C2 = [0,1]; Dx = 0;

[num1, den1]=ss2tf(A,B,C1,Dx);
[num2, den2]=ss2tf(A,B,C2,Dx);
G1 = tf(num1, den1);
G2 = tf(num2, den2);
G3 = tf(num2, num1);

Kpv = 1; Kiv = 1;
Gcv = Kpv*(s+Kiv)/s;
Kpi = 1; Kii = 1;
Gci = Kpi*(s+Kii)/s;

Vsa_ref= 650;
T = sisoinit(6);
T.F.Value = Vsa_ref;
T.G1.Value = G1;
T.C1.Value = Gcv;
T.C2.Value = Gci;
T.G2.Value = -G3;
sisotool(T);
```

with steady-state values:

$$D_{DQ} = \frac{V_{DQ}}{V_{SA}}, \qquad I_D = \frac{(-\alpha V_{SA} + \beta)}{3D_D} = \frac{I_{SA}}{3D_D}$$

The transfer function is derived using the MATLAB built-in function `ss2tf`, which converts the state-space model into its equivalent transfer function representation. The MATLAB m-file for the three-phase PCS controller is presented in Table 13.5.

13.2.2 Double-Loop *dq*-Controller Design

The solar PCS can also be implemented using a double-loop control structure in the synchronous *dq* reference frame. In this configuration, the previously designed solar-cell voltage controller is retained without modification, while the inner current controller is formulated in *dq* coordinates, following the same control strategy employed for the three-phase inverter discussed earlier.

The outer voltage controller regulates the PV capacitor voltage, and its output defines the *d*-axis reference current. Simultaneously, the *q*-axis reference current is set to zero, thereby preventing reactive power injection into the grid. This ensures

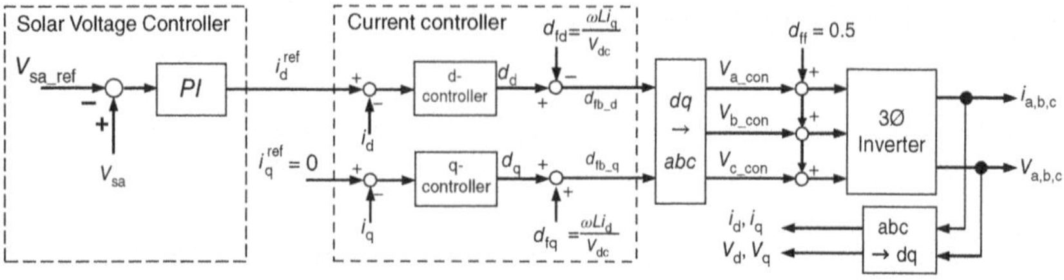

Figure 13.9 Block diagram of the solar PCS with double-loop control.

that the PCS transfers only the real power generated by the solar cells and operates at unity power factor, a key requirement for efficient and stable grid integration.

Figure 13.9 presents the block diagram of the solar PCS with double-loop control. Although not explicitly depicted, the MPPT algorithm continuously monitors the solar cell voltage and current to compute the reference voltage V_{sa_ref}, which enables maximum power extraction from the PV array.

The design specifications for the double-loop controllers are summarized in Table 13.6. For the voltage loop, the maximum achievable phase margin under the given parameters is 84°, which is adopted in the design.

Figure 13.10 illustrates the SISOTOOL auto-tuning results. Subfigure (a) presents the voltage loop Bode plot, while Subfigure (b) shows the current loop response. The design successfully satisfies the target specifications: 100 rad/sec bandwidth with an 84° phase margin for the voltage loop, and 4800 rad/sec bandwidth with a 60° phase margin for the current loop.

Using the auto-tuning feature in MATLAB's SISOTOOL, the double-loop controller was designed for a switching frequency of 20 kHz. The resulting continuous and discrete-time transfer function is summarized in Table 13.7.

Table 13.6 Design specification for double-loop controller.

Design specification	**Voltage controller (C_1)**	**Current controller (C_2)**
Bandwidth [rad/sec]	100	4800
Phase margin [deg]	**84**	60

Table 13.7 Controller gains in continuous-time and discrete-time formats.

	Voltage controller (C_1)	**Current controller (C_2)**
Continuous time	$\frac{0.143(s+11)}{s}$	$\frac{0.0063(s+2670)}{s}$
Discrete time	$\frac{0.144(z-0.999)}{z-1}$	$\frac{0.007(z-0.906)}{z-1}$

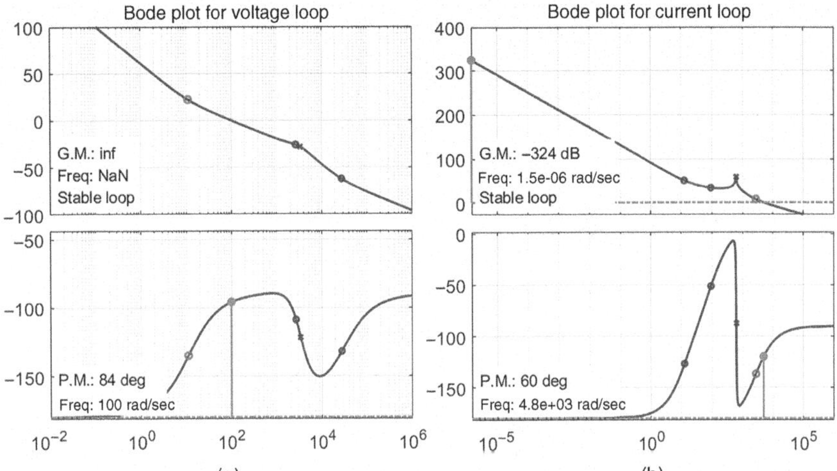

Figure 13.10 Bode plots of the double-loop controllers. (a) Voltage controller and (b) Current controller.

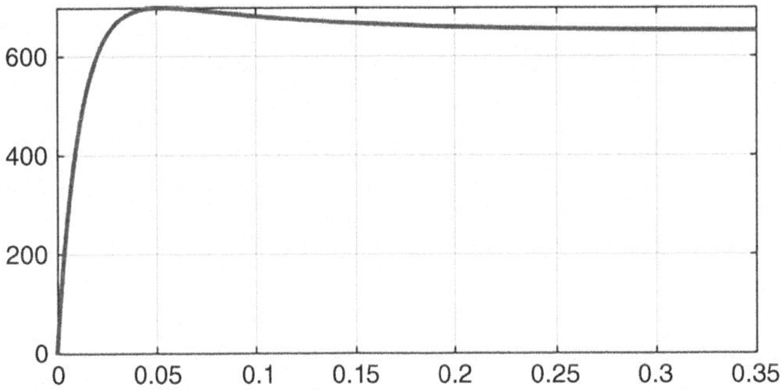

Figure 13.11 Step response of the double-loop controllers.

The closed-loop step response of the designed controllers is shown in Figure 13.11. The output voltage accurately tracks the reference, achieving the specified performance. No steady-state error is observed, and overshoot is 6.75%, and settling time is approximately 170 ms.

PSIM Simulation The complete PSIM schematic for three-phase solar PCS is shown in Figure 13.12.

The control logic is implemented in a C-block module labeled *Solar_PCS_DQ_V*, which facilitates the evaluation of the controller's dynamic performance. In this test, the reference voltage V_{sa_ref} is programmed to step from 670 V down to 650 V and then return to 670 V. The *C*-code for this three-phase PCS controller *C*-block is provided in Table 13.8.

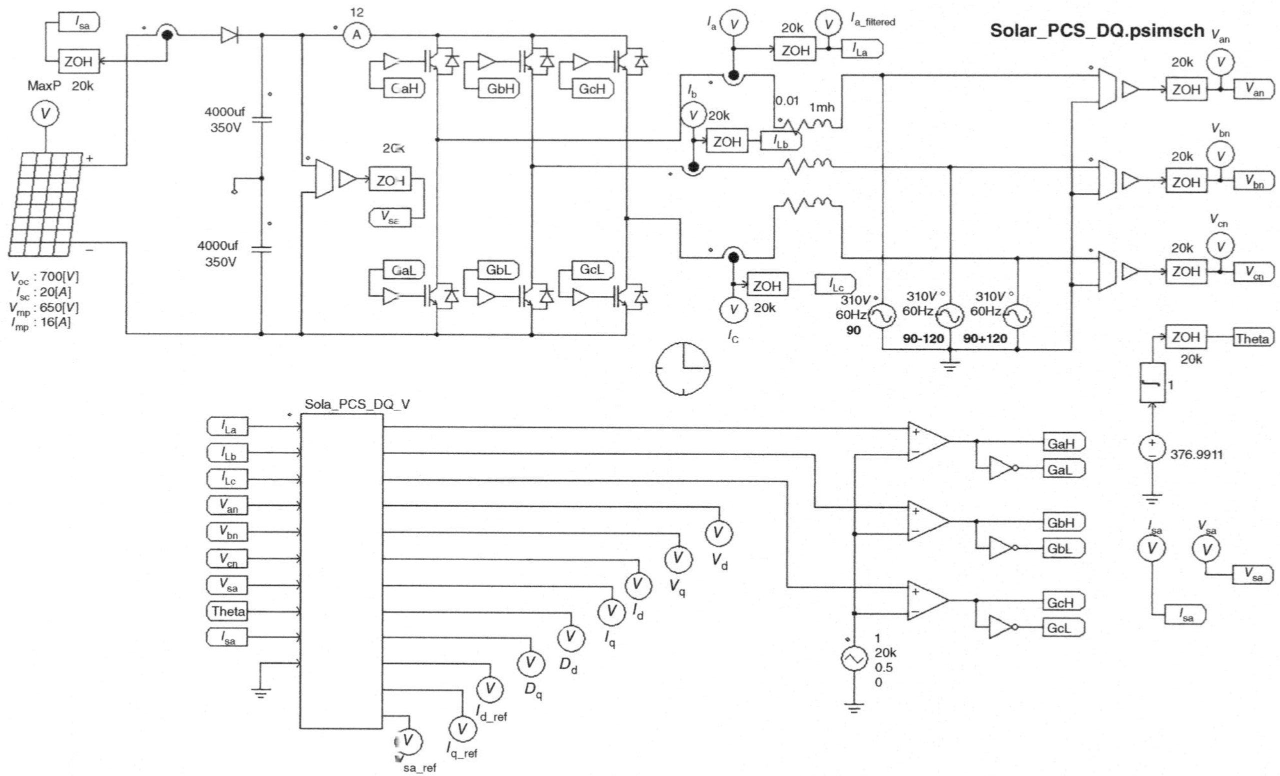

Figure 13.12 Schematic of the three-phase solar PCS.

Table 13.8 C-block code: Solar_PCS_DQ_V.

```
#define Lf      0.001      // 1 [mh]
#define rL      0.01       // 10 [mohm]
#define Omega 377          // 60 [Hz]

#define Kvp  0.144  // voltage Kp , 100 rad/s
#define Kvi  0.999  //Kiz
#define Kip  0.007  // current  Kp. 4800 rad/s
#define Kii  0.906  //Kiz

double ILa,ILb,ILc,Van,Vbn,Vcn,Vsa,theta, Isa;
double Valpha, Vbeta,  Ialpha, Ibeta, Vd, Vq,
ILd, ILq, dd, dq; // DQ variable
double Vd_ref, Vq_ref, Va_con, Vb_con, Vc_con;
double duty_a, duty_b, duty_c, Vsa_ref, Verr;
double dfb_d, dfb_q, Id_ref,
Iq_ref,Id_err,Iq_err;
static double dd_old=0.0,
dq_old=0.0,Id_err_old=0.;
static double Iq_err_old=0., Id_old=0.,
Verr_old=0.;

/**** Input Variable   ****/
ILa    = in[0];
ILb    = in[1];
ILc    = in[2];
Van    = in[3];
Vbn    = in[4];
Vcn    = in[5];
Vsa    = in[6];
```

```
/*** D-Q Current Controller ******/
Id_err = Id_ref - ILd;
dd = dd_old + Kip*Id_err -Kip*Kii*Id_err_old;
dd_old=dd;
Id_err_old=Id_err;

Iq_ref = 0;
//Iq_ref = Omega*Cf*Vd;
Iq_err = Iq_ref - ILq;
dq = dq_old +Kip*Iq_err -Kip*Kii*Iq_err_old;
dq_old=dq;
Iq_err_old=Iq_err;

dfb_d = dd  -  Lf*Omega*ILq/(Vsa);
dfb_q = dq +  Lf*Omega*ILd/(Vsa);

Valpha = dfb_d*cos(theta) - dfb_q*sin(theta);
Vbeta  = dfb_d*sin(theta) + dfb_q*cos(theta);

Va_con = Valpha;
Vb_con = -0.5*Valpha + sqrt(3)*Vbeta/2;
Vc_con = -0.5*Valpha - sqrt(3)*Vbeta/2;

// duty = dff + dfb = 0.5 + dfb
duty_a =  0.5 + Va_con;
duty_b =  0.5 + Vb_con;
duty_c =  0.5 + Vc_con;
```

(Continued)

Table 13.8 (Continued)

```
Theta   = in[7];
Isa     = in[8];

if (Vsa ==0 ) Vsa =0.1;

Valpha =(2./3.)*Van -  (1/3.)*Vbn -(1/3.)*Vcn;
Vbeta = (Vbn-Vcn)/sqrt(3);
Vd =  Valpha*cos(theta) + Vbeta*sin(theta);
Vq = -Valpha*sin(theta) + Vbeta*cos(theta);

Ialpha =(2./3.)*ILa -  (1/3.)*ILb -(1/3.)*ILc;
Ibeta = (ILb-ILc)/sqrt(3);
ILd =  Ialpha*cos(theta) + Ibeta*sin(theta);
ILq = -Ialpha*sin(theta) + Ibeta*cos(theta);

/**** Solar Voltage Controller ******/
if (t  < 0.4)        Vsa_ref = 670.;
else if ( t< 0.7)    Vsa_ref = 650.;
else                 Vsa_ref =  670.;

Verr = Vsa - Vsa_ref;
Id_ref=Id_old+Kvp*Verr-Kvp*Kvi*Verr_old;
Id_old= Id_ref;
Verr_old= Verr;

if ( duty_a > 1.0 )          duty_a=1.0;
else if ( duty_a  < 0.0 )    duty_a=0.0;
if ( duty_b > 1.0 )          duty_b=1.0;
else if ( duty_b  < 0.0 )    duty_b=0.0;
if ( duty_c > 1.0 )          duty_c=1.0;
else if ( duty_c  < 0.0 )    duty_c=0.0;

out[0] = duty_a;    // Inverter R
out[1] = duty_b;    // Inverter S
out[2] = duty_c;    // Inverter T

out[3] = Vd;
out[4] = Vq;
out[5] = ILd;
out[6] = ILq;
out[7] = dd;
out[8] = dq;
out[9] = Id_ref;
out[10]= Iq_ref;
out[11]= Vsa_ref;
```

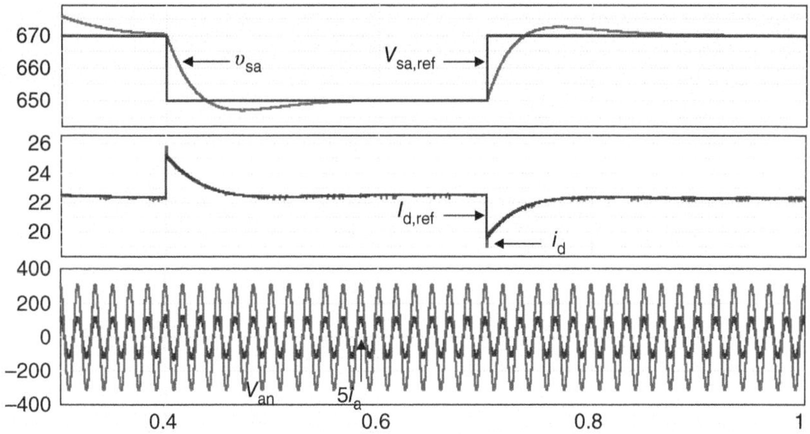

Figure 13.13 Simulation waveforms of the three-phase solar PCS.

The corresponding simulation waveforms are illustrated in Figure 13.13. The results verify that the solar cell voltage is precisely regulated to the commanded reference values. During transients, the controller exhibits fast tracking with no steady-state error, validating the robustness of the voltage regulation.

Furthermore, the current controller maintains accurate synchronization with the grid voltage, ensuring that the grid currents remain in phase with the grid voltages. This demonstrates that the system achieves unity power factor operation, thereby confirming efficient transfer of solar power from the PV array to the utility grid.

14

Single-Phase DQ Inverter

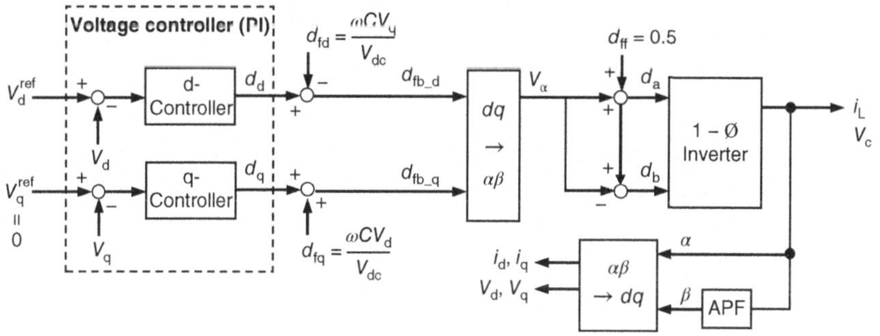

As explained in Chapter 11, the *dq* transformation converts the sinusoidal reference of a single-phase inverter into a step signal, allowing a *PI* controller to eliminate steady-state error and achieve zero steady-state deviation.

The single-phase *dq* inverter is an advanced control strategy that extends the *dq* transformation—traditionally used in three-phase systems—to single-phase inverter applications. This adaptation is made possible through a technique known as pseudo *dq* transformation or virtual two-phase *dq* control.

In single-phase systems, only a single AC signal (voltage or current) is available. To implement *dq* control, a virtual orthogonal signal must be generated. This is typically achieved using a Hilbert transform or by introducing a 90° phase-shifted sine wave. This process forms an artificial two-phase system, known as the α-β reference frame, which can then be transformed into *dq* coordinates using the Park transformation.

To generate the necessary 90° phase shift, an all-pass filter (APF) is commonly employed. The APF provides an efficient and practical means of enabling *dq* control in single-phase systems by creating the required virtual orthogonal signal without affecting signal amplitude.

14.1 APF (All Pass Filter)

The APF is a special type of filter designed to introduce a phase shift to the input signal while preserving a constant amplitude. The magnitude of the APF transfer function remains unity across all frequency bands, regardless of the introduced phase shift. The phase response of the filter depends on its order:

- In a first-order (one-pole) APF, the phase shift ranges from 0° to 180°.
- In a second-order (two-pole) APF, the phase shift extends from 0° to 360°.

In this chapter, only the first-order APF is considered.
The APF can be implemented using two fundamental configurations:

- High-pass filter (HPF)-based APF
- Low-pass filter (LPF)-based APF

First-Order APF Configurations Figure 14.1 illustrates the two primary configurations of the first-order APF:

a) High-pass-based APF: This configuration is derived from an RC HPF structure and provides a phase shift ranging from −180° to 0°. As the frequency increases, the phase shift asymptotically approaches 0°.
b) Low-pass-based APF: This configuration is derived from an RC LPF structure and provides a phase shift ranging from 0° to −180°. As the frequency increases, the phase shift asymptotically approaches −180°.

At the resonance frequency $\omega_0 = \frac{1}{RC}$, both configurations yield a phase shift of −90°, though the signal is inverted.

14.1.1 Transfer Function of the HPF-Based APF

Let V_p and V_n represent the voltages at the non-inverting and inverting terminals of the operational amplifier (Op-Amp), respectively. The transfer function for the

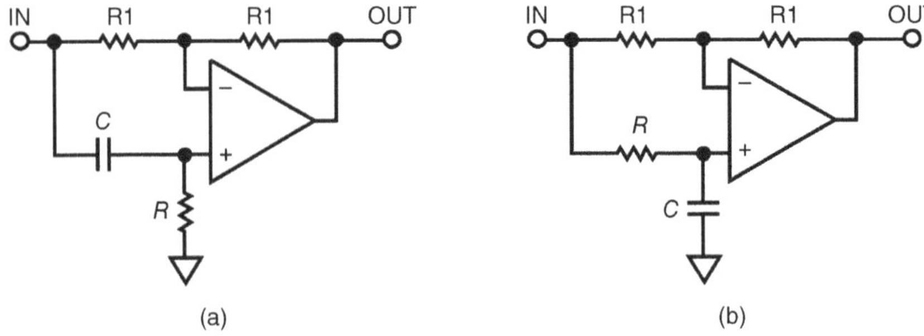

Figure 14.1 Configurations of the all-pass filter (APF). (a) High-pass based APF and (b) Low-pass based APF.

HPF-based APF is derived as follows:

$$V_p = \frac{R}{1/sC + R} V_{in}, \quad \frac{V_{in} - V_p}{R_1} = \frac{V_p - V_{out}}{R_1}$$

Solving for the transfer function:

$$\frac{V_{out}}{V_{in}} = \frac{s - \omega_0}{s + \omega_0} \tag{14.1}$$

where the cutoff frequency is:

$$\omega_0 = \frac{1}{RC}$$

To achieve a resonance frequency of $\omega_0 = 2\pi f = 120\pi = 377$, the following circuit parameters were selected:

$$R = 132.63[\Omega], \quad C = 20[\text{uF}]$$

Frequency Response Analysis Using MATLAB The APF frequency response is evaluated using the MATLAB script in Table 14.1.

The resulting Bode plot is shown in Figure 14.2, confirming that the magnitude remains constant across all frequencies, while the phase shift reaches 90° at the resonance frequency (377 rad/sec) and approaches 180° at higher frequencies.

PSIM Simulation of the HPF Circuit The actual circuit was implemented in PSIM, and the simulation results are shown in Figure 14.3. The input voltage V_{alpha} is represented as a cosine waveform, which is essential for subsequent *dq* transformations. The output voltage V_{beta}, obtained through the APF, leads the input phase, demonstrating the successful generation of the required 90° phase shift.

Please note that the V_{beta} is leading the V_{alpha}, which is *inconsistent* with the correct phase relationship required for the *dq*-transformation.

Discrete Transformation of the Transfer Function To implement the APF in a digital controller, the continuous-time transfer function is converted into discrete form

Table 14.1 MATLAB m-file for frequency response of APF.

```
R=132.63;
C=20e-6;
s=tf('s');
omega= 1/(R*C);
H=(s-omega)/(s+omega);
bode(H);
```

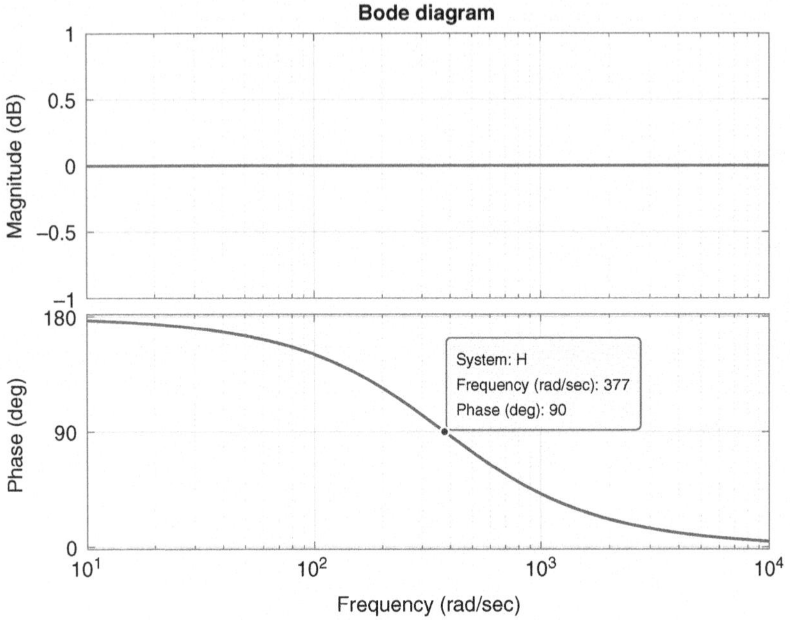

Figure 14.2 Bode plot of the HPF-based APF.

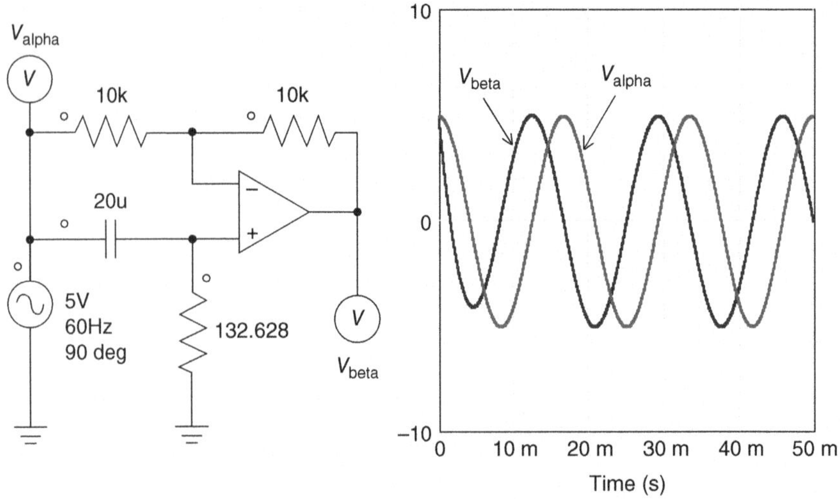

Figure 14.3 Circuit and simulation waveforms of the HPF-based APF.

using the bilinear (Tustin) transformation in MATLAB. The sampling time T_s is chosen to match the inverter switching period, typically $T_s = 1/20\,\text{kHz}$.

```
Ts = 1/(20000);
Hd= c2d(H,Ts,'tustin');
```

The resulting discrete transfer function is:

$$H_d = \frac{V_o(z)}{V_{in}(z)} = \frac{0.9813\,z - 1}{z - 0.9813} \tag{14.2}$$

Solving for the z-domain relationship yields:

$$z \cdot V_o - 0.9813 \cdot V_o = 0.9813\,z \cdot V_{in} - V_{in}$$

Expressed in time-domain difference equation:

$$V_o(k) = 0.9813 \cdot V_o(k-1) + 0.9813 \cdot V_{in}(k) - V_{in}(k-1) \tag{14.3}$$

This discrete implementation enables efficient real-time execution of the APF in digital controllers.

Transfer Function for the LPF Circuit Using the same method as the HPF case, the transfer function for the low-pass filter (*LPF*)-based *APF* is derived as follows:

$$\frac{V_o}{V_{in}} = \frac{-\left(s - \frac{1}{RC}\right)}{s + \frac{1}{RC}} = \frac{-(s - \omega_o)}{s + \omega_o} \tag{14.4}$$

Figure 14.4 presents the corresponding Bode plot for the LPF-based circuit, highlighting the characteristic phase response. The associated PSIM simulation is shown in Figure 14.5. As observed in the waveforms, the output signal V_{beta} lags behind the input signal V_{alpha}, confirming the expected phase shift introduced by the LPF configuration. The LPF configuration preserves the *correct* phase relationship required for an accurate *dq* transformation.

The discrete-time transfer function for the LPF-based APF, obtained using the bilinear (Tustin) transformation, is:

$$H_d = \frac{V_o(z)}{V_{in}(z)} = \frac{-0.9813\,z + 1}{z - 0.9813} \tag{14.5}$$

The corresponding time-domain difference equation is:

$$V_o(k) = 0.9813 \cdot V_o(k-1) - 0.9813 \cdot V_{in}(k) + V_{in}(k-1) \tag{14.6}$$

Summary of HPF and LPF APF Designs Thus far, both the high-pass-based APF in Figure. 14.1(a) and the low-pass-based APF in Figure. 14.1(b) have been examined.

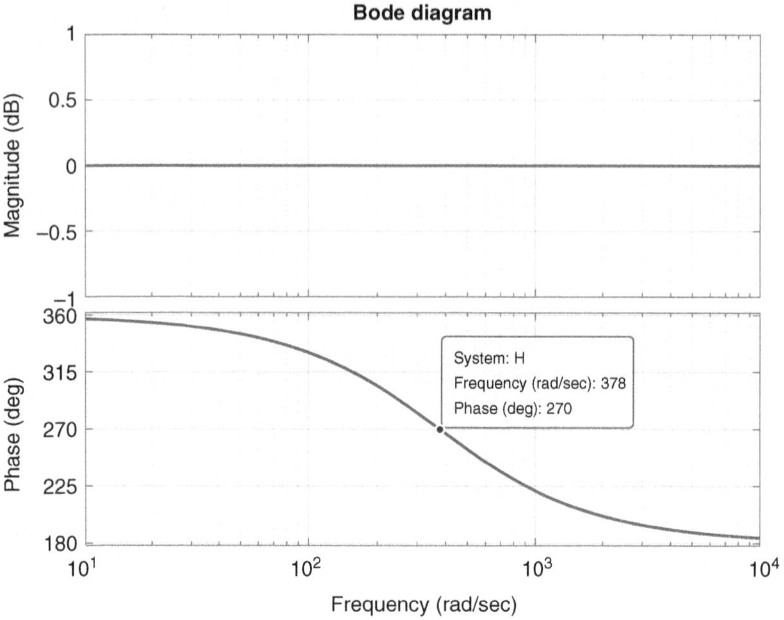

Figure 14.4 Bode plot of the LPF-based APF.

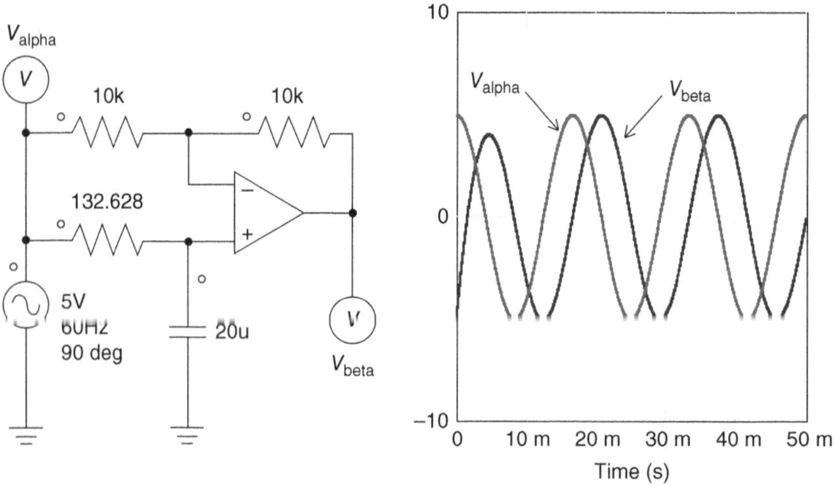

Figure 14.5 Circuit and simulation waveforms of the LPF-based APF.

Among these two configurations, the low-pass-based APF is typically preferred for applications such as phase-locked loop (PLL) implementation and single-phase *dq* control.

The key reason for selecting the LPF structure is that, in three-phase *dq* transformation, the virtual orthogonal component V_{beta} must lag the fundamental component V_{alpha} to maintain the correct rotational direction in the reference frame. The *LPF-based APF* naturally provides this phase relationship, making it essential for accurate *dq* transformations in single-phase systems.

14.2 Controller Design

14.2.1 Single-Phase Stand-Alone Inverter

The transfer functions of the single-phase inverter, as introduced in Chapter 11, are rewritten below for reference:

- Inductor Current to Duty Ratio:

$$G_1(s) = \frac{I_L(s)}{d(s)} = \frac{\frac{V_{in}}{L}\left(s + \frac{1}{RC}\right)}{s^2 + \left(\frac{1}{RC} + \frac{r_L}{L}\right)s + \frac{1}{LC}}$$

- Capacitor Voltage to Duty Ratio:

$$G_2(s) = \frac{V_C(s)}{d(s)} = \frac{\frac{V_{in}}{LC}}{s^2 + \left(\frac{1}{RC} + \frac{r_L}{L}\right)s + \frac{1}{LC}}$$

- Capacitor Voltage to Inductor Current:

$$G_3(s) = \frac{V_C(s)}{I_L(s)} = \frac{1/C}{s + (1/RC)}$$

Since I_L, V_C, and d are AC quantities, a conventional *PI* controller cannot be directly applied. To enable the use of a *PI* controller in single-phase systems, the decoupling control strategy commonly used in three-phase *dq*-control is adapted.

In a single-phase inverter, only one voltage and one current measurement are available, making direct application of the $abc \rightarrow dq$ transformation impossible. To address this limitation, an APF, similar to that used in PLL circuits, is employed to generate virtual α-variables by introducing a 90° phase shift. This allows the creation of β-variables, enabling the subsequent $\alpha\beta \rightarrow dq$ transformation required for *dq* control.

a) **Voltage Controller Design**

Figure 14.6 illustrates the configuration of the voltage controller for the single-phase *dq* control system. A MATLAB m-file example was employed to design the

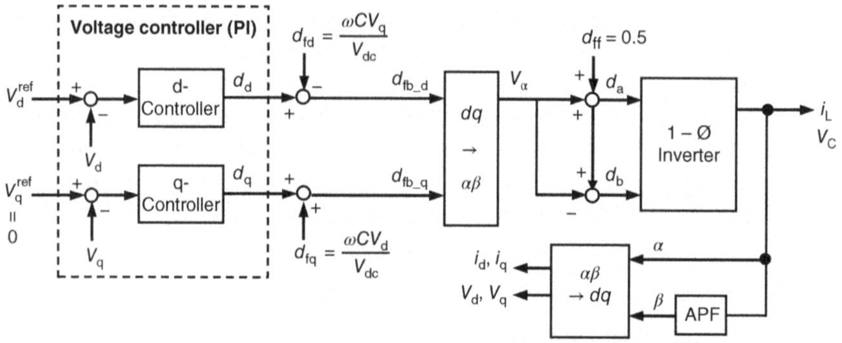

Figure 14.6 Voltage controller configuration for single-phase *dq* inverter.

Table 14.2 MATLAB m-file for voltage controller.

```
s=tf('s');

Vin = 360;   % Input voltage
rL=0.01;
L = 1e-3;    % L=1mh
C = 20e-6;   % C=20uf
R = 15;

Den = s^2+s*(1/(R*C)+rL/L)+1/(L*C);
G1 = (Vin/L)*(s+1/(R*C))/Den;
G2 = Vin/(L*C) / Den;
G3 = (1/C)/(s+1/(R*C));

Kpv = 1; Kiv=1;
Gcv = Kpv*(s+Kiv)/s;

Vref = 150;
T=sisoinit(1);
T.F.Value = Vref;
T.G.Value = G2;
T.C.Value = Gcv;
sisotool(T);
```

voltage controller according to the specified frequency-domain requirements. The corresponding MATLAB script for the voltage controller is provided in Table 14.2.

The frequency-domain design specifications for the voltage controllers are as follows:

- Bandwidth: 100 [rad/sec]
- Phase Margin: 90°

Using the SISOTOOL auto-tuning feature, the continuous-time *PI* controller is obtained as

$$G_{cv}(s) = \frac{0.000018574(s + 14950)}{s}$$

The discrete-time implementation, designed for a sampling time of 50 μs, is given by:

$$G_C(z) = \frac{3.246 \times 10^{-5}(z - 0.572)}{z - 1}$$

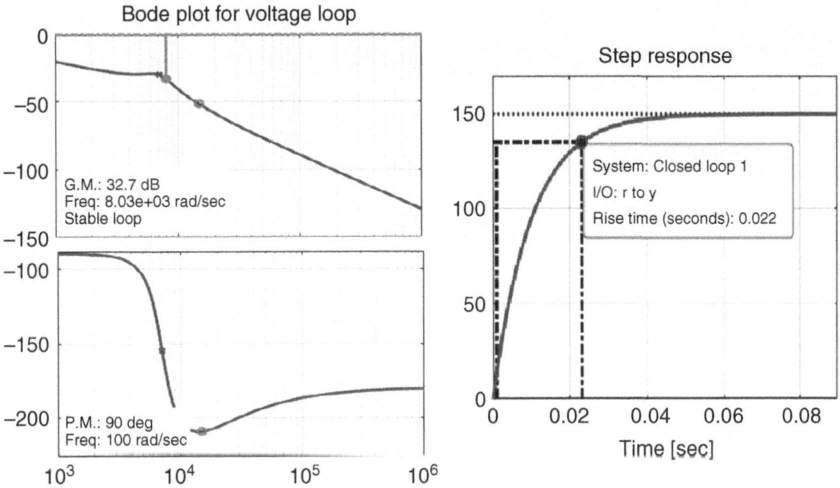

Figure 14.7 Bode plot and step response of the voltage controller.

The Bode plot and step response curves are shown in Figure 14.7, demonstrating a settling time of approximately 39 ms and a rise time of 22 ms, validating the controller's performance.

Simulation Results for Voltage Controller The PSIM simulation schematic and corresponding time-domain waveforms are presented in Figure 14.8.

The implementation of this voltage controller C-block is shown in Table 14.3.

Figure 14.9 shows the simulation waveforms for the single-phase *dq* control. When the reference voltage(V_{ref}) changes from 150 to 100 V, and back to 150 V, the output voltage exhibits smooth transitions without overshoot.

Key observations:

- The V_d voltage waveform has a settling time of approximately 40 ms.
- The simulation results obtained from MATLAB closely match those from PSIM, confirming the accuracy and effectiveness of the controller design.

b) **Current Controller Design**

Let's design the current controller. In this design example, since the output voltage frequency is 377 rad/sec, the cutoff frequency of the current control loop must be set lower than this value.

The frequency-domain design specifications for the current controller are as follows:

- Bandwidth: 200 rad/sec
- Phase Margin: 90°

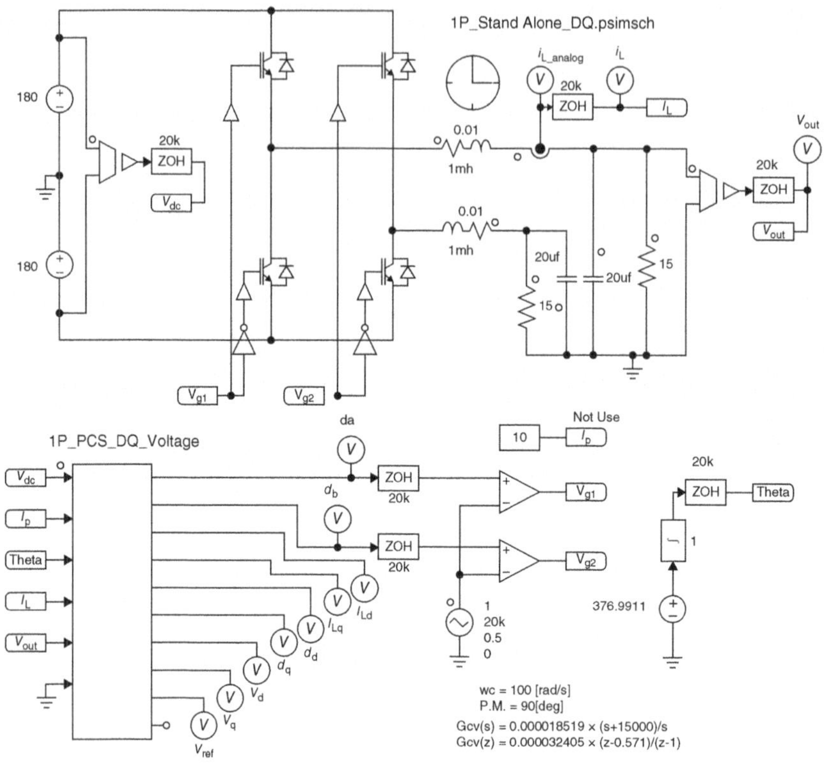

Figure 14.8 Schematic of the single-phase *dq* inverter voltage control system.

Based on these specifications, the PI controller for the current loop is given by:

$$G_{ci}(s) = \frac{K_{pi}(s + K_{ii})}{s} = \frac{8.31}{s}$$

Figure 14.10 illustrates the configuration of the current controller for the single-phase DQ stand-alone inverter.

The discrete-time of the current controller, with a sampling time of 50 μs, is expressed as:

$$G_C(z) = \frac{0.0002078(z + 1)}{z - 1}$$

The Bode plot and step response curve of the designed control system are presented in Figure 14.11. The Bode plot confirms that the frequency-domain specifications are satisfied. The step response characteristics show a settling time of approximately 20 ms and a rise time of approximately 11 ms, validating the dynamic performance of the controller.

Table 14.3 C-block code: 1P_PCS_DQ_Voltage.

```
#define Cf       0.00002    // 20[uf]
#define Lf       0.001      // 1 [mh]
#define rL       0.01       // 10  [mohm]
#define Omega  377          // 60[Hz]
#define Kvp      0.000032405 // Kp
#define Kvi      0.571        //Kiz

double Vdc, Ip, theta, IL, Vgrid;
double duty_a,duty_b, Va_con, Vb_con;

double Ialpha, Ibeta, ILd, ILq;
double Valpha, Vbeta, Vd, Vq;
double Vd_ref, Vd_err,Vq_ref, Vq_err;
double dd, dq, dfb_d, dfb_q;
static double Vbeta_old=0.,
Vgrid_old=0., Ibeta_old=0., IL_old=0.;
static double dd_old=0.0, dq_old=0.0,
Vd_err_old=0.0, Vq_err_old=0.0;

/****** start here ******/
Vdc      = in[0];
Ip       = in[1];
theta    = in[2];
IL       = in[3];
Vgrid    = in[4];

Valpha = Vgrid;
Vbeta = 0.9813*Vbeta_old - 0.9813*Vgrid
+ Vgrid_old;
Vbeta_old = Vbeta;
Vgrid_old = Vgrid;

Vd= cos(theta)*Valpha+sin(theta)*Vbeta;
Vq=-sin(theta)*Valpha+cos(theta)*Vbeta;

Ialpha = IL;
Ibeta=0.9813*Ibeta_old-0.9813*IL
+IL_old;
Ibeta_old = Ibeta;
IL_old = IL;

ILd=cos(theta)*Ialpha+sin(theta)*Ibeta;
ILq=-sin(theta)*Ialpha+cos(theta)*Ibeta;

if (t>0.1667) Vd_ref= 150;
else if (t > 0.08335) Vd_ref=100;
else Vd_ref = 150;

Vd_err = Vd_ref - Vd;

dd=dd_old+Kvp*Vd_err-Kvp*Kvi*Vd_err_old;
dd_old=dd;
Vd_err_old=Vd_err;

Vq_ref = 0;
Vq_err = Vq_ref -Vq;

dq=dq_old+Kvp*Vq_err-Kvp*Kvi*Vq_err_old;
dq_old=dq;
Vq_err_old=Vq_err;

dfb_d = dd - Cf*Omega*Vq/Vdc;
dfb_q = dq + Cf*Omega*Vd/Vdc;

Valpha=dfb_d*cos(theta)-dfb_q*sin(theta);
Vbeta =dfb_d*sin(theta)+dfb_q*cos(theta);

Va_con = Valpha;
Vb_con = Vbeta;

duty_a =0.5 + Valpha;
duty_b= 0.5 - Valpha;

out[0]= duty_a;
out[1]= duty_b;
out[2]= ILd;
out[3]= ILq;
out[4]= dfb_d;
out[5]= dfb_q;
out[6]= Vd;
out[7]= Vq;
out[8]= Vd_ref;
```

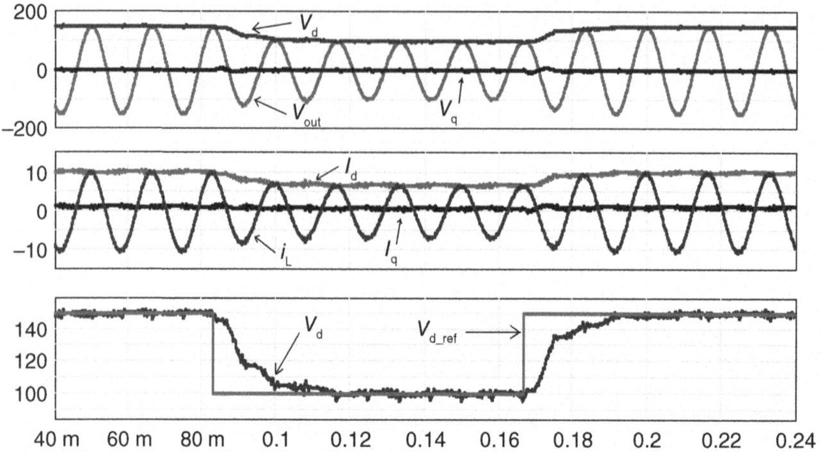

Figure 14.9 Simulation waveforms for single-phase *dq* control.

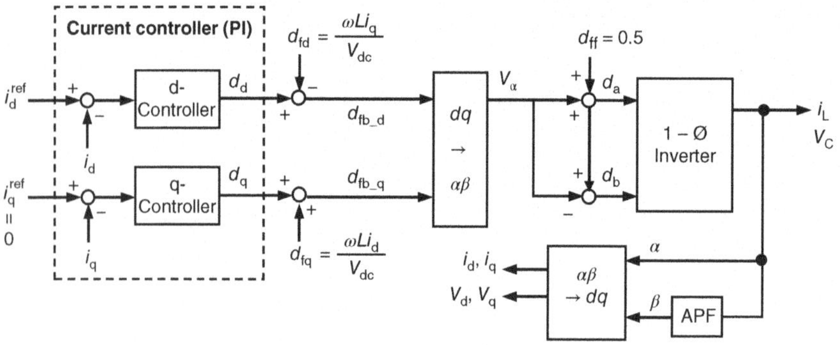

Figure 14.10 Current controller configuration for single-phase *dq* inverter.

The PSIM circuit for current control is identical to the voltage control schematic, except that the C-block is modified to implement the current controller block diagram. The corresponding C code for the current controller is provided in Table 14.4.

The PSIM simulation waveforms are shown in Figure 14.12.

In the simulation, the reference current (I_{ref}) changes from 10 to 5 A and back to 10 A. The output voltage and current waveforms demonstrate the following: It can be observed that the current waveform has no overshoot, and the settling time of the current waveform (I_d) is approximately 20 ms. These results confirm that the MATLAB step response of the mathematical model is consistent with the PSIM simulation of the actual circuit, thereby validating the accuracy and effectiveness of the designed current control system.

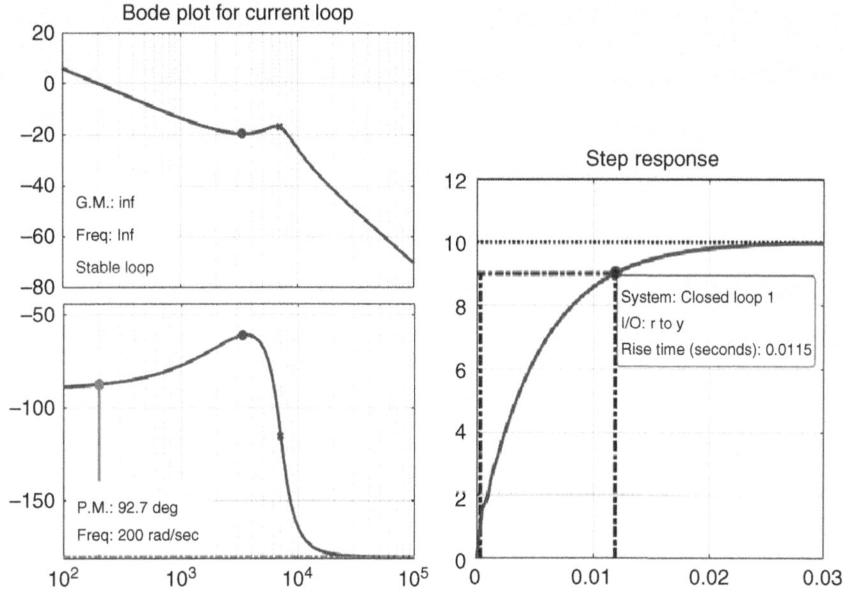

Figure 14.11 Bode plot and step response of the current controller.

c) Double-Loop Controller Design

The double-loop configuration integrates both the voltage and current controllers to achieve precise regulation of both output voltage and current. The block diagram of the double-loop control structure is shown in Figure 14.13. This configuration adopts the same control principles used in the three-phase stand-alone inverter, with identical voltage and current controller structures.

The design specifications for the double-loop controllers are summarized in Table 14.5.

The Bode plots of both the voltage and current control loops, obtained using MATLAB's SISOTOOL auto-tuning feature, are presented in Figure 14.14. These plots confirm that both controllers meet their respective design specifications in terms of bandwidth and phase margin.

The designed PI controller expressions for both loops are as follows:

$$G_{cv}(s) = \frac{0.074(s+90)}{s}, \quad G_{cv}(z) = \frac{0.074(z-0.996)}{z-1}$$

$$G_{ci}(s) = \frac{3.86}{s}, \quad G_{ci}(z) = \frac{9.68 \times 10^{-5}(z+1)}{z-1}$$

Step Response Analysis The step responses of the voltage and current controllers are presented in Figure 14.15. The results confirm that both controllers satisfy

Table 14.4 C-block code: 1P_PCS_DQ_Current.

```
#define Cf        0.00002    // 20[uf]
#define Lf        0.001      // 1  [mh]
#define rL        0.01       // 10  [mohm]
#define Omega     377        // 60  [Hz]
#define Kip       0.0002078 // Kp 200 rad/s
#define Kii       -1          //Kiz

double Vdc, Ip,  theta, IL, Vgrid;
double duty_a,duty_b, Va_con, Vb_con;

double Ialpha, Ibeta, ILd, ILq;
double Valpha, Vbeta, Vd, Vq;
double Id_ref, Id_err,Iq_ref, Iq_err;
double dd, dq, dfb_d, dfb_q;
static double Vbeta_old=0., Vgrid_old=0.,
Ibeta_old=0., IL_old=0.;
static double dd_old=0.0, dq_old=0.0,
Id_err_old=0.0, Iq_err_old=0.0;

/****** start here ******/
Vdc      = in[0];
Ip       = in[1];
Theta    = in[2];
IL       = in[3];
Vgrid    = in[4];

Valpha = Vgrid;
Vbeta=0.9813*Vbeta_old - 0.9813*Vgrid
+Vgrid_old;
Vbeta_old = Vbeta;
Vgrid_old = Vgrid;

Vd=cos(theta)*Valpha + sin(theta)*Vbeta;
Vq=-sin(theta)*Valpha +cos(theta)*Vbeta;

Ialpha = IL;
Ibeta=0.9813*Ibeta_old-0.9813*IL+IL_old;
Ibeta_old = Ibeta;
IL_old = IL;

ILd=cos(theta)*Ialpha + sin(theta)*Ibeta;
ILq=-sin(theta)*Ialpha +cos(theta)*Ibeta;

if (t>0.1667) Id_ref= Ip;
else if (t>0.08335) Id_ref=0.5*Ip;
else Id_ref = Ip;

Id_err = Id_ref - ILd;

dd = dd_old + Kip*Id_err -
Kip*Kii*Id_err_old;
dd_old=dd;
Id_err_old=Id_err;

Iq_ref = 0;
Iq_err = Iq_ref -ILq;

dq = dq_old +Kip*Iq_err -
Kip*Kii*Iq_err_old;
dq_old=dq;
Iq_err_old=Iq_err;

dfb_d = dd - Lf*Omega*ILq/Vdc;
dfb_q = dq + Lf*Omega*ILd/Vdc;

Valpha = dfb_d*cos(theta) -
dfb_q*sin(theta);
Vbeta  = dfb_d*sin(theta) +
dfb_q*cos(theta);

Va_con = Valpha;
Vb_con = Vbeta;

duty_a = 0.5 + Valpha;
duty_b = 0.5 - Valpha;

out[0]=duty_a;
out[1]=duty_b;
out[2]=ILd;
out[3]=ILq;
out[4]=dfb_d;
out[5]=dfb_q;
out[6]=Vd;
out[7]=Vq;
out[8]=Id_ref;
```

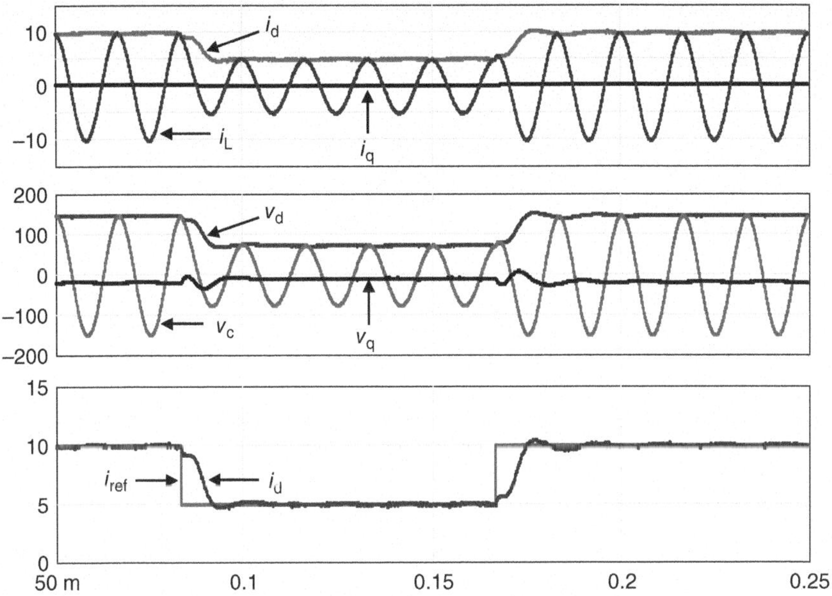

Figure 14.12 Simulation waveforms for current control.

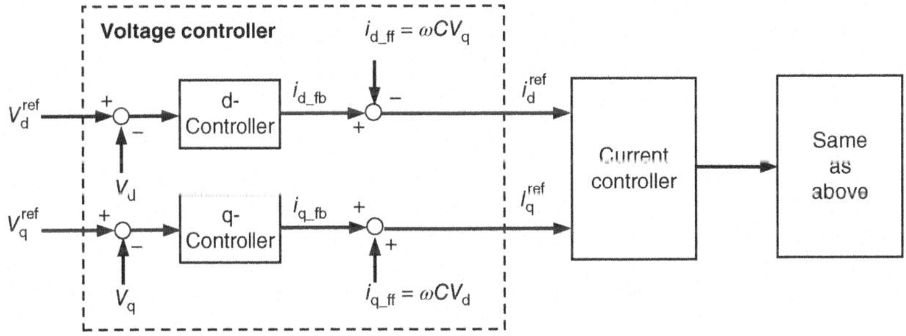

Figure 14.13 Double-loop configuration of the single-phase *dq* inverter.

Table 14.5 Design specification for double-loop controller.

	Bandwidth	Phase margin
Voltage controller (C_1)	100 rad/sec	90 deg
Current controller (C_2)	200 rad/sec	60 deg

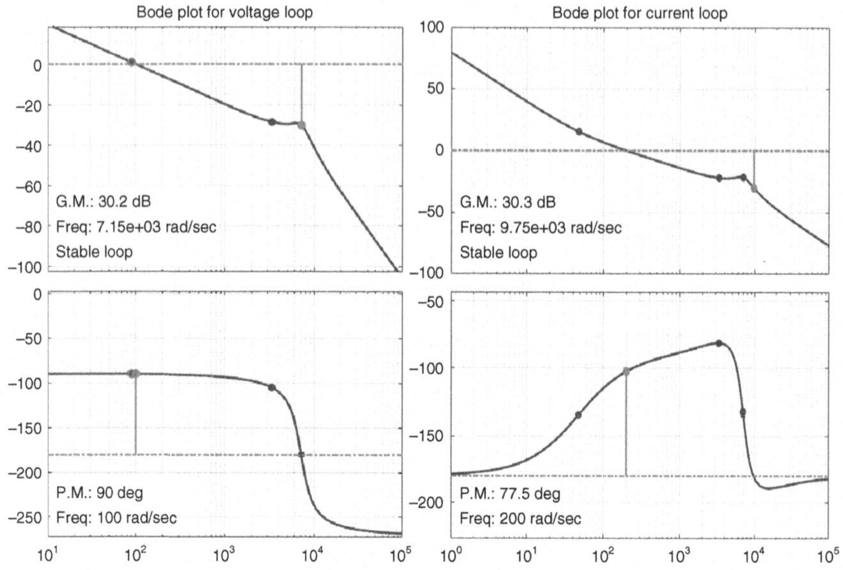

Figure 14.14 Bode plot of the double-loop controller. (a) Voltage loop and (b) Current loop.

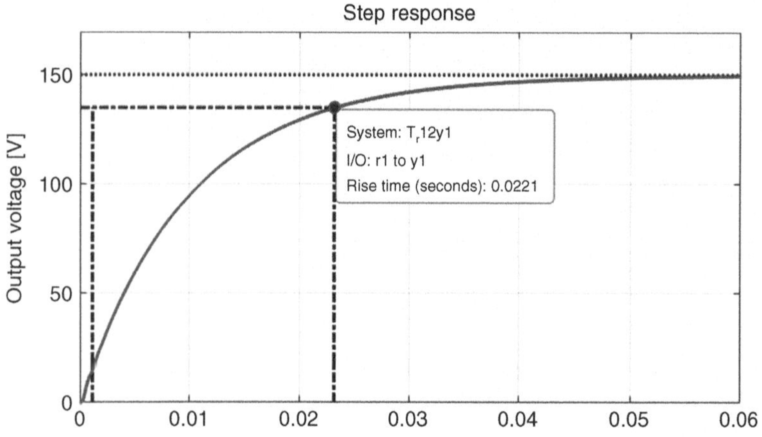

Figure 14.15 Step responses of the double-loop controller.

the specified transient performance requirements. In particular, the voltage loop exhibits a smooth response with no overshoot and an appropriate settling time.

PSIM Simulation and C-Code Implementation The PSIM circuit for the double-loop control system is identical to the previous voltage control schematic, except that the C-block is modified to implement both voltage and current controllers. The rest of the control structure, including the voltage and current feedback loops, follows the same design principles used in the three-phase stand-alone inverter.

The C-code implementation for the double-loop controller is provided in Table 14.6. Aside from the APF required for single-phase *dq* transformation, the control algorithm is consistent with the three-phase implementation.

The simulation waveforms obtained from PSIM are shown in Figure 14.16, confirming that the designed double-loop controller provides stable and accurate regulation of both voltage and current.

Table 14.6 is the C-code for the double-loop controller. This implementation ensures stable and accurate control of both the voltage and current loops in the single-phase DQ inverter system.

14.2.2 Single Phase Grid-Tied (PCS) DQ Inverter

The design of the single-phase grid-tied inverter, also known as a power conditioning system (PCS), using the DQ control method is presented in this section. The circuit configuration for the single-phase PCS is shown in Figure 14.17.

State-Space Equation Derivation The differential equation governing the inductor current i_L in the PCS circuit is given by:

$$L\frac{di_L}{dt} = \left(\frac{V_{dc}}{2} - V_{grid} - r_L i_L\right) d + \left(-\frac{V_{dc}}{2} - V_{grid} - r_L i_L\right)(1-d)$$

Simplifying this expression yields:

$$L\frac{di_L}{dt} = V_{dc}d - \frac{V_{dc}}{2} - V_{grid} - r_L i_L$$

This equation is inherently nonlinear due to the presence of a DC offset. To linearize the system, a feedforward compensation is introduced by defining the duty ratio as:

$$d = d_{ff} + d_{fb}$$

where:

$$d_{ff} = 0.5 + \frac{V_{grid}}{V_{dc}}$$

Substituting d_{ff} into the original equation, the linearized state equation becomes:

$$L\frac{di_L}{dt} = -r_L i_L + V_{dc} \cdot d_{fb} \tag{14.7}$$

Table 14.6 C-block code: 1P_PCS_DQ_Double.

```
#define Cf       0.00002   // 20[uf]
#define Lf       0.001     // 1 [mh]
#define rL       0.01      // 10  [mohm]
#define Omega    377       // 60 [Hz]
#define Kvp      0.074     // Voltage P
#define Kvi      0.996     // Voltage I
#define Kip      0.00009679// Current P
#define Kii     [illegible]   // Current I

double Vdc, Ip,  theta, IL, Vgrid;
double duty_a,duty_b, Va_con, Vb_con;
double Ialpha, Ibeta, ILd, ILq;
double Valpha, Vbeta, Vd, Vq;
double Vd_ref, Vd_err,Vq_ref, Vq_err;
double Id_ref, Iq_ref, Id_err, Iq_err;
double dd, dq, dfb_d, dfb_q;
double Id_fb, Iq_fb, Id_ff, Iq_ff;
static double Id_fb_old=0, Iq_fb_old=0.;
static double Vbeta_old=0., Vgrid_old=0.,
Ibeta_old=0., IL_old=0.;
static double dd_old=0.0, dq_old=0.0,
Vd_err_old=0.0, Vq_err_old=0.0;
static double Id_err_old=0.0,
Iq_err_old=0.0;
```

```
Vd_err = Vd_ref - Vd;
Id_fb=Id_fb_old+Kvp*Vd_err-Kvp*Kvi*Vd_err_old;
Id_fb_old=Id_fb;
Vd_err_old=Vd_err;

Id_ff =   Cf*Omega*Vq;
Id_ref = Id_fb-Id_ff;  // Id_ref

Vq_ref = 0.;
Vq_err = Vq_ref - Vq;
Iq_fb=Iq_fb_old+Kvp*Vq_err-Kvp*Kvi*Vq_err_old;
Iq_fb_old=Iq_fb;
Vq_err_old=Vq_err;

Iq_ff =  Cf*Omega*Vd;
Iq_ref = Iq_fb + Iq_ff;  //Iq_ref

/****** Current controller dq *******/
Id_err = Id_ref - ILd;
dd=dd_old+Kip*Id_err-Kip*Kii*Id_err_old;
dd_old=dd;
Id_err_old=Id_err;
```

```
/****** start here ******/
Vdc       = in[0];
Ip        = in[1];
theta     = in[2];
IL     = in[3];
Vgrid     = in[4];

Valpha = Vgrid;
Vbeta=0.9813*Vbeta_old-
0.9813*Vgrid+Vgrid_old;
Vbeta_old = Vbeta;
Vgrid_old = Vgrid;

Vd=cos(theta)*Valpha + sin(theta)*Vbeta;
Vq=-sin(theta)*Valpha +cos(theta)*Vbeta;

Ialpha = IL;
Ibeta=0.9813*Ibeta_old-0.9813*IL+IL_old;
Ibeta_old = Ibeta;
IL_old = IL;

ILd=cos(theta)*Ialpha + sin(theta)*Ibeta;
ILq=-sin(theta)*Ialpha +cos(theta)*Ibeta;

/****** Voltage controller dq *******/
if (t>0.1667) Vd_ref= 150;
else if (t > 0.08335) Vd_ref=100,
else Vd_ref = 150;
Iq_err = Iq_ref - ILq;
dq=dq_old+Kip*Iq_err-Kip*Kii*Iq_err_old;
dq_old=dq;
Iq_err_old=Iq_err;

dfb_d = dd  -  Lf*Omega*ILq/(Vdc);
dfb_q = dq +  Lf*Omega*ILd/(Vdc);
/****** End of controller ***********/

Valpha = dfb_d*cos(theta)- dfb_q*sin(theta);
Vbeta  = dfb_d*sin(theta)+ dfb_q*cos(theta);

Va_con = Valpha;
Vb_con = Vbeta;

duty_a =0.5 + Valpha;
duty_b= 0.5 - Valpha;

out[0]=duty_a;
out[1]=duty_b;
out[2] = ILd;
out[3] = ILq;
out[4]=dfb_d;
out[5]=dfb_q;
out[6]=Vd;
out[7]=Vq;
out[8]= Vd_ref;
```

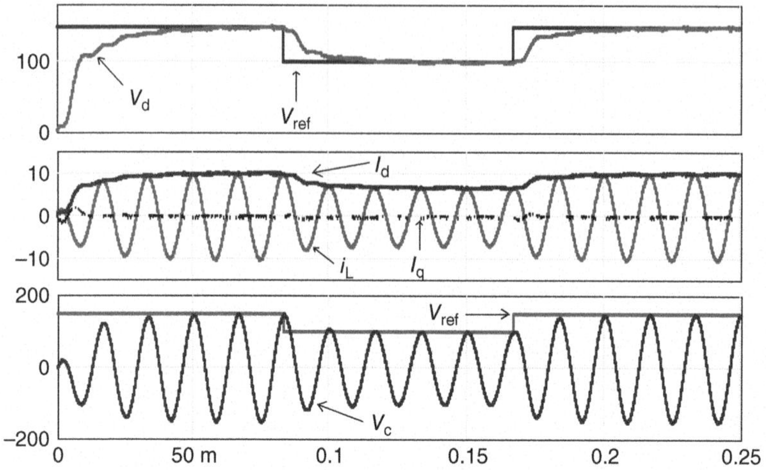

Figure 14.16 The simulation waveforms of the double-loop controller.

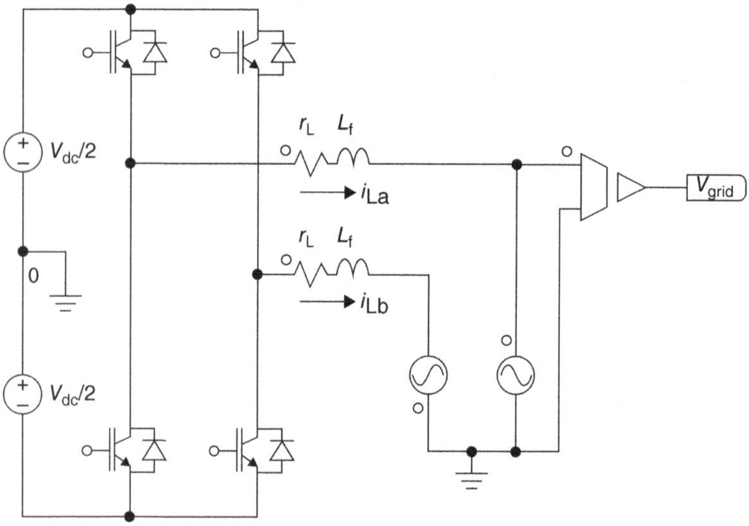

Figure 14.17 The configuration of the single-phase PCS.

The corresponding state-space representation is:

$$\dot{X} = AX + Bd_{\text{fb}}$$

where $X = i_{\text{L}}$

The transfer function derived from the state-space model is:

$$\frac{I_{\text{L}}(s)}{d(s)} = \frac{V_{\text{dc}}}{sL + r_{\text{L}}} = G_{1\text{dq}}(s) \tag{14.8}$$

Design of the Current Controller Since there is only one state variable—the inductor current i_L–only the current controller needs to be designed for the PCS system. The resulting transfer function is a first-order system, allowing for straightforward control design.

Two controller types are considered based on the desired dynamic response:

1) Proportional (P) Controller:
   - Suitable for overdamped or critically damped responses with no overshoot.
   - Simpler to implement but cannot eliminate steady-state error for a step reference.
2) Proportional–Integral (PI) Controller:
   - Suitable when some overshoot is acceptable or when zero steady-state error is required.
   - Places a pole at the origin to improve steady-state accuracy.

The choice between P and PI control depends on the specific application requirements, such as response speed, overshoot tolerance, and steady-state accuracy. The MATLAB code shown in Table 14.7 provides a basic framework for the design of the current controller in the single-phase PCS DQ inverter.

This code allows for frequency-domain design using the Bode plot and time-domain validation using the step response.

P Controller Design The design of a proportional (P) controller for the single-phase grid-tied (PCS) inverter focuses on achieving fast dynamic response without

Table 14.7 MATLAB m-file for current controller of single-phase PCS dq inverter.

```
s=tf('s');

Vin = 360;   % Input voltage
rL=0.01;
L = 1e-3;    % L=1mh

G1dq = (Vin/L) / ( s + rL/L );

Kpi = 1; Kii=1;
Gci = Kpi;                  % P-Controller
% Gci = Kpi*(s+Kii)/s;  % PI-Controller
Iref = 10;
T=sisoinit(1);
T.F.Value = Iref;
T.G.Value = G1dq;
T.C.Value = Gci;
sisotool(T);
```

overshoot, ensuring system stability and unity power factor operation. The target design specifications for the P controller are as follows

- Bandwidth = 4800 rad/sec

The P controller is represented in the frequency domain as:

$$G_{ci}(s) = 0.013$$

Figure 14.18 shows that the designed controller has no overshoot and a critically damped response.

The PSIM simulation results, illustrated in Figure 14.19, demonstrate that the current waveform follows the reference without overshoot and that unity

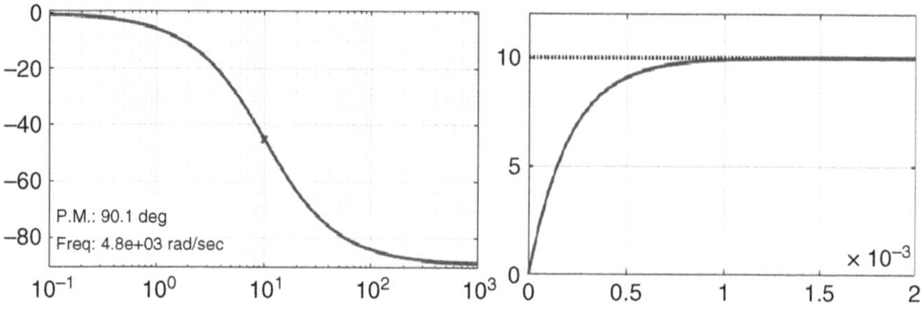

Figure 14.18 Bode plot and step response of the P controller.

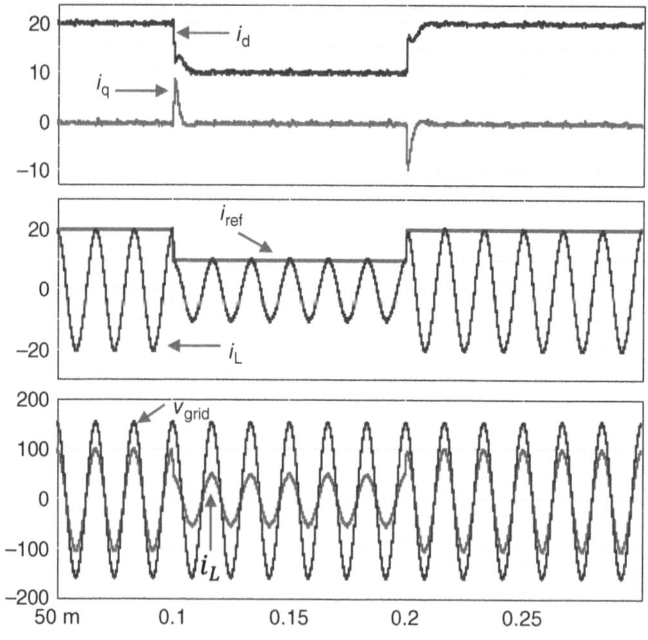

Figure 14.19 Simulation waveform for the *P* controller.

power factor control is achieved. These results validate the accuracy of the *P* controller design.

Table 14.8 presents the C-code implementation of the designed *P* controller. This proportional controller provides fast current tracking while maintaining both stability and power quality. Although a *P* controller alone cannot theoretically eliminate steady-state error, the inclusion of feedforward compensation,

Table 14.8 C-block code: 1P_PCS_DQ_Current.

```
#define Lf        0.001      // 1 [mh]
#define rL        0.01       // 10  [mohm]
#define Omega     377        // 60 [Hz]
#define Kip       0.013 // Current - Kp

double Vdc, Ip, theta, IL, Vgrid;
double duty_a,duty_b, flag;
double Ialpha, Ibeta, ILd, ILq;
double Valpha, Vbeta, Vd, Vq;
double Id_ref, Id_err,Iq_ref, Iq_err;
double dd, dq, dfb_d, dfb_q;
double Va_con, Vb_con;
static double Vbeta_old=0., Vg_old=0.,
Ibeta_old=0., IL_old=0.;
/****** start here ******/
Vdc     = in[0];
Ip      = in[1];
theta   = in[2];
IL    = in[3];
Vgrid   = in[4];

if (Vdc< 1.0 ) Vdc=1.0;
Valpha = Vgrid;
Vbeta=0.9813*Vbeta_old-0.9813*Vgrid+Vg_old;
Vbeta_old = Vbeta;
Vg_old = Vgrid;

Vd=cos(theta)*Valpha +sin(theta)*Vbeta;
Vq=-sin(theta)*Valpha+cos(theta)*Vbeta;

Ialpha = IL;
Ibeta=0.9813*Ibeta_old-0.9813*IL+IL_old;
Ibeta_old = Ibeta;
IL_old = IL;

ILd=cos(theta)*Ialpha+sin(theta)*Ibeta;
ILq=-sin(theta)*Ialpha+cos(theta)*Ibeta;

if (t>0.2) Id_ref= Ip;
else if (t > 0.1) Id_ref=0.5*Ip;
else Id_ref = Ip;

Id_err = Id_ref - ILd;
/**** P Controller *****/
dd = Kip*Id_err;
dq = Kip*Iq_err;

dfb_d = dd -  Lf*Omega*ILq/(Vdc);
dfb_q = dq +  Lf*Omega*ILd/(Vdc);

Valpha=dfb_d*cos(theta)-dfb_q*sin(theta);
Vbeta=dfb_d*sin(theta)+dfb_q*cos(theta);

Va_con = Valpha;
Vb_con = Vbeta;

duty_a = 0.5 + Valpha + Vgrid/Vdc;
duty_b=  0.5 - Valpha - Vgrid/Vdc;

out[0] = duty_a;
out[1] = duty_b;
out[2] = ILd;
out[3] = ILq;
out[4] = dd;
out[5] = dq;
out[6] = Vd;
out[7] = Vq;
out[8] = Id_ref;
out[9] = Va_con;
```

as discussed earlier, enables the system to achieve accurate control in practical applications.

PI Controller Design To enhance both steady-state accuracy and dynamic response, a PI controller is applied to the single-phase grid-tied PCS system. Since the design procedure of the PI controller follows the same methodology previously presented, it is not repeated here.

15

PLL (Phase-Locked Loop)

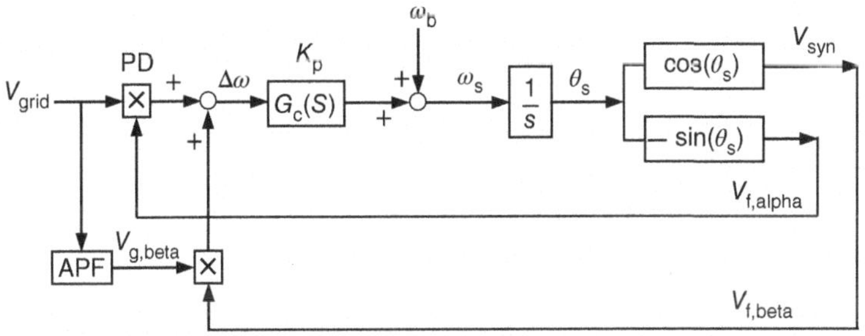

Phase-Locked Loop (PLL) in Inverters In inverters, the phase-locked loop (PLL) is a critical technology used to synchronize the phase of the inverter's output with that of the grid voltage. This synchronization is particularly essential in grid-connected inverters, as it contributes to the stability of the power system and ensures efficient power transfer.

The PLL continuously detects the phase of the grid voltage and adjusts the phase of the inverter's output accordingly. By minimizing the phase difference between the grid voltage and the inverter output current, the PLL enables stable power delivery and seamless grid integration.

Feedback Control Systems and Phase-Locked Loop (PLL) in Power Electronics The primary objective of a feedback control system is to ensure that the output voltage V_{out} accurately tracks the reference input voltage V_{in} under steady-state conditions. In other words, if the reference input is 1, the output should also be 1; if the input is 10, the output should likewise reach 10. This fundamental principle enables precise control and regulation in various applications.

To achieve this goal, the input signal must be a direct current (DC) signal—specifically, a positive signal without polarity reversal. Common examples of

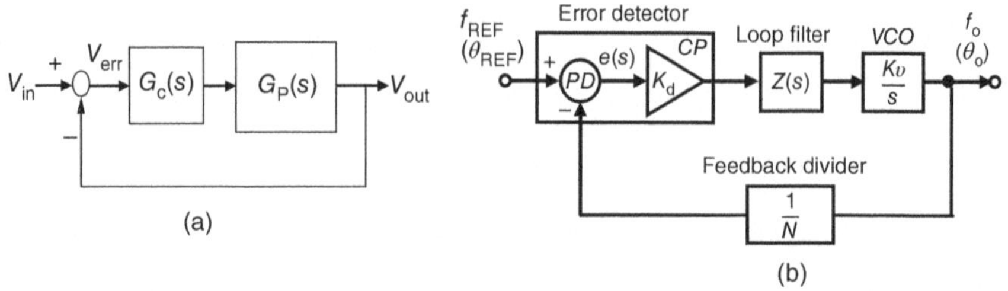

Figure 15.1 Block diagram of feedback system and PLL system. (a) Feedback control system and (b) Phase-locked loop system.

such inputs include step signals, ramp signals, and parabolic signals. Although these signals may vary in magnitude over time, they remain unidirectional and non-oscillatory, which makes them suitable for traditional feedback control systems.

As illustrated in Figure 15.1(a), a typical feedback control system comprises:

- The computation of the error, which is the difference between the reference input and the system output.
- A controller that processes this error and generates an appropriate control signal to drive the system towards the desired output.

Now consider a different scenario where the input is an alternating current (AC) sinusoidal signal that oscillates between positive and negative values. In this case, the system must be capable of reproducing the same sinusoidal signal at the output, matching both the frequency and phase of the input.

This requirement becomes particularly critical when the input sinusoidal signal contains high-frequency harmonics or distortions. In such cases, the system must not only follow the input waveform but also filter out distortions to produce a clean sinusoidal output. Achieving this demands a control system that can continuously synchronize with the frequency and phase of the input AC signal. Such a control system is known as a PLL. The basic structure of the PLL system is shown in Figure 15.1(b). In power electronics, the PLL plays a crucial role in grid-connected systems, where it extracts the phase and frequency of the grid voltage to ensure precise synchronization for stable and efficient power transfer.

15.1 Single-Phase PLL

Product PLL Structure for Frequency Detection Figure 15.2 illustrates the structure of a product-type PLL, which is commonly employed for general signal

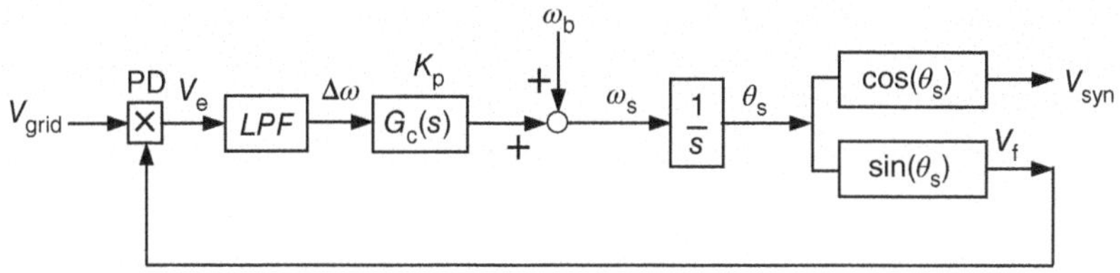

Figure 15.2 Block diagram of the product-type PLL.

frequency detection applications. This type of PLL comprises the following key components:

- A phase detector (PD) that multiplies the reference grid voltage by the feedback signal.
- A first-order low-pass filter (LPF) that suppresses high-frequency noise.
- An amplifier and integrator that processes the filtered signal to produce the phase angle output.

The output of this PLL structure is the phase angle, denoted as θ_s.

The mathematical expression for the input grid voltage V_g is given by:

$$V_g = V_P \cos(\omega_g t + \phi_g) = V_P \cos \theta_g$$

Similarly, the feedback signal V_f can be expressed as:

$$V_f = \sin(\omega_s t + \phi_s) = \sin \theta_s$$

The PD multiplies the grid voltage and the feedback signal, yielding the following output:

$$V_e = V_g \cdot V_f = V_P \cos(\omega_g t + \phi_g) \cdot \sin(\omega_s t + \phi_s)$$

When the PLL is locked, the grid frequency ω_g and the feedback signal frequency ω_s become equal to the reference frequency ω_b. In this synchronized condition, the PD output is given by:

$$V_e = \frac{V_P}{2} \sin(\phi_g - \phi_s) + \frac{V_P}{2} \sin(2\omega_b t + \phi_g + \phi_s) \tag{15.1}$$

The reference angular frequency is:

$$\omega_b = 2\pi \cdot 60 \text{ Hz} = 120 \text{ rad/ sec}$$

The following trigonometric identities are relevant for analyzing the PD output:

$$\sin(\alpha \pm \beta) = \sin\alpha \cdot \cos\beta \pm \cos\alpha \cdot \sin\beta$$

$$\cos(\alpha \pm \beta) = \cos\alpha \cdot \cos\beta \mp \sin\alpha \cdot \sin\beta$$

$$\sin\alpha \cdot \cos\beta = \frac{1}{2}[\sin(\alpha + \beta) + \sin(\alpha - \beta)]$$

The PD output V_e consists of two components:

- A DC component: $\frac{V_p}{2}\sin(\phi_g - \phi_s)$
- An AC component: A sinusoidal 120 Hz signal at twice the fundamental frequency

To attenuate the unwanted AC component in the PD output, a first-order LPF is employed as the loop filter. However, the AC component frequency (120 Hz) is relatively low compared to the fundamental grid frequency (60 Hz). For effective filtering, the LPF's 3 dB cutoff frequency should typically be at least ten times lower than the fundamental frequency.

Due to the low frequency of the AC ripple (120 Hz), the phase output θ_s inevitably contains significant ripple components. Consequently, this product-type PLL structure is rarely used in practical applications because it is difficult to achieve both system stability and distortion-free synchronization simultaneously.

PLL System Using an All-Pass Filter (APF) To design a PLL system without using LPF—which inherently introduces time delay—an all-pass filter (APF) can be employed. The PLL structure incorporating an APF is shown in Figure 15.3. This system is specifically designed for single-phase grid voltage synchronization and consists of the following key components:

- An APF
- A PD for detecting the phase of the grid voltage
- A proportional (P) controller
- An integrator for phase angle generation

Since the APF was explained in Chapter 14, detailed descriptions are omitted here. As discussed earlier, the APF in this context is implemented in a form similar to a LPF. To maintain consistency with the terminology used in

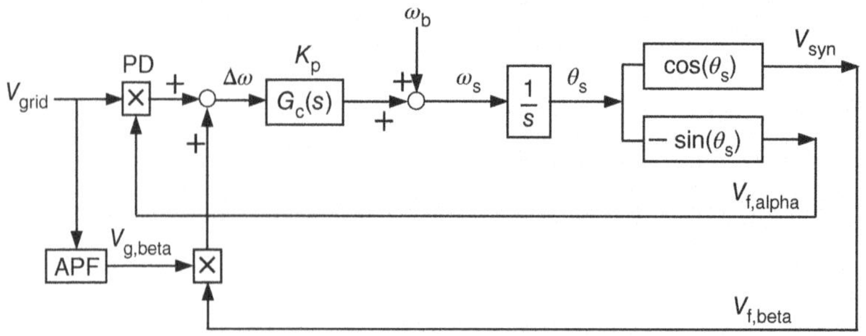

Figure 15.3 Configuration of single-phase PLL system using APF.

dq transformation for three-phase systems, the APF input signal V_{grid} in the single-phase system is denoted as $V_{g,alpha}$, while the APF output, which provides a 90-degree phase-shifted version of the input, is denoted as $V_{g,beta}$.

The mathematical expressions for these two signals are:

$$V_g = V_{g,alpha} = V_P \cos(\omega_g t + \phi_g) = V_P \cos\theta_g$$
$$V_{g,beta} = V_P \sin(\omega_g t + \phi_g) = V_P \sin\theta_g$$

The corresponding feedback signals are:

$$V_{f,alpha} = -\sin\theta_s$$
$$V_{f,beta} = \cos\theta_s$$

The output of the PD, representing the frequency variation $\Delta\omega$, is derived by multiplying the grid signals and feedback signals:

$$\begin{aligned}\Delta\omega &= V_{g,alpha} \cdot V_{f,alpha} + V_{g,beta} \cdot V_{f,beta} = -V_P \cos\theta_g \cdot \sin\theta_s + V_P \sin\theta_g \cdot \cos\theta_s \\ &= V_P \sin(\theta_g - \theta_s)\end{aligned} \tag{15.2}$$

Unlike conventional methods where the PD output contains both DC and AC components—requiring an LPF to extract the desired signal—this approach produces only a DC component. As a result, the need for an LPF is eliminated, reducing system delay and improving dynamic performance.

The P controller processes this PD output to generate the control signal $\Delta\omega_c$. By adding the nominal grid angular frequency ω_b to this control signal, the synchronized frequency ω_s is obtained:

$$\omega_s = \omega_b + \Delta\omega_c$$

By integrating this angular velocity, the phase angle θ_s is obtained.

The feedback signals are then recalculated using the updated phase angle:

$$V_{f,alpha} = -\sin\theta_s$$
$$V_{f,beta} = \cos\theta_s$$

Among these, $V_{f,beta}$ serves as the synchronized grid phase signal θ. By adjusting the phase difference around the nominal frequency ω_b, the PLL locks the phases of the grid and feedback signals.

This synchronized phase signal is essential in single-phase grid-connected inverters, where it is used to control active power. Specifically, the synchronized phase signal θ is multiplied by the desired grid current reference to generate the AC reference current for grid-connected operation.

Digital Implementation in C-Block To implement the PLL integrator $\frac{1}{s}$ in a C-block, it must first be converted into a discrete-time form suitable for digital control.

Table 15.1 MATLAB m-file for discrete-time PLL implementation.

```
s=tf('s');

Gc= tf(1,[1 0]); % integrator
Ts=0.00005;  % sampling 20 kHz
Gd = c2d(G,Ts, 'zoh')
```

This discretization is performed using MATLAB's ***c2d*** function with the zero-order hold (ZOH) method. The MATLAB script for this transformation is provided in Table 15.1.

The result of this discretization yields the following discrete-time integrator transfer function:

$$G_d(z) = \frac{\theta_s(z)}{\omega_s(z)} = \frac{T_s}{z-1}$$

In the time domain, this is expressed as:

$$\theta_s(k) = \theta_s(k-1) + T_s \cdot \omega_s(k)$$

The corresponding implementation in C code is straightforward:

```
theta_s = theta_s_old + Ts*wS;
theta_s_old = theta_s;
```

The schematic diagram for simulating the single-phase PLL system in PSIM is shown in Figure 15.4.

The C-block contains the digital implementation of the PLL, including the proportional controller, frequency computation, phase integration, and feedback signal generation. The MATLAB m-file for PLL is presented in Table 15.2.

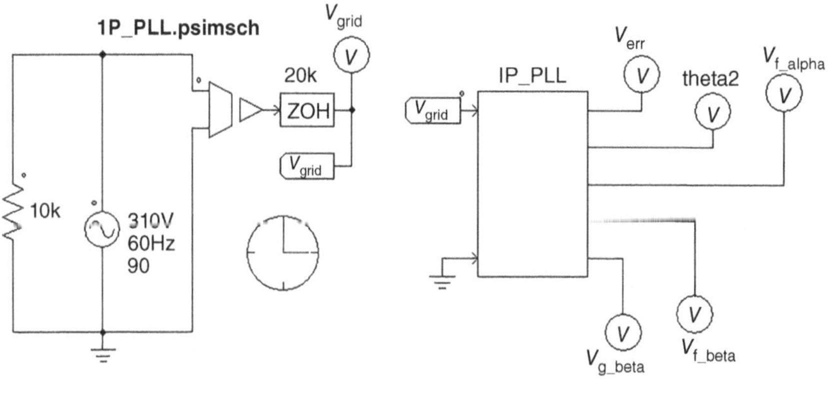

Figure 15.4 Circuit schematic of the single-phase PLL.

Table 15.2 C-block code: 1P_PLL.

```
#define wB 377
#define Kp 0.5
#define Ts 0.00005
#define PI 3.141592

double Vgrid, Vg_alpha, Vg_beta;
double Vf_alpha, Vf_beta;
double Verr, wS, theta2;
static double theta_s=0, theta_s_old=0;
static double Vgrid_old=0, Vg_beta_old=0;
/****** start here ******/
Vgrid = in[0];

Vg_alpha = Vgrid;
Vg_beta=0.9813*Vg_beta_old -
        0.9813*Vgrid+Vgrid_old;
Vg_beta_old = Vg_beta;
Vgrid_old = Vgrid;

Vf_alpha = -sin(theta_s);
Vf_beta  =  cos(theta_s);

Verr=Vg_alpha*Vf_alpha+Vg_beta*Vf_beta;
wS = wB + Kp*Verr;

theta_s = theta_s_old + Ts*wS;
theta_s_old = theta_s;

theta2=2*PI*(theta_s/(2*PI) -
        (int)( theta_s/(2*PI)));

out[0] = Verr;
out[1] = theta2;
out[2] = Vf_alpha;
out[3] = Vf_beta;
out[4] = Vg_beta;
```

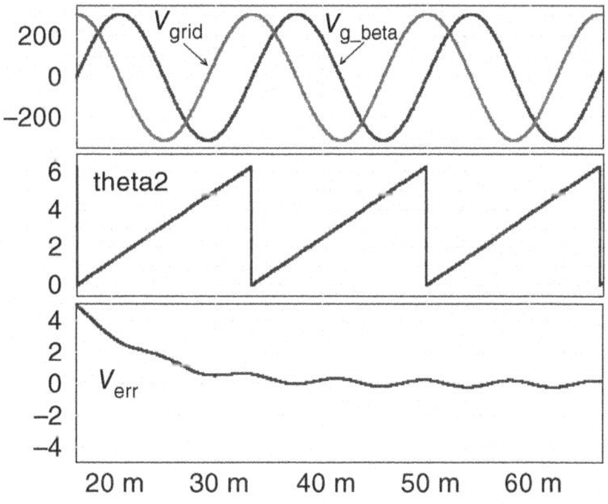

Figure 15.5 Simulation waveforms of the single-phase PLL.

By analyzing the simulation waveforms shown in Figure 15.5, the following observations can be made:

- The phase angle θ_s is successfully synchronized with the grid voltage V_{grid}.
- The PD output V_e converges to zero within one cycle, indicating proper locking of the PLL.

- The feedback signal $V_{f,beta}$ remains synchronized with the grid voltage, maintaining the same frequency and phase.

These results confirm the stable operation and effective phase synchronization capability of the implemented single-phase PLL system.

15.2 Three-Phase PLL

The structure of a three-phase PLL is fundamentally identical to that of a single-phase PLL, with the key difference lying in the method used to generate the α- and β-axis components(V_{alpha} and V_{beta}) from the grid voltage. In a single-phase system, these components are typically generated using an APF. In contrast, in a three-phase system, they can be obtained more directly and accurately using the $abc \rightarrow \alpha\beta$ transformation, owing to the inherent balance of three-phase systems.

When the three-phase circuit is configured using line-to-line voltages, it is necessary to compute the neutral point voltage in order to derive the corresponding phase voltages. The transformation formulas between line voltages (V_{ab}, V_{bc}, V_{ca}) and phase voltages (V_{an}, V_{bn}, V_{cn}) in a balanced system are expressed as follows:

$$V_{bn} = (V_{bc} - V_{ab})/3$$
$$V_{an} = V_{bn} + V_{ab}$$
$$V_{ca} = V_{ab} - 3V_{an}$$
$$V_{cn} = (V_{ca} - V_{bc})/3$$

Once the phase voltages are determined, the $abc \rightarrow \alpha\beta$ transformation for the three-phase inverter is given by:

$$V_{alpha} = V_{an}$$
$$V_{beta} = \frac{1}{\sqrt{3}}(V_{bn} - V_{cn})$$

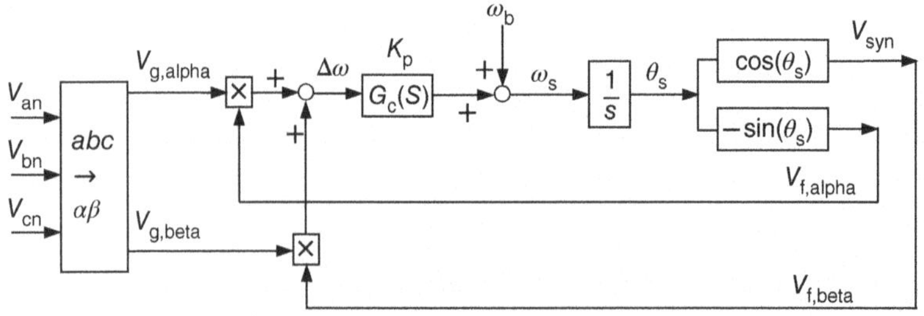

Figure 15.6 Configuration of the three-phase PLL system.

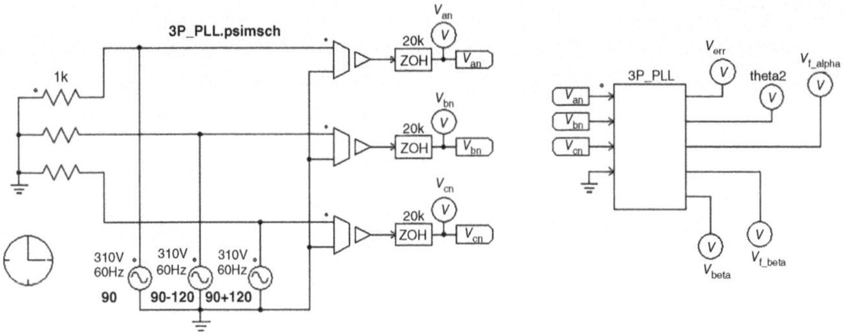

Figure 15.7 Circuit schematic of the three-phase PLL.

The complete structure of the three-phase PLL system is illustrated in Figure 15.6. After computing V_{alpha} and V_{beta} from the three-phase phase voltages, the subsequent PLL structure—including the PD, controller, and integrator—i s identical to that of the single-phase PLL. Therefore, further explanation of these common elements is omitted.

The three-phase PLL circuit implemented in PSIM is shown in Figure 15.7.

The corresponding C-block code implementing the digital PLL algorithm is provided in Table 15.3.

The simulation results are presented in Figure 15.8. From the simulation waveforms, it can be observed that the phase angle θ_s is successfully synchronized with the grid phase voltage V_{an}. The error signal V_e converges to zero within one cycle, indicating proper phase locking of the PLL system. Furthermore, the feedback signal $V_{\text{f,beta}}$ remains synchronized with the grid voltage V_{an}, maintaining the same frequency and phase throughout the operation. These results confirm the stable and accurate performance of the implemented three-phase PLL system.

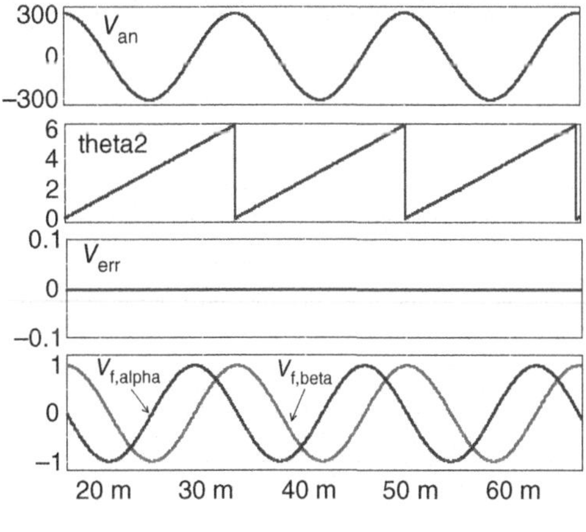

Figure 15.8 Simulation waveforms of the three-phase PLL.

Table 15.3 C-block code: 3P_PLL.

```
#define wB 377
#define Kp 10
#define Ts 0.00005
#define PI 3.141592

double Van, Vbn, Vcn;
double Valpha, Vbeta;
double Vf_alpha, Vf_beta;
double Verr, wS, theta2;
static double theta_s=0, theta_s_old=0;

/****** start here ******/
Van       = in[0];
Vbn       = in[1];
Vcn       = in[2];

Valpha=(2./3.)*Van-(1/3.)*Vbn-(1/3.)*Vcn;
Vbeta = (Vbn-Vcn)/sqrt(3);

Vf_alpha = -sin(theta_s);
Vf_beta =  cos(theta_s);

Verr = Valpha*Vf_alpha +
        Vbeta*Vf_beta;
wS = wB + Kp*Verr;

theta_s = theta_s_old + Ts*wS;
theta_s_old = theta_s;

theta2 = 2*PI*(theta_s/(2*PI) -
      (int)( theta_s/(2*PI)));

out[0] = Verr;
out[1] = theta2;
out[2] = Vf_alpha;
out[3] = Vf_beta;
out[4] = Vbeta;
```

16

Two-Stage Grid-Tied System

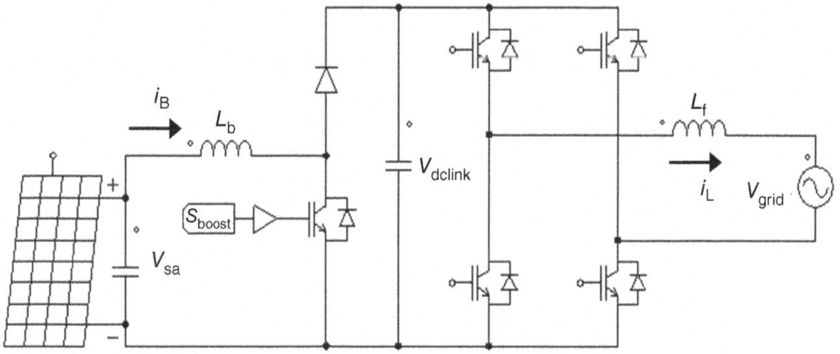

16.1 Solar Boost PCS

A two-stage grid-tied system, commonly employed in photovoltaic (PV) power generation, involves two distinct stages of power conversion to interface solar energy with the electrical grid. The first stage typically utilizes a DC–DC converter, often a boost converter, to maximize power extraction from the PV panels through maximum power point tracking (MPPT) and to step up the voltage to a suitable level for grid interfacing. The second stage uses a DC–AC inverter to convert the DC power into AC power compatible with the grid and to synchronize the inverter output with the grid voltage.

Figure 16.1 illustrates the circuit configuration of a two-stage solar PCS, consisting of a boost converter and a single-phase inverter connected to a

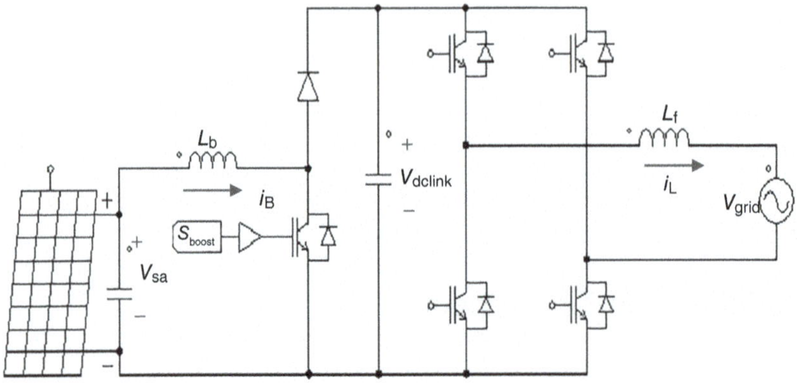

Figure 16.1 Solar boost PCS circuit schematic.

PV panel. The control system for this configuration requires the following controllers:

- A voltage controller for regulating the PV voltage V_{sa}
- A voltage controller for regulating the DC link voltage V_{dclink}
- A current controller for regulating the grid inductor current i_L

In a two-stage system where multiple controllers are required, an important design decision is determining which stage—either the converter or the inverter—will regulate the DC-link voltage.

In systems where the boost converter is responsible for controlling the PV voltage through MPPT, the output voltage $V_{dc-link}$ must remain stable, functioning effectively as a fixed voltage source. This stability is essential because the boost converter's control over the PV voltage is achieved by adjusting the duty cycle, which requires a constant output voltage.

Therefore, in such systems, the inverter must regulate the DC-link voltage while simultaneously controlling the inductor current to follow a sinusoidal reference that is in phase with the grid voltage. This dual objective is accomplished using a double-loop control structure.

Figure 16.2 shows the controller configurations for both the boost converter and the single-phase inverter:

a) The boost converter control employs the same voltage controller used in standalone solar boost applications.
b) The inverter control adopts a double-loop structure similar to that used in standard Solar PCS designs.

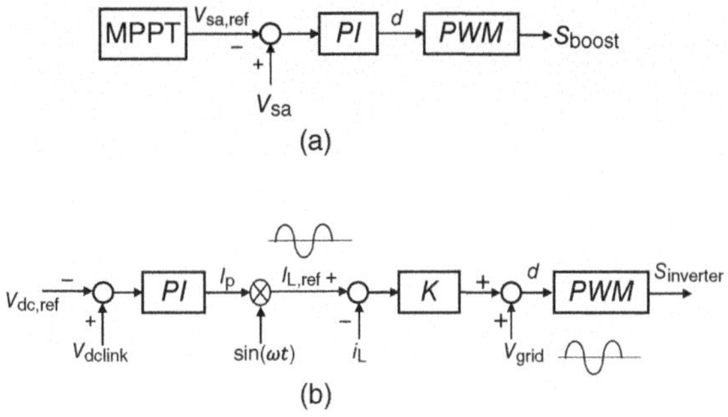

Figure 16.2 Controller configurations. (a) Solar boost voltage controller configuration and (b) Inverter double-loop controller configuration.

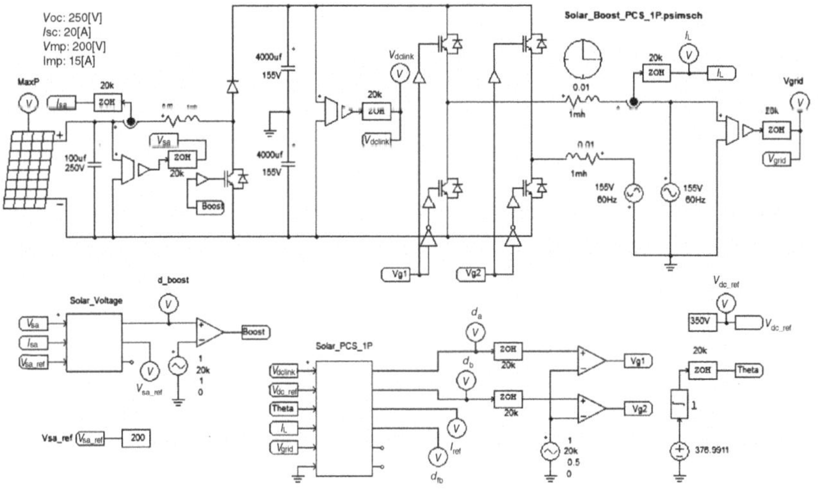

Figure 16.3 PSIM simulation schematic for solar boost PCS.

The PSIM simulation schematic for the Solar Boost PCS is shown in Figure 16.3. The system includes two distinct C-blocks:

- The ***Solar_Voltage*** C-block (Table 7.3), which implements the boost converter control by reusing the voltage controller code from the solar boost converter.
- The ***Solar_PCS_1P*** C-block (Table 13.4), which manages the single-phase inverter control by reusing the C-block code from the solar single-phase inverter design.

The corresponding simulation results are presented in Figure 16.4. The following observations can be made from the waveforms:

- The PV input voltage is regulated to the commanded setpoints of 200 V and 180 V; however, a 120 Hz ripple with accompanying harmonic distortion is

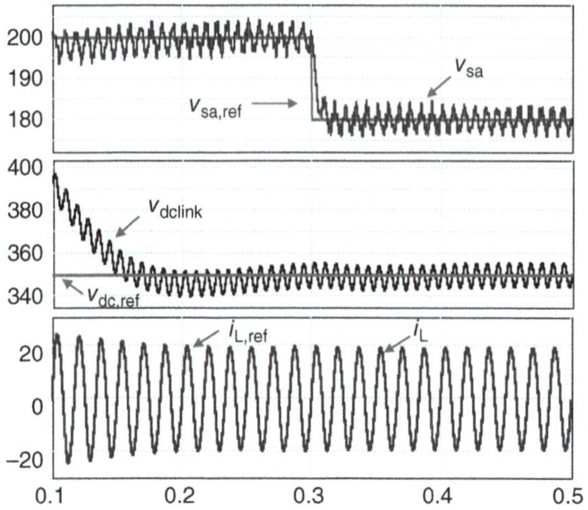

Figure 16.4 Simulation waveforms of the solar boost PCS.

observed. This arises from the 120 Hz component in the DC-link voltage (twice the 60 Hz grid frequency), which couples into the PV input and introduces harmonics.

- The DC-link voltage is maintained at its 350 V reference by the PCS double-loop controller.
- The inductor current is effectively controlled as a sinusoidal waveform, accurately tracking its reference and satisfying the controller's design specifications.

Discussion: Control Considerations for Standalone Operation Now consider an alternative scenario in which the system output is not connected to the grid but instead supplies power to a resistive load. In this case, the control structure poses a challenge because neither the PV input voltage nor the output voltage is inherently fixed, making conventional voltage regulation difficult.

One possible solution would involve measuring the load current and developing a control strategy based on this measurement. This approach could offer an alternative method of achieving stable operation under standalone conditions. Exploring such control strategies presents an interesting topic for further study and could be considered for a term project.

Design Project: Solar Boost Stand-Alone System Figure 16.5 illustrates the circuit configuration of a solar boost stand-alone inverter. This system is designed to supply AC power independently, without grid connection, and requires the

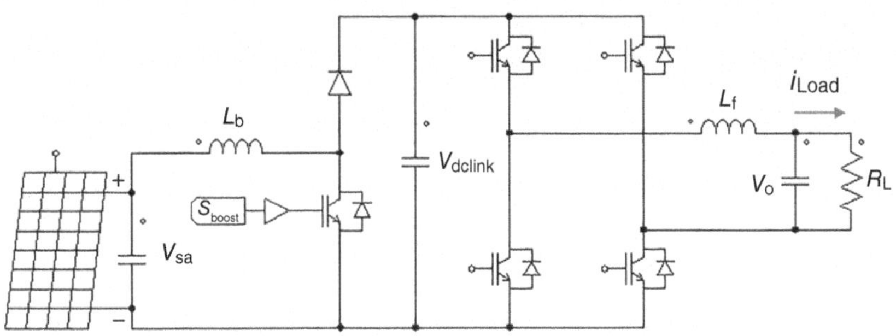

Figure 16.5 Solar boost stand-alone inverter.

implementation of multiple control loops to ensure stable and high-quality power delivery. The control system includes the following key controllers:

- PV Voltage Controller: Regulates the solar panel voltage to maximize power extraction through MPPT.
- DC-Link Voltage Controller: Maintains a stable DC-link voltage to provide a consistent power source for the inverter stage.
- Inverter Current Controller: Controls the inverter output current, ensuring proper waveform shaping and limiting current during load transients.
- AC Output Voltage Controller: Regulates the AC output voltage to maintain voltage stability and supply high-quality sinusoidal output to the load.

Additionally, the system allows for direct measurement of the load current i_{load}, enabling the possibility of implementing more advanced or adaptive control strategies based on real-time load feedback.

16.2 Two-Stage Energy Storage System (ESS)

Figure 16.6 illustrates the configuration of a two-stage energy storage system (ESS), which comprises a battery as the energy source, a boost converter, and a single-phase power conversion system (PCS). The primary distinction from the previously discussed solar boost PCS lies in the replacement of the PV panel with a battery. This change enables bi-directional current flow, allowing both charging and discharging operations.

Controller Design in Discharge Mode The control structure for discharge operation is first considered. The overall discharge-mode control configuration is depicted

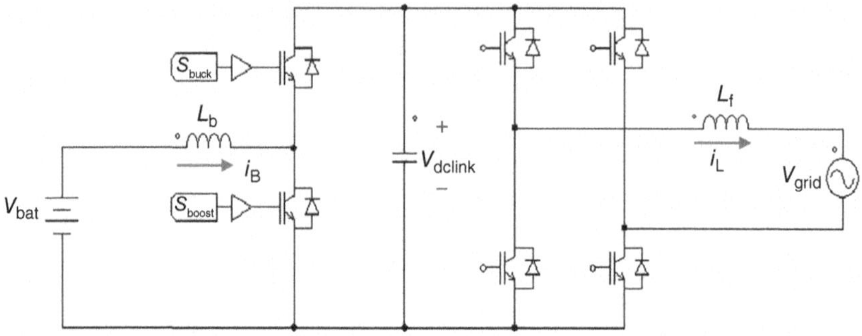

Figure 16.6 Two-stage ESS configuration.

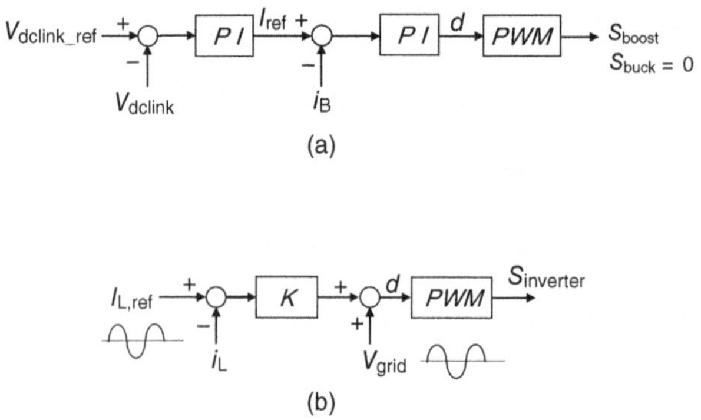

Figure 16.7 Controller configuration in discharge mode. (a) Boost converter double-loop control and (b) Inverter current controller.

in Figure 16.7. In this mode, power is transferred from the battery to the grid, and the battery voltage serves as the input while discharge current flows to the grid.

Since the battery acts as a fixed voltage source (V_{bat}), the output voltage of the boost converter ($V_{dc-link}$) becomes the control variable. Accordingly, the boost converter is designed with a double-loop controller to regulate the DC-link voltage. The inverter is governed by a single-loop current controller that ensures the inductor current follows a sinusoidal waveform synchronized with the grid voltage.

Controller Design in Charge Mode Next, the control strategy for charge mode is examined. Here, power flows from the grid to the battery. The grid voltage acts as the input, and the current flowing into the battery is the controlled output.

In charge mode, the inverter functions as a rectifier with its output ($V_{dc-link}$) being regulated using a double-loop control structure. The buck converter, with

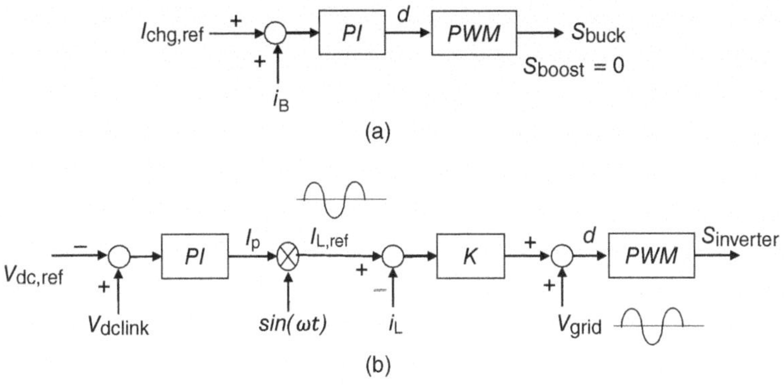

Figure 16.8 Controller configuration in charge mode. (a) Buck converter single-loop current controller and (b) Inverter (Rectifier) double-loop voltage controller.

$V_{dc-link}$ as its input and the battery (V_{bat}) as a fixed voltage load, cannot regulate voltage directly and therefore only controls the charging current.

An important consideration in bi-directional control is the fixed direction of the current sensor. In the simulation circuit, the sensor is placed to detect current flowing from the battery to the grid. During charging, the actual current direction is opposite to this configuration, resulting in negative sensor readings. To correct for this, the sensor output must be multiplied by −1 to reflect the actual charging current direction accurately. This adjustment is incorporated into the control structure illustrated in Figure 16.8.

Bi-Directional Simulation and Control Implementation Figure 16.9 shows the complete simulation schematic of the two-stage ESS with bi directional power flow. The system comprises a bi-directional buck–boost converter and a grid-connected PCS. Two C-blocks are employed: *Bi_Converter* for converter control (Table 4.2) and *Rectifier_Voltage* for single-phase rectifier/inverter control (Table 11.7). The *Bi_Converter* C-block is reused without modification, whereas the *Rectifier_Voltage* C-block is lightly revised to demonstrate bi-directional operation.

The bi-directional inverter control C-block is presented in Table 16.1.

The control performance of the system is validated through simulation waveforms. Figure 16.10 shows the system operating in discharge mode. The DC-link voltage is successfully regulated by the boost converter, and a 120 Hz ripple is observed due to the grid voltage. The grid voltage and current waveforms are in phase, confirming power transfer from the battery to the grid.

Figure 16.11 displays the charge mode operation. The rectifier effectively regulates the DC-link voltage, and the grid current is out of phase with the grid voltage,

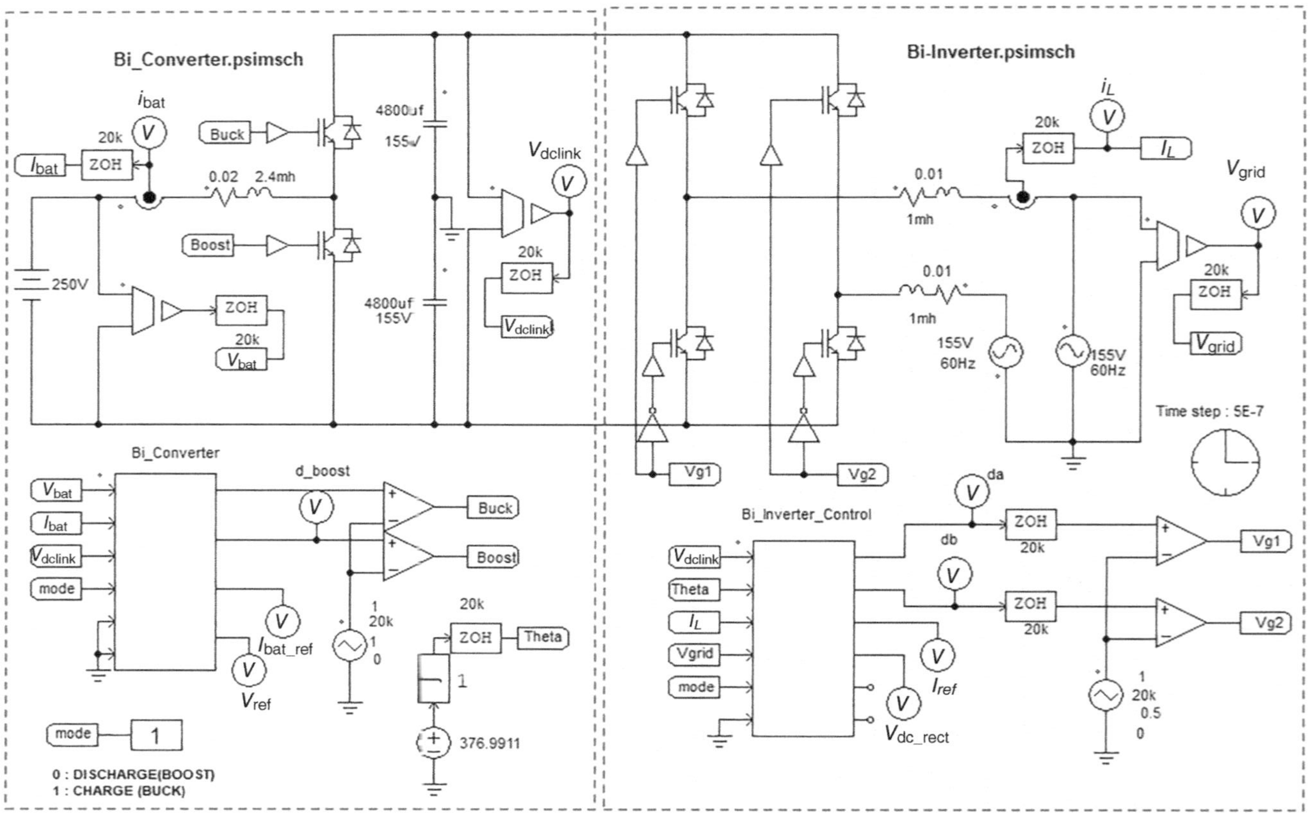

Figure 16.9 Simulation schematic for bi-directional two-stage ESS.

Table 16.1 C-block code: Bi_Inverter_Control.

```
double Vdc, Ip, Theta, IL, Vc;
double duty_a,duty_b, flag;
int mode;
double  Iref, Ierr, Vcon, Verr, Vdc_ref;
static double Verr_old =0.0, Ip_old=0.0;

/****** start here ******/
Vdc     = in[0];
Theta   = in[1];
IL      = in[2];
Vc      = in[3];
Mode    = in[4];

if (Vdc < 0.01) Vdc = 0.01;
if (mode == 0) { // discharge mode
     if (t<0.25) Ip = 10;
     else        Ip = 20;
}

else if (mode == 1) { // Charge mode
     if  (t>0.2) Vdc_ref = 550;
     else        Vdc_ref = 450;
     Verr =  -(Vdc_ref - Vdc);
     Ip = Ip_old +0.17946*Verr -
           0.17946*0.999*Verr_old;
     Ip_old = Ip;
     Verr_old = Verr;
}

/** Same as PCS Current Controller **/
Iref = Ip*sin(Theta);
Ierr = Iref - IL;

Vcon =20*Ierr + 0.01*IL +Vc;

if (Vcon >  Vdc/2)  Vcon =  Vdc/2;
if (Vcon < -Vdc/2)  Vcon = -Vdc/2;

duty_a =0.5 + Vcon/Vdc;
duty_b= 0.5 - Vcon/Vdc;

out[0] = duty_a;
out[1] = duty_b;
out[2] = Iref;
out[3] = Vdc_ref;
```

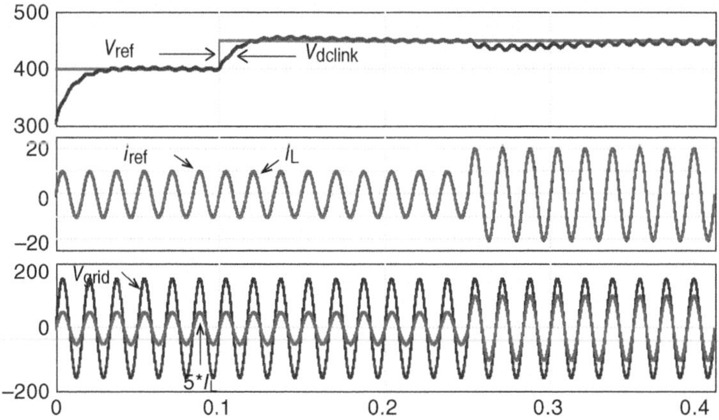

Figure 16.10 Waveforms in discharge mode.

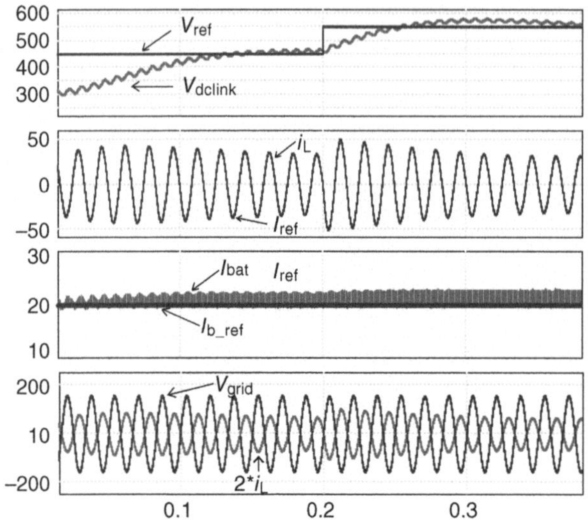

Figure 16.11 Waveforms in charge mode.

indicating power is flowing into the battery. The buck converter's current controller also functions correctly, ensuring stable battery charging.

16.3 Solar Energy Storage System (Solar ESS)

This system integrates a solar inverter (PCS) with an ESS, forming the configuration illustrated in Figure 16.12. In some references, this type of system is referred to as a Hybrid ESS.

The system consists of two main components:

- A solar PCS composed of four switches (*Q1*–*Q4*)
- A bi-directional converter composed of two switches (*Q5*, *Q6*)

The solar PCS performs MPPT by regulating the DC-link voltage ($V_{dc-link}$) and controlling the inductor current to maintain unity power factor. The bi-directional converter, connected in parallel with the solar PCS, manages the charging and discharging of the battery. These two controllers operate independently, performing the following functions:

a) Solar PCS controller:
   - MPPT control: Regulates the PV voltage (V_{sa}) to extract maximum solar power.

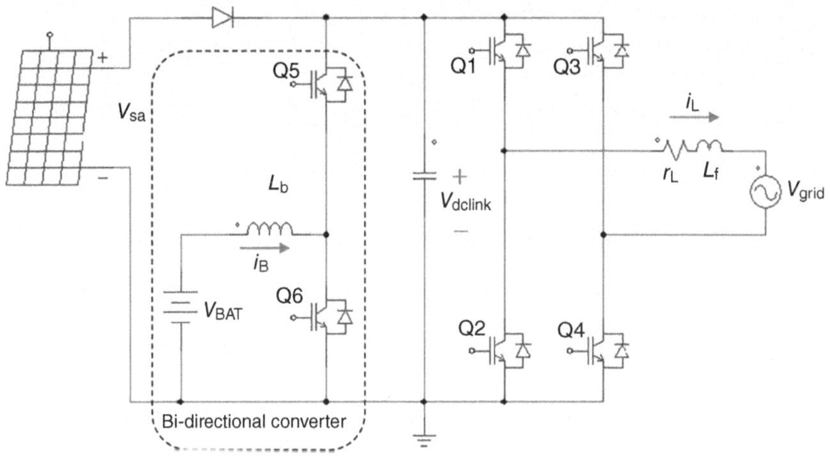

Figure 16.12 Solar ESS configuration.

- Inductor current control: Controls the grid current (i_L) to achieve a sinusoidal waveform in phase with the grid voltage, ensuring unity power factor. The magnitude of the current is determined by the power available from the solar array and is not directly regulated.

b) Bi-directional converter controller:
- Battery current control: Manages charging and discharging of the battery.
- Discharge operation: Supplies power when solar generation is insufficient.
- Charge operation: Stores excess solar power when available.
- DC-link voltage regulation: Regulates $V_{dc-link}$ in the absence of solar power by either sourcing or sinking current.

The control strategy of the solar PCS, which includes MPPT and current regulation, was discussed in detail in Chapter 13 and is therefore omitted here. This section focuses on the state-space modeling and control design of the bi-directional converter.

1) **Discharge Mode (Boost Converter Operation)**

In discharge mode, the bi-directional converter operates as a boost converter to transfer power from the battery to the DC link. The governing equations for the boost converter during the ON and OFF states are as follows:

- ON Time (Switch Closed, Duty Cycle = d):

$$V_{Bat} = L\frac{di_L}{dt} + r_L i_L$$

- OFF Time (Switch Open, Duty Cycle = 1 – *d*):

$$V_{\text{Bat}} = r_{\text{L}} i_{\text{L}} + L\frac{\mathrm{d}i_{\text{L}}}{\mathrm{d}t} + v_{\text{sa}}$$

By applying state-space averaging, the system equation becomes:

$$L\frac{\mathrm{d}i_{\text{L}}}{\mathrm{d}t} + r_{\text{L}} i_{\text{L}} = v_{\text{sa}} \cdot d + (V_{\text{Bat}} - v_{\text{sa}}) \tag{16.1}$$

This can be rewritten in matrix form as a nonlinear state-space model:

$$\dot{X} = AX + Bu + F$$

where d is defined as $d = d_{\text{fb}} + d_{\text{ff}}$. By introducing a feed-forward input (d_{ff}), the nonlinear term can be canceled out.

$$d_{\text{ff}} = -\frac{F}{B} = \frac{v_{\text{sa}} - V_{\text{Bat}}}{v_{\text{sa}}} \tag{16.2}$$

Substituting this into the original equation yields the state-space averaged equation:

$$L\frac{\mathrm{d}i_{\text{L}}}{\mathrm{d}t} + r_{\text{L}} i_{\text{L}} = v_{\text{sa}} \cdot d_{\text{fb}}$$

The corresponding transfer function from duty cycle to current is then expressed as:

$$G_1(S) = \frac{I_{L(S)}}{d(s)} = \frac{V_{\text{sa}}}{sL + r_{\text{L}}} \tag{16.3}$$

This transfer function provides the basis for the current controller design, which can be implemented using an m-file in MATLAB. The MATLAB m-file for current controller is presented in Table 16.2.

Table 16.2 MATLAB m-file for current controller.

```
%% Boost:Current Single %%
s=tf('s');

Vbat = 200;  % Input voltage
L = 3e-3;    % L=3mh
rL=0.02;     %rL = 20 mohm
Vsa = 350; % sa dclink voltage

G1 = Vsa/(s*L+rL);

Kpi =1; Kii =1; %temporarily value
Gci = Kpi*(s+Kii)/s;

IL_ref  = 8;
%sisotool Current Single loop
T=sisoinit(1);
T.F.Value = IL_ref;
T.G.Value = G1;
T.C.Value = Gci;
sisotool(T);
```

The design specifications define the key parameters required to achieve the desired control performance:

- Bandwidth: 4800 rad/sec
- Phase margin: 60 deg

Based on these specifications, the controller is auto-tuned to obtain optimized parameters. The resulting continuous-time controller is expressed as:

$$G_{Ci}(s) = 0.0356\frac{(s + 2780)}{s}$$

For real-time digital implementation, the controller is discretized using a sampling frequency of 20 kHz (T = 0.00005 s). The corresponding discrete-time controller is given by:

$$G_{Ci}(z) = 0.03967\frac{(z - 0.902)}{z - 1}$$

This digital controller is suitable for implementation in embedded systems or digital signal processors (DSPs) operating in real time.

2) **Charge Mode (Buck Converter Operation)**

In charge mode, the bi-directional converter operates as a buck converter, transferring power from the grid to the battery. The governing equations for this mode can be derived as follows:

- ON Time (Switch Closed, Duty Cycle = d):

$$v_{sa} = L\frac{di_L}{dt} + r_L i_L + V_{Bat}$$

- OFF Time (Switch Open, Duty Cycle = $1 - d$):

$$0 = L\frac{di_L}{dt} + r_L i_L + V_{Bat}$$

By applying state-space averaging, the following equation is obtained:

$$L\frac{di_L}{dt} + r_L i_L = v_{sa} \cdot d - V_{Bat} \tag{16.4}$$

To simplify control implementation, the duty cycle is expressed as the sum of feedback and feedforward components:

$$d = d_{fb} + d_{ff}$$

where the feedforward duty is defined as:

$$d_{ff} = \frac{V_{Bat}}{v_{sa}} \tag{16.5}$$

Substituting this into the averaged equation yields:

$$L\frac{di_L}{dt} + r_L i_L = v_{sa} \cdot d_{fb}$$

The corresponding transfer function from duty cycle to inductor current is:

$$G_1(S) = \frac{I_{L(S)}}{d(s)} = \frac{V_{sa}}{sL + r_L} \tag{16.6}$$

As a result, the same control code used in discharge mode can be applied in charge mode by simply adjusting the sign of the duty cycle. This unified approach simplifies implementation for both operating modes.

The simulation schematic for the bi-directional system is shown in Figure 16.13. The C-block for the *Solar_PCS_1P* remains identical to the example presented in Table 13.4, while the C-block for the bi-directional converter (*Bi_Converter*) is newly implemented for this simulation.

The newly implemented bi-directional converter C-block is shown in Table 16.3. The system operates in three distinct modes:

- Mode 0: Solar PCS and ESS operate simultaneously.
- Mode 1: Only Solar PCS operates.
- Mode 2: Only ESS operates.

Since Modes 1 and 2 were previously covered, the simulation focuses exclusively on Mode 0.

In ESS mode, the reference current smoothly transitions from a constant discharge current (+10 A) to a tapered-down profile, eventually stabilizing at a constant charge current (−10 A). The resulting waveforms—including the solar PCS PV voltage and the grid current—are presented in Figure 16.14.

Observations from the simulation results are as follows:

- When the battery discharges while the solar PCS controls the PV voltage, the voltage ripple increases.
- When the battery charges, the PV voltage ripple decreases.
- Consequently, the amplitude of the inductor AC current flowing into the grid is also reduced during charging.

These results demonstrate the effectiveness of the bi-directional converter control in coordinating with the solar PCS to ensure stable operation under varying power flow conditions.

Practical Exercise: Two-Stage PV-ESS Configuration The circuit shown in Figure 16.15 represents a two-stage PV-ESS configuration. This system integrates a two-stage solar PCS and a bi-directional converter, both interconnected through a shared DC link.

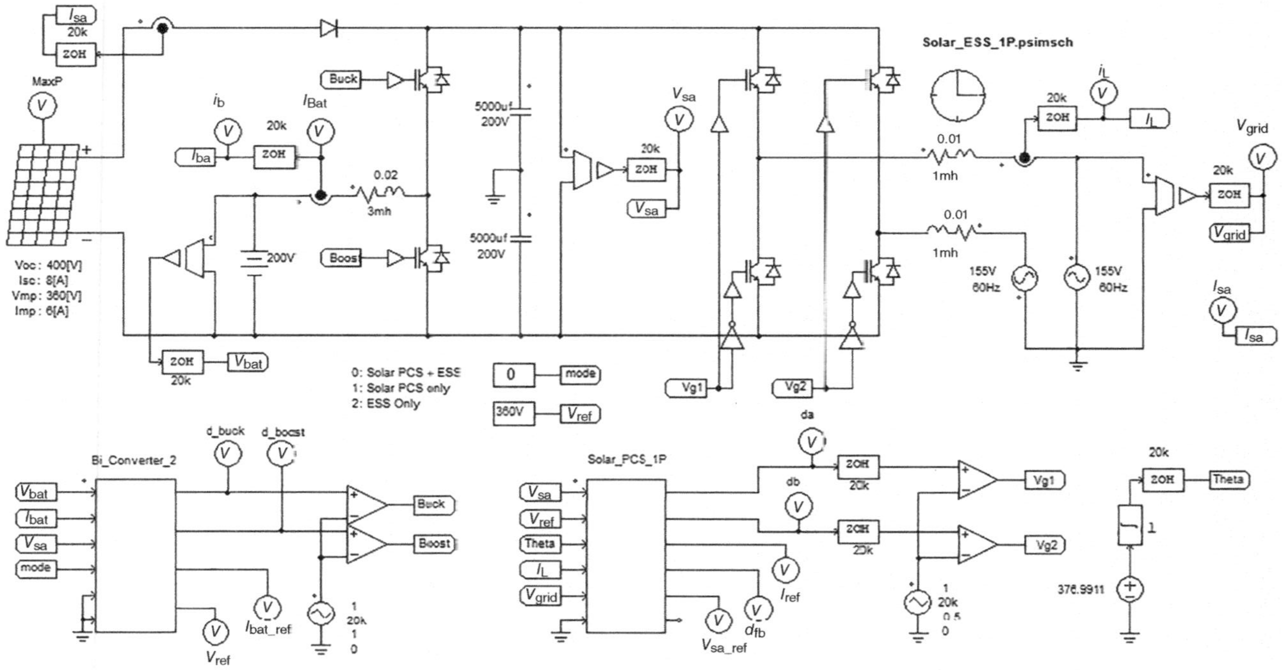

Figure 16.13 Simulation schematic for solar ESS with bi-directional converter.

Table 16.3 C-block code: Bi-Converter_2.

```
double Vbat, Ibat, Vlink,  Mode;
double Vdc_ref,  Vdc_err;
double IL_ref, IL_err, duty;
double Ichg_ref, Idchg_ref, Ibat_ref;

static double  IL_ref_old=0,
Vdc_err_old=0;
static double duty_old=0,
IL_err_old=0 ;

Vbat   = in[0];
Ibat    = in[1];
Vlink    = in[2];
Mode  =in[3];

if (Mode == 0)   { // ESS mode
 if       (t<0.2) Ibat_ref =  0.0;
 else if (t<0.5) Ibat_ref = +10.0;
 else if (t<0.7) Ibat_ref = -100*t+60;
 else             Ibat_ref = -10.0;

IL_err = Ibat_ref - Ibat;

duty = duty_old +0.03967*IL_err -
0.03967*0.902*IL_err_old;
duty_old=duty;
IL_err_old=IL_err;

if ( Ibat_ref > 0.0) {
  out[0]=0 ;        // buck
  out[1]=duty+(Vlink-Vbat)/Vlink;//boost
}
else if (Ibat_ref < 0.0) {
  out[0]=-duty + Vbat/Vlink ;    //buck
  out[1]= 0;    // boost
 }
}  // end of ESS mode

else if (Mode ==1) {    //Solar PCS Only
  out[0]=0 ;   // buck   oFF
  out[1]=0;    // boost oFF
}

out[2]  = Ibat_ref;
```

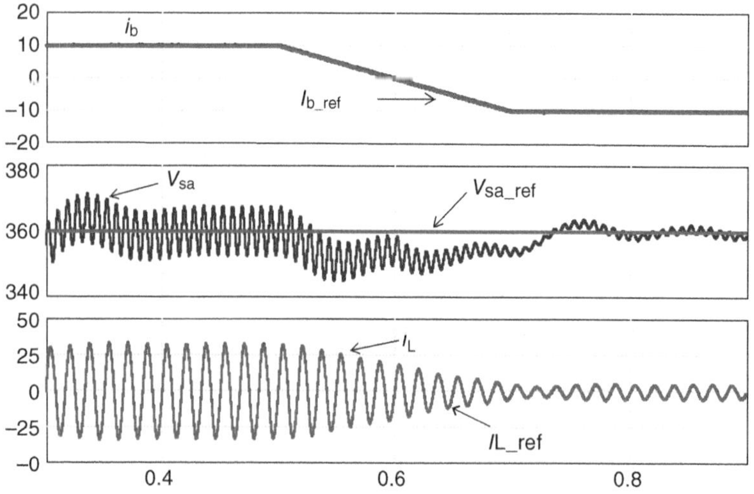

Figure 16.14 Simulation waveforms (Mode 0).

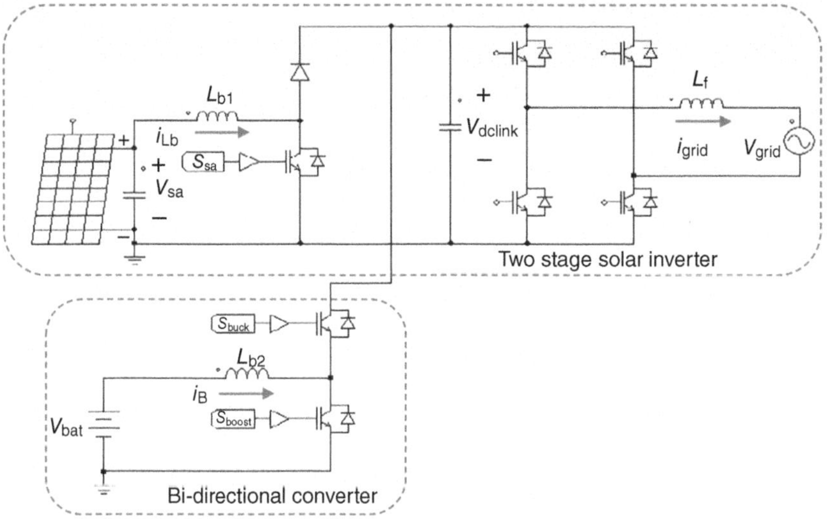

Figure 16.15 Two-stage PV-ESS configuration.

Table 16.4 Operational mode.

MODE	Power flow	Operation
0	Solar → Grid Battery → Grid	Solar + Battery Discharge
0	Solar → Grid Solar → Battery	Solar + Battery Charge
1	Grid → Battery	Battery Charge from Grid (ESS only)
2	Battery → Grid	Battery Discharge to Grid (ESS only)

The system operates in multiple modes depending on the direction of power flow, as summarized in Table 16.4.

Term Project As a term project, each student will individually design controllers for the two-stage PV-ESS System. The project will focus on:

- Developing control strategies tailored to each operating mode.
- Ensuring DC-link stability, efficient power distribution, and seamless grid integration.

Students are expected to:

- Simulate the proposed controllers.
- Validate system performance through waveform analysis.
- Document findings with professional reporting.

17

Frequency Analysis

17.1 Frequency Analysis Method

The controller design process involves constructing an equivalent circuit for the switch ON and OFF states, deriving state equations, and then applying the state averaging method to obtain the state-space representation. The Laplace transform is used to derive the frequency transfer function, followed by the application of the automated tuning method in SISOTOOL using the Bode plot to determine the gains of the *PI* controller. The controller structure consists of single-loop and double-loop configurations.

These processes rely on mathematical analysis, raising the question of whether the same frequency transfer function characteristics can be obtained in circuit simulations using actual hardware. The objective is to compare the Bode plot obtained from MATLAB's frequency transfer function with the Bode plot obtained from the AC sweep function in PSIM when implemented in real hardware. If both Bode plots are identical, it serves as a critical validation that the mathematical modeling and frequency transfer function calculations were correctly performed.

MATLAB, as a mathematical model-based tool, can generate a Bode plot using mathematical functions, whereas PSIM, which is based on real circuit simulations, requires an actual circuit implementation that meets design conditions to generate a Bode plot using AC sweep. The AC sweep library can be found under `Element` → `Other` → `AC Sweep` in PSIM. The help on the `AC Sweep` is presented in Table 17.1.

To use AC sweep in PSIM, an open-loop and closed-loop implementation must meet PSIM's required conditions. In the open-loop configuration, the feedback loop must be broken at a specific point to prevent forming a closed loop. Then AC probe must be inserted between two specific points, T_1 and T_2, and an AC signal, V_{sweep}, must be applied. The ratio of T_1 to T_2 represents the open-loop gain, which can be calculated using the AC sweep function in PSIM.

Table 17.1 Help on the AC sweep.

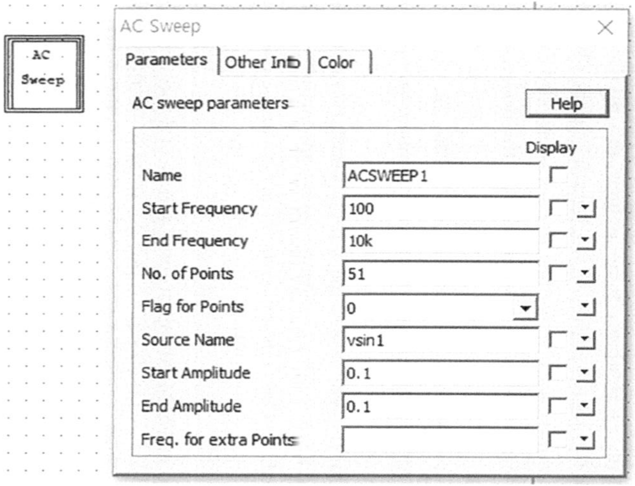

The parameters are defined as follows..

Start Frequency: Start frequency of the ac sweep, in Hz

End Frequency: End frequency of the ac sweep, in Hz

No. of Points: Number of data points

Flag for Points: Flag to define how the data points are generated.

Flag = 0: Points are distributed linearly in LOG10 scale

1: Points are distributed linearly in linear scale

Source Name: Name of the excitation source

Start Amplitude: Excitation source amplitude at the start frequency

End Amplitude: Excitation source amplitude at the end frequency

Table 17.2 Probes for closed-loop and open-loop measurement.

1. AC Sweep Probe	2. AC Sweep Probe (loop)
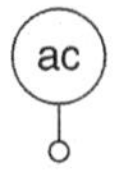	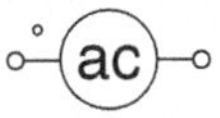
<Closed-loop gain measurement>	<Open-loop gain measurement>

In the closed-loop configuration, the AC signal, V_{sweep}, inserted to the reference input R in series, then measure the closed loop gain by connecting AC sweep probe to the output.

There are two probe libraries available for measuring loop gain: `Elements → Other → Probes → AC Sweep Probe and AC Sweep Probe(Loop)`. The figures of these two types of probes are presented in Table 17.2.

According to the PSIM operation manual, the process of conducting AC analysis follows specific steps. The following sections describe the methods to implement open-loop and closed-loop Bode plots in PSIM.

1. **Open-Loop Bode Plot Implementation**

   The structure for plotting an open-loop Bode plot is shown in Figure 17.1. Procedure for measuring open-loop gain is as follows:

   1) Break the feedback loop at the feedback loop.
   2) Inject an AC sweep signal at the open-loop insertion point.
   3) Connect the AC sweep probe (loop) across the two ends of the broken loop to measure loop gain.
   4) Use a sinusoidal excitation for the AC sweep, specifying both the start and end frequencies.
   5) As illustrated, place the AC sweep probe (Loop) between test nodes T_1 and T_2, and apply V_{sweep} between these nodes to obtain the frequency response.

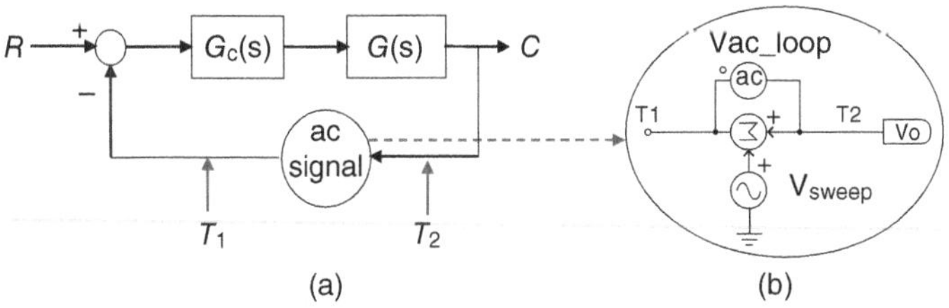

Figure 17.1 The configuration of open-loop gain simulation. (a) Configuration (b) AC sweep signal.

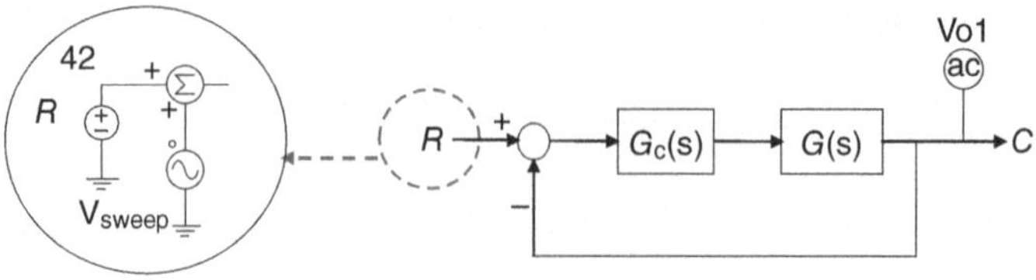

Figure 17.2 The configuration of closed-loop gain simulation.

6) Compute the open-loop gain $T_0(s)$ from the measured response by applying the corresponding transfer function, which is obtained using the following equation:

$$T_0(s) = \frac{T_1}{T_2} = G_c(s)G(s)$$

2. **Closed-Loop Bode Plot Implementation**
The configuration for plotting the closed-loop Bode plot is shown in Figure 17.2. The procedure for measuring the closed-loop gain is as follows:
1) Insert the AC sweep source in series with the reference input R.
2) Place the AC sweep probe at the output node C to record the closed-loop response V_{01}.
3) Set the sweep amplitude to approximately 2–5% of the reference (i.e., 1/50 to 1/20) to obtain a meaningful response without driving the system into nonlinearity.
4) The closed-loop gain, T_c, is calculated using the following equation:

$$T_c(s) = \frac{C}{R} = \frac{G_C(s)G(s)}{1 + G_C(s)G(s)}$$

By implementing these methods, the open-loop and closed-loop frequency responses can be analyzed in PSIM, allowing validation of the system's control design and ensuring accurate frequency-domain characteristics.

17.2 Buck Converter Frequency Analysis

In this chapter, the validity of the transfer function derived using the state-space averaging method is verified through an AC sweep analysis conducted in PSIM. The resulting frequency responses are compared with those obtained in Chapter 2 using MATLAB, thereby confirming the consistency of the mathematical model. Specifically, the open-loop and closed-loop Bode plots of the single-loop voltage controller for the buck converter are examined, followed by a comparison of the open-loop Bode plot of the double-loop controller. The close agreement between the PSIM and MATLAB results demonstrates the accuracy of the derived transfer function and validates the theoretical modeling approach.

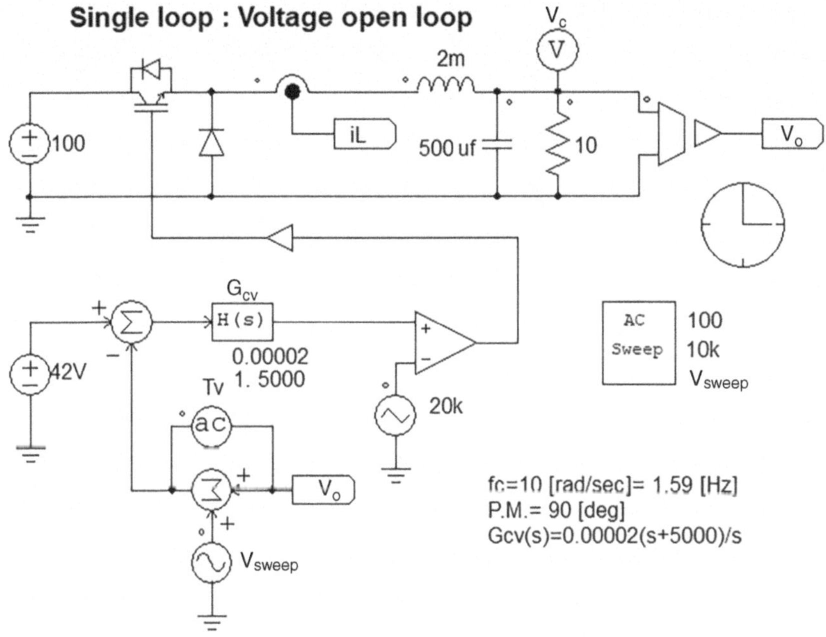

Figure 17.3 Schematic for measuring the open-loop response of the voltage controller.

a) **Voltage controller open-loop response (single-loop control)**

The PSIM schematic used to measure the open-loop Bode plot of the voltage controller is shown in Figure 17.3. The PI controller gains are identical to those introduced in Section 2.2.1 for the single-loop voltage control design.

The open-loop transfer function, including both the controller and the plant dynamics, is expressed as

$$T_{ov}(s) = \frac{T_1}{T_2} = G_{cv}(s)G_2(s)$$

The resulting Bode plots obtained from MATLAB and PSIM are presented in Figure 17.4. For comparison, the MATLAB Bode plot originally shown in Figure 2.12 has been redrawn using a frequency axis in Hertz to match the PSIM output format. As shown, the two plots exhibit an exact match, confirming the consistency between the analytical model and the simulation results.

b) **Voltage controller closed-loop response (single-loop control)**

The PSIM schematic used to evaluate the closed-loop Bode plot of the voltage controller is shown in Figure 17.5.

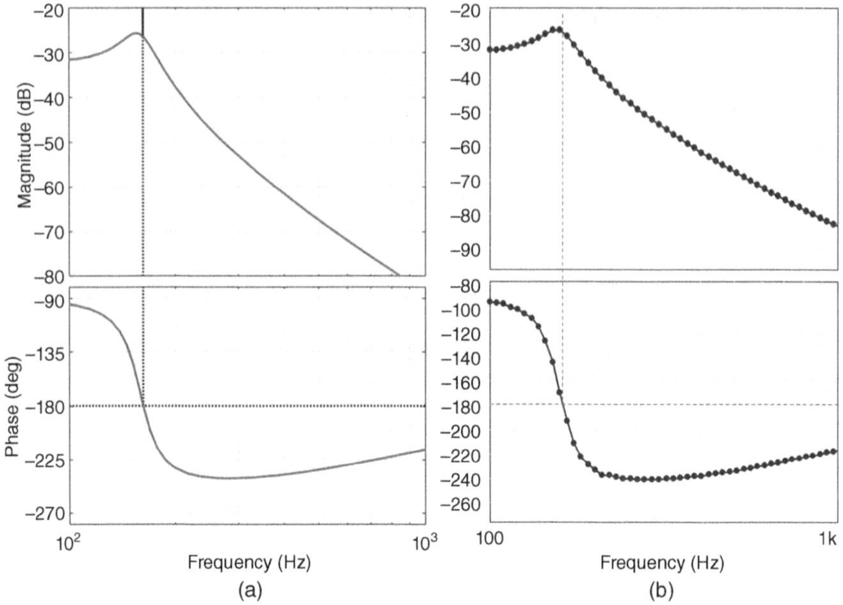

Figure 17.4 Comparison of open-loop gain $T_{ov}(s)$ from MATLAB and PSIM. (a) MATLAB (b) PSIM.

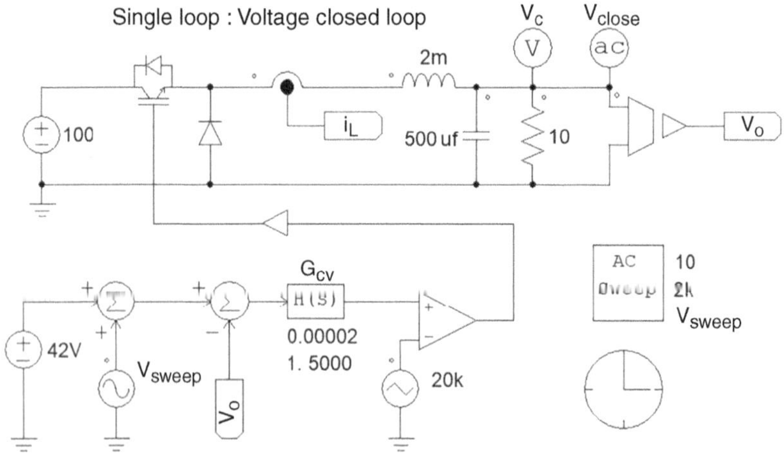

Figure 17.5 Schematic for measuring the closed-loop response of the voltage controller.

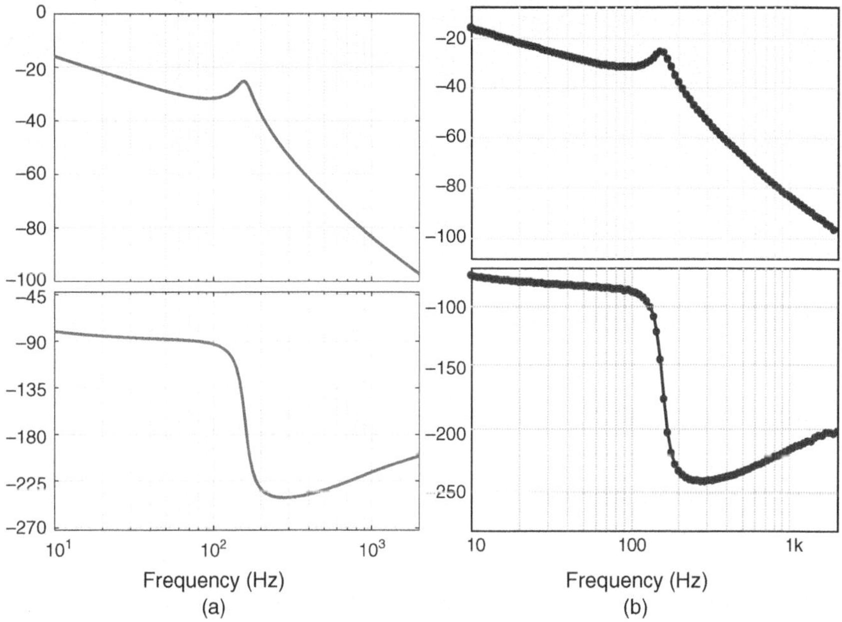

Figure 17.6 Comparison of closed-loop gain $T_{cv}(s)$ from MATLAB and PSIM. (a) MATLAB (b) PSIM.

The closed-loop transfer function for the voltage control system is given by

$$T_{cv}(s) = \frac{C(s)}{R(s)} = \frac{G_{cv}(s)G_2(s)}{1 + G_{cv}(s)G_2(s)}$$

The closed-loop Bode plots from MATLAB and PSIM are compared in Figure 17.6. Similar to the open-loop case, the MATLAB plot has been rescaled to a frequency axis in Hertz for consistency. The results from both MATLAB and PSIM show perfect agreement, further validating the accuracy of the derived transfer function and controller implementation.

c) **Double-loop controller**

The PSIM schematic used to evaluate the open-loop response of the double-loop controller is shown in Figure 17.7. The controller gains are identical to those specified in Table 2.8, ensuring consistency with the design parameters established in Chapter 2.

The open-loop Bode plot obtained from MATLAB (Figure 2.20) is compared with the corresponding PSIM result in Figure 17.8. Since the MATLAB frequency axis is originally expressed in radians per second, appropriate scaling was applied for comparison with the PSIM plot in Hertz. As observed, the two frequency responses exhibit excellent agreement, validating the accuracy of the derived transfer function and confirming the fidelity of the double-loop control model.

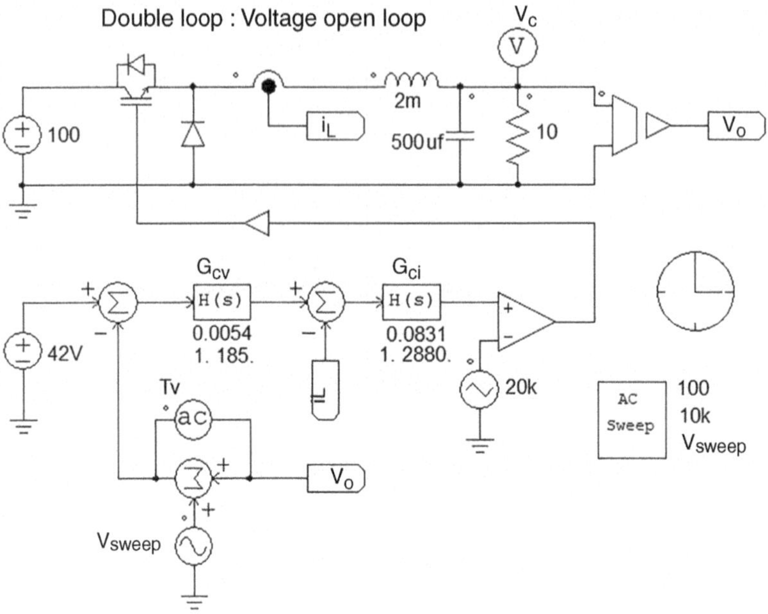

Figure 17.7 Schematic for measuring the open-loop response of the double-loop controller.

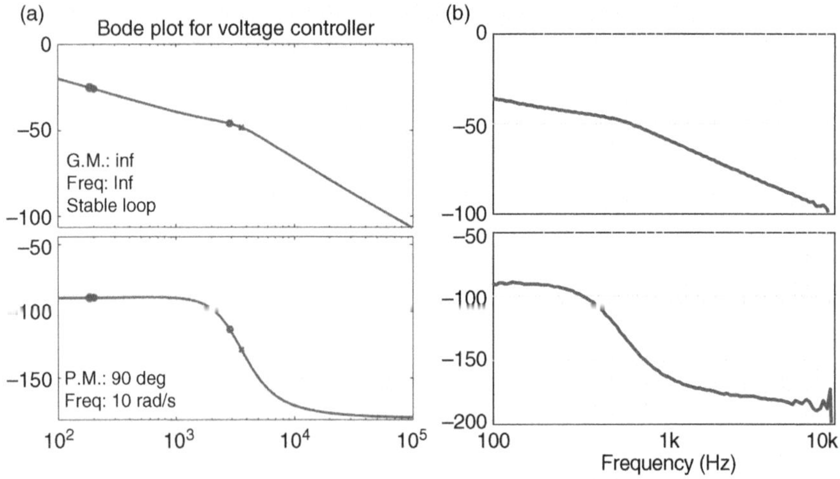

Figure 17.8 Comparison of double-loop open-loop gain from MATLAB and PSIM. (a) MATLAB (b) PSIM.

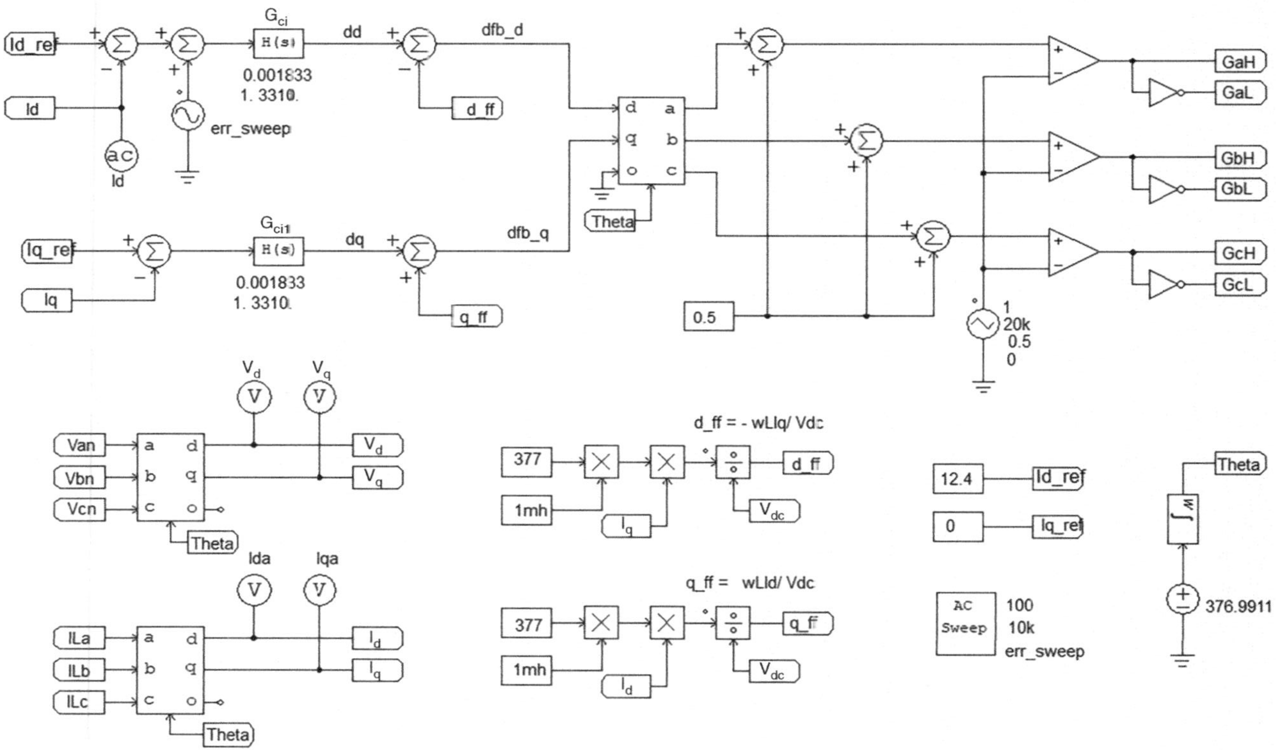

Figure 17.9 Schematic for measuring the open-loop response of the current controller.

17.3 Three-Phase Inverter Frequency Analysis

In this chapter, the frequency response characteristics of the three-phase stand-alone inverter are examined. We begin by considering the current controller design presented in Section 12.3.1. Figure 12.27 shows the MATLAB open-loop Bode plot of the designed current controller. To verify whether the same response can be obtained using the PSIM circuit, the schematic in Figure 12.28 was modified into an AC sweep configuration. The controller gains were configured to be identical to those used in the MATLAB design of Figure 12.27.

For the AC sweep analysis, the current controller implemented in the C-block was replaced by an analog decoupling model, as shown in Figure 17.9. Since the power circuit configuration is identical to that shown in Figure 12.28, the corresponding analog *dq*-model, including the PI controller, is illustrated in Figure 17.9. This figure represents the *d*-axis current control loop, where the *q*-axis current reference ($i_{q,\text{ref}}$) is set to zero.

The comparison between MATLAB and PSIM results is presented in Figure 17.10. The two plots exhibit excellent agreement, confirming that the mathematical model in MATLAB and the circuit model in PSIM produce consistent frequency response characteristics.

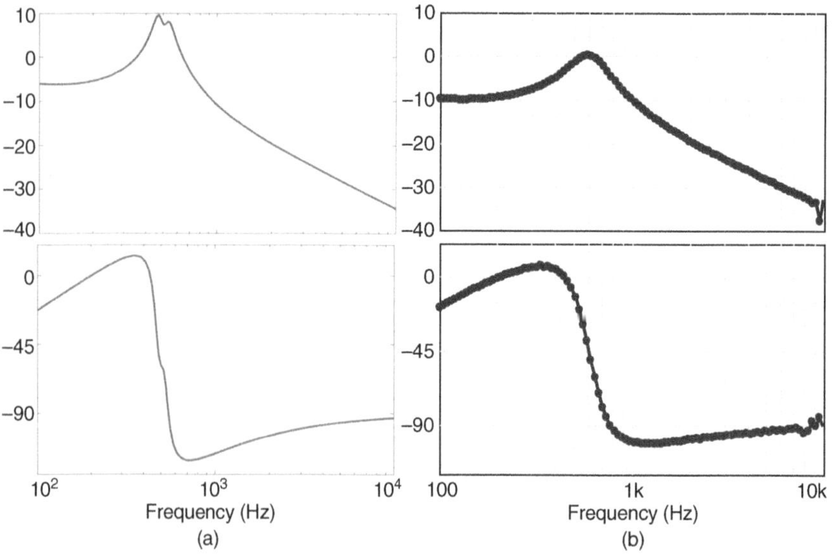

Figure 17.10 Comparison of current controller open-loop gain from MATLAB and PSIM. (a) MATLAB (b) PSIM.

18

Neutral-Point Clamped (NPC)

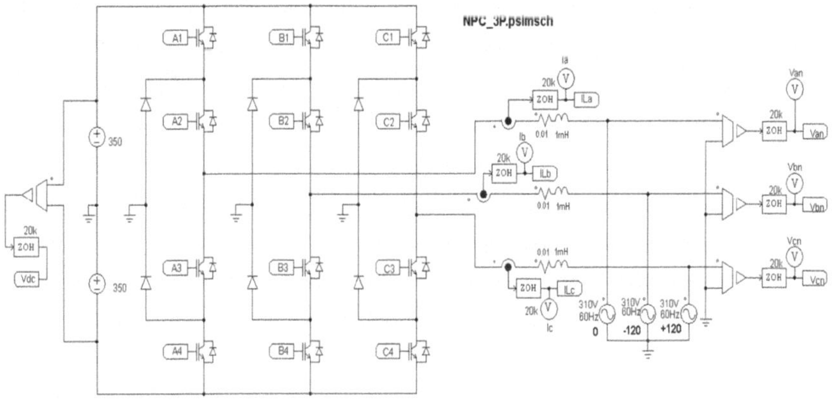

18.1 NPC Operation

Introduction to Neutral-Point Clamped (NPC) Inverter and Topology Power conversion devices are continuously evolving toward higher capacity and greater efficiency. In recent years, grid-tied inverters have advanced to the point where 1 MW-class products have become the industry standard. Achieving high efficiency in such systems requires the use of wide-bandgap devices, such as GaN transistors, which enable high switching frequencies. Additionally, increasing the system voltage reduces conduction and switching losses, prompting the development of high-voltage specifications.

For example, a 1500 V DC standard has recently been established for grid-tied solar inverters, while energy storage systems (ESS) are now commercializing battery packs exceeding 1000 V. In this high-voltage context, the most cost-effective semiconductor devices available are typically 1200 V-class IGBTs. Although higher-voltage devices rated at 1700 V or 2400 V exist, their cost is prohibitive for mainstream applications. Consequently, for systems operating in the 1000–1500 V range, the use of 1200 V-class IGBTs in series is the most economical and practical solution.

However, conventional two-level inverter topologies cannot accommodate such high-voltage applications using 1200 V IGBTs. This limitation has driven research and development of multilevel inverters (MLIs), where multiple switches are connected in series to distribute the voltage stress among devices. While the field of MLIs is still evolving with various new topologies and increasingly complex control strategies, the most practical and commercially viable solution today is the three-level inverter.

The primary reason for this is economic: two 1200 V IGBTs can be connected in series to safely handle system voltages in the 1000–1500 V range. Although this approach doubles the number of switches, it remains more cost-competitive than using higher-voltage devices. In this chapter, we focus on the neutral-point clamped (NPC) inverter, which is the most widely used form of a three-level inverter.

Overview of the NPC Inverter A conventional two-level inverter outputs either $\pm\frac{V_{dc}}{2}$. In contrast, an NPC inverter can additionally output 0 V, resulting in three distinct voltage levels. This additional voltage level reduces the voltage stress on each device and enables the use of lower-voltage IGBTs for high-voltage applications. Importantly, the overall operation of an NPC inverter is similar to that of a two-level inverter, with the main difference lying in the PWM modulation strategy. As a result, existing controller designs and gain settings used for two-level inverters can be directly adapted to the NPC topology.

Conventional Two-Level Inverter Figure 18.1(a) illustrates a three-phase voltage-source inverter, while Figure 18.1(b) presents the schematic for one inverter leg.

In Figure 18.1(b), the phase voltage V_{A0} determined by the switching states of S_1 and S_2, as summarized in Table 18.1.

Table 18.1 Phase voltage corresponding to switch states.

S_1	S_2	V_{A0}
1	0	$+\frac{V_{dc}}{2}$
0	1	$-\frac{V_{dc}}{2}$

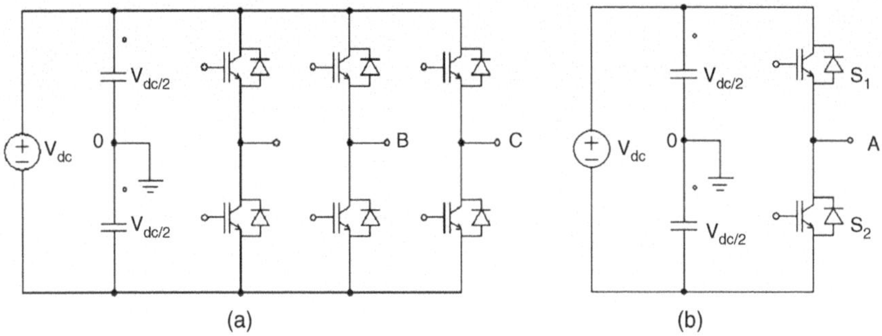

Figure 18.1 Schematic of a two-level inverter. (a) Three-leg configuration (b) Single-leg configuration.

The other two legs operate in the same manner, giving:

$$V_{AO} = \pm\frac{V_{dc}}{2},\ V_{BO} = \pm\frac{V_{dc}}{2},\ V_{CO} = \pm\frac{V_{dc}}{2}$$

The line-to-line voltage V_{AB} is expressed as follows. The same applies to V_{BC} and V_{CA}.

$$V_{AB} = V_{AO} - V_{BO} = \pm V_{dc}.0$$

The phase voltages with respect to the neutral point (N) are:

$$V_{AN} = \frac{V_{AB} - V_{CA}}{3} = \pm\frac{2V_{dc}}{3}, \pm\frac{V_{dc}}{3}$$

These voltage levels define the operation of a conventional two-level inverter.

Transition to Three-Level Inverter In contrast, a three-level inverter introduces a zero-voltage state, adding flexibility and reducing voltage stress across each switch. To simplify the explanation, Figure 18.2 depicts the circuit structures of both two-level and three-level inverters using equivalent mechanical switch representations.

The phase voltage for each leg in the three-level inverter is:

$$V_{AO} = 0, \pm\frac{V_{dc}}{2},\quad V_{BO} = 0,\ \pm\frac{V_{dc}}{2},\quad V_{CO} = 0,\ \pm\frac{V_{dc}}{2}$$

The line-to-line voltages become:

$$V_{AB} = V_{AO} - V_{BO} = \pm V_{dc},\quad \pm\frac{V_{dc}}{2},\quad 0$$

The phase voltages with respect to the neutral point are:

$$V_{AN} = \frac{V_{AB} - V_{CA}}{3} = \pm\frac{2V_{dc}}{3},\quad \pm\frac{V_{dc}}{2},\quad \pm\frac{V_{dc}}{3},\quad 0$$

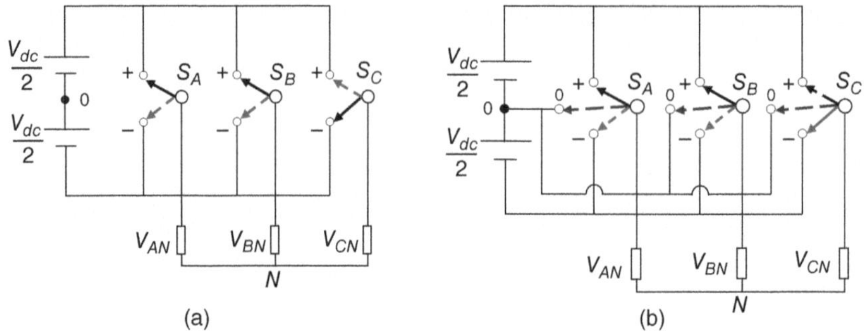

Figure 18.2 Simplified circuit diagrams of two-level and three-level inverters. (a) Two-Level circuit (b) Three-Level circuit.

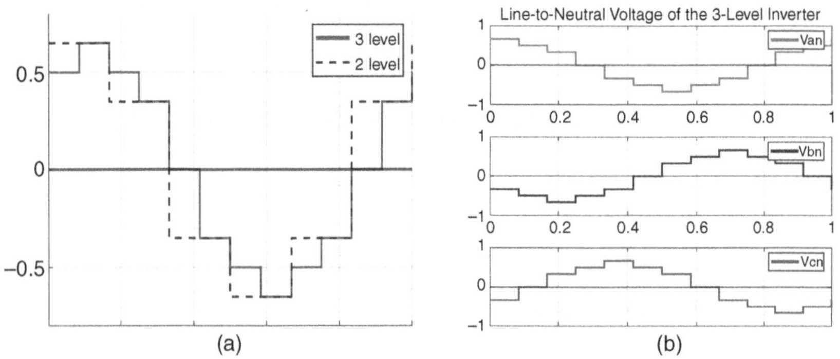

Figure 18.3 Voltage waveforms of two-level and three-level inverters. (a) Comparison of V_{AN} for two- and three-level (b) Line-to-neutral voltage of the three-level.

A comparison of the neutral point voltages (V_{AN}) between the two-level and three-level inverters is shown in Figure 18.3(a), while Figure 18.3(b) presents the line-to-neutral voltage waveform for the three-level inverter.

NPC Topologies The three-level inverter is commonly referred to as a neutral point clamped (NPC) inverter because it uses a clamping mechanism to stabilize the neutral point voltage. Several NPC topologies have been proposed and commercialized, the most notable being:

- Diode-Clamped NPC (Traditional NPC)
- T-type NPC (Improved efficiency and reduced loss)

The circuit diagrams for both topologies are illustrated in Figure 18.4.

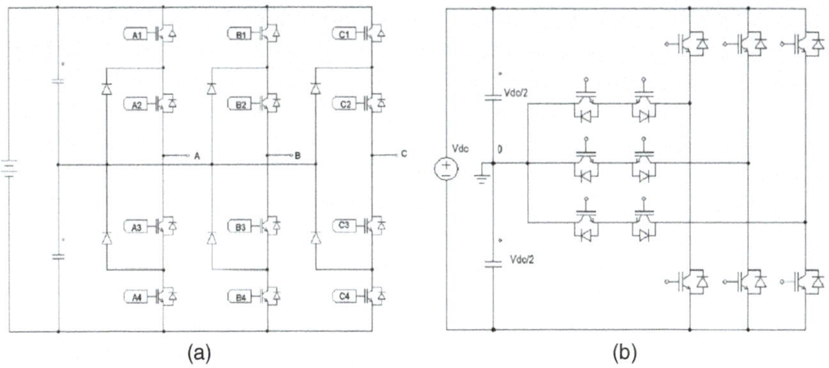

Figure 18.4 Common NPC inverter topology. (a) Diode-clamped NPC (b) T-type NPC.

These topologies form the foundation of modern three-level inverter designs and are widely used in industrial applications, including renewable energy systems and energy storage solutions.

18.2 Diode Clamped NPC

Figure 18.5(a) illustrates the single-leg circuit of a diode-clamped NPC inverter. In this configuration, switches S_1 through S_4 operate in complementary pairs: S_1 and S_3 function as one complementary pair, while S_2 and S_4 form the other. This coordinated switching enables the generation of three distinct voltage levels at the output.

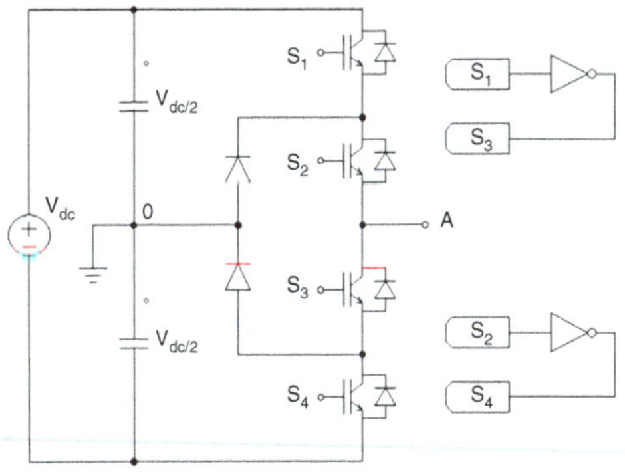

(a)

Mode	S_1	S_2	S_3	S_4	V_{A0}
(a)	1	1	0	0	$+\frac{V_{dc}}{2}$
(b)	0	1	1	0	0
(c)	0	0	1	1	$-\frac{V_{dc}}{2}$

(b)

Figure 18.5 Operation of the diode-clamped NPC inverter. (a) Single leg of diode-clamped NPC (b) Switching states for each operation mode.

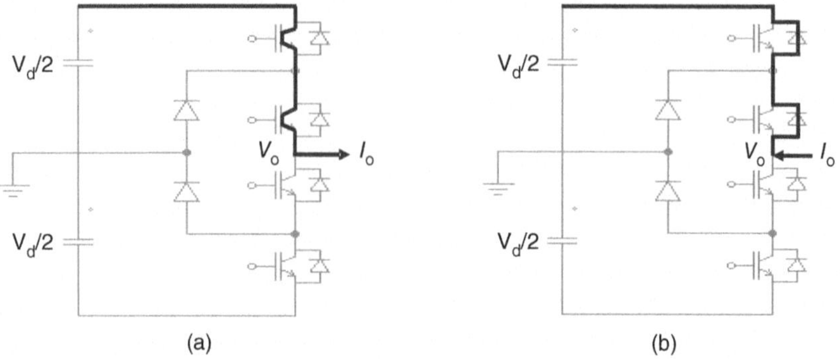

Figure 18.6 Operation during Mode (a). (a) I_o → Load (b) I_o → Input (regenerative mode)

The equivalent circuits, phase voltages, and current flow for each operation mode are as follows:

1) Operating Mode (a): Positive Voltage State
   In this mode, switches S_1 and S_2 are turned ON, while S_3 and S_4 remain OFF. Depending on the direction of the output current I_o, the current may either flow toward the load or return toward the input. The corresponding phase voltage is (Figure 18.6):

$$V_{A0} = +\frac{V_{dc}}{2}$$

2) Operating Mode (b): Zero Voltage State
   Here, switches S_2 and S_3 are ON, while S_1 and S_4 are OFF. Again, the output current I_o can flow either to the load or back toward the input. Since the output is clamped to the neutral point, the phase voltage is (Figure 18.7):

$$V_{A0} = 0$$

3) Operating Mode (c): Negative Voltage State
   In this mode, switches S_3 and S_4 are ON, and S_1 and S_2 are OFF. The current I_o flows either toward the load or toward the input, as in the previous modes. The phase voltage is (Figure 18.8):

$$V_{A0} = -\frac{V_{dc}}{2}$$

PWM Modulation Technique for NPC Pulse width modulation (PWM) techniques for MLIs are inherently more complex than those used in traditional two-level inverters. Numerous modulation methods have been proposed and continue to be actively researched. Among these, the most straightforward and widely used method for three-level inverters is an extension of conventional sinusoidal PWM.

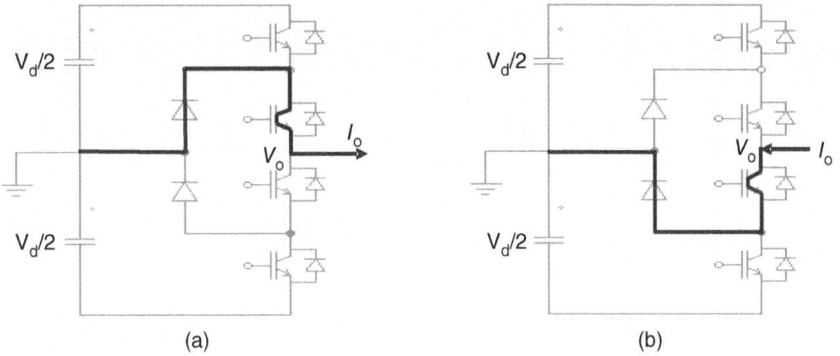

Figure 18.7 Operation during Mode (b). (a) I_o → Load (b) I_o → Input (regenerative mode).

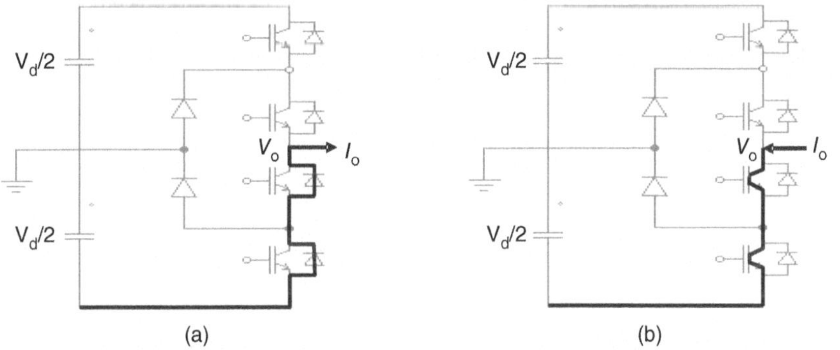

Figure 18.8 Operation during Mode (c). (a) I_o → Load (b) I_o → Input (regenerative mode).

Figure 18.9(a) shows the classic modulation waveform for a two-level inverter, which utilizes a single triangular carrier wave. In contrast, a three-level inverter, as shown in Figure 18.9(b), employs multiple carriers to generate the PWM signals. For example:

- Three-level inverter uses two carriers.
- Three-level inverter would use four carriers.

In the case of the three-level NPC inverter:

- The upper carrier controls the high-side switches (S_1 and S_3).
- The lower carrier controls the low-side switches (S_2 and S_4).

The circuit configurations used to generate the above PWM waveforms are illustrated in Figure 18.10. Figure 18.10(a) presents the circuit for two-level modulation, while Figure 18.10(b) shows the circuit for three-level modulation.

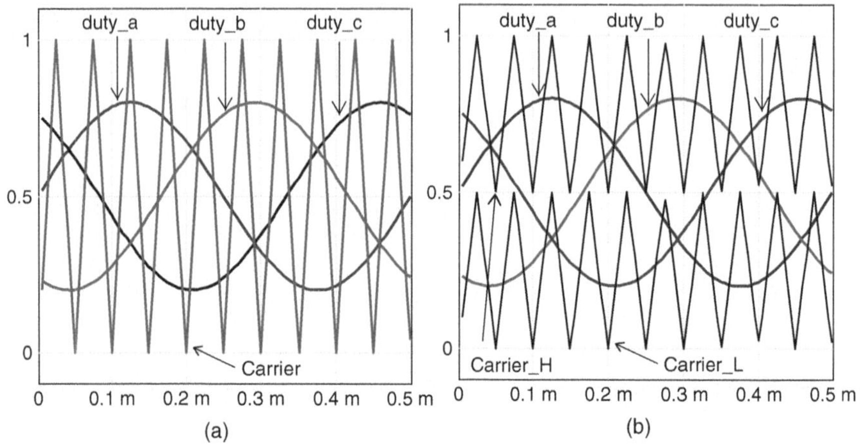

Figure 18.9 PWM modulation techniques for two-level and three-level inverters. (a) Two-level PWM modulation (b) Three-level PWM modulation.

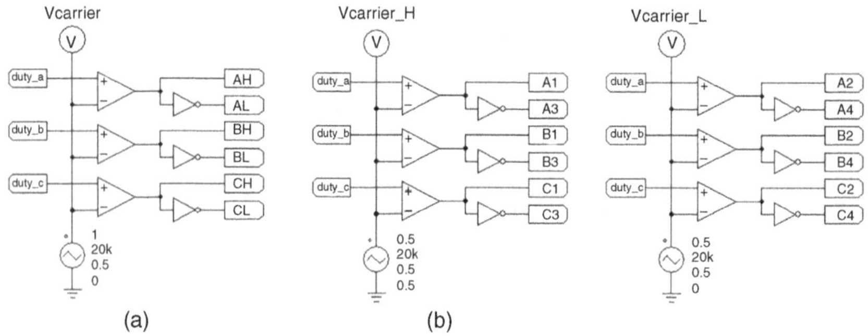

Figure 18.10 PWM circuit implementations for two-level and three-level NPC inverters. (a) PWM circuit for two-level inverter (b) PWM circuit for three-level inverter

It is important to note that both modulation circuits utilize the same control duty cycle provided by the controller; the difference lies only in the PWM generation method.

- In the two-level scheme, a single carrier wave with a magnitude of 1 is used.
- In the three-level scheme, two carriers are employed, each with a magnitude of 0.5. Additionally, an offset of 0.5 is applied to the high-side carrier to distinguish it from the low-side carrier.

This modulation strategy is known as phase disposition (PD) level-shifted PWM and is widely adopted in practical implementations. Alternative schemes, including phase opposition disposition (POD) and alternative phase opposition disposition (APOD), are not addressed here.

In summary, while both two-level and three-level inverters share similar control inputs, the modulation approach for three-level inverters enables the generation of multiple voltage levels, reducing device stress and improving output waveform quality—key advantages that make the NPC topology essential for modern high-voltage applications.

18.2.1 State Space Averaged Equation of Diode-NPC

The operation of the NPC inverter can be divided into three distinct switching intervals. The a-phase switching waveforms are shown in Figure 18.11 (b).

1) Current equation
   a) Upper-Switch ON Interval (d_p)
   When switches S_1, S_2 are ON and S_3, S_4 are OFF, the system equation is:

$$\frac{V_{\mathrm{dc1}}}{2} = r_L i_{La} + L\frac{di_{La}}{dt} + v_{ca}$$

   b) Center-Switch ON Interval ($1 - d_p - d_n$)
   When switches S_2, S_3 are ON and S_1, S_4 are OFF, the corresponding equation is:

$$0 = r_L i_{La} + L\frac{di_{La}}{dt} + v_{ca}$$

   c) Lower-Switch ON Interval (d_n)

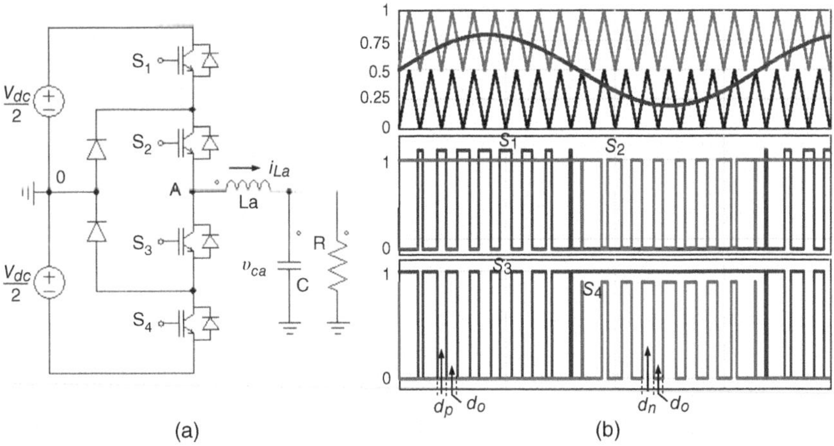

Figure 18.11 NPC circuits (a-phase example). (a) NPC a-phase circuit (b) a-phase switching waveform.

When switches S_1, S_2 are OFF and S_3, S_4 are ON, the system equation becomes:

$$-\frac{V_{dc2}}{2} = r_L i_{La} + L\frac{di_{La}}{dt} + v_{ca}$$

2) Voltage equation

$$C\frac{dv_{ca}}{dt} = i_{La} - \frac{v_{ca}}{R}$$

Duty Ratio Definition

Let the modulation index be defined as:

$$m = d_p - d_n \in [-1, 1]$$

Since the same control strategy is applied for both positive and negative half-periods, we assume:

$$d_p = d_n = d$$

Further, by means of a balancing controller, the split DC-link voltages can be maintained as:

$$\frac{V_{dc1}}{2} = \frac{V_{dc2}}{2} = \frac{V_{dc}}{2}$$

State-space average model

Averaging over one cycle, the state-space equation can be expressed as:

$$\begin{pmatrix} \frac{di_{La}}{dt} \\ \frac{dv_{ca}}{dt} \end{pmatrix} = \begin{pmatrix} -\frac{r_L}{L} & -\frac{1}{L} \\ \frac{1}{C} & -\frac{1}{RC} \end{pmatrix} \begin{pmatrix} i_{La} \\ v_{ca} \end{pmatrix} + \begin{pmatrix} \frac{V_{dc}}{2L} \\ 0 \end{pmatrix} m$$

Since $m \in [-1, 1]$, it can be expressed in terms of duty ratio $d \in [0, 1]$ as:

$$d = \frac{m+1}{2}$$

Substituting, we obtain:

$$\begin{pmatrix} \frac{di_{La}}{dt} \\ \frac{dv_{ca}}{dt} \end{pmatrix} = \begin{pmatrix} -\frac{r_L}{L} & -\frac{1}{L} \\ \frac{1}{C} & -\frac{1}{RC} \end{pmatrix} \begin{pmatrix} i_{La} \\ v_{ca} \end{pmatrix} + \begin{pmatrix} \frac{V_{dc}}{2L} \\ 0 \end{pmatrix} (2d - 1) \tag{18.1}$$

Comparison with Two-Level Inverter

For reference, the two-level inverter model is expressed in Eq. (12.2) and can be rewritten as:

$$\begin{pmatrix} \frac{di_{a}}{dt} \\ \frac{dv_{ca}}{dt} \end{pmatrix} = \begin{pmatrix} -\frac{r_L}{L} & -\frac{1}{L} \\ \frac{1}{C} & -\frac{1}{RC} \end{pmatrix} \begin{pmatrix} i_{a} \\ v_{Ca} \end{pmatrix} + \begin{pmatrix} \frac{V_{dc}}{L} \\ 0 \end{pmatrix} d + \begin{pmatrix} -\frac{1}{2L}V_{dc} \\ 0 \end{pmatrix} \tag{12.2}$$

The averaged state-space models of the three-level NPC inverter and the two-level inverter are shown to be exactly equivalent.

Extension to Three-Phase Representation

For the three-phase case, the system can be expressed as:

$$\begin{pmatrix} \dfrac{di_{L,abc}}{dt} \\ \dfrac{dv_{c,abc}}{dt} \end{pmatrix} = \begin{pmatrix} -\dfrac{r_L}{L} & -\dfrac{1}{L} \\ \dfrac{1}{C} & -\dfrac{1}{RC} \end{pmatrix} \begin{pmatrix} i_{L,abc} \\ v_{c,abc} \end{pmatrix} + \begin{pmatrix} \dfrac{V_{dc}}{L} \\ 0 \end{pmatrix} (d_{abc} - 0.5)$$

Applying the dq-transformation yields a decoupled model identical to that of the two-level inverter:

$$\begin{pmatrix} \dfrac{d}{dt} i_{dq} \\ \dfrac{d}{dt} v_{c,dq} \end{pmatrix} = \begin{pmatrix} -\dfrac{r_L}{L} & -\dfrac{1}{L} \\ \dfrac{1}{C} & -\dfrac{1}{RC} \end{pmatrix} \begin{pmatrix} i_{dq} \\ v_{c,dq} \end{pmatrix} + \begin{pmatrix} \dfrac{V_{dc}}{L} \\ 0 \end{pmatrix} d_{dq} \tag{18.2}$$

The averaged state-space model of the NPC inverter is mathematically equivalent to that of a conventional two-level inverter. Consequently, the transfer function of the NPC inverter is identical to the two-level case.

18.2.2 Controller Design

a) **Instantaneous AC Current Control (Single loop)**

A PSIM simulation was conducted on a three-phase power conditioning system (PCS) incorporating instantaneous AC current control for a diode-clamped NPC inverter. The simulation schematic is shown in Figure 18.12. The controller used in the simulation is the previously developed C-block (*Table 12.1 AC_3P_Control*) designed for conventional two-level three-phase PCS systems. In this study, the controller structure remains unchanged, with modifications applied only to the PWM generation circuit to accommodate the NPC topology.

This configuration supports bi-directional power flow, which is determined by the polarity of the input current I_p. The simulation results are presented in Figure 18.13. It can be observed that the waveform characteristics are essentially identical to those of a two-level inverter, indicating the functional compatibility of the controller with the NPC structure.

b) **DQ Current Control (Single loop)**

A separate simulation was performed using a dq transformation-based control strategy for the diode-clamped NPC inverter. The system schematic is shown in Figure 18.14. The controller utilized in this simulation is the C-block (*Table 12.5 DQ_CURRENT*), which was originally developed for a two-level three-phase PCS system and applied here without modification.

The simulation results are presented in Figure 18.15. The inverter system demonstrates stable and accurate tracking performance, consistent with the

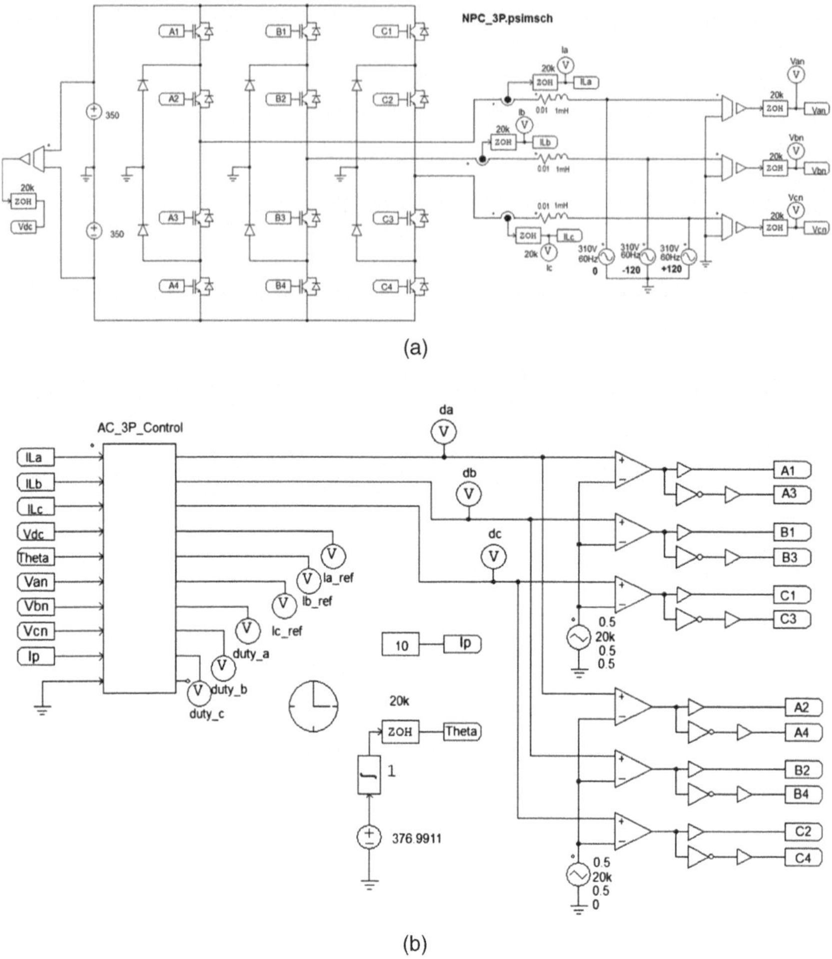

Figure 18.12 PSIM schematic of the diode NPC inverter with instantaneous AC control. (a) Diode-clamped NPC power stage (b) Controller with instantaneous AC current control.

design objectives. The output waveforms closely match those of a conventional two-level inverter, further validating the compatibility of the existing DQ control framework with the three level NPC structure.

These simulations confirm that both instantaneous and *DQ*-based control strategies—originally developed for two-level inverters—can be effectively applied to diode-clamped NPC inverters with minimal modifications. The key adaptation lies in the PWM generation scheme, while the core control logic remains unchanged. This compatibility enables straightforward

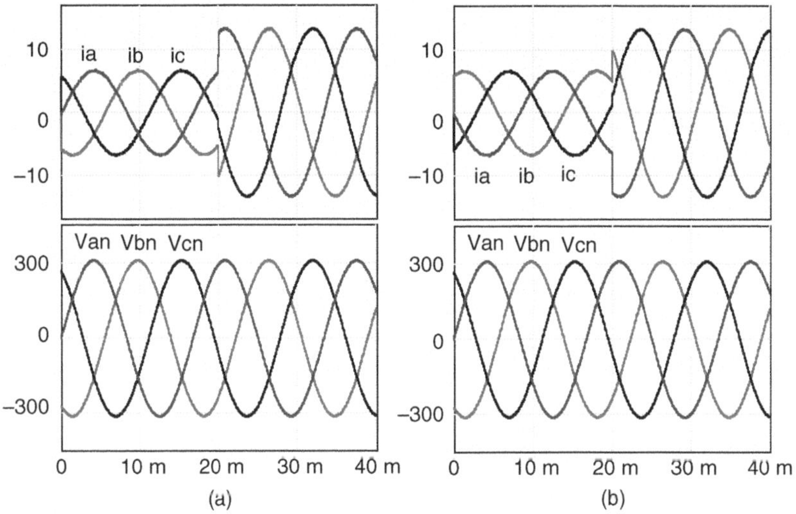

Figure 18.13 Simulation waveforms of AC current control using the diode NPC inverter. (a) Forward power flow ($I_p > 0$) (b) Reverse power flow ($I_p < 0$).

migration from two-level to three-level NPC systems, facilitating improved voltage handling and reduced switching stress without increasing controller complexity.

c) **Double-Loop Control for Stand-Alone Inverter**
The double-loop control system consists of inner current control loop and outer voltage loop to control output voltage and inductor current. Since the transfer functions of the NPC inverter are same as those of the two-level inverter, the same double loop controller having same control gains is used as those of the two-level double loop controller (*Table 12.11 DQ_Double*) as shown in Chapter 12. The design specifications are set to match the case of Chapter 12. Figure 18.16 shows the PSIM schematic of the NPC double-loop stand-alone inverter.

Figure 18.17 illustrates the simulation waveforms of the NPC double-loop inverter. The voltage and current waveforms are identical to those of the two-level double-loop inverter. From these waveforms, it is evident that the mathematical modeling of the NPC inverter and the two-level inverter yields identical results. Consequently, both inverter types can be controlled using the same control strategy and identical controller gains.

d) **AC/DC Rectifier (Power Factor Controller)**
In this section, the modeling and control method of the NPC Rectifier are presented. Similar to the two-level rectifier introduced in Chapter 12, the output stage consists of a capacitor and a resistive load. Figure 18.18 illustrates

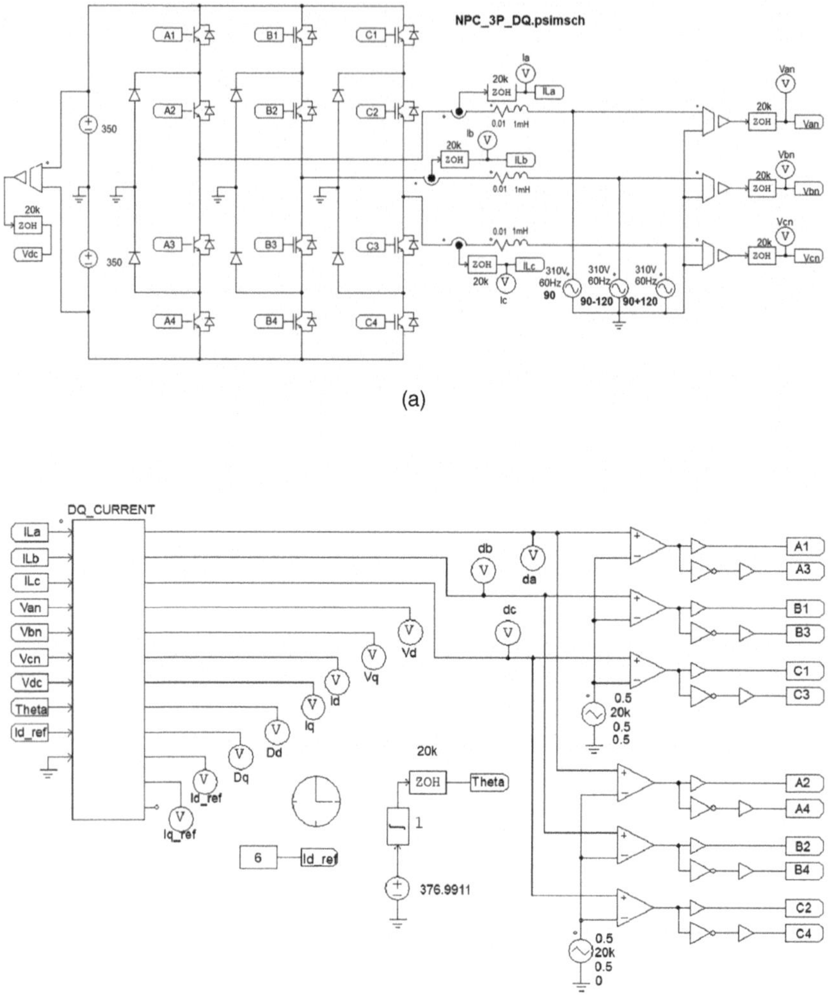

(a)

(b)

Figure 18.14 PSIM schematic of the diode-clamped NPC inverter using DQ control. (a) Power circuit schematic (b) Controller configuration.

the circuit diagram of the NPC Rectifier. For modeling purposes, the current direction is defined from the grid to the DC side, consistent with the convention used for the two-level rectifier. Please note that each capacitor in the series connection should have a capacitance of $\mathbf{2C}$ so that the resulting equivalent capacitance becomes $\boldsymbol{C}$.

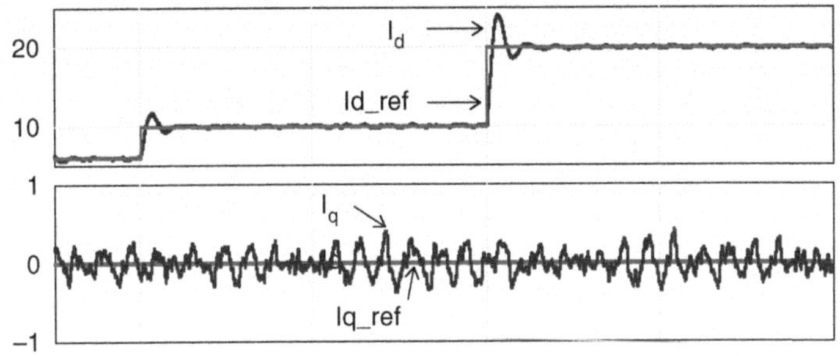

Figure 18.15 Simulation waveforms of DQ current control using the diode NPC inverter.

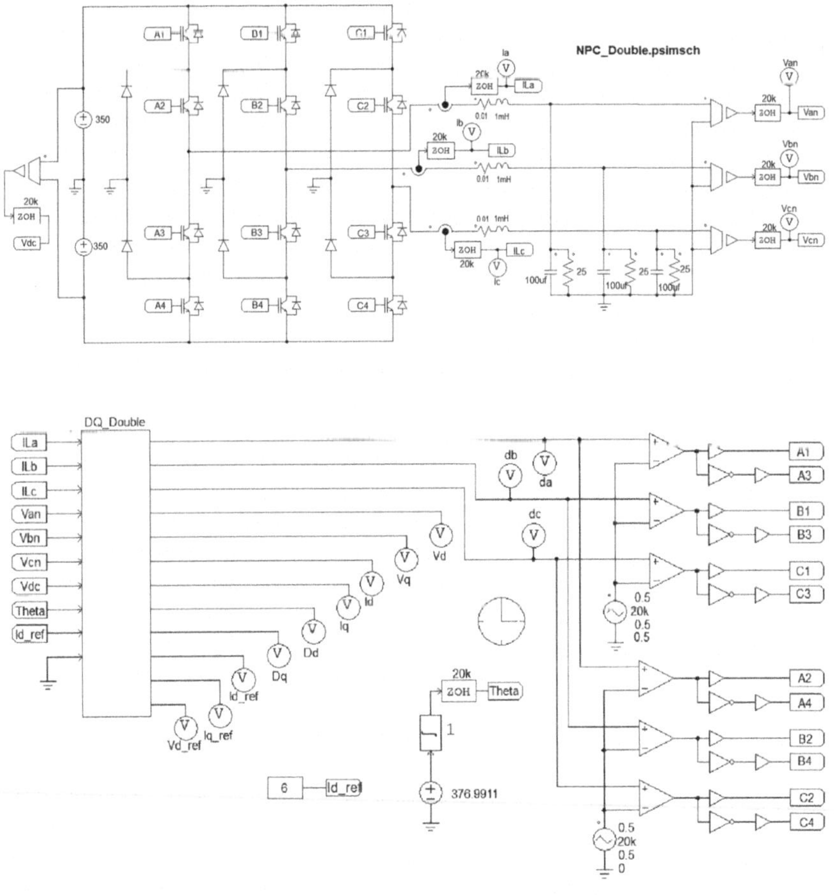

Figure 18.16 Schematic of the NPC double-loop inverter.

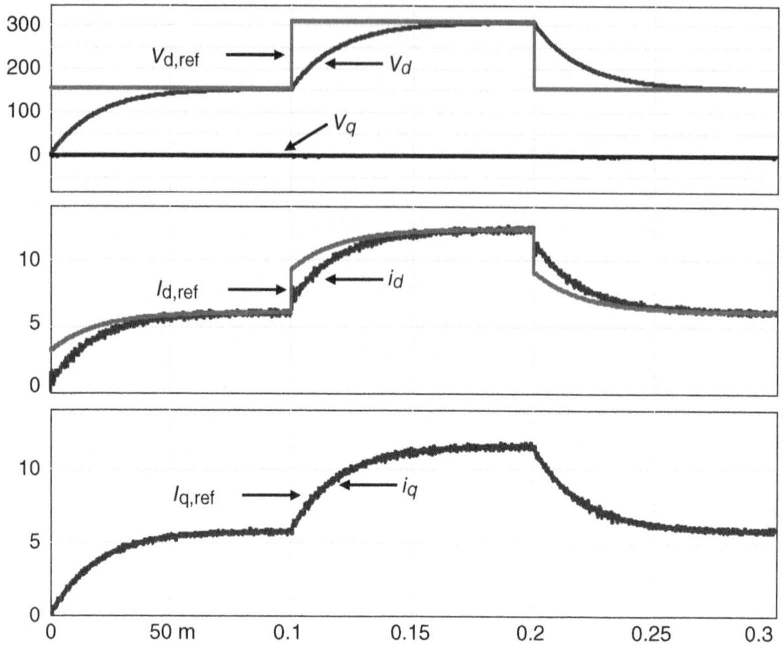

Figure 18.17 Simulation waveforms of the NPC double-loop inverter.

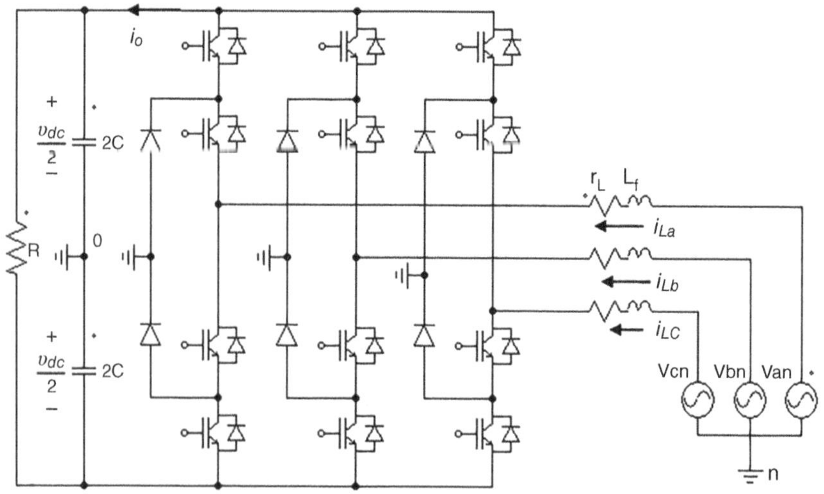

Figure 18.18 Schematic of NPC Rectifier.

1. State averaging equation and transfer function

1) Current Equation

For the a-phase, the state-space equation can be derived by considering three switching intervals:

a) Interval (a): Upper switch ON (d_p): S_1, S_2 : ON S_3, S_4 : OFF

$$V_{an} = r_L i_{La} + L\frac{di_{La}}{dt} + \frac{v_{dc}}{2}$$

b) Interval (b): Center switch ON ($1 - d_p - d_n$): S_2, S_3 : ON S_1, S_4 : OFF

$$V_{an} = r_L i_{La} + L\frac{di_{La}}{dt} + 0$$

c) Interval (c): Lower switch ON (d_n): S_1, S_2 : OFF S_3, S_4 : ON

$$V_{an} = r_L i_{La} + L\frac{di_{La}}{dt} - \frac{v_{dc}}{2}$$

Define the modulation index as:

$$m = d_p - d_n \in [-1, 1]$$

Since the same control strategy is applied to both the positive and negative half cycles,

$$d_p = d_n = d$$

The inductor current equation averaged over one cycle is expressed as:

$$\frac{di_{La}}{dt} = -\frac{r_L}{L} i_{La} - \frac{v_{dc}}{2L} m + \frac{V_{an}}{L}$$

Because $m \in [-1, 1]$, it can be normalized into the duty ratio duty $d \in [0, 1]$ as:

$$d = \frac{m+1}{2}$$

Substituting these yields,

$$\frac{di_{La}}{dt} = -\frac{r_L}{L} i_{La} - \frac{v_{dc}}{2L}(2d - 1) + \frac{V_{an}}{L}$$

In three-phase form, the expression becomes identical to that of the two-level AC/DC rectifier:

$$\frac{di_{L,abc}}{dt} = -\frac{r_L}{L} i_{L,abc} - \frac{v_{dc}}{2L}(2d_{abc} - 1) + \frac{V_{abc,n}}{L} \tag{18.3}$$

2) Voltage Equation

The DC-link voltage equation is also identical to that of the two-level PFC rectifier:

$$\frac{dv_{dc}}{dt} = \frac{1}{C}((2d_a - 1)i_{La} + (2d_b - 1)i_{Lb} + (2d_c - 1)i_{Lc}) - \frac{v_{dc}}{RC}$$

3) Final State-Averaged Equation

The combined state-averaged model is given as:

$$\begin{pmatrix} \dfrac{di_{L,abc}}{dt} \\ \dfrac{dv_{dc}}{dt} \end{pmatrix} = \begin{pmatrix} -\dfrac{r_L}{L} & -\dfrac{(2d_{abc}-1)}{2L} \\ \dfrac{(2d_{abc}-1)}{C} & -\dfrac{1}{RC} \end{pmatrix} \begin{pmatrix} i_{L,abc} \\ v_{dc} \end{pmatrix} + \begin{pmatrix} \dfrac{V_{abc,n}}{L} \\ 0 \end{pmatrix} \tag{18.4}$$

4) Small-Signal Model

Since the structure is identical, the small-signal model is the same as that of the two-level rectifier:

$$\begin{pmatrix} \dfrac{d\hat{i}_{dq}}{dt} \\ \dfrac{d(\hat{v}_{dc})}{dt} \end{pmatrix} = \begin{pmatrix} -\dfrac{r_L}{L} & -\dfrac{D_{DQ}}{L} \\ \dfrac{3D_{DQ}}{C} & -\dfrac{1}{RC} \end{pmatrix} \begin{pmatrix} \hat{i}_{dq} \\ \hat{v}_{dc} \end{pmatrix} + \begin{pmatrix} -\dfrac{V_{DC}}{L} \\ \dfrac{3I_{DQ}}{C} \end{pmatrix} \hat{d}_{dq,fb} \tag{18.5}$$

with steady-state values:

$$V_{DC} = \frac{V_{DQ}}{D_{DQ}}, \quad I_L = \frac{V_{DC}}{3RD_D}$$

5) PSIM Simulation

The PSIM simulation is carried out using the same control block and controller gains as those employed in the two-level rectifier. Please refer to Figure *12.46* for simulation circuit of AC/DC system for DC voltage control. The simulation circuit for the diode NPC rectifier is shown in Figure 18.19. The same C-block (*Table 12.15 DC_Voltage_Rectifier*) is used in the simulation and is not shown in the figure.

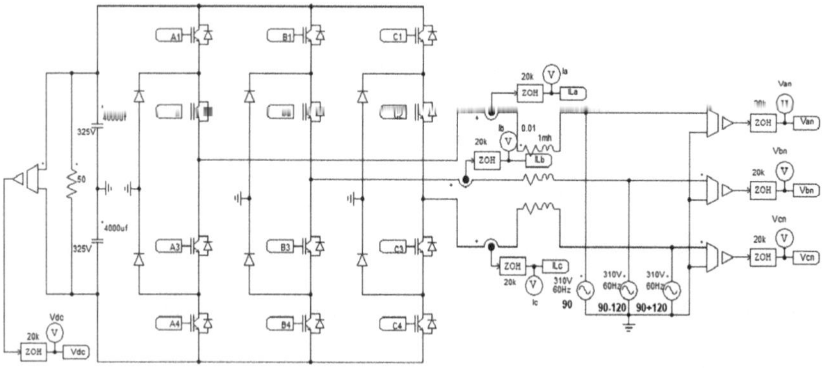

Figure 18.19 PSIM schematic of diode NPC Rectifier.

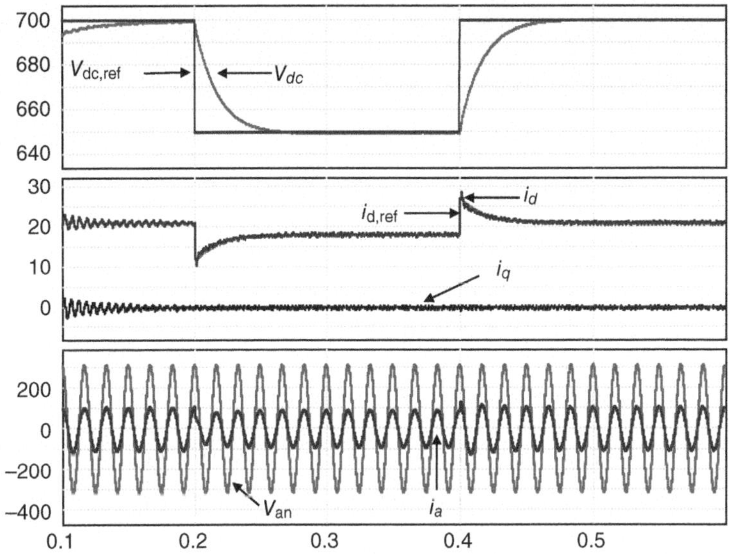

Figure 18.20 Simulation results of NPC Rectifier.

Simulation Results
The simulation results are illustrated in Figure 18.20. The obtained waveforms are identical to those of the two-level rectifier, confirming that the averaged model and small-signal equations are correct.

18.3 T-Type NPC Inverter

a) Instantaneous Current Control
An alternative three-level NPC inverter topology is the T-type NPC inverter, which differs structurally from the diode-clamped NPC. In the T-type configuration, instead of placing switches in series to divide voltage, a bidirectional switch is inserted in the ground path. While the total number of active switches remains the same as in the diode-clamped NPC, the elimination of clamping diodes makes the T-type NPC a more economical solution, particularly in high-efficiency applications.

The PSIM simulation schematic for the T-type NPC inverter is presented in Figure 18.21. This figure illustrates only the power stage circuit; the controller configuration is omitted, as it is identical to that shown in Figure 18.12(b). The control is implemented using instantaneous AC current control, employing the same C-block (*Table 12.1 AC_3P_Control*) previously used for the diode-clamped NPC.

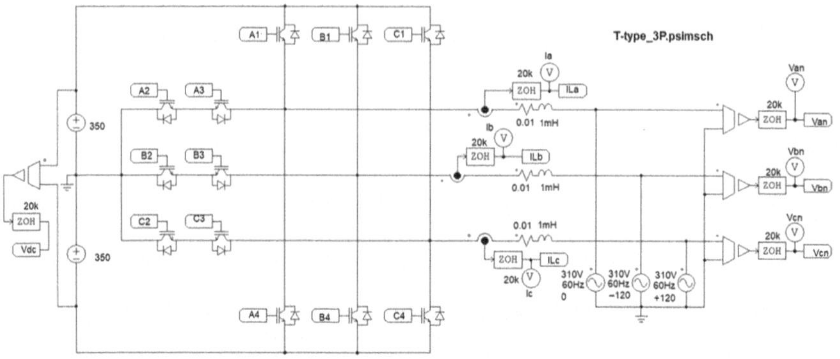

Figure 18.21 PSIM schematic of the T-type NPC inverter.

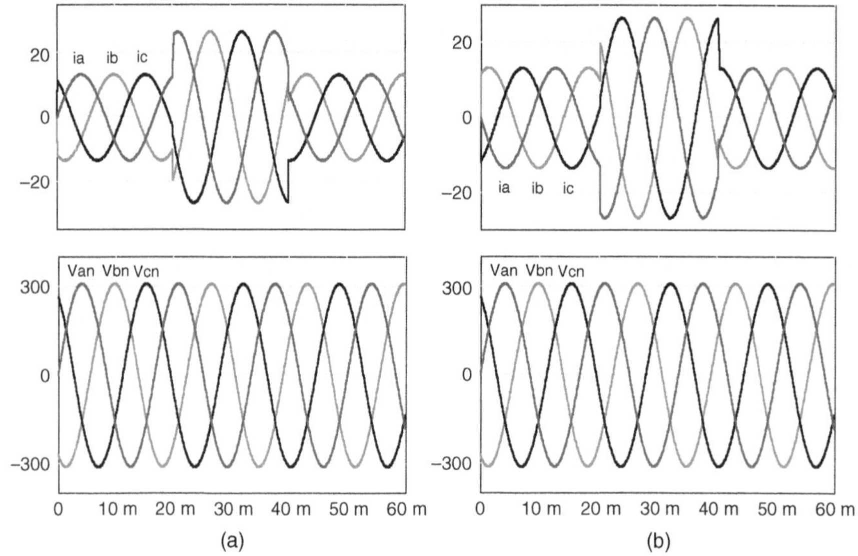

Figure 18.22 Simulation waveforms of the T-type NPC using instantaneous AC current control. (a) $I_p > 0$ (b) $I_p < 0$.

No modifications to the control logic were required, demonstrating the adaptability and versatility of the controller across different NPC topologies.

The simulation results are presented in Figure 18.22. As shown:

- Case (a): When $I_p > 0$, power flows from the DC input to the AC grid.
- Case (b): When $I_p < 0$, power flows from the AC grid to the DC input.

In both cases, the phase and magnitude of the output current are well-regulated, demonstrating precise current tracking. The inverter operation aligns with the design intent, exhibiting in-phase or out-of-phase operation with respect to the grid voltage depending on the power flow direction.

b) DQ Current Control (Omitted)

The T-type NPC inverter can also be controlled using DQ transformation-based current control, similar to the diode-clamped NPC. The same control code (*DQ_CURRENT*) can be directly applied without modification. As the implementation follows the same principles and structure, detailed discussion is omitted here. Students are encouraged to implement and simulate the T-type *DQ* control on their own to deepen their understanding.

The T-type NPC inverter provides a compact and cost-effective alternative to the diode-clamped topology while maintaining equivalent functional capabilities. Its compatibility with existing control structures—both instantaneous and *DQ*-based—makes it a practical solution for three-phase PCS applications requiring high performance and bidirectional power flow.

c) Double-Loop Control for Stand-Alone Inverter (Omitted)
d) AC/DC Rectifier (Omitted)

18.4 Solar NPC Inverter

This section examines the operation of a PCS that employs a photovoltaic (PV) source as the DC input to a three-level NPC inverter. Both the diode-clamped NPC and the T-type NPC configurations are considered. In each case, compatibility with existing controller structures developed for two-level solar inverters is demonstrated.

18.4.1 State-Averaged Equation and Transfer Function

The state-space behavior of the single-phase NPC inverter as shown in Figure 18.23 can be divided into three switching intervals within one cycle.

1) Current Averaging Equation
   - Interval (a): Upper switch ON (d_p) => S_1, S_2 : ON, S_3, S_4 : OFF

$$\frac{v_{sa}}{2} = r_L i_{La} + L\frac{di_{La}}{dt} + V_{an}$$

   - Interval (b): Center switch ON ($1 - d_p - d_n$) => S_2, S_3 : ON S_1, S_4 : OFF

$$0 = r_L i_{La} + L\frac{di_{La}}{dt} + V_{an}$$

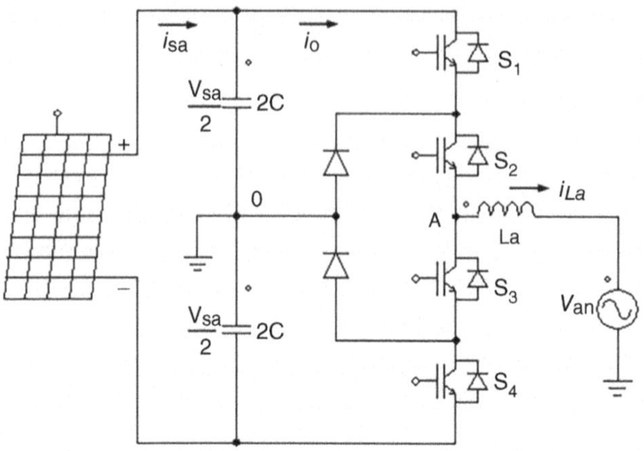

Figure 18.23 Circuit diagram of solar diode NPC inverter (a-phase only).

- Interval (c): Lower switch ON (d_n) => S_1, S_2 : OFF S_3, S_4 : ON

$$-\frac{v_{sa}}{2} = r_L i_{La} + L\frac{di_{La}}{dt} + V_{an}$$

Define the modulation variable as

$$m = d_p - d_n \in [-1, 1]$$

The inductor current dynamics can then be expressed as:

$$\frac{di_{La}}{dt} = -\frac{r_L}{L}i_{La} + \frac{v_{sa}}{2L}m - \frac{V_{an}}{L}$$

Since $m \in [-1, 1]$, it can be normalized into duty ratio duty $d \in [0, 1]$ as:

$$d = \frac{m+1}{2}$$

Substituting this into the current dynamics:

$$\frac{di_{La}}{dt} = -\frac{r_L}{L}i_{La} - \frac{v_{sa}}{2L}(1-2d) - \frac{V_{an}}{L}$$

In three-phase form, the averaged current equation becomes identical to that of a conventional two-level solar PCS converter:

$$\frac{di_{L,abc}}{dt} = -\frac{r_L}{L}i_{L,abc} - \frac{(1-2d_{abc})}{2L}v_{sa} - \frac{V_{abc,n}}{L} \tag{18.6}$$

2) Voltage Averaging Equation
The capacitor dynamics of the PV array are given by

$$\frac{dv_{sa}}{dt} = \frac{1}{C}(i_{sa} - i_o)$$

with i_{sa} linearized around the operating point as $i_{sa} = -\alpha v_{sa} + \beta$.
Accounting for the switch ON/OFF states,

$$\frac{dv_{sa}}{dt} = \frac{1}{C}((1-2d_a)i_{La} + (1-2d_b)i_{Lb} + (1-2d_c)i_{Lc} - \alpha v_{sa} + \beta)$$

Or, in compact form,

$$\frac{dv_{sa}}{dt} = \frac{(1-2d_{abc})i_{L,abc}}{C} + \frac{(-\alpha v_{sa} + \beta)}{C} \tag{18.7}$$

This formulation is also identical to the two-level PCS case.

Final small signal model Combining the current and voltage dynamics, the small-signal state-space model is given by:

$$\begin{pmatrix} \frac{d\hat{i}_{dq}}{dt} \\ \frac{d(\hat{v}_{sa})}{dt} \end{pmatrix} = \begin{pmatrix} -\frac{r_L}{L} & \frac{D_{DQ}}{L} \\ -\frac{3D_{DQ}}{C} & \frac{-\alpha}{C} \end{pmatrix} \begin{pmatrix} \hat{i}_{dq} \\ \hat{v}_{sa} \end{pmatrix} + \begin{pmatrix} \frac{V_{SA}}{L} \\ -\frac{3I_{DQ}}{C} \end{pmatrix} \hat{d}_{dq,fb}$$

where the steady-state relationships are:

$$D_{DQ} = \frac{V_{DQ}}{V_{SA}}, \quad I_D = \frac{(-\alpha V_{SA} + \beta)}{3D_D} = \frac{I_{SA}}{3D_D}$$

It is noteworthy that the derived equation is identical to that of the conventional two-level solar PCS. Accordingly, the transfer characteristics are identical to those of the two-level solar PCS.

18.4.2 Solar Diode-Clamped NPC

The PSIM simulation is carried out using the identical control block structure and controller gains as those employed in the two-level solar PCS. Figure 18.24 illustrates the circuit configuration of a solar-based diode-clamped NPC inverter. In this setup:

- Figure 18.24(a) shows the power stage, where the DC input is supplied by a solar panel array.
- Figure 18.24(b) shows the controller schematic, in which the C-block (*Table 13.8 Solar_PCS_DQ_V*), originally developed for a two-level solar inverter (see Figure 13.12), is reused without any modifications.

The simulation results are presented in Figure 18.25. It is observed that both the voltage and current controllers perform effectively, achieving stable operation. The output waveforms closely match those of a conventional two-level solar PCS system, confirming that the controller remains effective when applied to a three-level NPC topology.

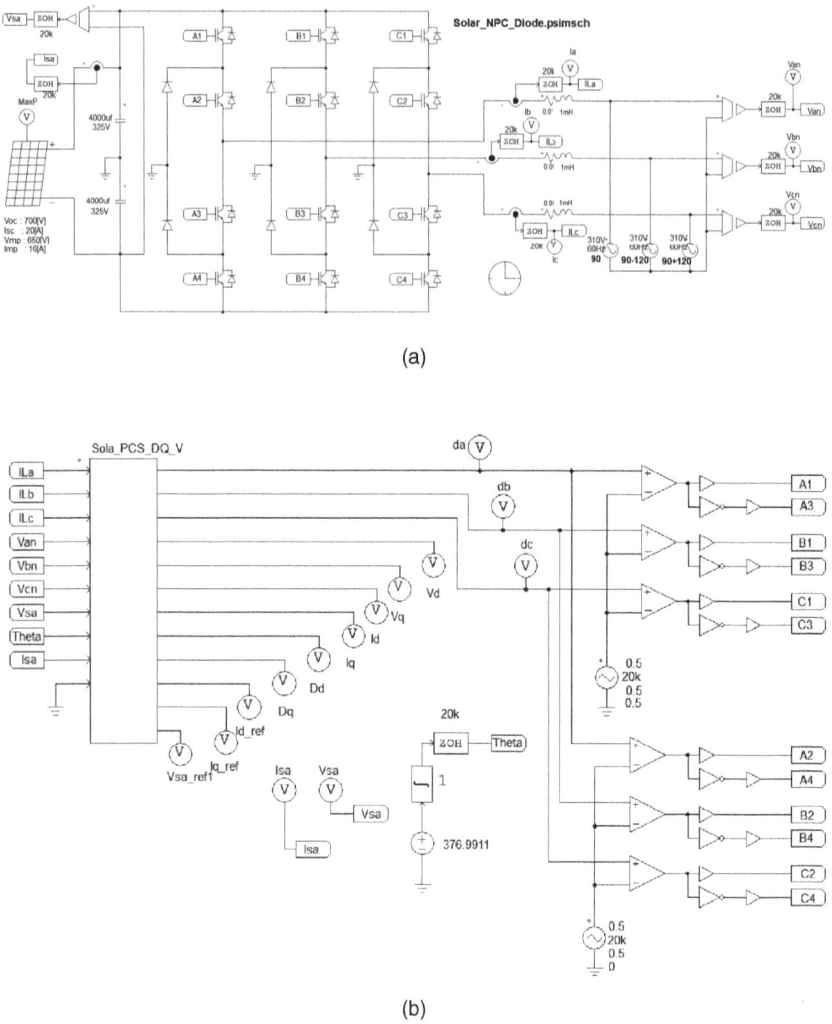

Figure 18.24 Schematic of solar diode-clamped NPC PCS (DQ control). (a) Power stage schematic (b) Controller schematic.

18.4.3 Solar T-Type NPC

Figure 18.26 shows the schematic of the solar T-type NPC PCS, in which a solar panel array serves as the input source and the output is connected to a three-phase grid. The controller section used for this simulation is identical to that applied in the diode-clamped configuration and previous two-level implementations. For this reason, it is omitted here.

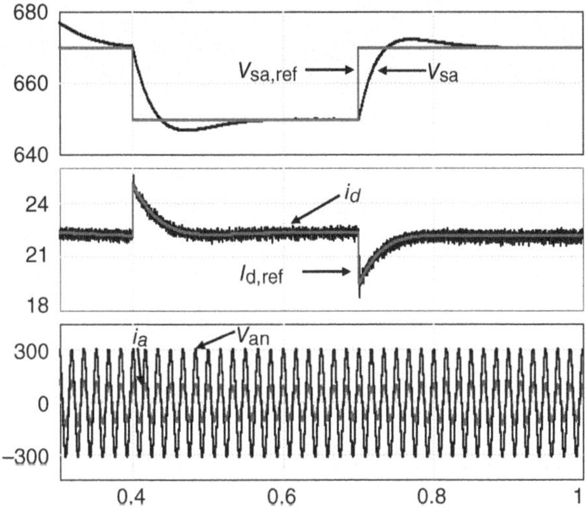

Figure 18.25 Simulation waveforms of solar diode-clamped NPC PCS.

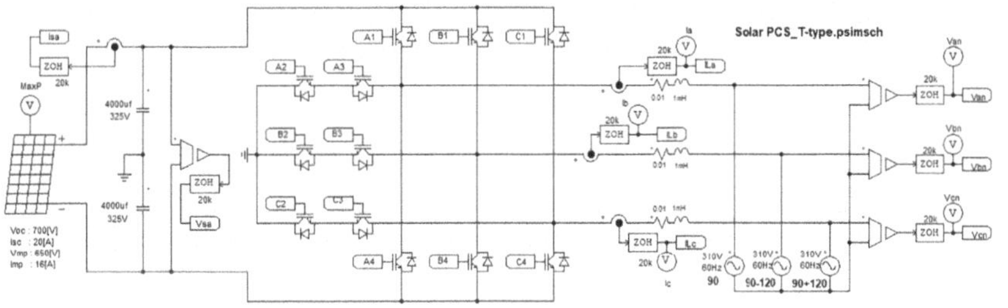

Figure 18.26 Simulation schematic of solar T-type PCS.

The simulation results are shown in Figure 18.27. The inverter output exhibits the expected waveform profiles, confirming that the system is being controlled according to the design objectives. As with the diode-clamped case, the waveform characteristics are consistent with those observed in the two-level solar PCS system.

Both diode-clamped and T-type NPC inverters are compatible with existing control structures developed for two-level solar PCS systems. The simulations confirm that controllers such as *Solar_PCS_PCS_DQ_V* can be reused without modification, enabling a smooth transition to higher-voltage three-level inverter topologies. This facilitates improved voltage handling, reduced switching stress, and better output quality—making NPC inverters a viable solution for next-generation solar energy systems.

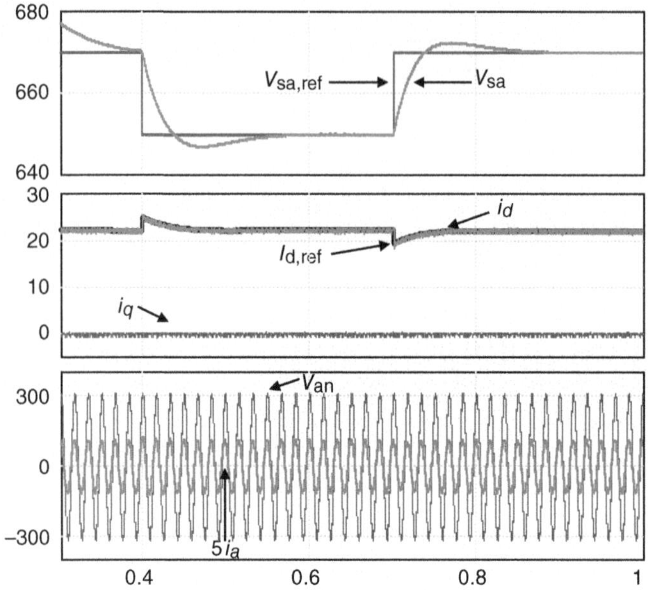

Figure 18.27 Simulation waveforms of solar T-type NPC inverter.

19

Permanent Magnet Synchronous Motor

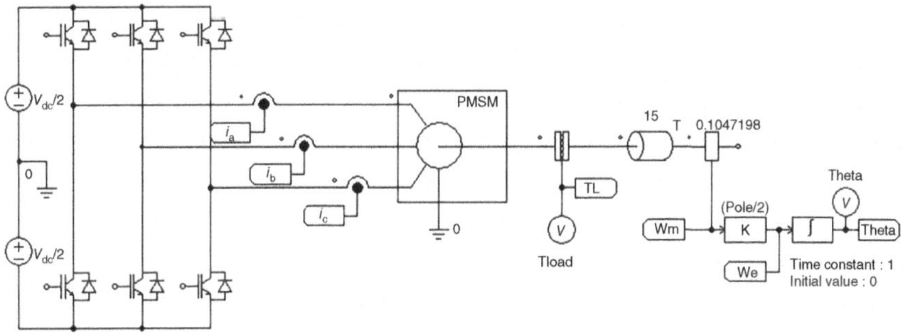

A permanent magnet synchronous motor (PMSM) consists of a permanent magnet rotor and a wound stator. PMSMs are widely used in traction applications such as electric vehicles (EVs) and in precision actuation systems, where they are often referred to as brushless servo motors. These motors typically exhibit a sinusoidal back electromotive force (EMF) profile, which minimizes undesirable mechanical and electrical current harmonics when compared to Brushless DC (BLDC) motors, which utilize a trapezoidal back EMF.

PMSMs form a broad category that includes both:

- SPMSM (Surface Permanent Magnet Synchronous Motor): Magnets are mounted on the rotor surface.
- IPMSM (Interior Permanent Magnet Synchronous Motor): Magnets are embedded within the rotor structure.

The IPMSM configuration provides higher torque and improved efficiency due to the additional reluctance torque generated by the rotor geometry. The various PMSM types are illustrated in Figure 19.1.

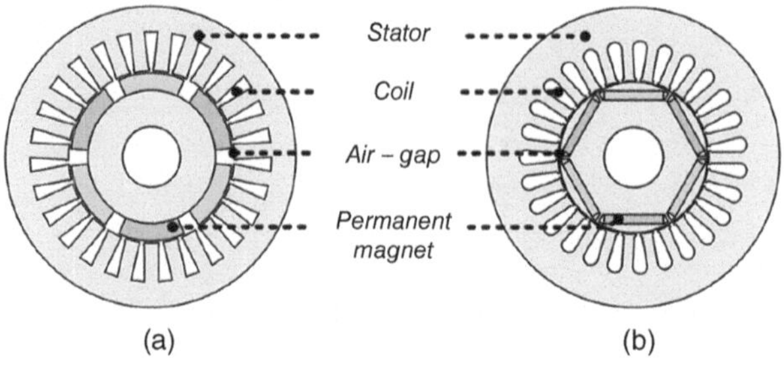

Figure 19.1 PMSM types. (a) SPMSM (b) IPMSM.

19.1 Motor Parameters

The fundamental parameters of a PMSM are identical to those of a DC motor, as discussed in *Chapter 10.1*. In addition, the following relationships are used to perform the dq transformation:

- The relationship between electric angle θ_e and mechanical angle θ_m:

$$\theta_e = \frac{P}{2}\theta_m$$

where P is the number of poles.

- The relationship between electrical angular velocity ω_e and mechanical velocity ω_m:

$$\omega_e = \frac{P}{2}\omega_m$$

A speed sensor measures the motor's angular velocity ω_m in revolutions per minute (rpm). The conversion from rpm to rad/s is given by:

$$\omega_m = \frac{2\pi}{60} \times [rpm] = 0.1047 \times [rpm]$$

For $abc - to - dq$ transformation, the electric angle θ_e must be derived from the mechanical speed. The relationship is:

$$\theta_e = \frac{P}{2}\theta_m = \frac{P}{2}\int \omega_m dt = \frac{P}{2}\int \left(\frac{2\pi}{60} \cdot \mathrm{rpm}\right) dt = \frac{P \cdot \pi \cdot rpm}{60} \cdot t \qquad (19.1)$$

a) **PMSM Equivalent Model**

The equivalent circuit of a PMSM is shown in Figure 19.2. Diagram (a) represents the phase model, while diagram (b) represents the dq model.

The variables of the phase model are defined as follows.

V_a, V_b, V_c: *Phase voltage,* E_a, E_b, E_c: *back EMF,*
L_a, L_b, L_c: *Inductance,* i_a, i_b, i_c: *Phase current.*

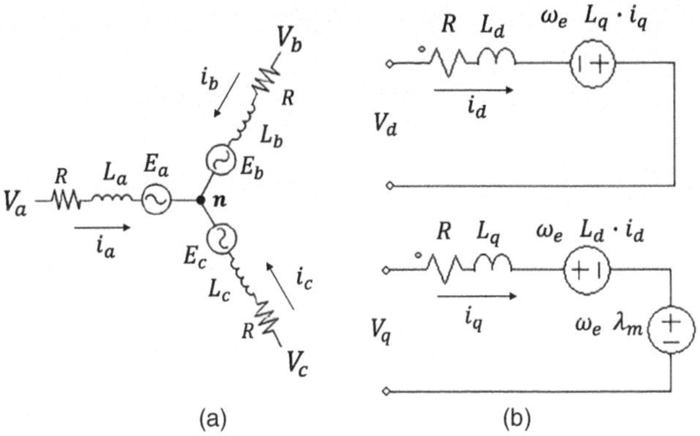

Figure 19.2 PMSM models. (a) Phase model. (b) dq model.

The electrical circuit equations for the phase model are:

$$v_{an} = Ri_a + L_a\frac{di_a}{dt} + E_a = Ri_a + L_a\frac{di_a}{dt} + \lambda_m \cos(\theta_e)$$

$$v_{bn} = Ri_b + L_b\frac{di_b}{dt} + E_b = Ri_b + L_b\frac{di_b}{dt} + \lambda_m \cos\left(\theta_e - \frac{2\pi}{3}\right)$$

$$v_{cn} = Ri_c + L_c\frac{di_c}{dt} + E_c = Ri_c + L_c\frac{di_c}{dt} + \lambda_m \cos\left(\theta_e + \frac{2\pi}{3}\right)$$

- SPMSM: The inductances are equal ($L_a = L_b = L_c$), resulting in $L_d = L_q$ in the *dq* model.
- IPMSM: The inductances vary sinusoidally, leading to $L_d \neq L_q$, typically with $L_d < L_q$.

dq *Model Equations* The *dq* axis voltage equations are:

$$V_d = Ri_d + L_d\frac{di_d}{dt} - \omega_e L_q \cdot i_q$$

$$V_q = Ri_q + L_q\frac{di_q}{dt} + \omega_e L_d \cdot i_d + \omega_e \lambda_m \tag{19.2}$$

The mechanical torque equation, which is common to both SPMSM and IPMSM, is:

$$T_e = \left(\frac{3}{2}\right)\left(\frac{P}{2}\right)[\lambda_m i_q + (L_d - L_q) i_q i_d] \tag{19.3}$$

The mechanical dynamics are described by:

$$J_m\left(\frac{2}{P}\right)\frac{d\omega_e}{dt} + B_m\left(\frac{2}{P}\right)\omega_e = T_e - T_L \tag{19.4}$$

This equivalent model serves as the foundation for simulation, control design, and performance analysis of both SPMSM and IPMSM motors.

Table 19.1 Comparison of DC motor and PMSM equations.

	DC motor	*PMSM*
Electrical	$V_f = R_f i_f + L_f \frac{di_f}{dt}$	$V_d = Ri_d + L_d \frac{di_d}{dt} - \omega_e L_q \cdot i_q$
	$V_t = R_a i_a + L_a \frac{di_a}{dt} + K_\varphi \cdot \omega_m$	$V_q = Ri_q + L_q \frac{di_q}{dt} + \omega_e L_d \cdot i_d + \lambda_m \cdot \omega_e$
Mechanical	$J_m \frac{d\omega_m}{dt} + B_m \omega_m = T_e - T_{\text{load}}$	$J_m \frac{d\omega_m}{dt} + B_m \omega_m = T_e - T_{\text{load}}$
Torque	$T_{em} = K_\varphi \cdot i_a$	$T_e = \left(\frac{3}{2}\right)\left(\frac{P}{2}\right)[\lambda_m i_q + (L_d - L_q) i_q i_d]$

Comparison with the DC Motor Model To gain deeper insight into the behavior of a PMSM, it is useful to compare its governing equations with those of a conventional DC motor. Table 19.1 presents a side-by-side comparison of the key equations.

a) Electrical Equation Comparison
   As explained in the three-phase inverter section, the PMSM electrical model introduces an additional decoupling term, specifically the feedforward components:
   - $-\omega_e L_q i_q$ in the d-axis voltage equation, and
   - $+\omega_e L_d i_d$ in the q-axis voltage equation.

   Apart from these feedforward terms, the fundamental structure of the field equation remains the same for both DC motors and PMSMs. The armature equation of the DC motor differs only in the absence of these cross-coupling terms and in the use of the back EMF constant K_φ instead of the PMSM's permanent magnet flux linkage λ_m.

   In DC motor modeling, the electrical angular velocity and the mechanical angular velocity are expressed as identical for simplicity, and thus ω_m is used interchangeably. However, strictly speaking, the notation ω_e is more appropriate in electrical equations since it reflects the electrical angular velocity. In PMSMs, this distinction is essential because the relationship between electrical and mechanical angular velocities is:

$$\omega_e = \frac{P}{2}\omega_m$$

b) Mechanical Equation Comparison
   The mechanical dynamics remain identical for both motor types:

$$J_m \frac{d\omega_m}{dt} + B_m \omega_m = T_e - T_{\text{load}}$$

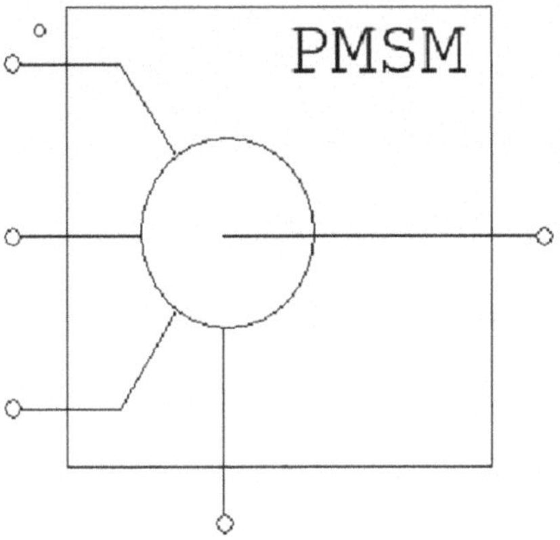

(a)

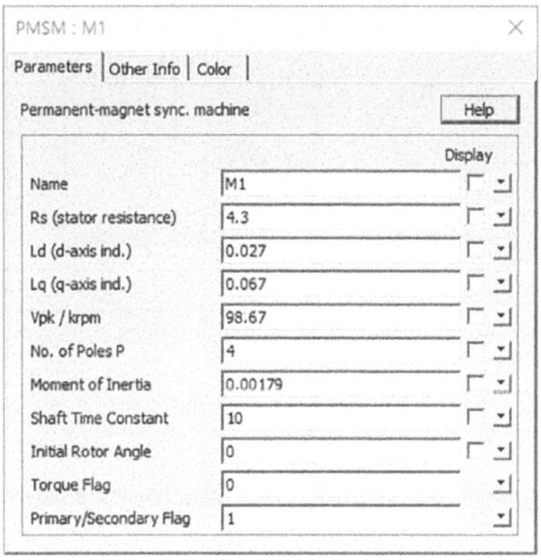

(b)

Figure 19.3 PMSM library model in PSIM.

In the torque equation, the term $\left(\frac{3}{2}\right) i_q$ in the PMSM corresponds to the phase current's contribution to torque generation, analogous to the armature current i_a in the DC motor. The DC motor's torque constant K_φ is equivalent to the PMSM term $\left(\frac{P}{2}\right) \lambda_m$, linking the electrical and mechanical aspects of torque production.

All simulation examples presented in this book are based on the PMSM library provided by PSIM. The key parameters of the dq model used for simulation are illustrated in Figure 19.3. The back EMF constant λ_m in the mathematical model can be calculated from the machine's rated values using the relation:

$$\lambda_m = V_{pk}/K_{\text{rpm}}$$

where

- V_{pk} is the peak phase voltage,
- K_{rpm} is the back EMF constant expressed in rpm.

Explanation of* V_{pk}/K_{rpm} *Parameter Clicking `Help` in the PMSM Library provides the following explanation for the parameter V_{pk}/K_{rpm}:

Peak line-to-line back emf constant, in V/K_{rpm} (mechanical speed). The value of V_{pk}/K_{rpm} should be available from the machine data sheet. If this data is not available, it can be obtained through an experiment by operating the machine as a generator at 1000 rpm and measuring the peak line-to-line voltage.

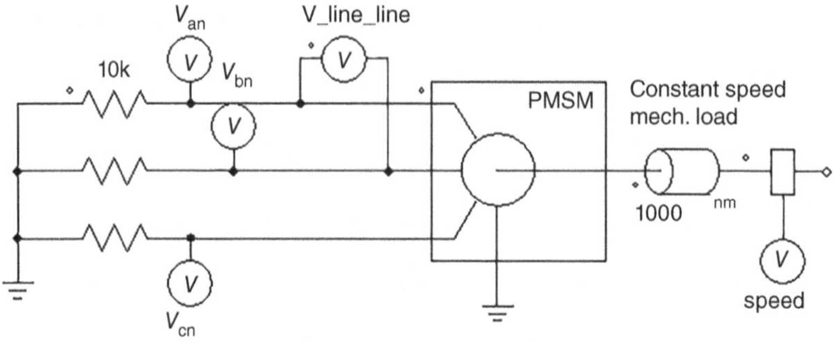

Figure 19.4 PSIM schematic for measuring the back EMF constant.

The parameter V_{pk}/K_{rpm} represents the peak value of the line-to-line back EMF voltage when the motor is rotating at 1000 rpm. This value is typically provided by the motor manufacturer. If it is not available, it can be experimentally obtained using a simple setup where the motor operates as a generator at a constant mechanical speed of 1000 rpm while the peak line-to-line voltage is measured. The default value of V_{pk}/K_{rpm} in the PSIM model is 98.67.

Figure 19.4 shows the PSIM schematic used to derive both the electrical position and the back EMF constant. In this setup:

- A constant mechanical load rotating the motor at 1000 rpm is applied.
- A large resistor (10 kΩ) is connected to the stator to ensure that no significant current flows, so that only the pure back EMF voltage is measured across the motor terminals.
- The phase voltage and line-to-line voltage are both monitored during the simulation.

The corresponding simulation waveforms are shown in Figure 19.5. The peak line-to-line voltage is observed to be approximately 98 V, which matches the default library value of $V_{pk}/K_{\text{rpm}}= 98$. This confirms the accuracy of the simulation.

In practical experiments, the back EMF constant can be measured using the same method, and the obtained value can then be used as the parameter input in the PSIM model.

The back EMF waveform of phase-a follows a sine wave starting from zero, indicating that the initial electrical position θ_e is 0 degrees. Since the motor in this simulation is externally rotated in the reverse direction, the resulting back EMF waveform is a negative sine wave. If, during an actual experiment, the measured waveform instead resembles a cosine wave, this indicates that the initial electrical position is shifted to +90 degrees.

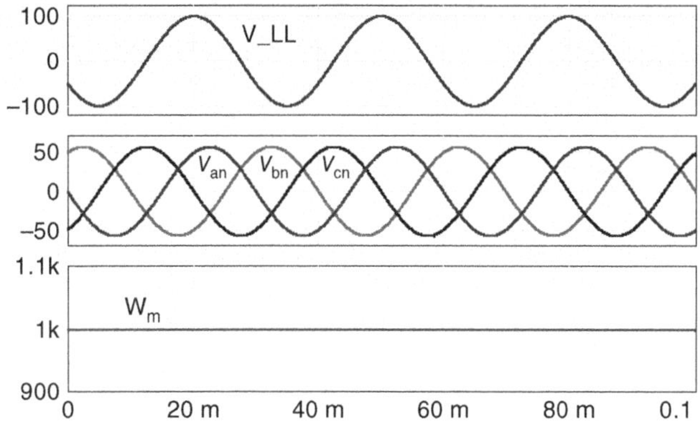

Figure 19.5 Back EMF voltage waveforms for constant load operation.

Calculation of Back EMF Constant λ_m The back EMF constant λ_m can be calculated from the V_{pk}/K_{rpm} value provided by the motor data or measurement. In the phase model, the back EMF is expressed as phase voltage. Since the given V_{pk}/K_{rpm} value is specified for line-to-line voltage, it must first be converted to phase voltage by dividing by $\sqrt{3}$.

The relationship between the back EMF and angular velocity is given by:

$$E_a = \lambda_m \cdot \omega_e$$

To compute the electrical angular velocity $\omega_{e,1000}$ at 1000 rpm, the following formula is used:

$$\omega_{e,1000} = \frac{P}{2} \cdot \frac{2\pi}{60} \times 1000[\text{rpm}] = 209.44 \left[\frac{\text{rad}}{\text{sec}}\right]$$

where:

- $P = 4$ poles (default value in the PSIM library).

The back EMF constant λ_m is then calculated as:

$$\lambda_m = \frac{E_a}{\omega_{e,1000}} = \frac{\frac{V_{pk}}{K_{rpm}}}{\sqrt{3}} \cdot \frac{1}{\omega_{e,1000}} = \frac{98.67}{\sqrt{3}} \cdot \frac{1}{209.44} = 0.272 \quad (19.5)$$

This value of λ_m can then be used in the PMSM *dq* model to achieve accurate simulation and controller design.

When developing a PSIM schematic for PMSM modeling and control, several important considerations must be taken into account to ensure accurate simulation results. One critical aspect is the definition of the electrical position θ_e, which varies depending on whether the back EMF of phase-*a* is synchronized to a cosine wave or a sine wave:

- In PSIM, the electrical position θ_e is synchronized with a *sine* wave.
- In contrast, the PMSM model in MATLAB Simulink aligns the back EMF of phase-*a* with a *cosine* wave.

If this difference is not properly addressed, it can lead to inaccuracies in the abc-to-dq transformation, potentially affecting control performance. Therefore, it is essential to perform a simple validation test to determine the correct electrical position alignment and apply initial position compensation as necessary. Careful attention to this detail ensures consistent behavior across different simulation platforms and prevents phase misalignment in the motor control algorithm.

dq-Transformation in PSIM Library vs. C-Block Implementation When performing the *dq* transformation in PSIM, two implementation methods are available:

- Using the *dq* transformation blocks provided in the PSIM Library.
- Implementing the transformation manually using a C-block.

It is well known that the dq transformation results obtained from the PSIM Library differ from those produced by a C-block-based dq transformation. To achieve the same results as the *dq* transformation implemented in the C-block, it is necessary to multiply the q-axis input or output by −1 when using the PSIM *dq* Library. This adjustment ensures consistency between the two transformation methods.

Figure 19.6 illustrates the PSIM schematic applying both the $abc \rightarrow dq$ and $dq \rightarrow abc$ transformations using the PSIM Library. As shown, applying a sign inversion on the q-axis aligns the transformation results with those of the C-block implementation. The detailed calculation process is explained in the following sections.

PSIM *dq* Libraries and their transformation equations are illustrated in Table 19.2.

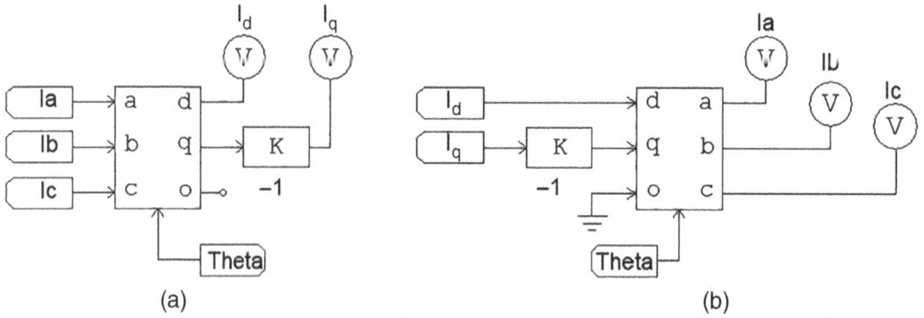

Figure 19.6 Schematic diagram using PSIM *dq* library. (a) abc → dq transformation (b) dq → abc transformation.

Table 19.2 PSIM *dq* libraries and transformation equations.

$abc \rightarrow dq$	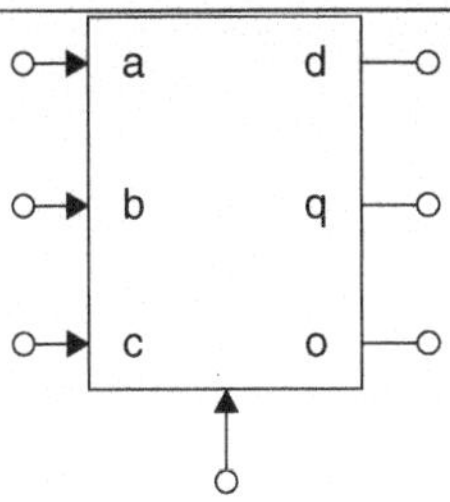	$\begin{bmatrix} v_d \\ v_q \\ v_0 \end{bmatrix} = \frac{2}{3} \begin{bmatrix} \cos(\theta_e) & \cos(\theta_e - 2\pi/3) & \cos(\theta_e + 2\pi/3) \\ \sin(\theta_e) & \sin(\theta_e - 2\pi/3) & \sin(\theta_e + 2\pi/3) \\ 1/2 & 1/2 & 1/2 \end{bmatrix} \begin{bmatrix} v_a \\ v_b \\ v_c \end{bmatrix}$
$dq \rightarrow abc$	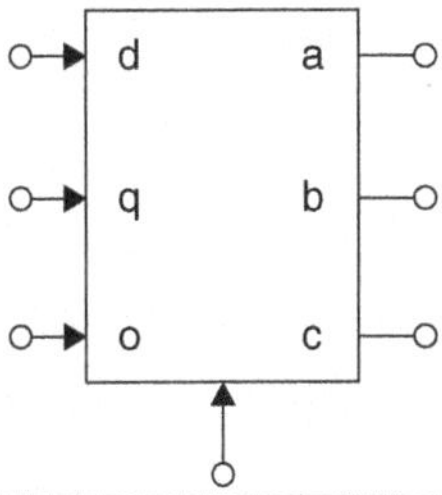	$\begin{bmatrix} v_a \\ v_b \\ v_c \end{bmatrix} = \begin{bmatrix} \cos(\theta_e) & \sin(\theta_e) & 1 \\ \cos(\theta_e - 2\pi/3) & \sin(\theta_e - 2\pi/3) & 1 \\ \cos(\theta_e + 2\pi/3) & \sin(\theta_e + 2\pi/3) & 1 \end{bmatrix} \begin{bmatrix} v_d \\ v_q \\ v_0 \end{bmatrix}$

Comparison with the C-Block-Based* dq *Transformation Let us derive and compare both approaches step-by-step.

Review of Trigonometric Identities:

$$\sin(\alpha \pm \beta) = sin\alpha \cdot cos\beta \pm cos\alpha \cdot sin\beta, \quad \cos(\alpha \pm \beta) = cos\alpha \cdot cos\beta \mp sin\alpha \cdot sin\beta$$

- **PSIM Library *abc* → *dq* Transformation:**

$$V_d = \frac{2}{3}\left\{cos\theta \cdot V_a + \cos\left(\theta - \frac{2}{3}\pi\right) \cdot V_b + \cos\left(\theta + \frac{2}{3}\pi\right) \cdot V_c\right\}$$

Expanding:

$$V_d = \left(\frac{2}{3} \cdot V_a - \frac{1}{3} \cdot V_b - \frac{1}{3} \cdot V_c\right) \cdot \cos\theta + \left(\frac{1}{\sqrt{3}} \cdot V_b - \frac{1}{\sqrt{3}} \cdot V_c\right) \cdot sin\theta$$

Thus:

$$V_d = V_{\text{alpha}} \cdot \cos\theta + V_{\text{beta}} \cdot sin\theta, \quad V_q = V_{\text{alpha}} \cdot \sin\theta - V_{\text{beta}} \cdot cos\theta$$

where

$$V_{\text{alpha}} = \frac{2}{3} \cdot V_a - \frac{1}{3} \cdot V_b - \frac{1}{3} \cdot V_c, \quad V_{\text{beta}} = \frac{1}{\sqrt{3}} \cdot V_b - \frac{1}{\sqrt{3}} \cdot V_c$$

- **C-Block abc → dq Transformation:**

In the C-block, the transformation equations are:

$$V_d = V_{\text{alpha}} \cdot \cos\theta + V_{\text{beta}} \cdot sin\theta, \quad V_q = -V_{\text{alpha}} \cdot \sin\theta + V_{\text{beta}} \cdot cos\theta$$

Therefore, when using the PSIM $abc \rightarrow dq$ Library, the *q-axis value must be multiplied by –1* to match the C-block results.

- **PSIM Library** $\boldsymbol{dq \rightarrow}$ **abc Transformation:**

$$V_a = cos\theta \cdot V_d + sin\theta \cdot V_q = V_{\text{alpha}}$$

$$V_b = -\frac{1}{2} \cdot V_{\text{alpha}} + \frac{\sqrt{3}}{2} \cdot V_{\text{beta}}$$

$$V_c = -\frac{1}{2} \cdot V_{\text{alpha}} - \frac{\sqrt{3}}{2} \cdot V_{\text{beta}}$$

with:

$$V_{\text{alpha}} = cos\theta \cdot V_d + sin\theta \cdot V_q$$

$$V_{\text{beta}} = sin\theta \cdot V_d - cos\theta \cdot V_q$$

- **C-Block** $\boldsymbol{dq \rightarrow abc}$ **Transformation:**

The corresponding equations in the C-block are:

$$V_{\text{alpha}} = cos\theta \cdot V_d - sin\theta \cdot V_q$$

$$V_{\text{beta}} = sin\theta \cdot V_d + cos\theta \cdot V_q$$

Therefore, when using the PSIM $dq \rightarrow abc$ Library, the *q-axis value must be multiplied by –1* to match the C-block results.

a) **Verification of Motor Parameter Values**

The accuracy of the calculated motor parameters can be verified using the PSIM *dq* Library through the following verification procedure:

1) Apply a voltage to the PMSM motor in simulation and measure the steady-state values of rpm, i_d, and i_q from the simulation waveforms.
2) Substitute the measured values into the motor modeling equations to calculate the steady-state values of electromagnetic torque T_e, V_d, and V_q.
3) Compare the calculated values with those obtained from the PSIM circuit to check for consistency.
4) In the steady state, the calculated electromagnetic torque T_e must match the applied load torque T_{load}.

The electrical position θ_e is obtained by integrating the mechanical angular velocity ω_m, measured using the speed sensor, after applying the appropriate conversion factor. In PSIM, the electrical position θ_e is inherently synchronized with the sine function, allowing the angle position to be used directly. However, in actual experiments, if the back EMF waveform appears as a cosine function, an additional phase shift of +90 degrees (equivalent to $\pi/2$ or 1.57) must be applied. This can be implemented by setting the initial value of the integrator to 1.57 instead of zero.

PSIM Schematic for Obtaining Steady-State Values The PSIM circuit used to calculate the electrical position θ_e from the mechanical angular velocity ω_m is shown in Figure 19.7.

To verify the steady-state motor parameters, the schematic shown in Figure 19.8 is used. For the IPMSM electrical model:

- The stator inductances L_a, L_b, L_c vary sinusoidally.
- In the dq reference frame, this results in unequal inductances: $L_d \neq L_q$

In the simulation:

- The PMSM model uses default parameter values: $L_d = 0.027$ and $L_d = 0.067$
- The applied motor terminal voltages are $V_q = 200[V]$ and $V_d = 0[V]$.
- A disturbance torque of $T_L = 15[Nm]$ is applied.

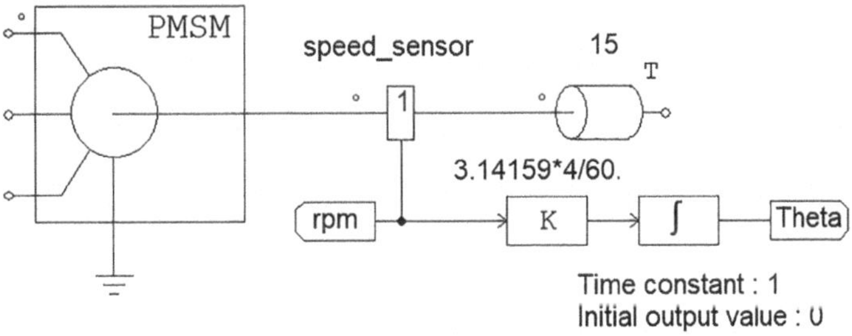

Figure 19.7 Conversion diagram from rpm to electrical angle θ_e.

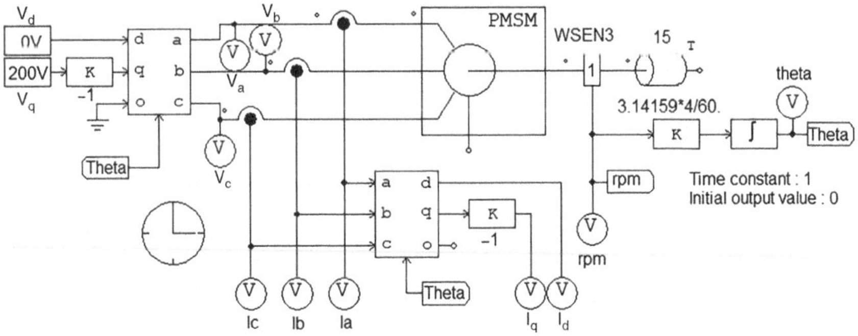

Figure 19.8 Schematic diagram for steady-state calculation.

- The speed sensor (WSEN) measures rpm, which is then converted into the electrical position θ_e.
- The dq transformation is used to obtain the motor currents I_d and I_q.

Using the measured values, the electromagnetic torque T_e can be calculated. If the motor parameters are correct, the calculated T_e should equal the applied torque T_L (15 Nm). Similarly, by substituting the steady-state values into the motor equations, the calculated values of V_q and V_d should match the applied values of $V_q = 200\ V$ and $V_d = 0\ V$. If both sets of values agree, this confirms the accuracy of the motor parameters and model constants.

Figure 19.9 displays the simulation results along with the m-file used to calculate the steady-state values.

From the simulation waveforms, the steady-state values are rpm=27.2, $I_q = 46$ [A], $I_d = 4.08$ [A]. Substituting these values into the motor state equations, the calculated results are: $T_e = 15.01\ [Nm]$, $V_q = 199.97\ [V]$, $V_d = -0.01[V]$. Comparing these calculated values with the schematic input values, they match exactly, confirming the accuracy of the modeling.

Next, the conditions were modified for a surface-mounted permanent magnet synchronous motor (SPMSM) model by setting $L_d = L_q = 0.027[H]$ and changing the disturbance torque $T_L = 10[Nm]$. The simulation results are shown in Figure 19.10. The steady-state values from the simulation waveform are:

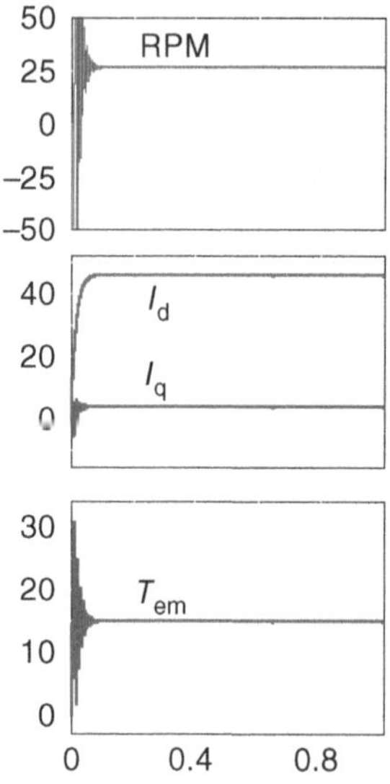


```
%% PMSM Steady-State Calculation %%
% Motor and Circuit Parameter
Rs = 4.3;                   % Stator Resistance
Ld = 0.027;                 % d-axis Inductance
Lq = 0.067;                 % q-axis Inductance
Vpk_krpm = 98.67;           % Back-emf constant
P=4;                        % Number of Poles
Jm = 0.00179;               % Moment of Inertia
Tau_shaft = 10;             % Shaft time constant
Bm = Jm/Tau_shaft;          % Friction Constant
TL = 15;                    % Load torque [Nm]
% Back-EMF Constant Calculation
We_1000 = (P/2)*(2*pi/60)*1000;
Lamda_m=Vpk_krpm/sqrt(3)/We_1000; %Lamda_m=0.272;
% Steady-State data from PSIM Simulation Waveform
rpm=27.2;
Iq = 46;
Id =4.08;
% Unit Conversion
Wm = (2*pi/60)*rpm; % 0.1047 * rpm;
We= (P/2)*Wm;
% Step 1: Electric Torque Equation: Te
Te = (3/2)*(P/2)*( (Lamda_m)*Iq + (Ld-Lq)*Iq*Id);
% Step 2: Vd, Vq Calculation
Vd = Rs*Id - We*Lq*Iq;
Vq = Rs*Iq + We*Ld*Id + We*Lamda_m;
```

Figure 19.9 Simulation waveforms and steady-state calculation m-file.

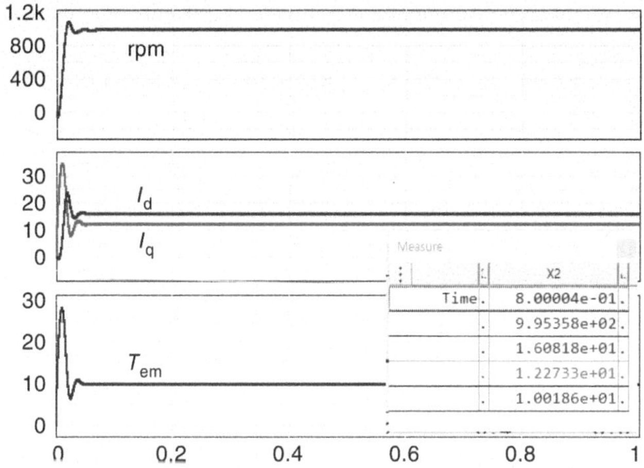

Figure 19.10 Waveforms of the surface-mounted PMSM.

$$\text{rpm}=994,\ I_d = 16.1\ [\text{A}],\quad I_q = 12.28\ [\text{A}],\quad T_{em\_PMSM} = 10[Nm].$$

These simulation results match the applied inputs, confirming that the back EMF constant, unit conversions, and state-space equations have been correctly implemented.

19.2 State-Space Averaged Equation and Controller Design

Figure 19.11 presents the schematic of the PMSM drive circuit. The power stage consists of a three-phase inverter composed of three bridge legs. A speed sensor is

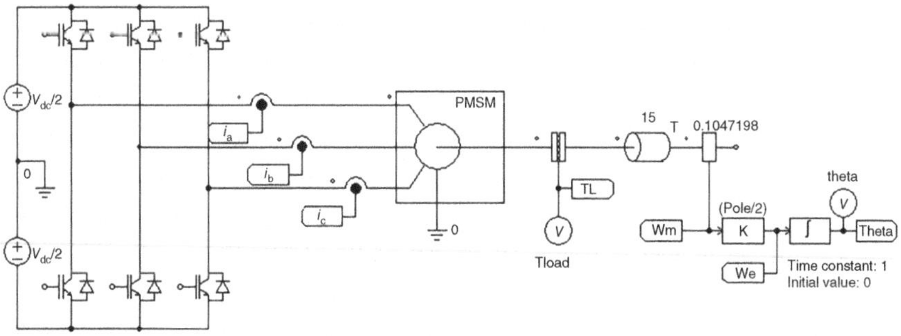

Figure 19.11 PMSM driving circuit schematic.

connected to measure the motor's angular velocity, and a torque sensor is used to measure the external load torque.

Since the speed sensor provides the output in rpm, a conversion factor of 0.1047 is applied to obtain the mechanical angular velocity ω_m in rad/s. The electrical angular velocity ω_e is then calculated and integrated to obtain the electrical position angle θ_e. Using the same approach as for the three-phase inverter, the state-space averaged equations for the PMSM can be derived.

The dynamic equations for the dq-axes of the PMSM are given by:

$$V_q = Ri_q + L_q\frac{di_q}{dt} + \omega_e L_d \cdot i_d + \omega_e \lambda_m$$

$$V_d = Ri_d + L_d\frac{di_d}{dt} - \omega_e L_q \cdot i_q$$

Applying the state-space averaging method, as used in the three-phase inverter, the following derivations are made:

a) ***q*-Axis State-Space Averaged Expression**

The inverter phase voltage V_q behaves as follows:

- Switch ON (duration d_q): $+\frac{V_{\mathrm{dc}}}{2}$
- Switch OFF (duration $1 - d_q$): $-\frac{V_{\mathrm{dc}}}{2}$

Thus, the averaged voltage equation becomes:

$$+\left(\frac{V_{\mathrm{dc}}}{2}\right)(d_q) - \left(\frac{V_{\mathrm{dc}}}{2}\right)(1 - d_q) = Ri_q + L_q\frac{di_q}{dt} + \omega_e L_d \cdot i_d + \omega_e \lambda_m$$

Simplifying:

$$V_{\mathrm{dc}} \cdot d_q - \frac{V_{\mathrm{dc}}}{2} = Ri_q + L_q\frac{di_q}{dt} + \omega_e L_d \cdot i_d + \omega_e \lambda_m$$

To eliminate the offset voltage $-\frac{V_{\mathrm{dc}}}{2}$, an additional offset of +0.5 is added to the q-axis duty cycle d_q. The duty cycle can be decomposed into:

$$d_q = d_{q\mathit{ff}} + d_{q\mathit{fb}}$$

Feed-forward duty is expressed as follows.

$$d_{q\mathit{ff}} = \frac{\omega_e L_d i_d}{V_{\mathrm{dc}}}$$

The resulting q-axis state-space averaged equation is:

$$\frac{di_q}{dt} = -\frac{R}{L_q}i_q - \frac{\lambda_m}{L_q}\left(\frac{P}{2}\right)\omega_m + \frac{V_{dc}}{L_q}d_{q\mathit{fb}} \tag{19.6}$$

b) ***d*-Axis State-Space Averaged Expression**

Using the same approach for the *d*-axis:

$$V_{dc} \cdot d_d - \frac{V_{dc}}{2} = Ri_d + L_d \frac{di_d}{dt} - \omega_e L_q \cdot i_q$$

The duty cycle for the *d*-axis is similarly split:

$$d_d = d_{d,ff} + d_{d,fb}$$

with:

$$d_{d,ff} = -\frac{\omega_e L_q \cdot i_q}{V_{dc}}$$

Thus, the *d*-axis state-space averaged equation is:

$$\frac{di_d}{dt} = -\frac{R}{L_d} i_d + \frac{V_{dc}}{L_d} d_{d,fb} \tag{19.7}$$

c) **Mechanical Variable State-Space Averaged Equation**

The mechanical dynamics are independent of the inverter duty cycles and are described by:

$$J_m \frac{d\omega_m}{dt} + B_m \omega_m = T_e - T_{load}$$

which gives:

$$\frac{d\omega_m}{dt} = \left(\frac{3}{2}\right)\left(\frac{P}{2}\right)\frac{\lambda_m}{J_m} i_q - \frac{B_m}{J_m}\omega_m - \frac{T_{load}}{J_m} \tag{19.8}$$

Thus, the state-space averaged equation for the PMSM is given as follows.

$$\begin{pmatrix} \dfrac{di_q}{dt} \\ \dfrac{d\omega_m}{dt} \end{pmatrix} = \begin{pmatrix} -\dfrac{R}{L_q} & -\dfrac{\lambda_m}{L_q}\left(\dfrac{P}{2}\right) \\ \left(\dfrac{3}{2}\right)\left(\dfrac{P}{2}\right)\dfrac{\lambda_m}{J_m} & -\dfrac{B_m}{J_m} \end{pmatrix} \begin{pmatrix} i_q \\ \omega_m \end{pmatrix} + \begin{pmatrix} \dfrac{V_{dc}}{L_q} \\ 0 \end{pmatrix} d_{q,fb} + \begin{pmatrix} 0 \\ -\dfrac{T_{load}}{J_m} \end{pmatrix}$$

$$\frac{di_d}{dt} = -\frac{R}{L_d} i_d + \frac{V_{dc}}{L_d} d_{d,fb} \tag{19.9}$$

For reference, the state-space equation of a DC motor is presented below for comparison with the PMSM model:

$$\begin{pmatrix} \dfrac{di_a}{dt} \\ \dfrac{d\omega_m}{dt} \end{pmatrix} = \begin{pmatrix} -\dfrac{R_a}{L_a} & -\dfrac{K_\varphi}{L_a} \\ \dfrac{K_\varphi}{J_m} & -\dfrac{B_m}{J_m} \end{pmatrix} \begin{pmatrix} i_a \\ \omega_m \end{pmatrix} + \begin{pmatrix} \dfrac{V_t}{L_a} \\ 0 \end{pmatrix} d + \begin{pmatrix} 0 \\ -\dfrac{T_{load}}{J_m} \end{pmatrix} \tag{10.13}$$

The difference between the DC motor and PMSM models lies solely in the numerical values of the system matrix A, while the overall state-space structure remains identical. Therefore, the same methodology used to derive the transfer function for the DC motor can be directly applied to the PMSM model.

Transfer Function Derivation for PMSM The inverse of the system matrix is computed as follows.

$$(sI - A)^{-1} = \begin{pmatrix} s + \frac{R}{L_q} & \frac{\lambda_m}{L_q}\left(\frac{P}{2}\right) \\ -\left(\frac{3}{2}\right)\left(\frac{P}{2}\right)\frac{\lambda_m}{J_m} & s + \frac{B_m}{J_m} \end{pmatrix}^{-1}$$

$$= \frac{\begin{pmatrix} s + \frac{B_m}{J_m} & -\frac{\lambda_m}{L_q}\left(\frac{P}{2}\right) \\ \left(\frac{3}{2}\right)\left(\frac{P}{2}\right)\frac{\lambda_m}{J_m} & s + \frac{R}{L_q} \end{pmatrix}}{s^2 + \left(\frac{R}{L_q} + \frac{B_m}{J_m}\right)s + \left(\frac{RB_m}{L_qJ_m} + \left(\frac{3}{2}\right)\left(\frac{P}{2}\right)^2\frac{\lambda_m^2}{L_qJ_m}\right)}$$

The transfer function is then expressed as follows. For convenience, the feedback duty cycles of each phase are denoted as $d_{d,fb} = d_d$ and $d_{q,fb} = d_q$.

1) q-axis current to duty cycle transfer function:

$$G_{1q}(s) = \frac{I_q(s)}{d_q(s)} = \frac{\frac{V_{dc}}{L_q}\left(s + \frac{B_m}{J_m}\right)}{s^2 + \left(\frac{R}{L_q} + \frac{B_m}{J_m}\right)s + \left(\frac{RB_m}{L_qJ_m} + \left(\frac{3}{2}\right)\left(\frac{P}{2}\right)^2\frac{\lambda_m^2}{L_qJ_m}\right)} \tag{19.10}$$

2) Mechanical speed to duty cycle transfer function:

$$G_{2q}(s) = \frac{\omega_m(s)}{d_q(s)} = \frac{\left(\frac{3}{2}\right)\left(\frac{P}{2}\right)\frac{\lambda_m}{J_m}\frac{V_{dc}}{L_q}}{s^2 + \left(\frac{R}{L_q} + \frac{B_m}{J_m}\right)s + \left(\frac{RB_m}{L_qJ_m} + \left(\frac{3}{2}\right)\left(\frac{P}{2}\right)^2\frac{\lambda_m^2}{L_qJ_m}\right)} \tag{19.11}$$

3) Mechanical speed to q-axis current transfer function:

$$G_{3q}(s) = \frac{\omega_m(s)}{I_q(s)} = \frac{\left(\frac{3}{2}\right)\left(\frac{P}{2}\right)\frac{\lambda_m}{J_m}}{\left(s + \frac{B_m}{J_m}\right)} \tag{19.12}$$

4) d-axis current to duty cycle transfer function:

$$G_{1d}(s) = \frac{I_d(s)}{d_d(s)} = \frac{V_{dc}}{L_d s + R} \tag{19.13}$$

19.2.1 Controller Design

a) Current Controller Design

The duty cycles for each phase of current controller are given as follows:

$$d_q = d_{q,ff} + d_{q,fb} = \frac{\omega_e L_q \cdot i_q}{V_{dc}} + d_{q,fb}$$

$$d_d = d_{d,ff} + d_{d,fb} = -\frac{\omega_e L_q \cdot i_q}{V_{dc}} + d_{d,fb}$$

The terms $d_{q,fb}$ and $d_{d,fb}$ represent the controller outputs responsible for regulating the q-axis and d-axis currents, respectively.

The current control block diagram based on these transfer functions is shown in Figure 19.12.

b) Double-Loop Controller Design

The double-loop controller structure is shown in Figure 19.13 (a). The structure consists of an outer speed controller and an inner q-axis current controller, which is almost identical to the structure used for a DC motor. The d-axis current control follows a single-loop configuration, as shown in Figure 19.13 (b), where the reference current is typically set to zero.

MATLAB m-file for current and double loop controller is presented in Table 19.3.

The frequency design specifications for the speed controller and current controller are summarized in Table 19.4.

After auto-tuning with SISOTOOL, the following controller transfer functions were obtained:

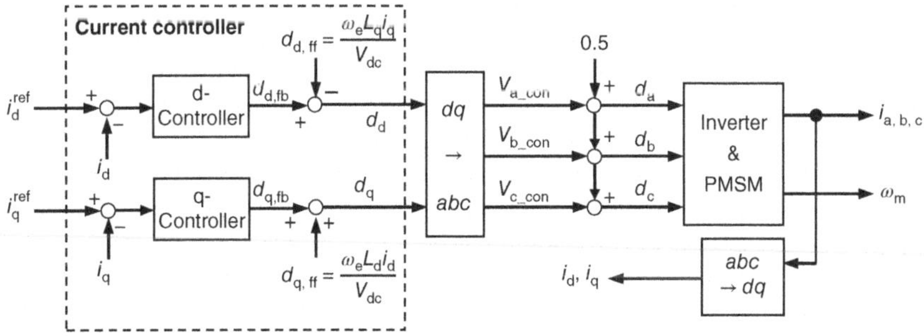

Figure 19.12 Current controller configuration (single-loop).

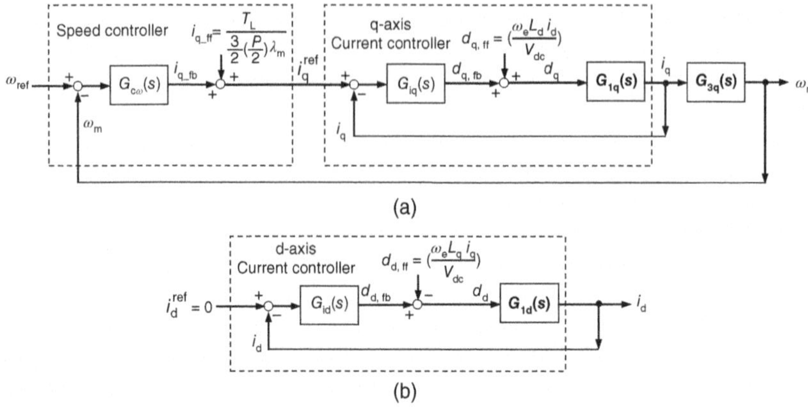

Figure 19.13 The double-loop controller configuration. (a) Double-loop controller (q-axis) (b) Single-loop current controller (d-axis).

Table 19.3 MATLAB m-file for current and double loop controllers.

```
s=tf('s');
%%% PMSM Motor Parameter%%%
Vdc=500;
Vq=Vdc/2;
R = 4.3;
Ld = 0.027;
Lq = 0.027;
Vpk_krpm =98,67;
P=4;
Jm = 0.00179;
Tau_shaft=10;

%%%% Motor Constant Calculation %%%%
We_1000 = (P/2)*(2*pi/60)*1000;
Lamda_m= Vpk_krpm / (sqrt(3)*We_1000);
Bm=Jm/Tau_shaft;
TL = 5; % Load torque TL = 5 [Nm]

G1d= Vdc/(Ld*s +R);

Delta =s^2 + (R/Lq+Bm/Jm)*s + (R*Bm +
   (3/2)*(P/2)^2*Lamda_m^2 )/(Lq*Jm);

G1q=(Vdc/Lq)*(s+Bm/Jm)/Delta; % Ia/d
G2q=(3/2)*(P/2)*Lamda_m*Vdc/(Lq*Jm)/Delta;
G3q = (3/2)*(P/2)*Lamda_m/Jm/(s+Bm/Jm);
```

```
Kpw=1; Kiw=1;
Gcw=Kpw*(s+Kiw)/s;
Kpi=1; Kii=1;
Gci=Kpi*(s+Kii)/s;

Wref = 100;

%%%% Double loop %%%%
% Speed:Wc=10 P.M=90[deg]
% Current:Wc=2400, P.M.=60
T=sisoinit(6);
T.F.Value = Wref;
T.G1.Value = G1q;
T.C2.Value = Gci;
T.G2.Value = G3q;
T.C1.Value = Gcw;
sisotool(T);

% Single loop:Current-id %
% Current:Wc=2400, P.M.=60
% Iref= 10;
% T=sisoinit(1);
% T.F.Value = Iref;
% T.G.Value = G1d;
% T.C.Value = Gci;
% sisotool(T);
```

Table 19.4 Design specifications for double loop controller.

	Bandwidth	*Phase margin*
Speed Controller (C_1)	*10 rad/s*	*90 degree*
Current Controller (C_2)	*2400 rad/s*	*60 degree*

- Speed Controller:

$$C_1 = G_{cw}(s) = \frac{0.0221(s + 0.175)}{s}$$

- Current Controller:

$$C_2 = G_{cl}(s) = \frac{0.108(s + 1500)}{s}$$

For the d-axis, the same current controller gains as the q-axis are applied.

Using a switching frequency of 5 kHz, the discrete-time controller expressions are:

- Speed Controller:

$$C_1 = G_{cw}(z) = \frac{0.0221(z - 0.99996)}{z - 1}$$

- Current Controller:

$$C_2 = G_{ci}(z) = \frac{0.13164(z - 0.838)}{z - 1}$$

The Bode plots of the designed speed and q-axis current controllers using SISO-TOOL are shown in Figure 19.14. The target phase margin for the speed controller was 90°, with the design achieving 89.5°, resulting in a minor phase margin deviation of 0.5°, which causes a slight overshoot. The current controller meets its design specifications.

The step response of the speed control loop is presented in Figure 19.15. The response shows a rise time of approximately 0.22 seconds, indicating satisfactory control performance.

The simulation schematic for the double-loop controller implemented using a C-block is shown in Figure 19.16. This configuration integrates both the speed and current controllers into a single control structure using digital logic.

The simulation results are illustrated in Figures 19.17 and 19.18. The waveform analysis confirms the correct operation of the controller:

- Motor speed accurately tracks the reference value without overshoot, demonstrating a well-tuned speed control loop.

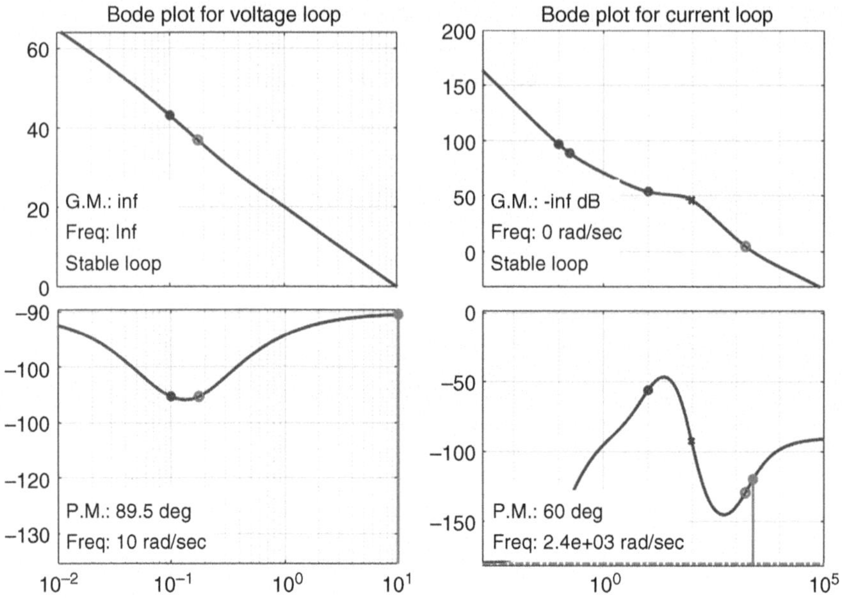

Figure 19.14 Bode plot for double-loop controller.

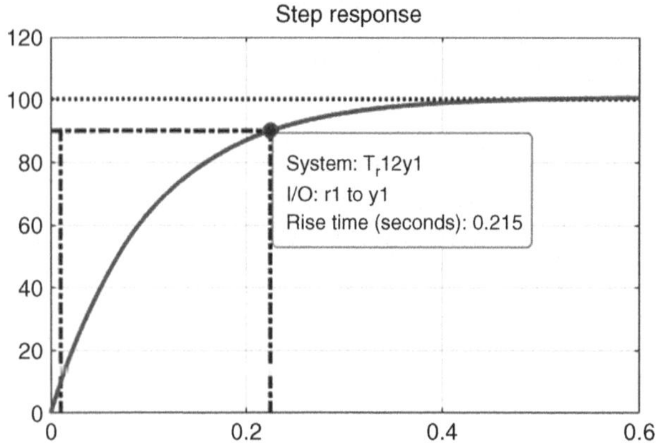

Figure 19.15 Step response of speed in double-loop controller.

- The current controller operates as expected, maintaining stability and responsiveness.
- The second waveform displays the q-axis current error, providing insight into controller performance in real time.

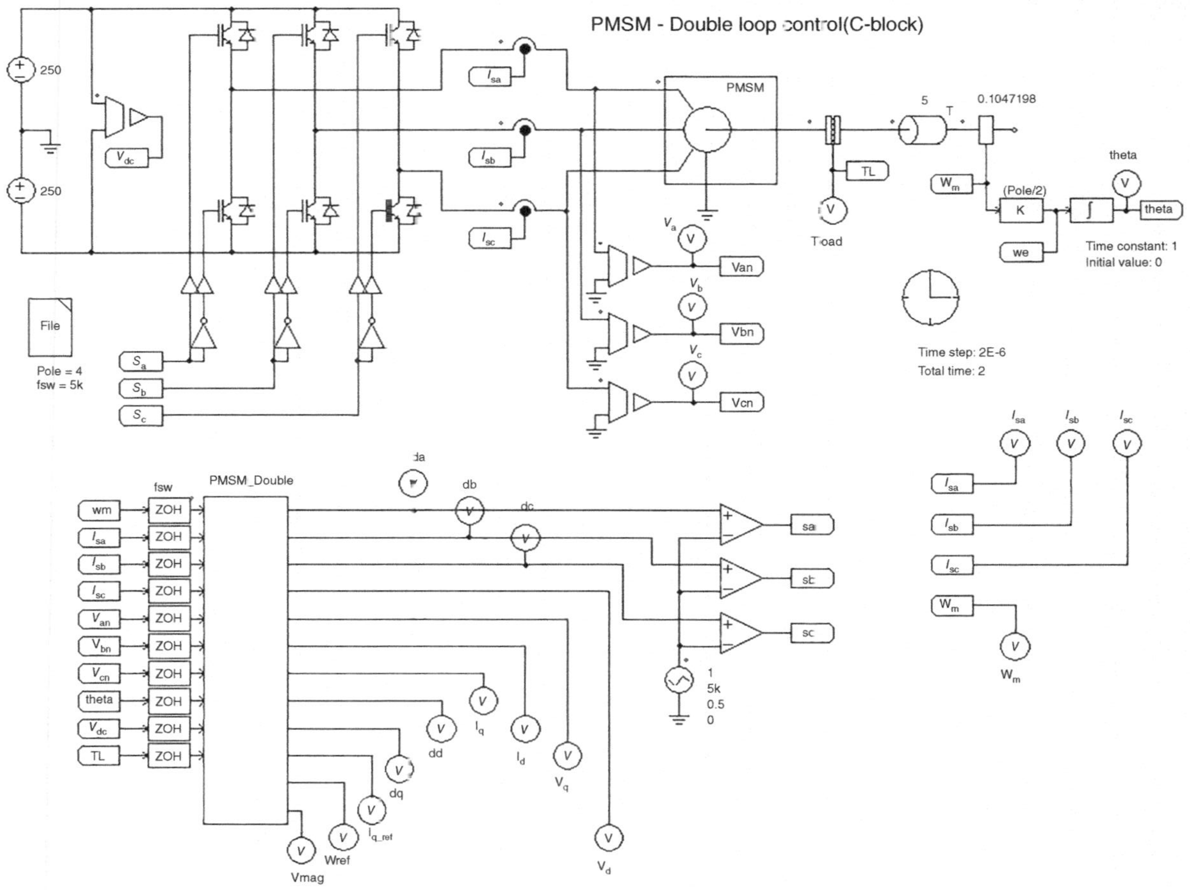

Figure 19.16 Simulation schematic for the double-loop controller.

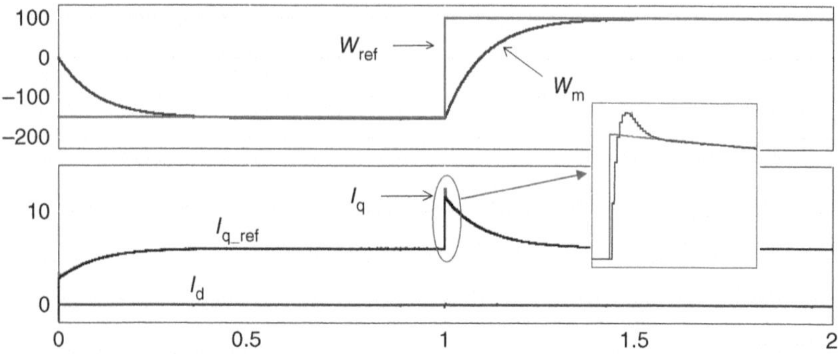

Figure 19.17 Waveforms of the double-loop controller.

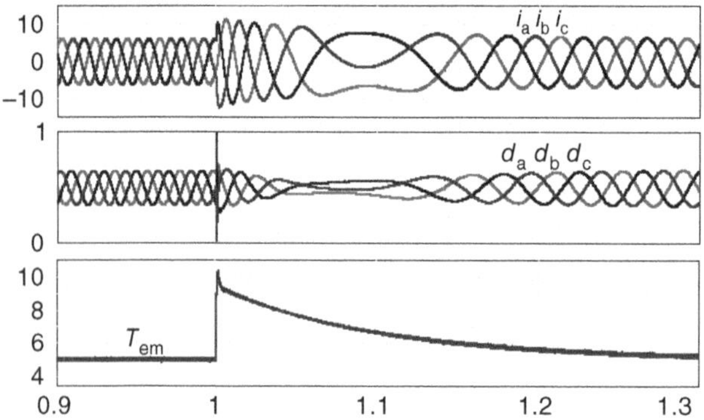

Figure 19.18 Waveforms of the double-loop controller (continued).

- The fourth waveform shows the duty cycle commands applied to each phase of the inverter.
- The fifth waveform represents the motor torque, confirming that the current is properly regulated to generate the required torque.

The C-block code used to implement this control logic is presented in Table 19.5.

Figure 19.19 shows the PSIM schematic of the double-loop controller implemented using analog circuits. As the power circuit remains unchanged, only the analog controller section is presented for clarity.

The corresponding simulation waveforms are displayed in Figure 19.20. It can be observed that the results are identical to those obtained with the digital controller, demonstrating that both the analog and digital implementations achieve equivalent control performance.

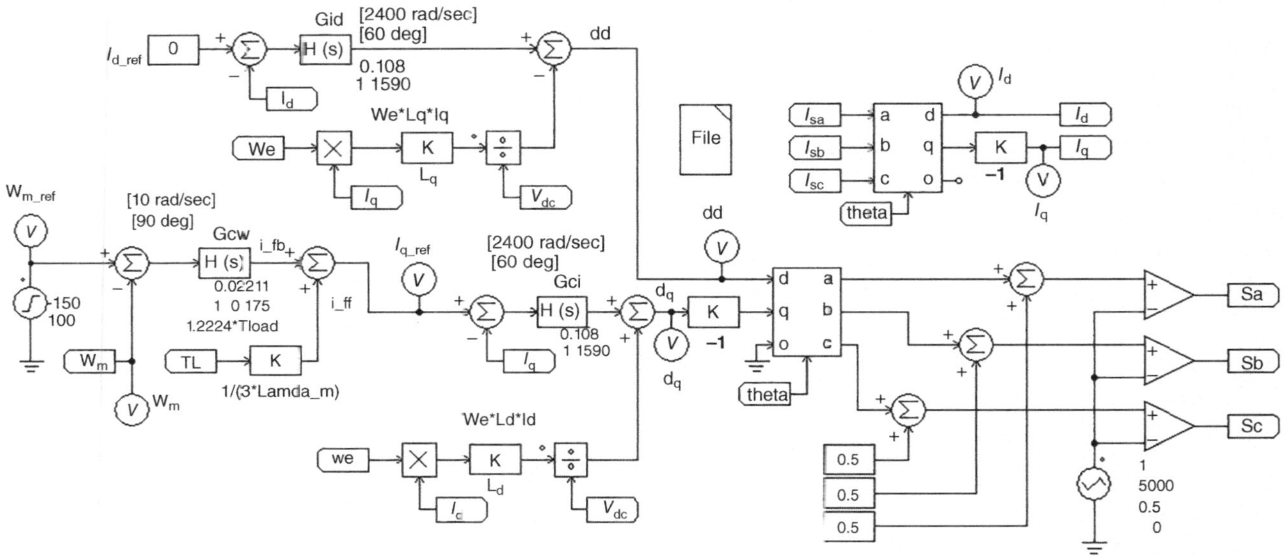

Figure 19.19 Double-loop controller implemented with analog circuits.

Table 19.5 C-block code: PMSM_ Double.

```
#define Ld       0.027 //d-axis Inductance
#define Lq       0.027 //q-axis Inductance
#define Lamda_m 0.273 //Back-emf constant
#define Pole 4         //4-pole motor

#define KWp 0.02218  //Speed P: BW: 10[rad]
#define KWi 0.99996  //Speed I: 90 [deg]

#define Kip_d 0.1317 // d-axis Current P
#define Kii_d 0.837  // d-axis CurrentrI

#define Kip_q 0.1317 // q-axis Current P
#define Kii_q 0.837  // q-axis CurrentrI

double Wm, Isa, Isb, Isc, Van, Vbn, Vcn,
theta, Vdc, TL;
double Valpha, Vbeta, Ialpha, Ibeta, Vd, Vq,
Id, Iq, Id_ref, Iq_ref;
double We, Wm_ref, Wm_err, Iq_fb, Iq_ff;
double Id_err, Iq_err, dd, dq, dfb_d, dfb_q;
double Va_con, Vb_con, Vc_con, duty_a,
duty_b, duty_c, Vmag;

static double Wm_err_old=0.0, Iq_fb_old=0.0,
Id_err_old=0.0, Iq_err_old=0.0;
static double dd_old=0.0, dq_old=0.0;

/**** Input Variable ****/
Wm = in[0];
Isa = in[1];
Isb = in[2];
Isc = in[3];
Van = in[4];
Vbn = in[5];
Vcn = in[6];
theta = in[7];
Vdc = in[8];
TL = in[9];

/**** Speed Controller ****/
Wm_err = Wm_ref - Wm;
Iq_fb = Iq_fb_old + KWp*Wm_err-
        KWp*KWi*Wm_err_old;
Iq_fb_old = Iq_fb;
Wm_err_old = Wm_err;

Iq_ff = TL/(3.* Lamda_m);
Iq_ref = Iq_fb + Iq_ff;

/*** Current controller dq ***/
Id_ref = 0.0;
Id_err = Id_ref - Id;
dd_fb = dd_fb_old + Kip_d*Id_err
         -Kip_d*Kii_d*Id_err_old;
dd_fb_old = dd_fb;
Id_err_old = Id_err;

Iq_err = Iq_ref - Iq;
dq_fb = dq_fb_old +Kip_q*Iq_err
        -Kip_q*Kii_q*Iq_err_old;
dq_fb_old = dq_fb;
Iq_err_old = Iq_err;

if (Vdc < 0.1) Vdc = 0.1;
dd = dd_fb - (We*Lq*Iq) / Vdc ;
dq = dq_fb + (We*Ld*Id) / Vdc;

Vmag=Vdc*sqrt(dd*dd+dq*dq);
Valpha=dd*cos(theta)-dq*sin(theta);
Vbeta =dd*sin(theta)+dq*cos(theta);

Va_con= Valpha;
Vb_con=-0.5*Valpha+sqrt(3)*Vbeta/2;
Vc_con=-0.5*Valpha-sqrt(3)*Vbeta/2;

// duty = dff + dfb = 0.5 + dfb
duty_a = 0.5 + Va_con;
duty_b = 0.5 + Vb_con;
```

Table 19.5 (Continued)

```
if (t < 1.0) Wm_ref = -150;
else Wm_ref = 100;

Ialpha =(2./3.)*Isa-(1./3.)*Isb-(1./3.)*Isc;
Ibeta = (Isb-Isc)/sqrt(3);
Id = Ialpha*cos(theta) + Ibeta*sin(theta);
Iq = -Ialpha*sin(theta) + Ibeta*cos(theta);

Valpha =(2./3.)*Van-(1./3.)*Vbn-(1./3.)*Vcn;
Vbeta = (Vbn-Vcn)/sqrt(3);
Vd = Valpha*cos(theta) + Vbeta*sin(theta);
Vq = -Valpha*sin(theta) + Vbeta*cos(theta);

We = Wm*(Pole/2.);
```

```
duty_c = 0.5 + Vc_con;

out[0] = duty_a;    // Inverter R
out[1] = duty_b;    // Inverter S
out[2] = duty_c;    // Inverter T
out[3] = Vd;
out[4] = Vq;
out[5] = Id;
out[6] = Iq;
out[7] = dd;
out[8] = dq;
out[9] = Iq_ref;
out[10]= Wm_ref;
out[11]= Vmag;
```

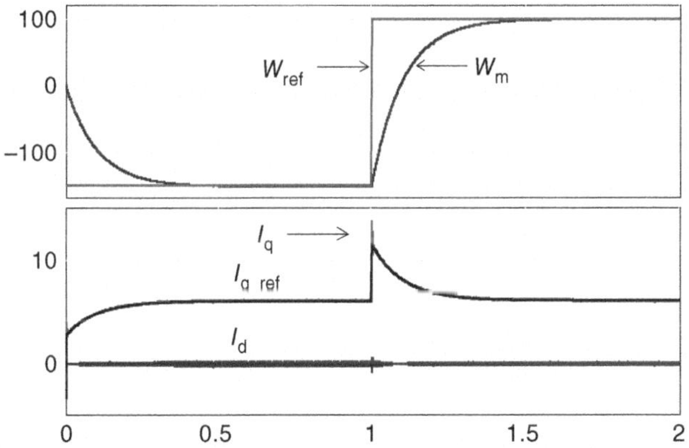

Figure 19.20 Simulation waveforms of the analog double-loop controller.

19.3 MTPA

Permanent magnet synchronous motors (PMSMs) can be classified into two types based on the placement of their magnets:

- Surface-mounted PMSM (SPMSM)
- Interior-mounted PMSM (IPMSM)

The key distinction between these two configurations lies in their inductance characteristics. In SPMSMs, the direct-axis and quadrature-axis inductances are equal, i.e., $L_d = L_q$. In contrast, IPMSMs exhibit unequal inductances, specifically $L_d < L_q$.

The general torque equation for a PMSM is expressed as:

$$T_e = \left(\frac{3}{2}\right)\left(\frac{P}{2}\right)[\lambda_m i_q + (L_d - L_q)i_q i_d]$$

In this equation:

T_e: Electromagnetic torque
P: Number of poles
λ_m: Permanent magnet flux linkage
$i_{q,}$ i_d: q-axis and d-axis current components
L_d, L_q: q-axis and d-axis inductances

For SPMSM, where $L_d = L_q$, the second term vanishes, and torque is generated solely by the q-axis current. However, in an IPMSM, the torque comprises two components:

- permanent magnet torque
- reluctance torque due to the inductance difference ($L_d < L_q$)

By setting the y-axis as i_q and the x-axis as i_d, and rearranging the torque equation for a given constant torque T_{e1}, the expression becomes:

$$i_q = \frac{T_{e1}}{i_d + k}$$

where

$$k = \frac{\lambda_m}{(L_d - L_q)}, \quad T_{e1} = \frac{T_e\left(\frac{2}{P}\right)\left(\frac{2}{3}\right)}{(L_d - L_q)}$$

This equation reveals a hyperbolic relationship between i_d and i_q for a given torque T_e. That is, multiple combinations of i_d and i_q satisfy the same torque constraint, which forms a family of MTPA curves on the i_d–i_q plane.

Using this relationship, MTPA trajectories were plotted by varying i_d from −20 A to +2 A. The results are illustrated in Figure 19.21 (a). Each curve corresponds to a different torque level (e.g., 10 Nm, 20 Nm, 30 Nm). The curves demonstrate that for higher torque, a greater magnitude of i_q is required, and the optimal i_d shifts accordingly.

The topmost curve corresponds to a torque output of 30 Nm, followed by curves for 20 Nm and 10 Nm, respectively. These curves demonstrate that a constant torque can be achieved through various i_q–i_d combinations, with the shape and position of each curve varying according to the target torque level.

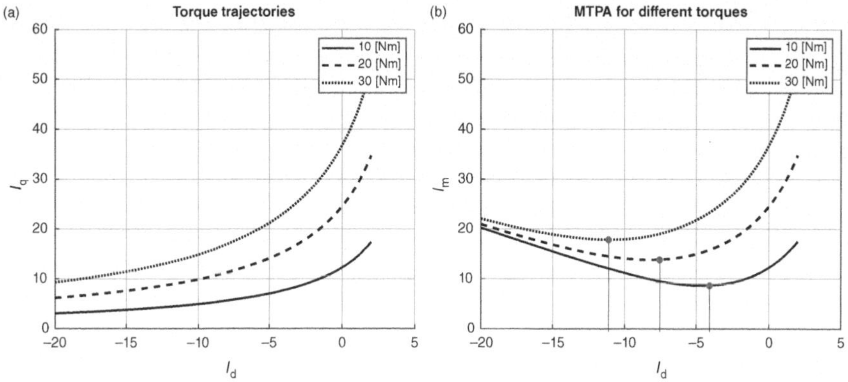

Figure 19.21 Torque and MTPA trajectories. (a) Various torque trajectories. (b) MTPA for various torques.

Let the magnitude of the stator current be

$$I_m = \sqrt{i_d^2 + i_q^2}$$

By computing I_m for each combination, we can evaluate the current required to generate the target torque. The graph in Figure 19.21 (b) illustrates the relationship between i_d and I_m for various torque levels. From the graph, it is evident that there exists a unique i_d value that minimizes the stator current I_m for each torque level. It can be observed that the MTPA i_d values for 10, 20, and 30 Nm are −4.6 A, −8.2 A, and −11.0 A, respectively.

This means that for each torque curve, there exists a specific combination of q-axis and d-axis currents that allows torque generation with the minimum stator current ($I_{m,\min}$). This point is called maximum torque per ampere (MTPA) and controlling the motor using such a *q*-axis and *d*-axis current combination is referred to as MTPA control.

So far, we have explained the concept of determining the MTPA point using an offline method based on torque curves for *q*-axis and *d*-axis currents. In the next chapter, we will discuss the design method for a controller that enables MTPA control.

MTPA Controller Design Figure 19.22 illustrates the torque trajectories of surface permanent magnet synchronous motors (SPMSM) and IPMSMs. In SPMSM, where $L_d = L_q$, torque is generated exclusively through the i_q component. In contrast, IPMSMs, characterized by $L_d < L_q$, produce torque through a combination of both i_d and i_q currents.

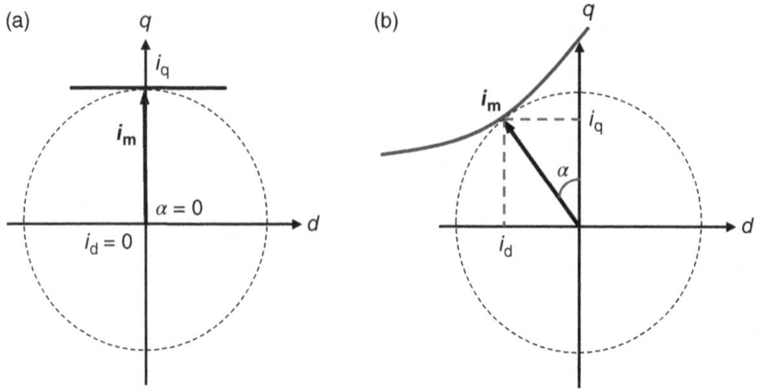

Figure 19.22 Torque trajectory. (a) SPMSM. (b) IPMSM.

The MTPA control strategy aims to generate the desired torque with the minimum possible stator current magnitude I_m. Define the current angle α such that:

$$i_d = I_m sin\alpha, \ \ i_q = I_m cos\alpha$$

The torque equation becomes:

$$T_e = \left(\frac{3}{2}\right)\left(\frac{P}{2}\right)\left[\lambda_m I_m cos\alpha + (L_d - L_q)I_m^2 sin\alpha cos\alpha\right] \tag{19.14}$$

Using the identity $sin2\alpha = 2sin\alpha \cdot cos\alpha$, this can be rewritten as:

$$T_e = \left(\frac{3}{2}\right)\left(\frac{P}{2}\right)\left[\lambda_m I_m cos\alpha + \frac{(L_d - L_q)}{2} I_m^2 sin2\alpha\right]$$

To maximize torque, take the derivative with respect to α and set it to zero:

$$\frac{\partial T_e}{\partial \alpha} = \left(\frac{3}{2}\right)\left(\frac{P}{2}\right)\left[-\lambda_m I_m sin\alpha + (L_d - L_q)I_m^2 cos2\alpha\right] = 0 \tag{19.15}$$

Substitute the trigonometric identity $cos2\alpha = 1 - 2sin^2\alpha$ and define $sin\alpha = x$. The equation becomes:

$$-\lambda_m \cdot x + (L_d - L_q)I_m \cdot (1 - 2x^2) = 0$$

Solving for x, we obtain a quadratic equation. The α value satisfying this equation corresponds to the stator current components i_d and i_q at the MTPA point. By applying trigonometric identities, particularly:

$$2(L_d - L_q)I_m \cdot x^2 + \lambda_m \cdot x - (L_d - L_q)I_m = 0$$

Solving for x gives:

$$x = \frac{-\lambda_m + \sqrt{\lambda_m^2 + 8 \cdot I_m^2 \cdot (L_d - L_q)^2}}{4(L_d - L_q)I_m} \tag{19.16}$$

$$\Rightarrow \quad \alpha = sin^{-1}(x)$$

Using the derived angle α, the optimal i_d and i_q currents components corresponding to MTPA can be computed. Figure 19.23 illustrates the characteristics of MTPA control based on the IPMSM library parameters:

a) MTPA point vs. T_e: This plot shows the trajectory of the MTPA operating point as the torque level varies.
b) (b) Angle α vs. I_m: This plot depicts the variation of the current angle with respect to the stator current magnitude.

These plots are fundamental for understanding the optimal current control strategy in IPMSM drives, as they clearly demonstrate how MTPA control enables efficient operation through proper current vector selection.

The block diagram of the MTPA controller is shown in Figure 19.24.

This block diagram builds upon the conventional double-loop control architecture by incorporating MTPA functionality. The speed and current controllers maintain the same gain structure as in the original double-loop design.

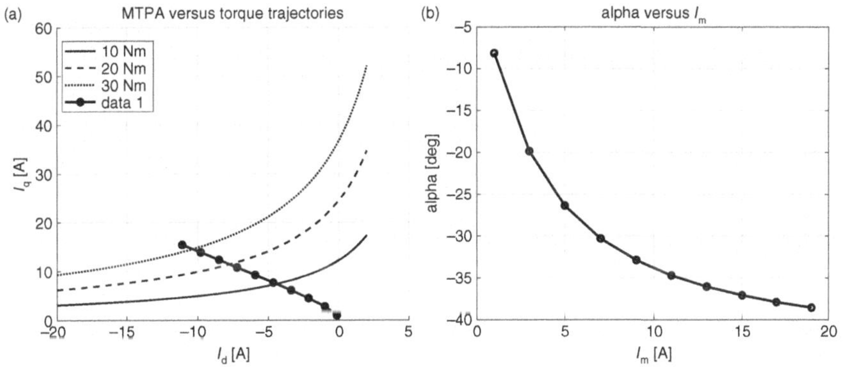

Figure 19.23 MTPA curves. (a) MTPA point vs. Torque (b) α vs. I_m.

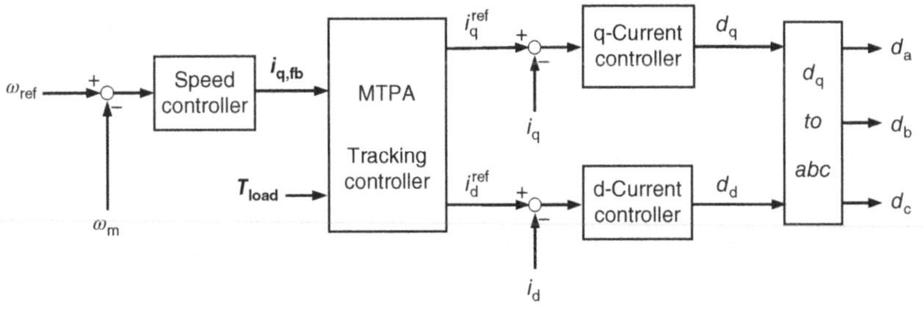

Figure 19.24 Block diagram of the MTPA controller.

There are two primary approaches for implementing MTPA control:

- Off-line Method: This method utilizes a precomputed look-up table that stores optimal stator current magnitudes I_m corresponding to various external torque values. These values are retrieved as needed during operation.
- On-line Method: In this approach, the external torque is measured in real time, and the control algorithm dynamically calculates the corresponding I_m required to achieve MTPA.

For this implementation, the off-line method is adopted for its simplicity. The external torque is sequentially set to 10, 20, and 30 Nm, with the corresponding feedforward current values $I_{q,ff}$ defined as 8.6 A, 13.8 A, and 17.8 A, respectively, based on the predefined look-up table. The feedback current $I_{q,fb}$, obtained from the output of the speed controller, is added to the feedforward current to calculate the total reference stator current magnitude:

$$I_m = I_{q,fb} + I_{q,ff}$$

Using this I_m, the current angle α is computed according to the MTPA control equation. The resulting current components (i_d, i_q) are then used as reference inputs to the current controller. For simulation purposes, the PSIM schematic remains unchanged. The existing C-block code from the double-loop controller only requires the addition of the MTPA tracking controller segment. Since the overall schematic is structurally identical to the previously implemented double-loop control circuit, it is not repeated here. Instead, only the additional MTPA control logic is described. The implementation of the MTPA controller C-block is provided in Table 19.6. Functionally, the MTPA controller is positioned between the speed controller and the current controller, allowing real-time adjustment of the reference current to achieve optimal torque efficiency.

MTPA Control Simulation Results The simulation results for a load torque of 10 Nm are presented in Figure 19.25. It can be observed that both the motor speed and current are effectively regulated. The current angle $\alpha = -32.4\ deg$ aligns precisely with the corresponding value in the look-up table, confirming the correct operation of the MTPA controller.

The simulation results for a load torque of 30 Nm are shown in Figure 19.26. As in the previous case, both speed and current exhibit stable and accurate control performance. Additionally, the current angle α matches the designed MTPA value, further validating the effectiveness of the control strategy.

Table 19.6 MTPA tracking controller C-block.

```
/****** Speed Controller *******/
---- Same as the speed controller code ----
/********** MTPA *****************/
// from table : 10N -> 8.6A, 20N -> 13.8A, 30N -> 17.8A
if (TL > 21)       Iq_ff = 17.8; // TL=30
else if (TL > 11) Iq_ff = 13.8; // TL=20
else if (TL > 1)  Iq_ff = 8.6;  // TL=10
else               Iq_ff = 8;

Im = Iq_fb + Iq_ff;

ang=(-Lamda_m+sqrt(Lamda_m*Lamda_m+8*Im*Im*(Ld-Lq)*(Ld-Lq)))/(4*(Ld-Lq)*Im);
alpha = asin(ang);

Iq_ref = Im*cos(alpha);
Id_ref = Im*sin(alpha);

/****** Current controller dq *******/
//Id_ref = 0.0;
----- Same as the current controller code ----
```

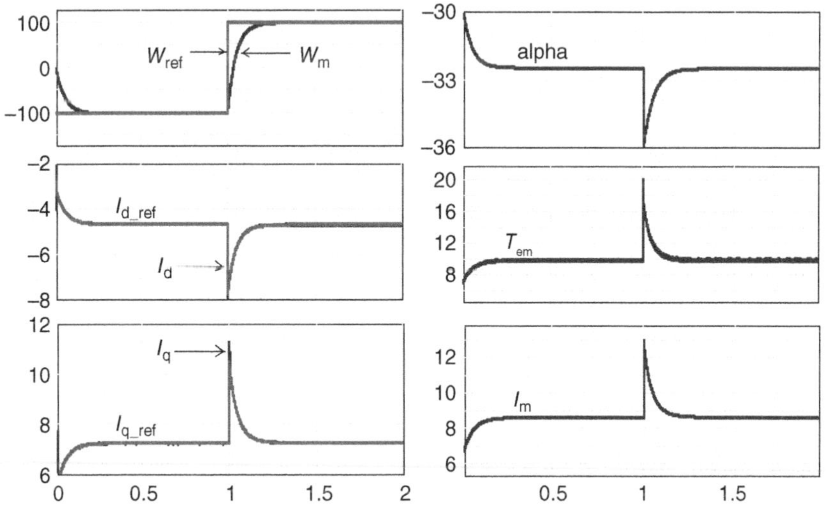

Figure 19.25 Waveform of MTPA control (T_L=10 *Nm*).

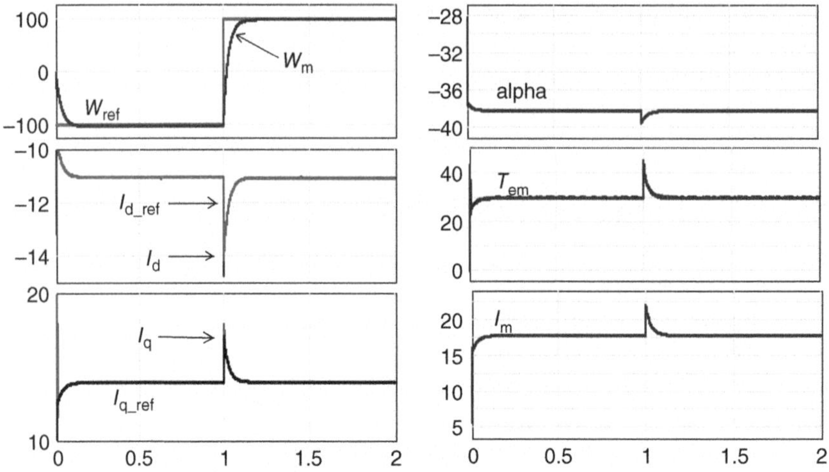

Figure 19.26 Waveform of MTPA control (T_L=30 *Nm*).

19.4 Field Weakening Control

Let us consider the following question:

"What is the maximum speed a motor can achieve under MTPA control, and how can the motor operate beyond this speed limit ?"

To answer this, we begin by revisiting the steady-state voltage equations of a PMSM. Neglecting stator resistance due to its relatively small influence at high speeds, the equations simplify to:

$$V_d = -\omega_e L_q \cdot i_q$$

$$V_q = \omega_e L_d \cdot i_d + \omega_e \lambda_m$$

These represent the d-axis and q-axis voltage components under steady-state conditions. The maximum available voltage from the inverter imposes a voltage constraint, expressed as:

$$V_d^2 + V_q^2 \leq V_{\max}^2$$

Where $V_{\max}$ depends on the PWM method and a safety margin η (typically 0.9 to 0.95) that accounts for dead time and control voltage limitations:

$$V_{\max} = \frac{V_{dc}}{2} = \frac{1}{2} V_{dc} \cdot \eta, \text{ for sinusoidal PWM}$$

$$V_{\max} = \frac{V_{dc}}{\sqrt{3}} \eta \approx 0.577 \cdot V_{dc} \cdot \eta, \text{ for space vector PWM}$$

Expressing the voltage constraint in terms of current variables, we obtain:

$$\left(\frac{L_q}{L_d}\right)^2 \cdot i_q^2 + \left(i_d + \frac{\lambda_m}{L_d}\right)^2 \leq \frac{V_{max}^2}{\omega_e^2 \cdot L_d^2}$$

In parallel, the current constraint is defined by the maximum allowable stator current:

$$i_q^2 + i_d^2 \leq i_{\max}^2$$

In field weakening (FW) control, the motor operates at speeds beyond the base speed where MTPA is no longer valid due to voltage limitations. In this region, the optimal operating point lies at the intersection of the voltage and current constraint boundaries. By equating these conditions, the following quadratic equation in i_d is derived:

$$\left(L_q^2 - L_d^2\right) i_d^2 - 2L_d\lambda_m i_d + \left(\frac{V_{\max}}{\omega_e}\right)^2 - (L_q I_{\max})^2 - \lambda_m^2 = 0 \tag{19.17}$$

Solving this using the quadratic formula gives:

$$i_d = \frac{1}{\left(L_d^2 - L_q^2\right)} \times \left[-L_d\lambda_m + \sqrt{(L_d\lambda_m)^2 - \left(L_d^2 - L_q^2\right) \cdot \left(\lambda_m^2 + (L_q I_{max})^2 - \left(\frac{V_{max}}{\omega_e}\right)^2\right)}\right] \tag{19.18}$$

The boundary speed ω_b, which separates the MTPA region from the FW region, can be calculated using the stator current values i_{q1} and i_{d1} obtained at the intersection of the load torque and MTPA trajectory:

$$\omega_b = \frac{V_{\max}}{\sqrt{(L_q i_{q1})^2 + (\lambda_m + L_d i_{d1})^2}} \tag{19.19}$$

This boundary defines the maximum speed achievable using MTPA control. Beyond this speed, the motor enters the FW region, where i_d is actively reduced (more negative) to limit the back-EMF and allow higher speed operation within the voltage constraints.

Field Weakening Operation with PMSM Library Parameters Using the PMSM library values directly and applying them to the circuit described in the previous chapter, the simulation results for speed variation under a voltage limit condition are illustrated in Figure 19.27. The speed ω_e is varied from 200 to 600 rad/s, and the voltage constraint is defined as $V_{\max} = 250 \cdot 0.95 = 237.5\ V$. The figure includes the MTPA curve, the load torque curve for $T_L = 10\ Nm$, and the voltage constraint ellipses.

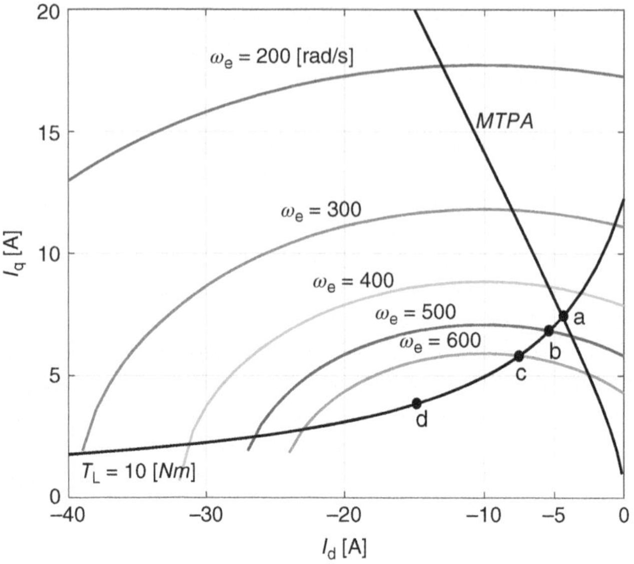

Figure 19.27 MTPA, load torque, and voltage constraint curves for varying speeds.

It is evident from the figure that as the speed increases, the size of the voltage constraint ellipse progressively shrinks. Starting from standstill, the motor accelerates along the MTPA trajectory until it reaches point ***a***, where it intersects with the load torque curve. This point represents the boundary speed ω_b, which marks the transition from MTPA to FW operation. The corresponding current values $i_d = -4.4\ A$ and $i_q = 7.5\ A$ at this point serve as the reference values for current control. In this case, the boundary speed is approximately 452 rad/s, which corresponds to about 226 rad/s in mechanical speed.

To achieve speeds beyond the boundary speed, the motor must deviate from the MTPA condition and enter the FW region, shifting its operating point along the load torque curve from point ***b*** to point ***c*** and ***d***. During this transition, the speed increases from approximately 500 to 600 rad/s and more. In this region:

- The reference i_d becomes more negative, increasing the demagnetizing (reactive) component of the stator current.
- The reference i_q is reduced, lowering the torque-producing (active) component of the stator current.

This coordinated adjustment reduces the back-EMF and allows higher operating speed under the voltage constraint—hence the term *field weakening control.*

It is crucial to note that points ***a***, ***b***, ***c***, and ***d*** all lie on the voltage constraint boundary, meaning that the inverter is operating at its maximum voltage output

$V_{\max}$. Thus, the inverter voltage remains constant at V_{max} during the entire transition from MTPA to FW operation.

Maximum Torque Per Voltage (MTPV) Maximum torque per voltage (MTPV) refers to the operating condition where the maximum torque is achieved relative to the available input voltage from the inverter. It is particularly relevant in the FW region, where voltage becomes the limiting factor for motor operation at high speeds.

The magnitude of the stator voltage vector under steady-state conditions is given by:

$$V_{op} = \sqrt{V_d^2 + V_q^2} = \sqrt{(-\omega_e L_q i_q)^2 + (\omega_e L_d i_d + \omega_e \lambda_m)^2}$$

The electromagnetic torque is expressed as:

$$T_e = \left(\frac{3}{2}\right)\left(\frac{P}{2}\right)[\lambda_m i_q + (L_d - L_q) i_q i_d]$$

To determine the optimal value of i_q for MTPV operation, we apply the condition:

$$\frac{\partial T_e}{\partial V_{op}} = 0$$

Solving this condition yields:

$$i_q = \frac{L_d}{L_q}\sqrt{\left(i_d - \frac{\lambda_m}{(L_q - L_d)}\right)\left(i_d + \frac{\lambda_m}{L_d}\right)} \tag{19.20}$$

This equation provides the current trajectory that maximizes torque while satisfying the voltage constraint, forming the MTPV trajectory.

Operating Behavior and Trajectory Figure 19.28 illustrates the operating trajectories for MTPA, FW, and MTPV under a constant load torque of $T_L = 10\ Nm$. For reference, the current-limit circles are also depicted for $I_{\max} = 20$ A and $I_{\max} = 13$ A, providing a clear visualization of the boundaries imposed by current constraints. The key points are summarized as follows.

- Point ***A*** (MTPA boundary). This is the intersection of the load-torque line with the MTPA trajectory. It marks the base-speed boundary where the stator-voltage magnitude (V_s) first reaches the limit $V_{\max}$. Below this speed the drive follows MTPA.
- Point ***B*** (FW → MTPV transition). From A to B the drive operates in field-weakening: negative i_d is increased to keep $|V_s| = V_{\max}$ while sustaining $T_e = T_L$ as speed rises. Point B is the intersection of the load-torque line with the MTPV trajectory and indicates the onset of voltage-limited operation

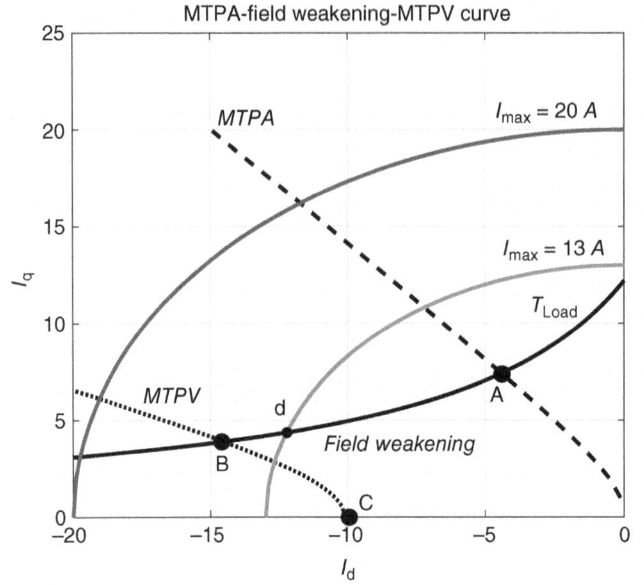

Figure 19.28 Trajectories of MTPA, FW, and MTPV at $T_L = 10\ Nm$.

where the commanded torque can no longer be maintained by FW alone; control transitions to MTPV.

- Point ***C*** (high-speed MTPV). Beyond B the operating point follows the MTPV path with $|V_s| = V_{\max}$. As speed increases toward C, the current magnitude decreases and the achievable torque progressively drops, reflecting the trade-off among current, torque, and speed under the voltage constraint.

Simulation Verification A PSIM simulation was carried out using the schematic shown in Figure 19.16 to validate the preceding analysis. The schematic and C-code framework are identical to those employed in the MTPA example of the previous chapter, with the exception that the stator resistance is modified from $R_s = 4.3\ \Omega$ to $0.5\ \Omega$.

The only functional change is in the current-reference generator, as illustrated in Figure 19.24 and Table 19.6. The reference current pair (i_{dref}, i_{qref}) is generated according to the selected operating region—MTPA, field-weakening, or MTPV—while the remainder of the control structure remains unchanged. The load torque is fixed at $T_L = 10\,\text{Nm}$ for all cases, providing a consistent basis for performance comparison between MTPA and MTPV. The maximum stator voltage magnitude is set to $V_{\max} = 237.5\,\text{V}$, and the maximum stator current is limited to $I_{\max} = 30\,\text{A}$.

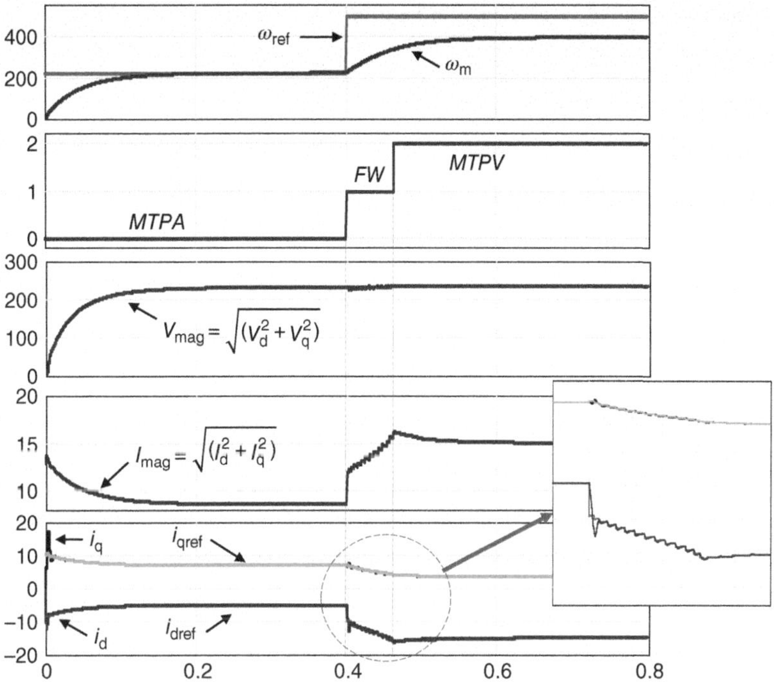

Figure 19.29 Waveforms of MTPA, FW, and MTPV at $T_L = 10\ Nm$.

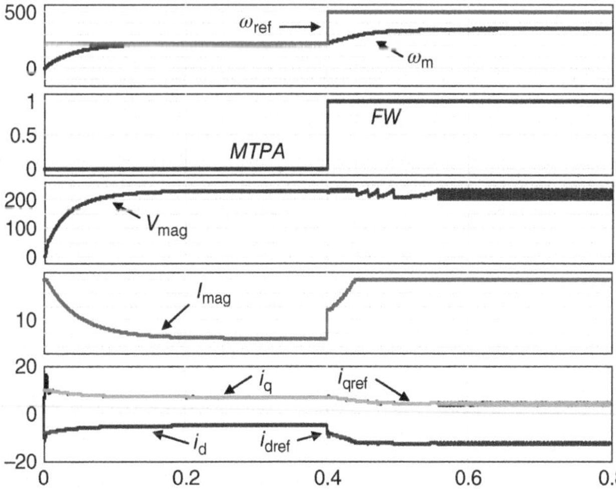

Figure 19.30 Waveforms of current-limit conditions.

Table 19.7 C-block code-1: Field_Weakening_Sub_Function.

```
double clamp(double x, double lo, double hi){ return (x<lo)?lo:((x>hi)?hi:x); }
double sqr(double x){ return x*x; }
double signnz(double x) {return (x >= 0.0) ? 1.0: -1.0; }

double voltage_mag(double id, double iq, double we)
{
    double Vd = Rs*id - we*Lq*iq;
    double Vq = Rs*iq + we*(Ld*id + Lamda_m);
    return sqrt(Vd*Vd + Vq*Vq);
}

int solve_quadratic_stable(double a, double b, double c, double* x1, double* x2)
{
    if (fabs(a) < eps)
    {
        if (fabs(b) < eps) return 0; /* Linear bx + c = 0 */
        *x1 = *x2 = -c / b;
        return 1;
    }
    double D = b*b - 4.0*a*c;
    if (D < 0.0) return 0;
    double s = sqrt(D);
    double q = -0.5 * (b + signnz(b) * s);
    double r1 = q / a;
    double r2 = (fabs(q) > eps) ? (c / q) : r1;
    if (r1 <= r2){ *x1 = r1; *x2 = r2; } else { *x1 = r2; *x2 = r1; }
    return (D > 0.0) ? 2: 1;
}

int apply_field_weakening(double Te, double we, double *id, double *iq)
{
    double idk = *id;
    double iqk = *iq;
    double Vmag = voltage_mag(idk, iqk, we);
    if (Vmag <= V_lim){
       *id = idk; *iq = iqk; return 0; /* Not needed */ }
```

Table 19.7 (Continued)

```
    /* Iteratively push i_d negative to reduce back EMF */
    int t;
    for (t =0; t < 60; ++t)
    {
        idk = fmax(idk - 0.02 * I_max, -I_max);
    /* Keep torque: recompute iq from Te = 1.5*P*(lambda_m+(Ld-Lq)*id)*iq */
        double psi = Lamda_m + (Ld - Lq) * idk;
        if (fabs(psi) < eps) psi = (psi >= 0.0 ? eps : -eps);
        iqk = Te / (1.5 * Pair * psi);

        /* current circle */
        double max_iq = sqrt(fmax(0.0, I_max*I_max - idk*idk));
        iqk = clamp(iqk, -max_iq, max_iq);

        Vmag = voltage_mag(idk, iqk, we);
        if (Vmag <= V_lim * 1.0005){ *id = idk; *iq = iqk; return 1; }
        if (idk <= -I_max + 1e-9) break;              /* out of room */
    } // end of for int t

    *id = idk; *iq = iqk;
    return 1; /* FW attempted;may still violate V->MTPV caller should check */
}

int mtpv_id_iq(double we, double Te_desired, double *id, double *iq)
{
     *id = 0.0; *iq = 0.0;

    if (we < eps || V_lim < eps || I_max < eps) return;

    const int N = 200;
    const double sgnT = (Te_desired >= 0.0) ? 1.0: -1.0;

    double best_absTe_match = -1.0;
    double best_absTe_any = -1.0;
    double best_id_match=0.0, best_iq_match=0.0;
    double best_id_any=0.0, best_iq_any=0.0;
```

(Continued)

Table 19.7 (*Continued*)

```
    const double A = we*we*Lq*Lq + Rs*Rs; /* constant in iq^2 */
    int k;
    for (k = 0; k <= N; ++k){
        double idk = - (double)k / (double)N * I_max; /* sweep id ∈ [0, -Imax] */
        double Ld_id_lam = Ld*idk + Lamda_m;

        /* Quadratic in iq: A iq^2 + B iq + C = 0 */
        double B = 2.0*Rs*we*(Lamda_m + (Ld-Lq)*idk ); /* simplified */
        double C = (Rs*idk)*(Rs*idk) + (we*Ld_id_lam)*(we*Ld_id_lam)-V_lim*V_lim;

        double q1=0.0, q2=0.0;
        int roots = solve_quadratic_stable(A, B, C, &q1, &q2);
        if (!roots) continue;
        int r;
        for (r = 0; r < roots; ++r){
            double iqk = (r == 0) ? q1 : q2;

            /* current circle clip at fixed idk */
            double max_iq = sqrt(fmax(0.0, I_max*I_max - idk*idk));
            iqk = clamp(iqk, -max_iq, max_iq);

            /* torque */
            double Te = 1.5*Pair*(Lamda_m*iqk + (Ld-Lq)*idk*iqk);
            double absTe = fabs(Te);
            /* track best overall */
            if (absTe > best_absTe_any){
                best_absTe_any = absTe; best_id_any = idk; best_iq_any = iqk; }
            /* track best with desired sign */
            if (signnz(Te) == signnz(sgnT) && absTe > best_absTe_match){
                best_absTe_match = absTe; best_id_match = idk; best_iq_match = iqk;
            }
        }
    }

    if (best_absTe_match > 0.0) { *id = best_id_match; *iq = best_iq_match;}
    else                        { *id = best_id_any; *iq = best_iq_any; }
    return 2;
}
```

The simulation results are presented in Figure 19.29. The mechanical speed reference was initially set to $\omega_{ref} = 225$ rad/s, which corresponds to the boundary speed ω_b in the MTPA region as shown in Figure 19.27. At $t = 0.4$ s, the speed reference steps up to 500 rad/s to evaluate maximum-speed operation. At this instant, the operating mode transitions to FW, corresponding to point A in Figure 19.28. At $t = 0.46$ s, the current values are measured as $i_d = -15.8$ A and $i_q = 4.5$ A, which corresponds to point B. Subsequently, the controller shifts into MTPV operation, guiding the trajectory toward point C. By the end of the simulation, the currents are $i_d = -14$ A and $i_q = 3.9$ A. This operating point remains close to point B because the controller is constrained by the constant load-torque curve. Throughout the simulation, both $i_d$and $i_q$are well regulated to their respective references.

The stator voltage magnitude V_{mag} reaches the maximum limit V_{max}, whereas the stator current magnitude I_{mag} remains below I_{max}. This indicates that the drive is operating under a voltage-constrained condition rather than a current-constrained condition. The current magnitude is minimized during the MTPA period, increases during the FW period, and then decreases during the MTPV period. This trend is also clearly observed in the individual current waveforms, where i_d and i_q shift in the negative direction during FW and move in the opposite direction under MTPV operation.

To investigate the influence of current constraints, the maximum current I_{max} was reduced to 13 A, and the corresponding simulation results are presented in Figure 19.30. The MTPA period remains unchanged, as the required current magnitude during this region is below the current limit. However, during the FW period, the direct-axis current decreases to approximately $i_d = -11.5$ A and remains clamped at point $\boldsymbol{d}$ due to the imposed current limit. Under this condition, the operating trajectory is unable to progress toward point $\boldsymbol{C}$, where the MTPV region would normally begin. Consequently, when the system operates under current-limited conditions, the transition into the MTPV region cannot occur.

It is also observed that the voltage-limit period exhibits small ripples. This behavior arises because the speed controller attempts to push the operating point into the high-speed region, while the FW controller counteracts this action and forces the trajectory back onto the load-torque curve.

The implementation of the controller C-block is summarized in Tables 19.7 and 19.8. Table 19.7 lists the subfunctions used in the program, while Table 19.8 presents the main function routines that integrate these subfunctions to realize the complete controller.

Table 19.8 C-block code-1: Field_Weakening_Main_Function.

```
#define Rs        0.5     // stator resistance 0.5 ohm
#define Ld        0.027   // d-axis Inductance
#define Lq        0.067   // q-axis Inductance
#define Lamda_m   0.273   // Back-emf constant
#define Pole      4
#define Pair      Pole/2

#define KWp       0.02218 // Speed P-gain
#define KWi       0.99996 // Speed I-gain

#define Kip_d     0.13167 // d-axis Current P-gain
#define Kii_d     0.837   // d-axis Currentr I-gain

#define Kip_q     0.13167 // q-axis Current P-gain
#define Kii_q     0.837   // q-axis Currentr I-gain

#define V_lim     250*0.95
#define I_max     30      // 13 for current limit
#define eps       1e-9

double Wm, Isa, Isb, Isc, Van, Vbn, Vcn, theta, Vdc, TL;
double Valpha, Vbeta, Ialpha, Ibeta, Id, Iq, Id_ref, Iq_ref;
double We, Wm_ref, Wm_err, Iq_fb, Iq_ff;
double Id_err, Iq_err, dd_fb, dq_fb, dd, dq;
double Va_con, Vb_con, Vc_con, duty_a, duty_b, duty_c;
double Im, ang, alpha, mi,Imag, WeB, WmB;
static double Wm_err_old=0.0, Iq_fb_old=0.0, Id_err_old=0.0;
static double Iq_err_old=0.0, dd_fb_old=0.0, dq_fb_old=0.0;

/**** Input Variable ****/
Wm      = in[0];
Isa     = in[1];
Isb     = in[2];
Isc     = in[3];
Van     = in[4];
Vbn     = in[5];
Vcn     = in[6];
theta   = in[7];
Vdc     = in[8];
TL      = in[9];
```

Table 19.8 (Continued)

```
if (t<0.4) Wm_ref = 225;
else       Wm_ref = 500;

Ialpha =(2./3.)*Isa - (1./3.)*Isb -(1./3.)*Isc;
Ibeta = (Isb-Isc)/sqrt(3);
Id = Ialpha*cos(theta) + Ibeta*sin(theta);
Iq = -Ialpha*sin(theta) + Ibeta*cos(theta);

/****** Speed Controller *******/
Wm_err = Wm_ref - Wm;
Iq_fb = Iq_fb_old + KWp*Wm_err -KWp*KWi*Wm_err_old;
Iq_fb_old=Iq_fb;
Wm_err_old=Wm_err;

/********** MTPA Operation **********/
if (TL > 21)        Iq_ff = 17.8;     // TL = 30 N
else if (TL > 11)   Iq_ff = 13.8;     // TL = 20 N
else if (TL > 1)    Iq_ff = 8.659;    // TL = 10 N

Im = Iq_fb + Iq_ff;
Im = clamp(Im, -I_max, I_max);

ang=(-Lamda_m+sqrt(Lamda_m*Lamda_m+8.*Im*Im*(Ld-Lq)*(Ld-Lq))) / (4.*(Ld-Lq)*Im);
alpha = asin(ang);
Iq_ref = Im*cos(alpha);
Id_ref = Im*sin(alpha);

/* Enforce current circle (safety) */
double I2 = Id_ref*Id_ref + Iq_ref*Iq_ref;
if (I2 > I_max*I_max)
{
          double scale = I_max / sqrt(I2);
          Id_ref *= scale;
          Iq_ref *= scale;
}

We = Wm*2;                       // 4-pole motor
WeB = V_lim /sqrt(sqr(Ld*Id_ref+Lamda_m)+sqr(Lq*Iq_ref)); // Boundary We
WmB=WeB/2.0; // Convert We to Wm
```

(Continued)

Table 19.8 (Continued)

```
/* Check voltage limit and apply FW if needed */
static int mode = 0; /* 0=MTPA, 1=FW, 2=MTPV */
double Vmag = voltage_mag(Id_ref, Iq_ref, We);

if (Vmag > V_lim * 1.0005)
{
    mode = apply_field_weakening(Im, We, &Id_ref, &Iq_ref);
    /* Re-check voltage */
    Vmag = voltage_mag(Id_ref, Iq_ref, We);
    if (Vmag > V_lim * 1.0005)
    {      /* Fall back to MTPV if still violating */
          mode = mtpv_id_iq(We, Im, &Id_ref, &Iq_ref);
          double I = sqrt(Id_ref*Id_ref + Iq_ref*Iq_ref);
          if (I > I_max && I > 1e-12)
          {
                double sc = I_max / I;
                Id_ref *= sc; Iq_ref *= sc;
          }
    /* Very rare: tiny pullback if still a hair over the ellipse due to rounding*/
         Vmag = voltage_mag(Id_ref, Iq_ref, We);
         if (Vmag > V_lim * 1.0005)
         {
             double scv = V_lim / Vmag;
             Id_ref *= scv; Iq_ref *= scv;
         }
    }
}

Imag = sqrt(Id_ref*Id_ref + Iq_ref*Iq_ref);

/****** Current controller dq *******/
Id_err = Id_ref - Id;
dd_fb = dd_fb_old + Kip_d*Id_err -Kip_d*Kii_d*Id_err_old;
dd_fb_old=dd_fb;
Id_err_old=Id_err;

Iq_err = Iq_ref - Iq;
dq_fb = dq_fb_old+Kip_q*Iq_err -Kip_q*Kii_q*Iq_err_old;
dq_fb_old=dq_fb;
Iq_err_old=Iq_err;
```

Table 19.8 (*Continued*)

```
if (Vdc < 0.1) Vdc = 0.1;
dd =     dd_fb - We*Lq*Iq / Vdc ;
dq =     dq_fb + We*Ld*Id / Vdc;

Valpha = dd*cos(theta) - dq*sin(theta);
Vbeta  = dd*sin(theta) + dq*cos(theta);

Va_con = Valpha;
Vb_con = -0.5*Valpha + sqrt(3)*Vbeta/2;
Vc_con = -0.5*Valpha - sqrt(3)*Vbeta/2;

duty_a = 0.5 + Va_con;
duty_b = 0.5 + Vb_con;
duty_c = 0.5 + Vc_con;

out[0] = duty_a;   // Inverter R
out[1] = duty_b;   // Inverter S
out[2] = duty_c;   // Inverter T

out[3] = Id_ref;
out[4] = Iq_ref;
out[5] = Id;
out[6] = Iq;
out[7] = Vmag;
out[8] = Im;

out[9] = mode;
out[10]= Wm_ref;
out[11]= WmB;
out[12]= Imag;

}
```

20

Induction Motor

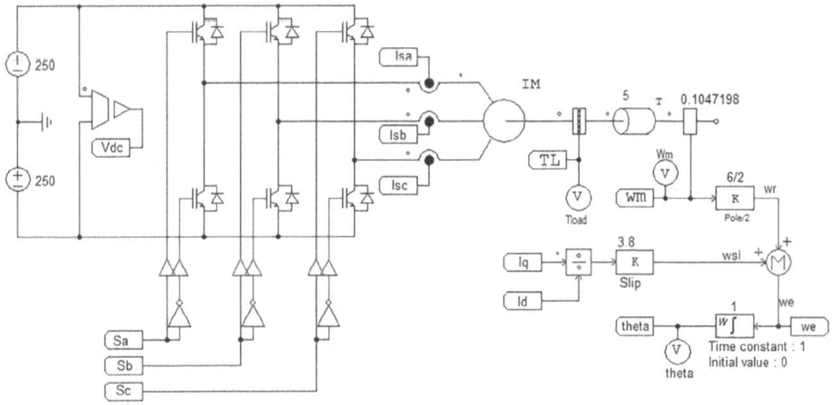

The induction motor is one of the most widely used electric machines in industrial and commercial high-power applications. Its popularity stems from its simple and rugged construction, high reliability, low manufacturing cost, and minimal maintenance requirements. Unlike synchronous machines, an induction motor does not require an external field excitation. Instead, rotor currents are induced by the rotating magnetic field produced by the stator windings, and electromagnetic torque is generated through the principle of electromagnetic induction.

Induction motors are manufactured in several configurations, most notably the squirrel-cage and wound-rotor types. They are extensively employed in applications such as pumps, fans, compressors, conveyors, and traction systems. With the development of power-electronic converters and advanced control strategies—particularly field-oriented control—induction motors are now capable of high-performance operation comparable to that of DC motors and permanent-magnet synchronous machines. In this chapter, the discussion is restricted to squirrel-cage induction motors.

To understand the operating principle of an induction motor, it is essential to introduce the concept of *slip*, which represents the relative speed difference between the rotating magnetic field and the rotor. Slip is a fundamental quantity in induction machines, as it explains both the mechanism of electromagnetic torque production and the reason why the rotor speed can never reach the synchronous speed of the rotating magnetic field.

A detailed derivation of the induction-machine operating principles is beyond the scope of this section. Instead, the basic concept of slip is briefly introduced. Let the mechanical speed of the rotor (or disk) be denoted by n, and let the synchronous speed of the rotating magnetic field be n_s. The slip S is defined as the normalized difference between these two speeds:

$$S = \frac{n_s - n}{n_s}$$

Slip represents the essential condition for torque production in an induction motor. When the mechanical load increases, the rotor speed n decreases, resulting in an increase in slip and a corresponding increase in induced rotor current and electromagnetic torque. Conversely, under light-load conditions, the rotor speed approaches the synchronous speed, and the slip decreases. At no load, the motor operates with a very small but nonzero slip, which is necessary to sustain the induced rotor currents and maintain torque production.

20.1 *d-q* Dynamic Equation for IFOC

This section derives the d–q dynamic equations of the induction motor in a synchronous reference frame and shows how they are simplified under indirect field-oriented control (IFOC). The derivation remains exact until the field-orientation assumptions are explicitly introduced.

Flux Linkage Relations In the synchronous d–q reference frame, the stator flux linkages (λ_{ds}, λ_{qs}) and rotor flux linkages (λ_{dr}, λ_{qr}) are expressed as

$$\lambda_{ds} = L_s i_{ds} + L_m i_{dr}, \quad \lambda_{qs} = L_s i_{qs} + L_m i_{qr}$$

$$\lambda_{dr} = L_r i_{dr} + L_m i_{ds}, \quad \lambda_{qr} = L_r i_{qr} + L_m i_{qs}$$

where

$$L_s = L_{ls} + L_m, \quad L_r = L_{lr} + L_m, \quad \sigma = 1 - \frac{L_m^2}{L_s L_r}$$

Eliminating the rotor currents yields the exact relations

$$\lambda_{ds} = \sigma L_s\, i_{ds} + k\, \lambda_{dr}, \quad \lambda_{qs} = \sigma L_s\, i_{qs} + k\, \lambda_{qr}$$

with $k = \frac{L_m}{L_r}$.

Differentiating,

$$\frac{d\lambda_{ds}}{dt} = \sigma L_s \frac{di_{ds}}{dt} + k\frac{d\lambda_{dr}}{dt}, \quad \frac{d\lambda_{qs}}{dt} = \sigma L_s \frac{di_{qs}}{dt} + k\frac{d\lambda_{qr}}{dt}$$

Thus, the stator current derivatives are obtained as

$$\frac{di_{ds}}{dt} = \frac{1}{\sigma L_s}\left(\frac{d\lambda_{ds}}{dt} - k\frac{d\lambda_{dr}}{dt}\right), \quad \frac{di_{qs}}{dt} = \frac{1}{\sigma L_s}\left(\frac{d\lambda_{qs}}{dt} - k\frac{d\lambda_{qr}}{dt}\right)$$

Stator Voltage Equations in the Synchronous Frame In a synchronous reference frame rotating at electrical speed ω_e, the stator voltage equations are:

$$v_{ds} = R_s i_{ds} + \frac{d\lambda_{ds}}{dt} - \omega_e\lambda_{qs}, \quad v_{qs} = R_s i_{qs} + \frac{d\lambda_{qs}}{dt} + \omega_e\lambda_{ds}$$

Rearranging,

$$\frac{d\lambda_{ds}}{dt} = v_{ds} - R_s i_{ds} + \omega_e\lambda_{qs}, \quad \frac{d\lambda_{qs}}{dt} = v_{qs} - R_s i_{qs} - \omega_e\lambda_{ds}$$

Complete Stator Current Dynamics (Before Field Orientation) Substituting the flux relations into the stator current derivatives yields the exact current dynamics with full d–q coupling:

$$\begin{aligned} \frac{d}{dt}i_{ds} &= \frac{1}{\sigma L_s}(v_{ds} - R_s i_{ds}) + \omega_e i_{qs} + \frac{\omega_e k}{\sigma L_s}\lambda_{qr} - \frac{k}{\sigma L_s}\frac{d\lambda_{dr}}{dt} \\ \frac{d}{dt}i_{qs} &= \frac{1}{\sigma L_s}(v_{qs} - R_s i_{qs}) - \omega_e i_{ds} - \frac{\omega_e k}{\sigma L_s}\lambda_{dr} - \frac{k}{\sigma L_s}\frac{d\lambda_{qr}}{dt} \end{aligned} \tag{20.1}$$

These equations represent the exact stator current dynamics with d–q coupling, prior to applying any IFOC assumptions.

Rotor Flux Dynamics For a squirrel-cage rotor, the rotor voltages satisfy $v_{dr} = v_{qr} = 0$. The rotor flux equations are

$$\frac{d\lambda_{dr}}{dt} = -\alpha\lambda_{dr} + \alpha L_m i_{ds} + \omega_{sl}\lambda_{qr}$$

$$\frac{d\lambda_{qr}}{dt} = -\alpha\lambda_{qr} + \alpha L_m i_{qs} - \omega_{sl}\lambda_{dr} \tag{20.2}$$

Where, $\alpha = \frac{R_r}{L_r}$, $\omega_{sl} = \omega_e - \omega_r$. ω_{sl}: slip frequency

Application of Indirect Field-Oriented Control (IFOC) Under rotor-flux orientation, the d-axis is aligned with the rotor flux:

$$\lambda_{qr} = 0, \quad \lambda_{dr} = \lambda_r, \quad \frac{d\lambda_{dr}}{dt} \approx 0.$$

The stator current dynamics reduce to

$$\begin{aligned} \frac{d}{dt}i_{ds} &= \frac{1}{\sigma L_s}(v_{ds} - R_s i_{ds}) + \omega_e i_{qs} \\ \frac{d}{dt}i_{qs} &= \frac{1}{\sigma L_s}(v_{qs} - R_s i_{qs}) - \omega_e i_{ds} - \frac{\omega_e k}{\sigma L_s}\lambda_r \end{aligned} \tag{20.3}$$

Linearization for Stator Current Applying state averaging equations to the inverter control input v_{ds} and v_{qs},

$$v_{ds} = +\left(\frac{V_{dc}}{2}\right)(d_{ds}) - \left(\frac{V_{dc}}{2}\right)(1 - d_{ds}) = \left(\frac{V_{dc}}{2}\right)(2d_{ds} - 1)$$

$$v_{qs} = +\left(\frac{V_{dc}}{2}\right)(d_{qs}) - \left(\frac{V_{dc}}{2}\right)(1 - d_{qs}) = \left(\frac{V_{dc}}{2}\right)(2d_{qs} - 1)$$

By adding a constant offset of 0.5, the control voltages can be written as

$$v_{ds} = V_{dc}\, d_{ds}, \quad v_{qs} = V_{dc}\, d_{qs}$$

Each duty ratio is decomposed into feedback and feed-forward components:

$$d_{ds} = d_{ds\_fb} + d_{ds\_ff}, \quad d_{qs} = d_{qs\_fb} + d_{qs\_ff}.$$

The feed-forward terms are selected to cancel nonlinear coupling terms:

$$d_{ds\_ff} = \frac{-\omega_e\, \sigma L_s\, i_{qs}}{V_{dc}}, \quad d_{qs\_ff} = \frac{\omega_e\, \sigma L_s\, i_{ds} + \omega_e\, k\, \lambda_r}{V_{dc}}$$

The resulting *linearized stator current dynamics* are

$$\frac{d}{dt}i_{ds} = \frac{1}{\sigma L_s}(V_{dc}d_{ds\_fb} - R_s i_{ds}), \quad \frac{d}{dt}i_{qs} = \frac{1}{\sigma L_s}(V_{dc}d_{qs\_fb} - R_s i_{qs})$$

Taking the Laplace transform yields the current-to-duty transfer functions:

$$G_{1d}(s) = \frac{I_{ds}(s)}{d_{ds}(s)} = \frac{V_{dc}}{\sigma L_s s + R_s}, \quad G_{1q}(s) = \frac{I_{qs}(s)}{d_{qs}(s)} = \frac{V_{dc}}{\sigma L_s s + R_s} \tag{20.4}$$

Rotor Flux Equation and Slip Frequency Under IFOC, the rotor flux dynamics reduce to

$$\frac{d}{dt}\lambda_r = -\alpha\lambda_r + \alpha L_m i_{ds}$$

The ideal slip frequency law is

$$\omega_{sl} = \alpha \frac{L_m i_{qs}}{\lambda_r}$$

For steady-state operation, the rotor flux is often approximated as $\lambda_r \approx L_m i_{ds}$, yielding

$$\omega_{sl} \approx \alpha \frac{i_{qs}}{i_{ds}}$$

Mechanical speed dynamics The mechanical dynamics are identical to those of a PMSM and are given by

$$J_m \frac{d\omega_m}{dt} = T_e - T_L - B_m \omega_m$$

where ω_m is mechanical speed (rad/s), J_m is the rotor inertia, B_m is viscous friction coefficient, T_e is electromagnetic torque, and T_L is load torque.

The electromagnetic torque is expressed as

$$T_e = \frac{3}{2}\left(\frac{P}{2}\right)\frac{L_m}{L_r}(\lambda_{dr} i_{qs} - \lambda_{qr} i_{ds})$$

Using $\lambda_{qr} = 0$, this simplifies to

$$T_e = \frac{3}{2}\left(\frac{P}{2}\right)\frac{L_m}{L_r}\lambda_r i_{qs}$$

Defining the torque constant for a given flux level,

$$K_T \triangleq \frac{3}{2}\left(\frac{P}{2}\right)\frac{L_m}{L_r}\lambda_r \Rightarrow T_e = K_T i_{qs}$$

Mechanical dynamics can be rewritten as

$$\frac{d\omega_m}{dt} = \frac{K_T}{J_m} i_{qs} - \frac{T_{load}}{J_m} - \frac{B_m}{J_m}\omega_m$$

To compensate for load torque using feed-forward control,

$$i_{qs} = i_{qs\_fb} + i_{qs\_ff}, \quad i_{qs\_ff} = \frac{T_{\text{load}}}{K_T}$$

The resulting linear mechanical dynamics are

$$\frac{d\omega_m}{dt} = \frac{K_T}{J_m} i_{qs\_fb} - \frac{B_m}{J_m}\omega_m$$

Taking the Laplace transform gives the speed-to-current transfer function:

$$G_{3q}(s) = \frac{\omega_m(s)}{I_{qs}(s)} = \frac{K_T}{J_m s + B_m} \tag{20.5}$$

Comparison with the PMSM Equations When the d–q state equations of an induction motor are compared with those of a permanent magnet synchronous motor (PMSM) presented in Chapter 19, it becomes evident that the two machine models are mathematically very similar. The overall structure of both electrical and mechanical dynamics is nearly identical, providing valuable insight into unified modeling and control approaches for AC machines.

In particular, the electromagnetic torque expression and the d-axis current equation have the same mathematical form in both machines. The primary difference appears in the q-axis current equation, where the induction motor includes additional terms associated with rotor-flux dynamics. This distinction arises because torque production in an induction motor relies on rotor currents induced by the stator magnetic field, whereas in a PMSM, the rotor magnetic field is established directly by permanent magnets.

From a control and linearization perspective, the feed-forward compensation terms required for decoupling are also essentially the same for both machines. The only additional element that must be considered in the induction motor is the rotor-flux dynamics compensation term, which accounts for transient rotor behavior. Apart from this contribution, the resulting linearized models and decoupling structures are equivalent.

Finally, while the electrical angle in a PMSM is obtained directly from the mechanical rotor position, the corresponding electrical angle in an induction motor is determined by adding the slip frequency to the mechanical speed. When this relationship is considered, the strong equivalence between the two machine models is clearly evident from the governing equations. This strong similarity explains why control strategies originally developed for PMSMs can be systematically extended to induction motors with only minor modifications.

20.2 Double-Loop Controller Design

The double-loop controller structure for the induction motor under IFOC is shown in Figure 20.1(a). The configuration consists of an outer speed control loop and an inner q-axis current control loop. This structure is almost identical to that used for a PMSM, reflecting the close similarity between the dynamic models of the two machines.

The only distinction from the PMSM control structure lies in the feed-forward compensation terms, which, in the induction motor case, must include rotor-flux dynamic compensation. Aside from these additional terms, the overall double-loop control configuration remains the same as that of the PMSM.

The d-axis current controller follows a single-loop configuration, as shown in Figure 20.1(b). Under rotor-flux-oriented control, the d-axis current reference is

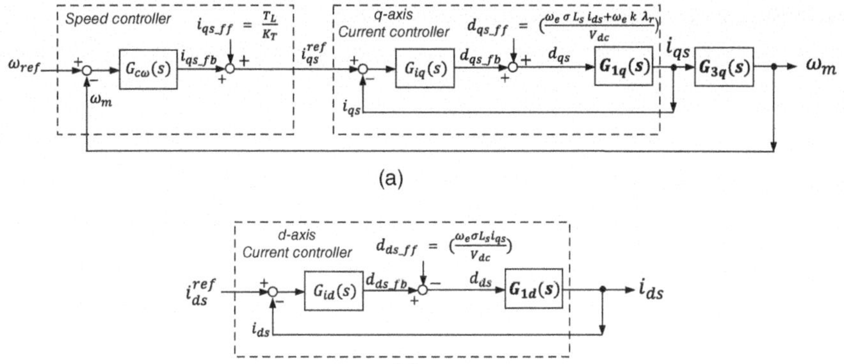

Figure 20.1 The double-loop controller configuration. (a) Double-loop controller (q-axis) (b) Single-loop current controller (d-axis)

Table 20.1 Design specifications for double loop controller.

	Bandwidth	Phase margin
Speed Controller (C_1)	100 rad/s	84 degree
Current Controller (C_2)	2400 rad/s	60 degree

typically selected to regulate the rotor flux magnitude and is given by

$$i_{ds}^{ref} = \frac{\lambda_r}{L_m} \tag{20.6}$$

It should be noted that, in contrast, the *d*-axis reference current in a PMSM controller is usually set to zero, since the rotor flux is established by permanent magnets rather than stator current.

The frequency design specifications for double-loop controller are summarized in Table 20.1.

The phase margin of the speed controller is set to 84°, which is the maximum achievable value given the selected motor parameters. This choice results in a small transient overshoot in the speed response; however, the steady-state error is eliminated within the specified settling time. MATLAB m-file for double-loop controller is presented in Table 20.2.

After applying the automated tuning procedure in SISOTOOL, the following continuous-time controller transfer functions were obtained:

- Speed Controller:

$$C_1 = G_{cw}(s) = \frac{0.219(s+10)}{s}$$

Table 20.2 MATLAB m-file for double loop controllers.

```
%% Double-Loop Design for Induction Motor using IFOC
s = tf('s');
%% --- PSIM Induction Motor Parameters (given) ---
Rs = 0.294; % [Ohm] stator resistance
Lls = 0.00139; % [H] stator leakage
Rr = 0.156; % [Ohm] rotor resistance
Llr = 0.00074; % [H] rotor leakage
Lm = 0.041; % [H] magnetizing inductance
P = 6; % [-] number of poles
pp = P/2; % pole pairs
J = 0.002; % [kg*m^2] inertia
B = 0.0; % [N*m*s] viscous friction
Vin = 500; % [V] input voltage
%% --- Choose Flux(id) operating point
id0 = 5; % [A] chosen flux current (d-axis)

%% Compute Kt from rotor-flux-oriented IFOC model
Lr_tot = Llr + Lm; % [H] rotor total inductance
sigma_r = Lm/Lr_tot; % Lm/(Lr+Lm)
Ls_tot = Lls + Lm; % [H] stator total inductance
sig = 1-(Lm^2 / (Ls_tot*Lr_tot) );

% Approx rotor flux magnitude at steady state
psi_r0 = Lm * id0;                    % [Wb]

Kt = (3/2)*pp*sigma_r*psi_r0;     % Te = Kt * iq
G1q = Vin / (sig*Ls_tot*s + Rs); % Iq(s)/dq(s)
G3q = Kt / (J*s + B);             % Wm(s)/Iq(s)

Kpw=1; Kiw=1;
Gcw=Kpw*(s+Kiw)/s;
Kpi=1; Kii=1;
Gci=Kpi*(s+Kii)/s;

Wref = 100;

%%%% Double loop %%%%
% Speed:Wc=100
% P.M=84[deg]
% Current:Wc=2400,
% P.M.=60

T=sisoinit(6);
T.F.Value = Wref;
T.G1.Value = G1q;
T.C2.Value = Gci;
T.G2.Value = G3q;
T.C1.Value = Gcw;
sisotool(T);
```

- Current Controller:

$$C_2 = G_{ci}(s) = \frac{0.0087(s + 1440)}{s}$$

The same current controller gains are applied to both the d-axis and q-axis current control loops.

The Bode plots of the designed speed and q-axis current controllers are shown in Figure 20.2. The speed controller achieves the target phase margin of 84°, resulting in a small transient overshoot, while the current controller satisfies its bandwidth and phase margin requirements.

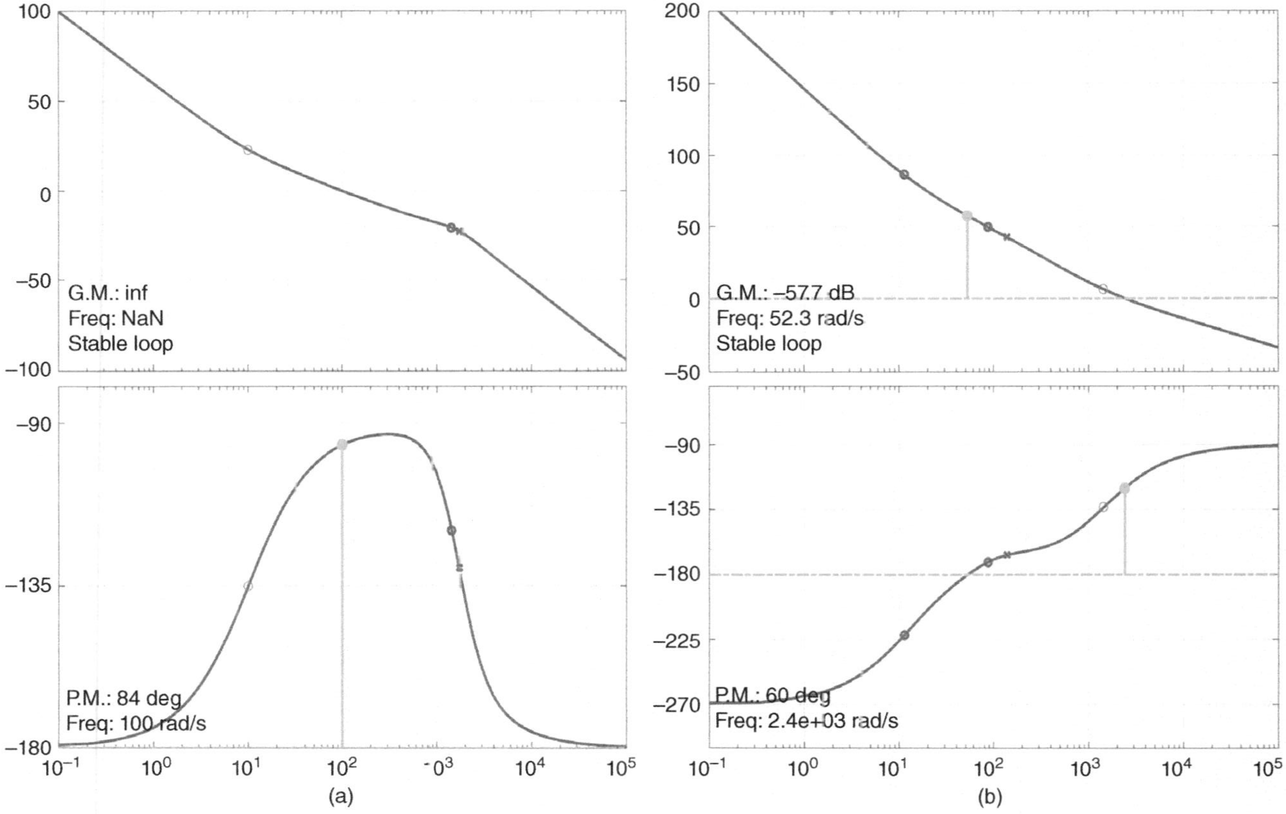

Figure 20.2 Bode plot for double-loop controller. (a) Speed loop (b) Current loop.

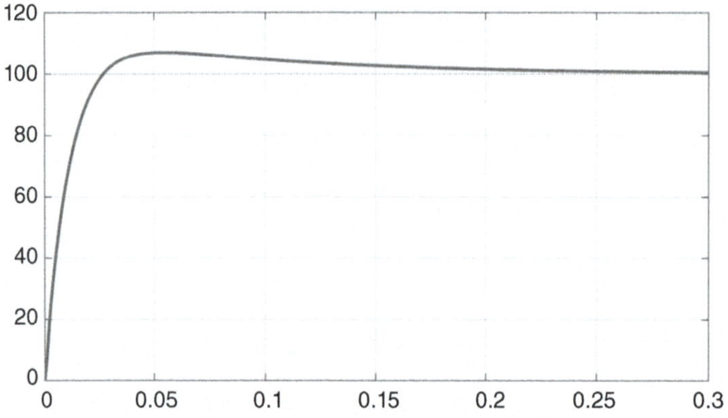

Figure 20.3 Step response of speed in double-loop controller.

The step response of the speed control loop is presented in Figure 20.3. The response exhibits an overshoot of approximately 7% and a rise time of 18 ms, demonstrating satisfactory dynamic performance for the double-loop control structure.

PSIM Verification The simulation schematic for the double-loop controller implemented using an *analog control structure* is shown in Figure 20.4. This configuration integrates both the outer speed controller and the inner current controllers using analog logic blocks. One important advantage of the analog model is that it allows the control logic to be understood intuitively and with minimal abstraction. Although identical results can be achieved by implementing the controller using a C-block, the analog model is employed in this chapter to clearly illustrate the control structure and signal flow. Readers are encouraged to implement the equivalent C-block controller as an exercise.

The overall schematics closely resemble the double-loop controller used for the PMSM. The principal difference lies in the generation of the electrical angle ω_e, which is obtained by combining the mechanical rotor speed ω_m and the slip speed ω_{sl}. Under steady-state rotor-flux-oriented control, the slip speed is approximated as

$$\omega_{sl} \approx \alpha \frac{i_{qs}}{i_{ds}}, \quad \alpha = \frac{R_r}{L_r} = 3.74$$

The speed controller is implemented using a PI regulator, and its output serves as the reference input for the q-axis current controller, together with the appropriate feed-forward compensation terms. The d-axis current reference is set to 5 A. This value is selected based on Eq. (20.6),

$$i_{ds}^{\text{ref}} = \frac{\lambda_r}{L_m} \approx \frac{V_{\text{dc}}}{\omega_e \sqrt{3}} \frac{\gamma}{L_m}$$

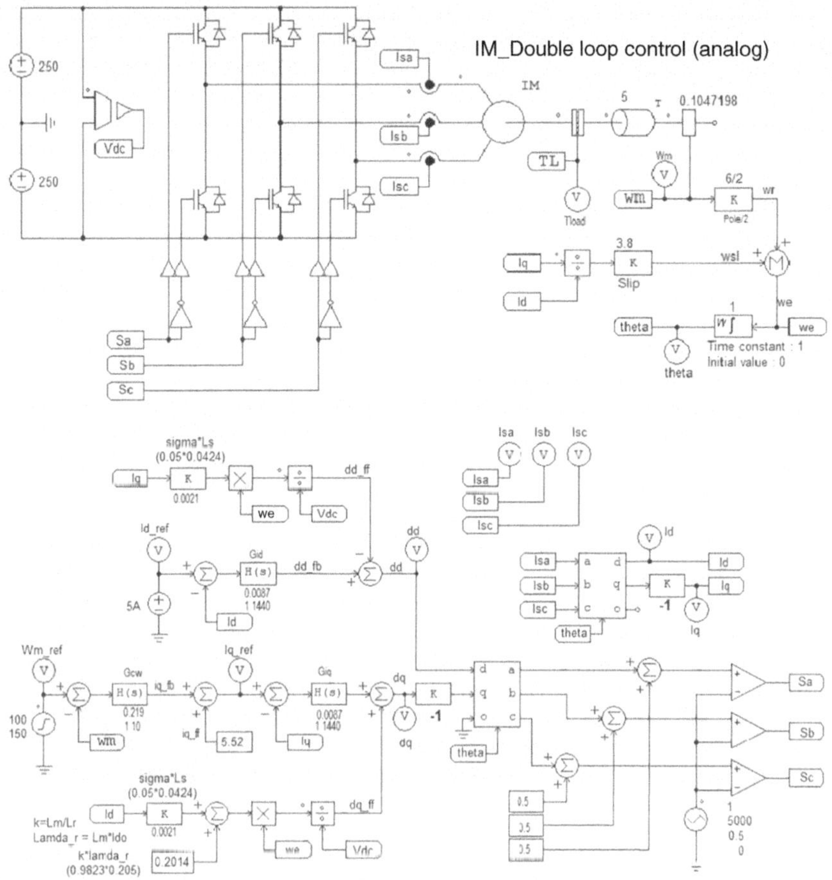

Figure 20.4 Simulation schematic for the double-loop controller.

where γ is a derating factor typically chosen in the range of 0.2–0.6. This choice ensures an appropriate rotor flux level while avoiding magnetic saturation.

The simulation schematic is constructed to exactly match the mathematical equations derived in the preceding sections. As a result, the simulation waveforms closely correspond to the controller characteristics predicted by the MATLAB-based design.

The simulation results are presented in Figures 20.5 and 20.6. The waveform analysis in Figure 20.5 confirms the correct operation of the double-loop controller:

- The motor speed accurately tracks the reference value with a small overshoot, consistent with the MATLAB step-response prediction.

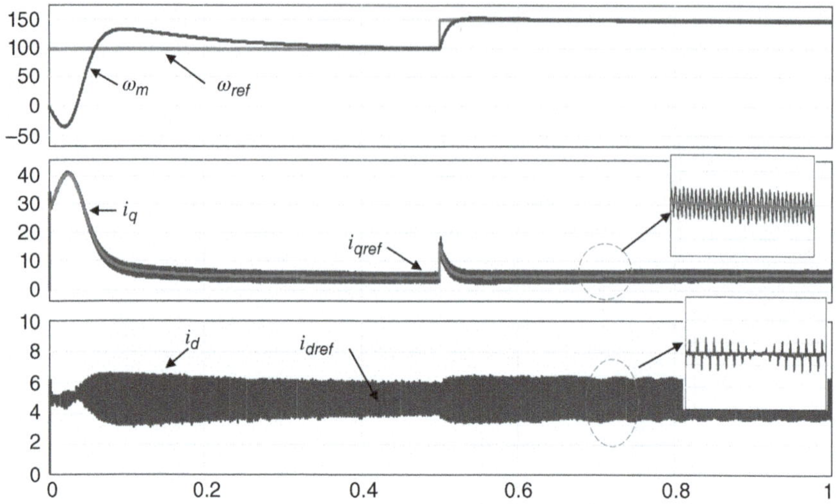

Figure 20.5 Waveforms of the double-loop controller.

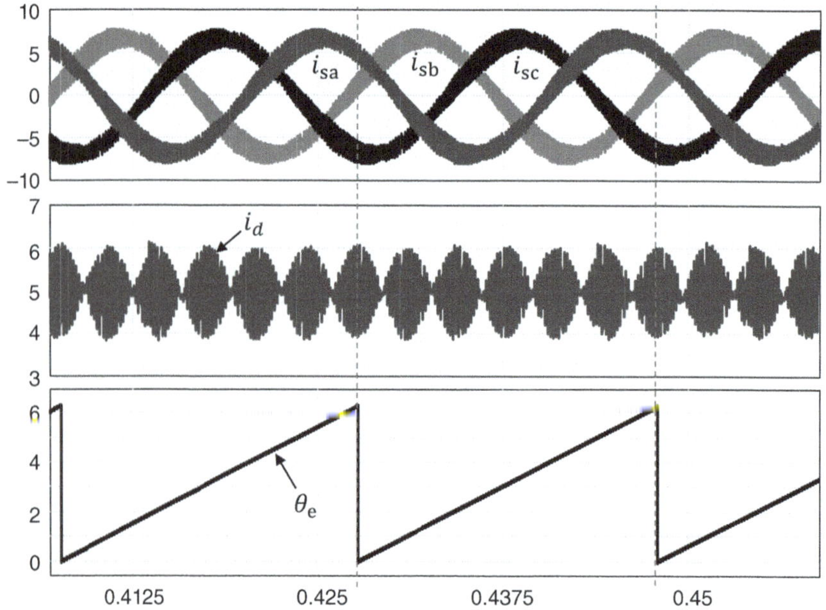

Figure 20.6 Waveforms of the double-loop controller.

- The second waveform shows the q-axis current, providing real-time insight into torque-producing current behavior.
- The third waveform shows the d-axis current, which exhibits ripple around the 5 A reference value.

The waveforms shown in Figure 20.6 further validate the controller's performance:

- The first waveform represents the three-phase stator currents, confirming proper current regulation in all phases.
- The second waveform shows the *d*-axis current ripple. Since the induction motor has six poles, six current ripple components per mechanical revolution are observed, as expected.
- The third waveform illustrates the electrical angle θ_e, confirming correct angle generation based on rotor speed and slip frequency.

Part III

Special Session

21. Phase Shift Full-Bridge (PSFB) Converter
22. Dual Active Bridge (DAB) Converter
23. LLC Converter
24. Discontinuous Current Mode
25. Multi-Modular Converter

Part I of this book focuses on DC power conversion, while Part II is devoted to AC power conversion. In both parts, a unified and systematic control-design methodology is adopted. Specifically, state-space averaging techniques are employed to derive small-signal transfer functions of power converter circuits, and classical control theory is applied to design controllers based on these models. This approach enables readers to establish a clear relationship between circuit dynamics and controller behavior, facilitating intuitive and systematic controller design.

The methodologies presented in Parts I and II are particularly effective for power conversion systems operating under continuous conduction conditions, where linear PWM modulation can be assumed and averaged models accurately represent the system dynamics. As a result, these methods have been widely used in both academic research and industrial practice for the design of voltage and current controllers in conventional DC–DC converters and grid-connected inverters.

However, not all power conversion systems satisfy the assumptions required for conventional state-space averaging. In many practical converters, the current may become discontinuous, zero-current intervals may exist within a switching period, or the control variable may not be a linear duty ratio. In addition, resonant and phase-shift–controlled converters exhibit strongly nonlinear characteristics, operating-point–dependent gains, or frequency- or phase-based control inputs. In such cases, the direct application of conventional averaged modeling techniques becomes challenging or even invalid.

To address these challenges, a substantial body of research has been devoted to developing alternative modeling and control techniques for such systems, and active research in this area continues. These studies often rely on harmonic approximations, large-signal energy balance, discrete-time modeling, or numerical approaches, which can be mathematically involved and difficult to apply systematically from a practical control-design perspective.

In Part III (Special Session), this book extends the core design philosophy developed in Parts I and II to these more complex power conversion systems. Rather than introducing entirely new control paradigms, the emphasis is placed on adapting and reinterpreting the familiar transfer-function–based and PI-controller–based design framework so that it can be applied effectively to nonconventional converters. This approach allows readers to leverage previously acquired knowledge while gaining insight into advanced converter operations and control.

The converters addressed in Part III include the phase-shift full-bridge converter, the dual active bridge converter, the LLC resonant converter, discontinuous current mode converters, and modular multilevel converters. These topologies have attracted significant attention in recent years due to their high efficiency, bidirectional power flow capability, high-power scalability, and suitability for applications such as renewable energy systems, energy storage systems, electric vehicles, and high-voltage power conversion.

Although the specific operating principles of these converters differ, many of their control challenges can be systematically addressed by carefully defining equivalent control variables, identifying dominant dynamic characteristics, and constructing appropriate small-signal or quasi-linear models. By following the step-by-step procedures presented in this book, readers can design stable and robust controllers for these systems without relying exclusively on trial-and-error tuning or overly complex analytical tools.

Beyond the specific converter topologies explicitly discussed in this book, a wide variety of power conversion systems exist in modern power electronics. The control design methodologies presented herein are intentionally developed in a general and extensible framework, rather than being tailored to individual circuits. As a result, with appropriate interpretation of the underlying circuit dynamics and control variables, these methodologies can be systematically applied to a broad range of converter topologies. This enables engineers and researchers to design effective and reliable control strategies not only for established converters, but also for newly emerging power electronic systems.

21

Phase-Shift Full Bridge Converter

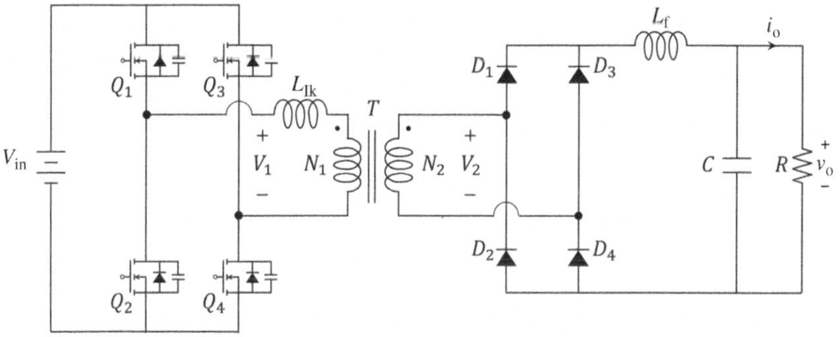

Phase Shift Full Bridge (PSFB) Converter The phase-shift full-bridge (PSFB) converter is an isolated high-power DC–DC conversion topology widely adopted in off-line power supplies, telecommunications rectifiers, and electric-vehicle auxiliary power systems. Owing to its high efficiency, fixed-frequency operation, and reduced switching stress, it is especially suitable for medium- and high-power conversion stages.

In the PSFB topology, four primary-side power switches are arranged in a full-bridge configuration to apply a high-frequency square-wave voltage to the transformer primary. Output voltage regulation is achieved by controlling the phase shift between the two bridge legs, rather than by varying the duty ratio of the individual switches. This phase-shift modulation effectively adjusts the applied transformer voltage and the transferred power while maintaining a constant switching frequency, which simplifies magnetic design and EMI filtering.

A key advantage of the PSFB converter is its ability to achieve zero-voltage switching (ZVS) over a wide load range. By utilizing the transformer leakage inductance and/or an external resonant inductor, the energy stored in the circuit is used to discharge the switch output capacitances prior to turn-on. This soft-switching mechanism significantly reduces switching losses, mitigates electromagnetic interference, and enhances overall efficiency. On the secondary side, a full-wave rectifier and output filter convert the high-frequency AC voltage into a regulated DC output, while the transformer provides galvanic isolation and voltage scaling. Due to its high efficiency, scalability, and compatibility with digital control platforms, the PSFB converter remains a standard solution for high-performance isolated DC–DC power conversion systems.

21.1 Operation of the PSFB Converter

The schematic diagram of the PSFB converter is shown in Figure 21.1. The full-bridge switches Q_1–Q_4 are operated with a fixed 50% duty ratio, while a controlled phase shift is introduced between the left and right bridge legs. This phase shift determines the effective duty cycle applied to the transformer primary and, consequently, the power transferred to the load.

Zero-voltage turn-on of the active switches is achieved by exploiting the energy stored in the transformer leakage inductance. During each commutation interval, the leakage inductance current discharges the output capacitance of the incoming switch before it is turned on, thereby enabling ZVS operation. Representative voltage and current waveforms of the PSFB converter are illustrated in Figure 21.2, highlighting the phase-shifted bridge voltages and the resonant behavior during switching transitions.

In practical PSFB operation, the effective duty applied to the transformer primary is reduced relative to the commanded phase-shift duty. This reduction, referred to as duty loss (ΔD), arises from the finite time required for ZVS commutation. Specifically, the leakage inductance current cannot reverse instantaneously

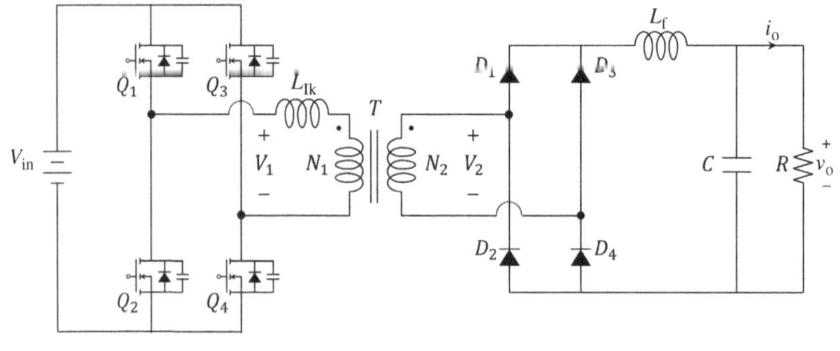

Figure 21.1 Phase-shift full bridge converter.

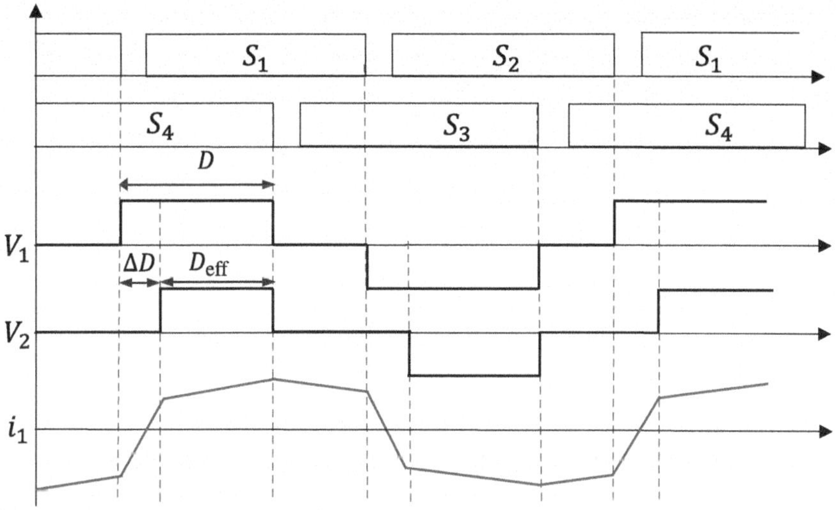

Figure 21.2 Voltage and current waveforms.

during the transition interval, resulting in a period during which no voltage is applied to the transformer primary. Consequently, power transfer is reduced for a portion of each switching half-cycle.

This section derives an analytical expression for the duty loss ΔD and explains its physical origin based on current waveforms, transformer operation, and output inductor ripple characteristics. Understanding this phenomenon is essential for accurate modeling, control design, and performance optimization of the PSFB converter.

Duty Loss (ΔD) Calculation The following derivation quantifies ΔD based on the output inductor current ripple and the leakage inductance dynamics.

1) Output Inductor Current Ripple During Freewheeling Interval
During the freewheeling interval of each half-cycle, the secondary-side output inductor L_f is subjected to a negative voltage equal to the output voltage.

$$v_{L_f} = -V_{\text{out}}$$

The duration of the freewheeling interval is

$$t_{\text{freewheel}} = \frac{(1-D)T_s}{2}$$

where D is the commanded phase-shift duty and T_s is the switching period. The corresponding current slope of the output inductor L_f is

$$\frac{di_{L_f}}{dt} = \frac{v_{L_f}}{L_f} = -\frac{V_{\text{out}}}{L_f}$$

Therefore, the peak-to-peak ripple current of the output inductor is given by

$$\Delta I = \left|\frac{di}{dt}\right| t_{\text{freewheel}} = \frac{V_{\text{out}}(1-D)T_s}{2L_f}$$

If the DC output current is denoted by I_{load}, the minimum (valley) inductor current at the end of the freewheeling interval becomes

$$I_{\text{valley}} = I_{\text{load}} - \frac{\Delta I}{2}$$

In order to achieve ZVS in the primary bridge, the current is required to reverse its direction during the commutation interval. As a result, the primary bridge must commutate a current equal to twice the valley current.

$$2I_{\text{valley}} = 2I_{\text{load}} - \Delta I = 2I_{\text{load}} - \frac{V_{\text{out}}(1-D)T_s}{2L_f}$$

2) Reflection to Primary and Required Leakage Inductor Current Swing
Reflecting the secondary current to the primary side using the transformer turns ratio $n = \frac{N_s}{N_p}$, the corresponding primary-side valley current is

$$I_{\text{p,valley}} = n\, I_{\text{valley}}$$

During each commutation interval, the leakage inductor current must reverse from $+I_{\text{p,valley}}$ to $-I_{\text{p,valley}}$. Hence, the total required current change in the leakage inductor is

$$\Delta I_{\text{lk}} = 2I_{\text{p,valley}} = 2nI_{\text{valley}} = n\left(2I_{\text{load}} - \frac{V_{\text{out}}(1-D)T_s}{2L_f}\right)$$

This represents the minimum leakage current swing required to guarantee ZVS during each switching transition.

3) Commutation Interval and Leakage Inductor Current Slew
The rate at which the leakage inductor current can change is limited by the applied voltage and the leakage inductor L_{lk}. During commutation, the full DC bus voltage V_{in} is applied across L_{lk}, resulting in a current slew rate

$$\frac{di_{\text{lk}}}{dt} = \frac{V_{\text{in}}}{L_{\text{lk}}}$$

The commutation interval corresponds to the duty loss ΔD multiplied by half of the switching period,

$$t_{\Delta D} = \frac{\Delta D T_s}{2}$$

Accordingly, the maximum achievable change in the leakage inductor current during this interval is

$$\Delta I_{\text{lk}} = \frac{V_{\text{in}}}{L_{\text{lk}}} \cdot \frac{\Delta D T_s}{2}$$

Equating the required current reversal to the achievable current change yields

$$\frac{V_{\mathrm{in}}}{L_{\mathrm{lk}}} \cdot \frac{\Delta D T_{\mathrm{s}}}{2} = n\left(2I_{\mathrm{load}} - \frac{V_{\mathrm{out}}(1-D)T_{\mathrm{s}}}{2L_{\mathrm{f}}}\right)$$

Solving for duty loss ΔD, the final expression becomes

$$\Delta D = \frac{n\left(2I_{\mathrm{load}} - \frac{V_{\mathrm{out}}(1-D)T_{\mathrm{s}}}{2L_{\mathrm{f}}}\right)}{\frac{V_{\mathrm{in}}T_{\mathrm{s}}}{2L_{\mathrm{lk}}}} \tag{21.1}$$

with

n: transformer turns ratio (secondary/primary as you defined)
I_{load}: DC load current
V_{out}: output voltage
D: commanded phase-shift duty (0.0 ~ 1.0)
T_{s}: switching period
L_{f}: secondary-side output inductor
L_{lk}: primary-side leakage (or series resonant) inductance
V_{in}: DC bus voltage

The duty loss ΔD represents the portion of each switching half-cycle during which the converter cannot apply voltage to the transformer primary. During this interval, the leakage inductor current is reversing to satisfy the ZVS condition, and no power is transferred to the secondary. A larger leakage inductance L_{lk} or a lower load current I_{load} increases the commutation time, thereby enlarging ΔD. Additionally, output inductor ripple further reduces the available valley current, exacerbating the duty loss under light-load conditions. Understanding and accounting for ΔD is therefore essential for accurate modeling and control design of PSFB converters.

21.2 Controller Design

From a control perspective, the operation of the PSFB converter is analogous to that of a buck converter, with the primary distinction arising from the shape of the applied input voltage. In a conventional buck converter, the input voltage V_{in} is constant DC value. In contrast, the effective input voltage applied to the equivalent buck stage of the PSFB converter is a high-frequency rectangular waveform whose effective duty ratio is denoted by D_{eff}.

The relationship between the commanded phase-shift duty D, the effective duty D_{eff}, and the duty loss ΔD is expressed as

$$D = D_{\mathrm{eff}} + \Delta D$$

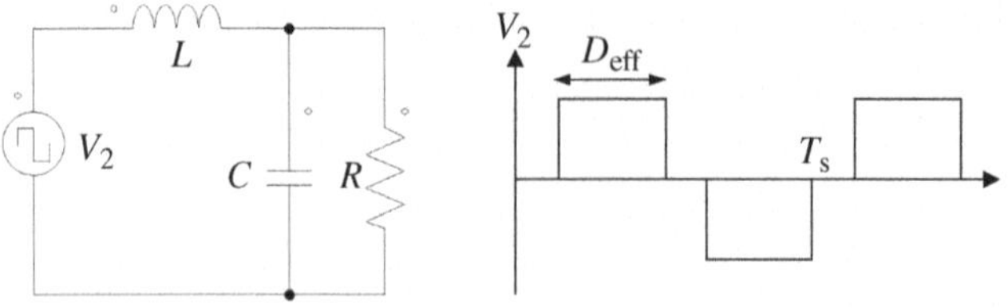

Figure 21.3 The equivalent circuit of the PSFB converter.

An equivalent circuit representation of the PSFB converter, which highlights its correspondence to the buck converter, is shown in Figure 21.3.

Because the equivalent circuit of the PSFB converter is identical to that of a buck converter, the small-signal transfer functions derived for the buck converter can be directly employed for controller design. The only required modification is that the control input is the effective duty ratio D_{eff}, rather than the commanded duty D. The duty loss ΔD, which arises from the ZVS commutation interval, is treated as a feedforward term and is precomputed prior to controller synthesis. Consequently, the voltage controller regulates D_{eff}, while the actual duty applied to the switches is given by $D_{\text{eff}} + \Delta D$.

The controller design procedure begins with a single-loop voltage controller and is subsequently extended to a double-loop control structure by incorporating an inner current loop.

21.2.1 Voltage Controller (Single Loop)

The single-loop voltage controller regulates the output capacitor voltage v_C using the same control configuration as that of a buck converter. The proportional and integral gains, K_{pv} and K_{iv}, are designed using MATLAB's SISOTOOL based on the following frequency-domain specifications:

- Bandwidth: 100 rad/sec
- Phase Margin: 90 deg

Using the Automated Tuning feature in SISOTOOL, the optimized PI controller is obtained as:

$$G_{cv}(s) = \frac{8.0513 \cdot 10^{-7}(s + 207000)}{s}$$

Step Response of Capacitor Voltage The closed-loop step response of the capacitor voltage is shown in Figure 21.4. With the phase margin set to 90°, the system

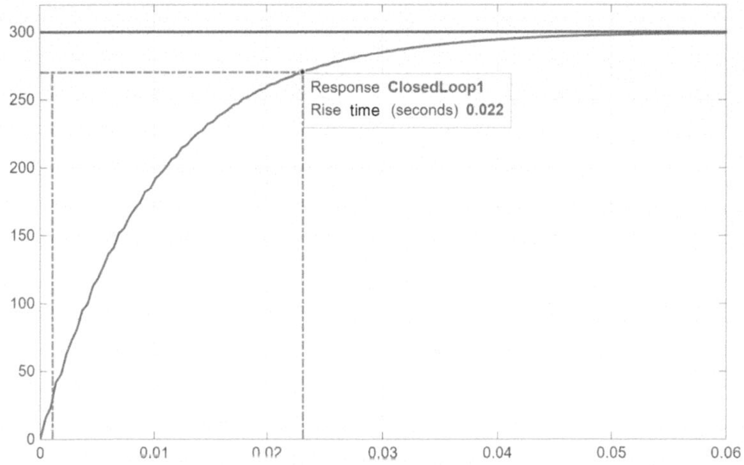

Figure 21.4 Step response of the voltage controller.

exhibits a well-damped transient response with no overshoot, confirming the robustness and stability of the voltage control loop.

PSIM Verification To validate the MATLAB-based controller design, a time-domain simulation was performed using PSIM. The PSIM schematic of the single-loop voltage controller is illustrated in Figure 21.5, and the corresponding voltage and current waveforms are shown in Figure 21.6. The simulation results closely match the MATLAB predictions, thereby confirming the correctness and consistency of the designed controller.

Double-Loop Controller Design To improve dynamic performance and enhance disturbance rejection, a double-loop control structure is adopted, consisting of an outer voltage loop and an inner current loop. The frequency-domain design specifications for the two controllers are summarized in Table 21.1.

Table 21.1 Frequency domain design specifications for double loop controller.

Controller type	Bandwidth [rad/sec]	Phase Margin [deg]
Voltage controller (C_1)	100	90
Current controller (C_2)	20,000	60

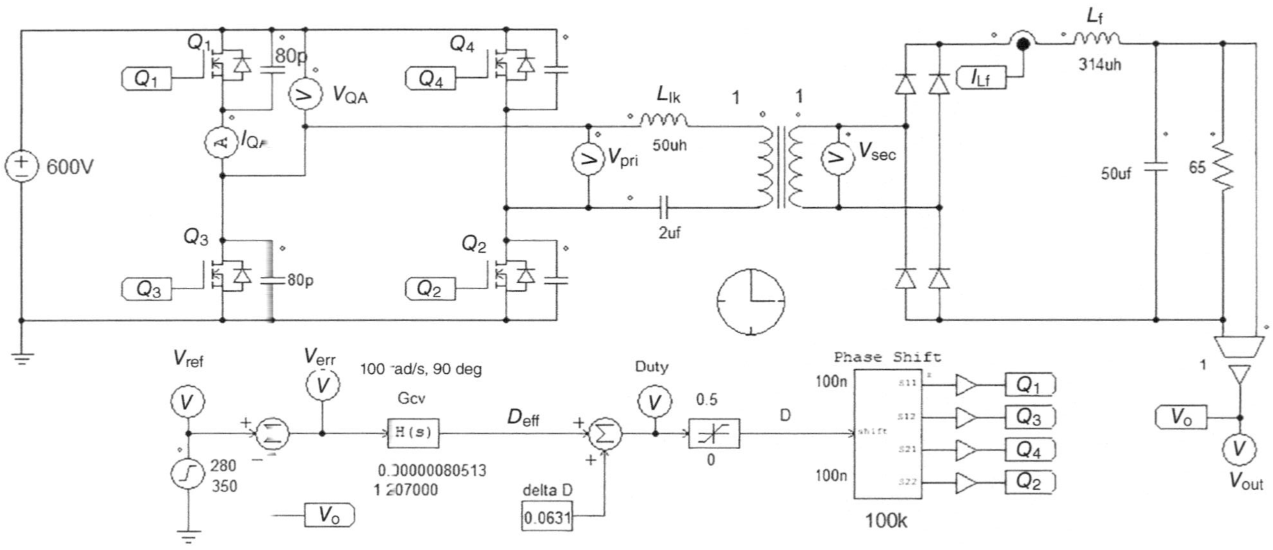

Figure 21.5 PSIM schematic of the single-loop voltage controller.

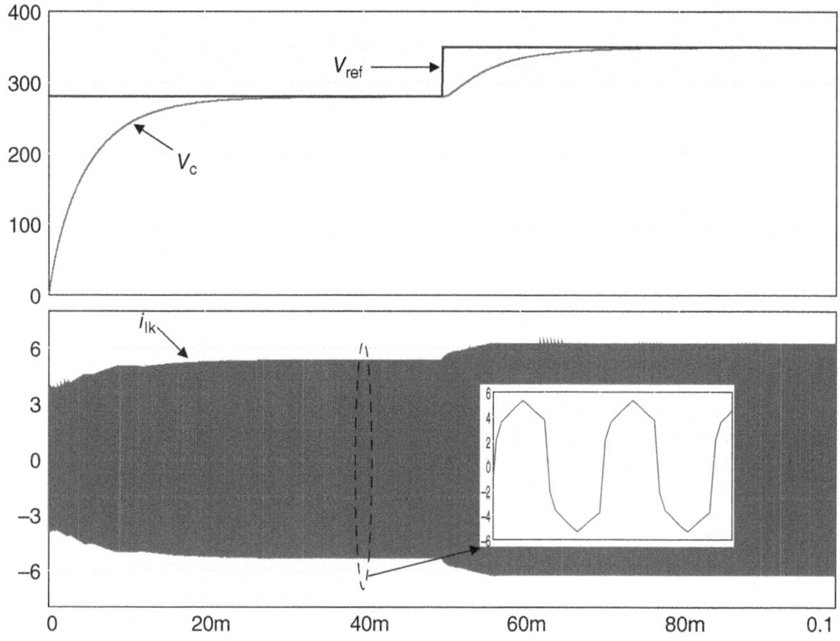

Figure 21.6 Voltage and current waveforms of the single-loop voltage controller.

The controllers are tuned using MATLAB's automated tuning capability, resulting in the following transfer functions:

$$C_1 = G_{cv}(s) = \frac{0.0069(s + 223)}{s}$$

$$C_2 = G_{ci}(s) = \frac{0.00764(s + 11400)}{s}$$

The PSIM schematic of the double-loop control system is shown in Figure 21.7, and the corresponding voltage and current waveforms are presented in Figure 21.8. The simulation results demonstrate excellent agreement with the MATLAB design, confirming that the double-loop controller achieves the desired dynamic performance and stability characteristics.

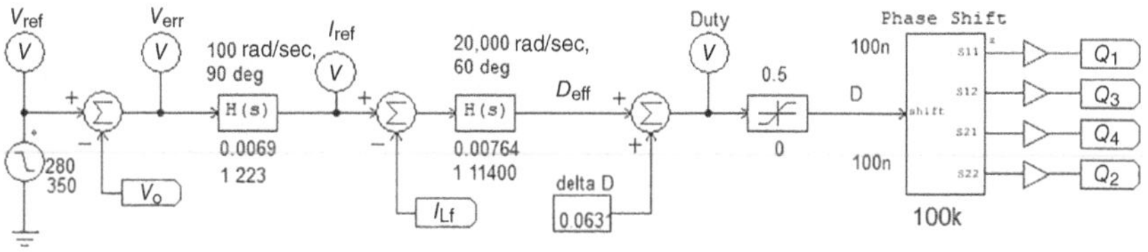

Figure 21.7 The PSIM schematic of the double-loop controller.

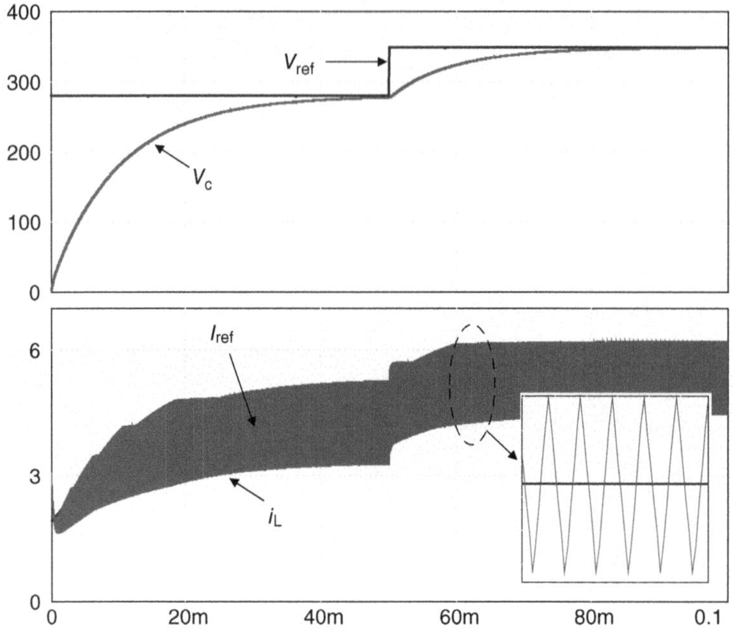

Figure 21.8 Voltage and current waveforms of the double-loop controller.

22

Dual Active Bridge Converter

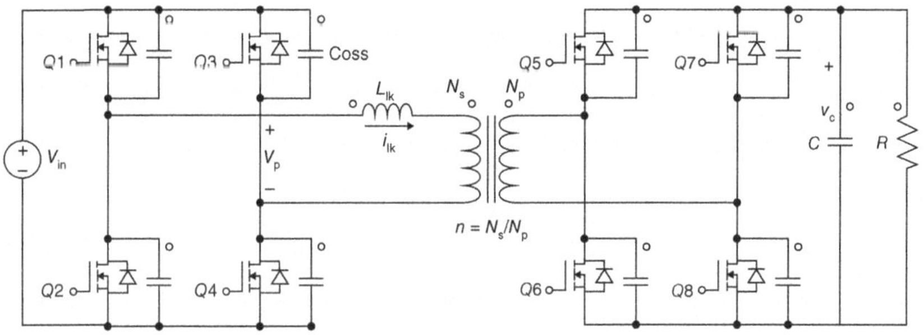

Dual Active Bridge (DAB) Converter The dual active bridge (DAB) Converter is a high-performance isolated DC–DC conversion topology widely employed in renewable-energy systems, energy-storage interfaces, electric-vehicle chargers, and solid-state transformers. Its widespread adoption is primarily attributed to the following key characteristics:

- Bi-directional power flow with simple control, enabling charging and discharging operations
- Galvanic isolation provided by a high-frequency transformer
- Soft-switching capability, which allows high efficiency over a wide operating range

A DAB converter consists of two actively controlled full bridges—one on the primary side and the other on the secondary side—interconnected by a high-frequency isolation transformer. The leakage inductance of the transformer (or an intentionally added series inductance) serves as the main energy-transfer element. The circuit schematic is shown in Figure 22.1. It should be noted that no output inductor is employed on the secondary side.

Unlike PWM-based converters, where power regulation relies on duty-ratio averaging and an output inductor to maintain continuous current, the DAB

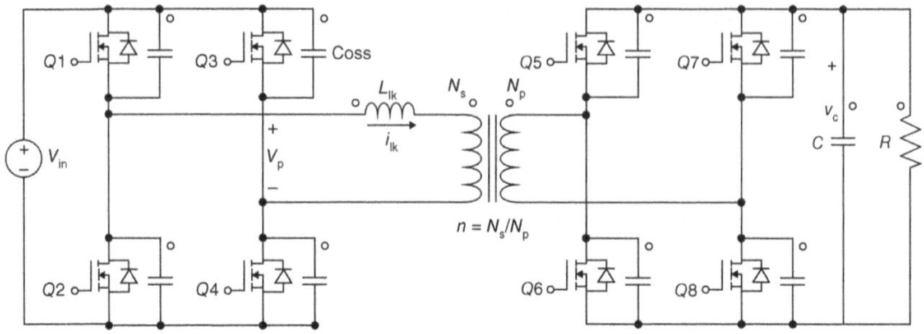

Figure 22.1 Schematic of the dual active bridge (DAB) converter.

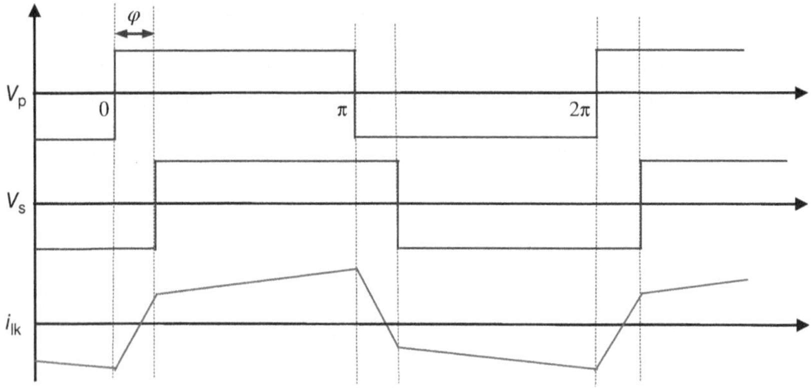

Figure 22.2 Voltage and current waveforms of the DAB converter.

converter regulates power flow through *phase-shift control* between the primary- and secondary-side bridge voltages. The phase difference generates a controlled voltage across the transformer leakage inductance, which in turn determines the magnitude and direction of the power transfer. Consequently, the leakage inductance inherently performs the role of an energy-storage and transfer element.

Since the existing inductance already dictates the current dynamics, no additional output inductor is required. The output capacitor alone provides adequate voltage smoothing, and adding an output inductor would only increase losses, size, and control complexity, especially under bi-directional power flow.

Figure 22.2 illustrates the typical voltage and current waveforms of the DAB converter. The bridge voltages on both sides of the transformer are square waves with fixed duty ratio (typically 50%), while the transformer current exhibits a quasi-triangular waveform determined by the applied phase shift and leakage inductance. By adjusting the phase shift, the converter achieves precise and fast control of power flow in both directions while maintaining soft-switching operation over a wide range of load and voltage conditions.

22.1 Basic Operating Principle

In the DAB converter, both the primary- and secondary-side full bridges are operated with a fixed duty ratio of 50%. As a result, each bridge generates a bipolar square-wave voltage at the switching angular frequency $\omega = 2\pi f_s$ where f_s is a switching frequency. The primary-side bridge produces a square-wave voltage $v_p(t) = \pm V_{in}$, while the secondary-side bridge generates $v_s(t) = \pm V_C$ where V_{in} is the input DC voltage and V_C is the DC-link (or output) voltage on the secondary side. The transformer turns ratio is defined as $n = \frac{N_s}{N_p}$.

By introducing a phase shift φ between the switching functions of the two bridges, a controlled AC voltage difference is established across the transformer leakage inductance L. This voltage difference is given by

$$v_L(t) = v_p(t) \quad nv_s(t)$$

The applied voltage $v_L(t)$ alternates in polarity over each switching interval, forcing a nearly triangular current to flow through the leakage inductance. This current represents the energy-transfer mechanism between the two DC ports.

Power Transfer Using Phase Shift The fundamental operating principle of the DAB converter is that the average power flow is governed exclusively by the phase shift φ between the two bridge voltages, while the duty ratios remain fixed. When the primary-side voltage leads the secondary-side voltage ($\varphi > 0$), power flows from the primary to the secondary side, corresponding to operating modes such as battery charging or DC-bus power export. Conversely, when the secondary-side voltage leads ($\varphi < 0$), power flows from the secondary to the primary side, enabling battery discharge or bi-directional energy-storage operation.

Under the fundamental-frequency approximation (FFA), which retains only the fundamental components of the square-wave voltages, the average transferred power can be expressed as

$$P(\varphi) = \frac{nV_{in}V_C}{\omega L}\left(\varphi - \frac{\varphi^2}{\pi}\right), \quad |\varphi| < \pi \tag{22.1}$$

This relationship reveals several important characteristics of DAB operation:

1) Zero power at zero phase shift: When $\varphi = 0$, the primary and secondary voltages are aligned, resulting in no net voltage across the leakage inductance and, therefore, zero average power transfer.
2) Approximately linear behavior for small phase shifts: For small values of φ, the quadratic term is negligible, and the transferred power is approximately proportional to φ, which simplifies control design.

3) Maximum power at intermediate phase shift: The power reaches its maximum near $\varphi = \pi/2$, beyond which further increases in phase shift reduce the effective power transfer.

Because the power flow is regulated solely by phase shift rather than duty-ratio modulation, the DAB converter is classified as a phase-shift–controlled DC–DC converter. In this sense, its operating principle is analogous to that of the phase-shift full-bridge (PSFB) converter. However, unlike the PSFB, the DAB inherently supports bi-directional power flow, making it particularly suitable for modern energy-storage systems, electric vehicles, and DC microgrids.

22.2 Modeling and Controller Design

In conventional PWM-based converters and inverters, the inductor current within a single switching period contains a ripple component whose average value is non-zero. This property allows the instantaneous switching behavior to be replaced by an averaged representation over one switching interval. As a result, the state-space averaging method is widely used to derive small-signal models and control-oriented transfer functions.

In the DAB converter, however, the situation is fundamentally different. As illustrated in Figure 22.2, the inductor (transformer leakage) current exhibits a *bipolar waveform whose average over one switching period is zero*. This is a direct consequence of the bi-directional power-transfer mechanism and the fixed 50% duty operation of both bridges. Because the averaged inductor current is identically zero, the classical state-space averaging technique breaks down and cannot be used to obtain a meaningful small-signal model.

Therefore, alternative modeling approaches are required to accurately describe the dynamics of the DAB converter. Commonly used methods include energy-balance modeling, describing-function analysis, and frequency-domain perturbation techniques. Among these, energy-based modeling is particularly attractive for controller design because it yields a low-order, physically intuitive representation of the output-voltage dynamics.

Output-Side Energy-Balance Equation Consider the secondary-side DC bus, where a DC-link capacitor C supplies a resistive load R. Let $v_c(t)$ denote the output voltage. The input side is assumed to be fed by a stiff DC source, allowing V_{in} to be treated as constant.

The large-signal energy-balance equation on the output DC bus can be written as

$$C\frac{dv_c}{dt} = \frac{P_{in} - P_{load}}{v_c}$$

where the input and load powers are given by

$$P_{in} \approx \frac{nV_{in}v_c}{\omega L}\left(\varphi - \frac{\varphi^2}{\pi}\right)$$

$$P_{\text{load}} = \frac{v_c^2}{R}$$

This formulation highlights that the output-voltage dynamics are governed by the power imbalance between the phase-shift–controlled input power and the load power absorbed at the DC bus.

Small-Signal Linearization To derive a control-oriented model, small perturbations around a steady-state operating point are introduced:

$$v_c(t) = V_{c,0} + \hat{v}_c(t), \quad \varphi(t) = \varphi_0 + \hat{\varphi}(t)$$

Linearizing the input power around $(V_{c,0}, \varphi_0)$ yields

$$P_{\text{in}}(\varphi) = P_o + \hat{P}_{\text{in}}(\varphi) = P_o + \left(\frac{\partial P}{\partial \varphi}\right)_{\varphi_0, V_{c,0}} \cdot \hat{\varphi} = P_o + \frac{nV_{\text{in}}V_{c,0}}{\omega L}\left(1 - \frac{2\varphi_0}{\pi}\right) \cdot \hat{\varphi}$$

Similarly, the load power perturbation is

$$P_{\text{load}} = P_o + \hat{P}_{\text{load}} = P_o + \frac{2V_{c,0}}{R}\hat{v}_c$$

For the left-hand side of the energy-balance equation,

$$C\frac{dv_c}{dt} \cdot v_c = C\frac{d(V_{c,0} + \hat{v}_c)}{dt} \cdot (V_{c,0} + \hat{v}_c) \approx V_{c,0} \cdot C\frac{d\hat{v}_c}{dt}$$

where second-order perturbation terms have been neglected.

Substituting these expressions into the energy-balance equation gives

$$V_{c,0} \cdot C\frac{d\hat{v}_c}{dt} = \frac{nV_{\text{in}}V_{c,0}}{\omega L}\left(1 - \frac{2\varphi_0}{\pi}\right) \cdot \hat{\varphi} - \frac{2V_{c,0}}{R}\hat{v}_c$$

Dividing both sides by $V_{c,0}$ results in the linearized voltage-dynamics equation

$$C\frac{d\hat{v}_c}{dt} = \frac{nV_{\text{in}}}{\omega L}\left(1 - \frac{2\varphi_0}{\pi}\right) \cdot \hat{\varphi} - \frac{2}{R}\hat{v}_c$$

Taking the Laplace transform, the small-signal control-to-output transfer function is obtained as,

$$G_{v\phi}(s) \equiv \frac{\hat{v}_c(s)}{\hat{\varphi}(s)} = \frac{\frac{nV_{\text{in}}}{\omega L}\left(1 - \frac{2\varphi_0}{\pi}\right)}{Cs + \frac{2}{R}} \tag{22.2}$$

This result reveals that the DAB output-voltage dynamics are first-order, with a dominant pole determined by the DC-link capacitance and load resistance. The control gain depends explicitly on the operating phase shift φ_0, implying that controller performance varies with operating point. This characteristic necessitates the adoption of gain scheduling or robust PI controller designs in practical DAB voltage-control implementations.

Voltage Controller Design (Single Loop) The single-loop output-voltage controller is designed using a proportional–integral (PI) compensator. The controller gains, K_{pv} and K_{iv}, are determined in the frequency domain using MATLAB SISOTOOL to ensure robust closed-loop performance. The design specifications are selected as follows:

- Bandwidth: 100 rad/sec
- Phase Margin: 90 deg

A relatively low bandwidth is chosen to guarantee sufficient separation from the switching frequency and to maintain the validity of the small-signal model, while the large phase margin ensures a well-damped transient response with strong robustness against parameter variations. The MATLAB m-file is presented to design single-loop voltage controller in Table 22.1.

Table 22.1 MATLAB m-file for single-loop voltage controller design.

```
s=tf('s');
Vin = 400;
N  = 2;          % transformer turn ratio Np/Ns
fs = 20000;      % switching frequency 20 kHz
L = 70e-6;       % leakage inductance
C   = 1000e-6; % output capacitor
R   = 4;         % load resistance
Td  = 200e-9;   % dead time
Ws = 2*pi*fs;

%%%%% DC Operating Point %%%%%%
phi0 = pi/5;    % phase shift of 0.6283
Vc0 = Vin*N*R/(Ws*L)*(phi0- phi0^2 / pi);

Kg = N*Vin/(Ws*L)*(1-2*phi0/pi);
G2 = Kg/(C*s + 2/R)

Kpv=1; Kiv=1;
Gcv = Kpv*(s+Kiv)/s;

Vref = int16(Vc0);
%sisotool Single loop-voltage
T=sisoinit(1);
T.F.Value = Vref;
T.G.Value = G2;
T.C.Value = Gcv;
sisotool(T);
```

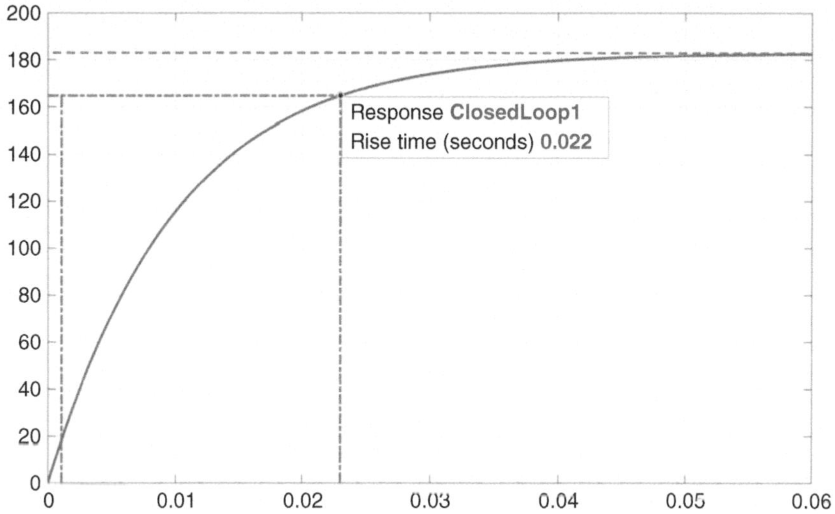

Figure 22.3 Step response of the voltage controller.

Using the Automated Tuning feature of SISOTOOL to meet the specified bandwidth and phase-margin requirements, the optimized PI voltage controller is obtained as:

$$G_{cv}(s) = \frac{0.00183(s + 500)}{s}$$

This controller provides integral action to eliminate steady-state voltage error and sufficient proportional gain to achieve the desired dynamic response.

Step Response of Capacitor Voltage The closed-loop step response of the capacitor voltage is shown in Figure 22.3. With the phase margin set to 90°, the response exhibits smooth and well-damped transient behavior without overshoot. This confirms the robustness and stability of the single-loop voltage control design and validates the suitability of the chosen bandwidth for DAB operation.

PSIM Verification To verify the MATLAB-based controller design in the time domain, a simulation was performed using PSIM. The PSIM schematic of the single-loop voltage controller is shown in Figure 22.4. In the PSIM implementation, the output of the voltage controller is divided by 2π to match the input range [0, 1] of the *phase-shift* control library.

The corresponding output-voltage response and control input are illustrated in Figure 22.5, demonstrating close agreement with the MATLAB simulation results.

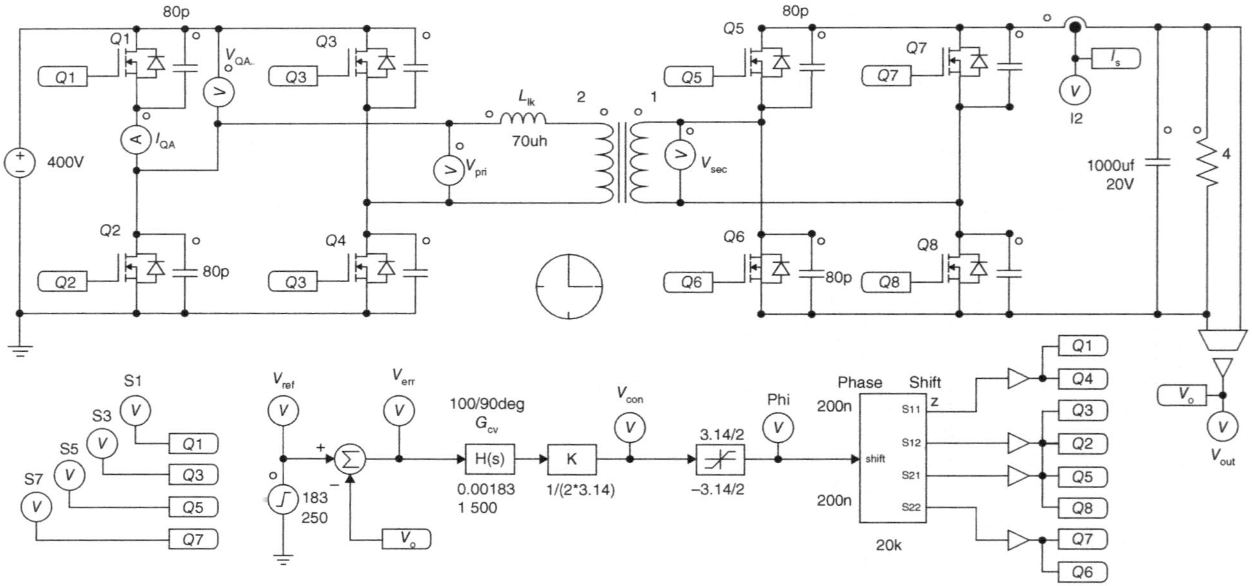

Figure 22.4 PSIM schematic of the single-loop voltage controller.

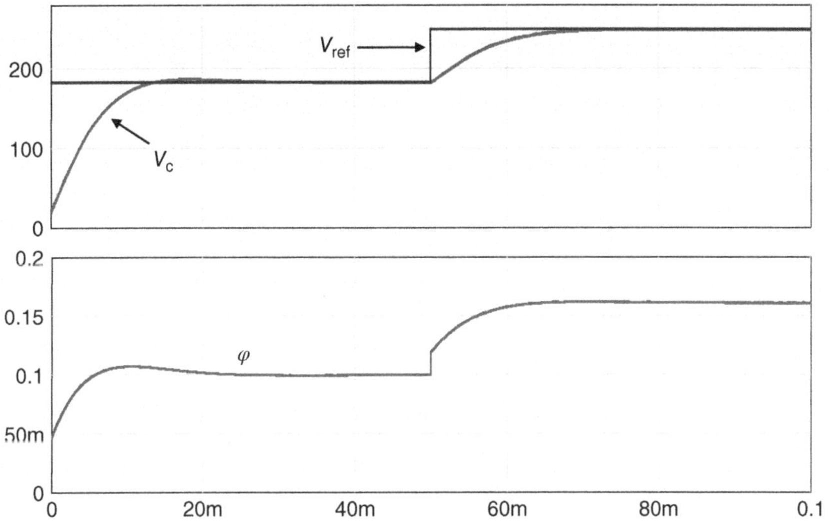

Figure 22.5 Step response waveforms of output voltage and control input.

The primary- and secondary-side voltage and current waveforms are presented in Figure 22.6, confirming proper phase-shift modulation, correct power transfer, and stable bi-directional operation. The strong correlation between MATLAB and PSIM results validates both the derived small-signal model and the effectiveness of the designed single-loop voltage controller.

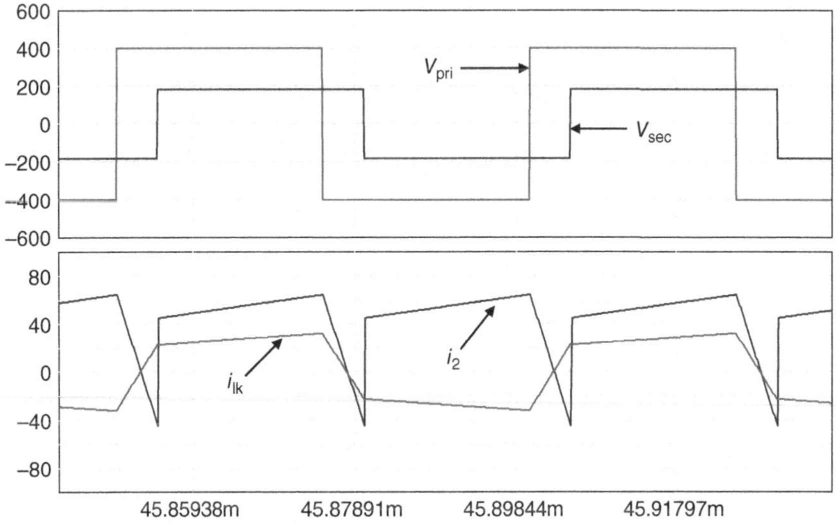

Figure 22.6 Voltage and current waveforms on the primary and secondary sides.

23

LLC Resonant Converter

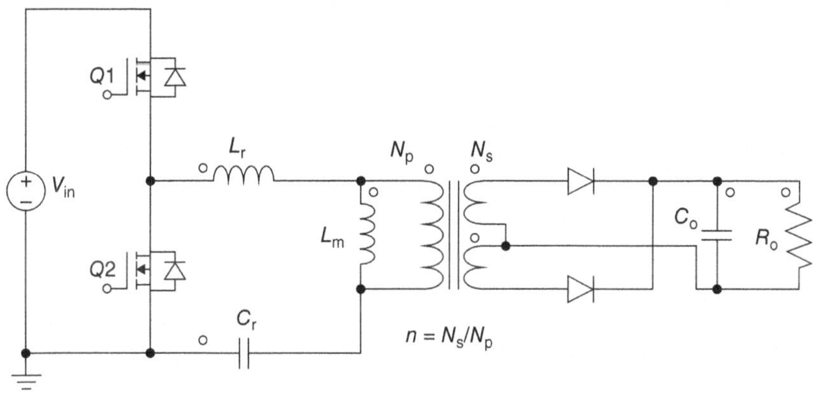

23.1 Basic Operation of the LLC Resonant Converter

The LLC resonant converter is a high-efficiency isolated DC–DC power conversion topology that is widely employed in applications such as server power supplies, telecommunications systems, and electric-vehicle chargers. Its widespread adoption is primarily attributed to its ability to achieve soft-switching operation over a wide load range, resulting in reduced switching losses, high efficiency, and increased power density.

Unlike conventional PWM converters, which regulate the output voltage by varying the duty ratio, the LLC resonant converter controls the output predominantly through switching-frequency modulation while maintaining a fixed duty ratio, typically 50%. Power transfer is achieved through a resonant tank composed of a series resonant inductor, a resonant capacitor, and the magnetizing inductance of the isolation transformer. This resonant network transforms the applied square-wave voltage into a near-sinusoidal current, thereby enabling zero-voltage switching (ZVS) of the primary-side switches and significantly reducing switching stress.

In a typical half-bridge configuration, two primary switches alternately apply a high-frequency square-wave voltage to the resonant tank. The resulting resonant current flows through the transformer, providing galvanic isolation and voltage scaling according to the transformer turn ratio. On the secondary side, a full-wave rectifier and an output filter capacitor convert the high-frequency AC voltage into a regulated DC output supplied to the load. The output voltage is governed by the switching frequency, resonant tank parameters, and load conditions.

A defining characteristic of the LLC resonant converter is its frequency-dependent voltage gain. Near the resonant frequency, the converter achieves maximum efficiency but exhibits limited voltage-regulation capability. Consequently, practical LLC designs operate above the resonant frequency, where the voltage gain decreases monotonically as the switching frequency increases. This operating region enables stable output-voltage regulation through frequency modulation while preserving soft-switching operation over a wide load range. Owing to its resonant energy transfer mechanism and frequency-based control strategy, the LLC resonant converter provides an efficient and reliable solution for isolated DC–DC power conversion. The circuit configuration of the LLC converter is illustrated in Figure 23.1. It should be noted that, like the dual active bridge (DAB) converter, no output inductor is employed on the secondary side.

The voltage and current waveforms of the LLC converter operating at the resonant frequency are shown in Figure 23.2. The primary switches Q_1 and Q_2 are operated with a fixed 50% duty ratio, including a short dead time. At the instant when each switch is turned on, the corresponding switch current is negative,

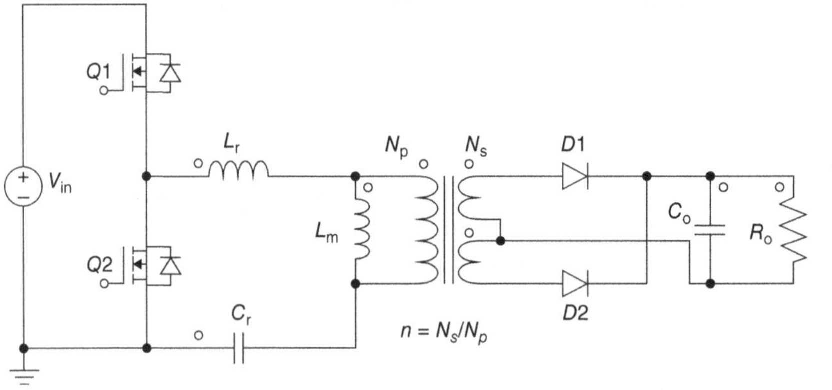

Figure 23.1 The LLC circuit configuration.

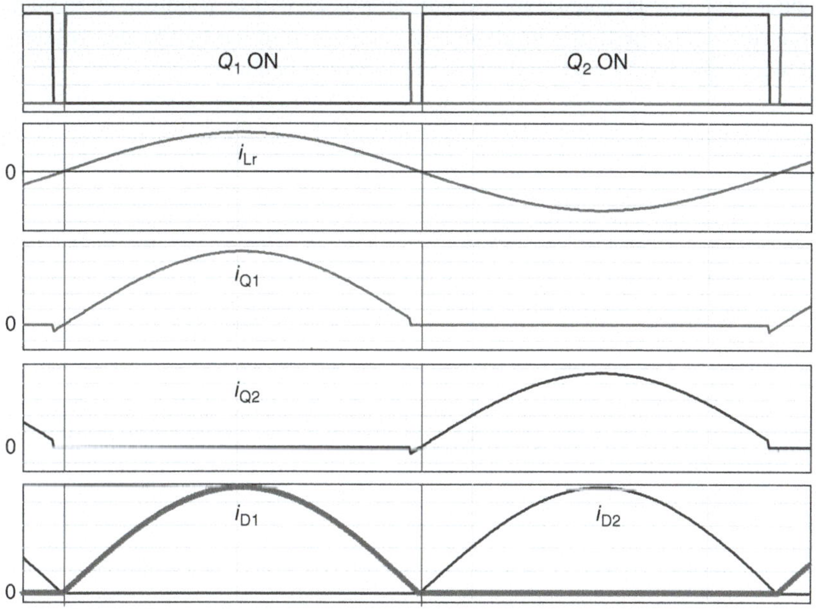

Figure 23.2 Voltage and current waveforms of the LLC converter operating at the resonant frequency.

enabling ZVS. On the secondary side, the rectifier diode currents commutate at half the switching period, consistent with the resonant operation of the converter.

Resonant Tank Model (FHA Equivalent Circuit) Consider a conventional half-bridge LLC resonant converter analyzed using the fundamental harmonic approximation (FHA). Under this approximation, only the fundamental component of the applied square-wave voltage is considered, and the resonant tank is modeled as a linear time-invariant AC network. The resonant tank consists of the following elements:

- Series resonant branch: Resonant inductor L_r and resonant capacitor C_r.
- Parallel branch: Magnetizing inductance L_m in parallel with the reflected load resistance R_{ac}, where the load is referred to the primary side
- Input excitation: Fundamental component of the half-bridge square-wave voltage
- Output stage: Full-wave rectifier followed by an output capacitor supplying a load R_o

Under FHA, the primary-side resonant tank is excited by a sinusoidal voltage source corresponding to the fundamental component of the half-bridge voltage, enabling frequency-domain analysis of the converter behavior.

FHA Primary-Side AC Model The impedance of the series resonant branch is given by

$$Z_s = Z_{L_r} + Z_{C_r} = j\omega L_r + \frac{1}{j\omega C_r}$$

The parallel branch, consisting of the magnetizing inductance and the reflected load resistance, has an equivalent impedance expressed as

$$Z_p = \left(\frac{1}{j\omega L_m} + \frac{1}{R_{ac}}\right)^{-1}$$

The total impedance of the resonant tank as seen by the fundamental input voltage source is therefore

$$Z_{tank} = Z_s + Z_p$$

The AC voltage transfer function from the fundamental input voltage to the resonant-tank output node is defined as

$$H(j\omega) \triangleq \frac{\hat{V}_p}{\hat{V}_{in,1}} = \frac{Z_p}{Z_s + Z_p}$$

where $\hat{V}_{in,1}$ denotes the amplitude of the fundamental component of the half-bridge input voltage, and $\hat{V}_p$ is the corresponding voltage across the parallel branch.

FHA Secondary-Side DC Output Model On the secondary side, a full-wave bridge rectifier followed by a large output capacitor forms a DC bus. Under FHA, the DC load can be represented by an equivalent AC resistance on the secondary side, given by

$$R_{ac,sec} = \frac{8}{\pi^2} R_o$$

Referring this resistance to the primary side yields

$$R_{ac} = \frac{R_{ac,sec}}{n^2} = \left(\frac{8}{n^2\pi^2} R_o\right)$$

where

$$n \equiv \frac{N_s}{N_p}$$

and N_p and N_s denote the number of turns of the primary and secondary windings, respectively.

Based on the average-rectified sinusoidal assumption, the DC output voltage gain $M$is often expressed as

$$M = (4/\pi^2) \cdot n \mid H \mid$$

However, when a diode bridge rectifier is combined with a large output capacitor, the rectifier current becomes highly pulsed, and the simple equivalent-resistance mapping is no longer exact. In this practical operating condition, a more accurate expression for the DC voltage gain under FHA is given by

$$M = (n/2) \cdot \mid H \mid$$

This expression more accurately captures the relationship between the resonant-tank AC behavior and the DC output voltage for LLC converters operating with capacitive output filtering.

23.2 Small-Signal Modeling

Under the FHA and assuming a constant input voltage V_{in}, the DC output voltage of the LLC resonant converter can be expressed as

$$V_{\text{o}} = M(\omega_{\text{s}})\, V_{\text{in}}$$

where $M(\omega_{\text{s}}) = V_{\text{o}}/V_{\text{in}}$ is the DC voltage gain of the LLC converter as a function of the switching angular frequency ω_{s}, and ω_{s} serves as the control input.

The output-stage dynamics, consisting of the output capacitor C_{o} and load resistance R_{o}, are approximated by the following DC-side equation:

$$C_{\text{o}} \frac{\text{d}v_{\text{o}}}{\text{d}t} = \frac{V_{\text{src}} - v_{\text{o}}}{R_{\text{o}}} = \frac{M(\omega_{\text{s}})\, V_{\text{in}} - v_{\text{o}}}{R_{\text{o}}}$$

Small-Signal Linearization Define small perturbations around a steady-state operating point as

$$\omega_{\text{s}} = \omega_{\text{s0}} + \hat{\omega}_{\text{s}}, \quad v_{\text{o}} = V_{\text{o0}} + \hat{v}_{\text{o}}, \quad M = M(\omega_{\text{s0}}) + \hat{M}$$

Using a first-order Taylor expansion of the gain function yields

$$\hat{M} \approx \left.\frac{\partial M}{\partial \omega_{\text{s}}}\right|_{\omega_{\text{s0}}} \cdot \hat{\omega}_{\text{s}}.$$

Substituting these expressions into the output dynamic equation gives

$$C_{\text{o}} \frac{\text{d}(V_{\text{o0}} + \hat{v}_{\text{o}})}{\text{d}t} = \frac{\left(M(\omega_{\text{s0}}) + \left.\frac{\partial M}{\partial \omega_{\text{s}}}\right|_{\omega_{\text{s0}}} \cdot \hat{\omega}_{\text{s}} \right) V_{\text{in}} - (V_{\text{o0}} + \hat{v}_{\text{o}})}{R_{\text{o}}}$$

At steady state, the DC operating-point condition is

$$V_{o0} = M(\omega_{s0})V_{in}$$

Subtracting the steady-state equation yields the small-signal perturbation model:

$$C_o \frac{d\hat{v}_o}{dt} = \frac{F_m \cdot V_{in}}{R_o}\hat{\omega}_s - \frac{\hat{v}_o}{R_o}$$

where the frequency sensitivity is defined as

$$F_m \equiv \left.\frac{\partial M}{\partial \omega_s}\right|_{\omega_{s0}}$$

Control-to-Output Transfer Function Taking the Laplace transformation of the small-signal equation results in

$$\left(C_o s + \frac{1}{R_o}\right)\hat{v}_o(s) = \frac{V_{in}F_m}{R_o}\hat{\omega}_s(s)$$

Therefore, the control-to-output voltage transfer function of the LLC converter is obtained as

$$G_{vd}(s) = \frac{\hat{v}_o(s)}{\hat{\omega}_s(s)} = \frac{V_{in}F_m}{1 + sR_oC_o} \tag{23.1}$$

Equation (23.1) reveals that the resonant-tank dynamics do not explicitly appear in the small-signal model; instead, their effect is captured implicitly through the frequency sensitivity F_m. When the operating point is close to the resonant frequency, the slope of the gain characteristic satisfies

$$\frac{dM}{d\omega_s} \approx 0,$$

resulting in a very small F_m. Consequently, the loop gain becomes low, leading to weak control authority and poor output voltage regulation.

In contrast, when the converter operates above the resonant frequency, the gain characteristic exhibits a large negative slope. In this region, F_m attains a sufficiently large magnitude, enabling effective control of the output voltage through switching-frequency modulation and ensuring stable closed-loop operation.

Because the sensitivity term F_m is a highly nonlinear function of frequency, load, and resonant parameters, it is typically evaluated numerically. In this work, F_m is calculated using MATLAB-based numerical analysis. The corresponding MATLAB code used for controller design is presented in Table 23.1.

Table 23.1 MATLAB m-file for LLC small-signal transfer function derivation (frequency control).

```
%% LLC PI controller design (frequency control)
%% Parameters
Vin = 400;          % [V]
n   = 4;            % turns ratio, n = Ns/Np
Lr  = 20e-6;        % [H]
Cr  = 47e-9;        % [F]
Lm  = 80e-6;        % [H]
Ro  = 20;           % [ohm]
Co  = 470e-6;       % [F]

%% Operating point
fn0 = 1.10;                 % fs0/f0 (set operating point)
w0  = 1/sqrt(Lr*Cr);        % [rad/s]
ws0 = fn0*w0;               % [rad/s]
fs0 = ws0/(2*pi);           % [Hz]

%% Frequency sweep for numerical derivative
fn_vec = linspace(0.6, 2.0, 2000);
w_vec  = fn_vec*w0;

%% AC load mapping (FHA resistive mapping)
Rac_sec = (8/pi^2)*Ro;
Rac     = Rac_sec / n^2;

%% Tank impedances and H(jw)
Zs = 1j*w_vec*Lr + 1./(1j*w_vec*Cr);
Zp = 1 ./ ( 1./(1j*w_vec*Lm) + 1./Rac );
H  = Zp ./ (Zs + Zp);

%% DC gain mapping
// M = (4/pi^2)*n.*abs(H); % average-rectified mapping
M = (n/2).*abs(H);          % capacitor-input (peak) approximation

%% Fm = dM/dw at operating point
Fm_vec = gradient(M, w_vec);       % per (rad/s)

[~, idx0] = min(abs(w_vec - ws0));
M0  = M(idx0);
Fm  = Fm_vec(idx0);         % dM/dw at ws0
K   = Vin * Fm;             % plant DC gain (V per rad/s)
```

(Continued)

Table 23.1 (Continued)

```
%% Small-signal plant: Gvd(s)=(Vin*Fm)/(1+ s*Ro*Co)
s   = tf('s');
Gvd = K / (1 + s*Ro*Co);

fprintf('Operating fs0 = %.1f kHz (fn0=%.3f)\n', fs0/1e3, fn0);
fprintf('M0 = Vout/Vin ≈ %.4f\n', M0);
fprintf('Fm = dM/dw ≈ %.3e per(rad/s)\n', Fm);

%% Bode plot of Gvd(s)
figure;
bode(Gvd); grid on;
title('Small-signal transfer function Gvd(s) of LLC(frequency control)');

%% Plot M(fn) for reference (DC gain vs frequency)
figure;
plot(fn_vec, M, 'LineWidth', 1.5); hold on;
plot(fn0, M0, 'ro', 'MarkerSize', 8, 'LineWidth', 1.5);
xlabel('Normalized frequency fn = fs/f0');
ylabel('DC gain M(fn) = Vo/Vin');
title('LLC DC gain vs normalized frequency(FHA)'); grid on;
legend('M(fn)',sprintf('Operating point fn=%.2f',fn0),'Location','Best');
```

MATLAB Results Using the converter parameters listed in Table 23.1 and selecting the operating point at $f_{n0} = f_{s0}/f_0 = 1.10$, the numerical evaluation produces the following results:

- Operating switching frequency: $f_{s0} \approx 180.6$ kHz
- DC gain at the operating point: $M_0 = V_o/V_{in} \approx 0.4967$
- Frequency sensitivity at the operating point:

$$F_m = \left.\frac{dM}{d\omega_s}\right|_{\omega_{s0}} \approx -4.308 \times 10^{-6}\ (\text{rad/sec})^{-1}$$

- Small-signal plant gain: $K = V_{in}F_m \approx -1.723 \times 10^{-3}$ V/(rad/sec)

Because $F_m < 0$ in the above-resonance region, an increase in switching frequency reduces the DC gain M, and thus decreases V_o. This negative slope provides effective control authority for closed-loop output-voltage regulation using frequency modulation.

The Bode plot of the control-to-output transfer function $G_{vd}(s)$, together with the DC gain at the selected operating frequency, is illustrated in Figures 23.3 and 23.4.

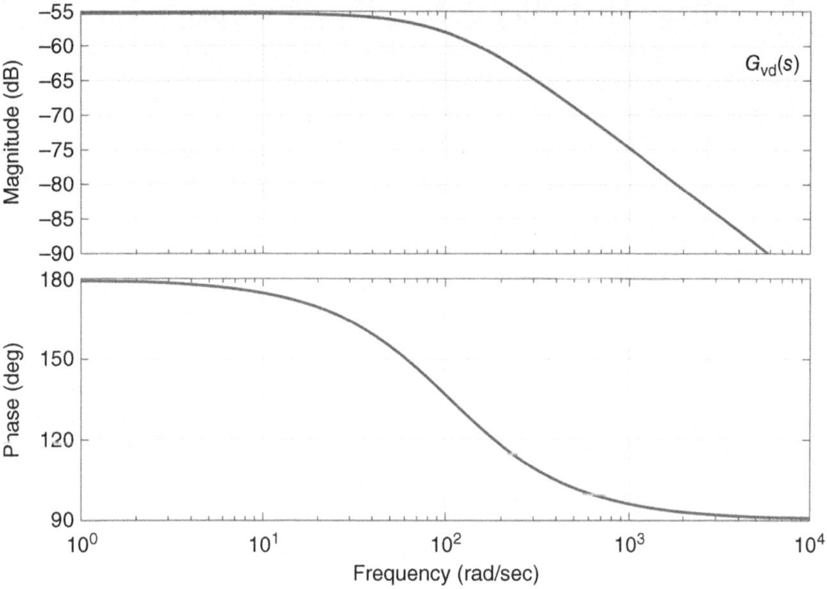

Figure 23.3 Small-signal transfer function of $G_{vd}(s)$ of LLC converter.

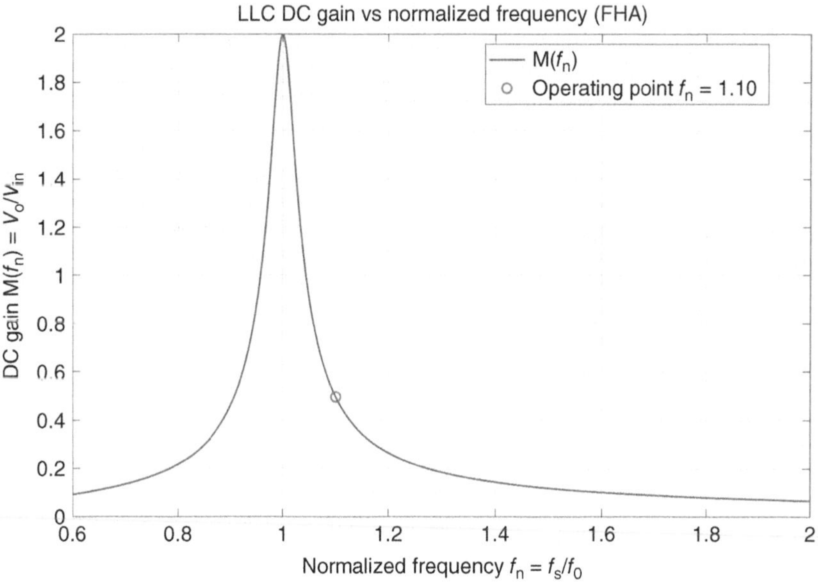

Figure 23.4 LLC DC gain vs. normalized frequency graph.

23.3 Voltage Controller Design (Single Loop)

Based on this small-signal model, a single-loop voltage controller is designed using a proportional–integral (PI) structure. The proportional and integral gains, K_{pv} and K_{iv}, are determined using the MATLAB SISOTOOL according to the following frequency-domain specifications:

- Bandwidth: 100 rad/sec
- Phase Margin: 90 deg

These specifications provide a conservative design margin that ensures stable operation despite modeling uncertainties and nonlinearities inherent to the LLC resonant converter. The MATLAB m-file using SISOTOOL is illustrated in Table 23.2.

Using the automated tuning feature in SISOTOOL, the optimized PI controller is obtained as:

$$G_{cv}(s) = \frac{-554(s + 106)}{s}$$

The corresponding closed-loop step response of the output voltage is shown in Figure 23.5, demonstrating a fast transient response with negligible overshoot and robust stability margins.

PSIM Verification Because the control-to-output transfer function $G_{vd}(s)$ exhibits a negative gain, the output of the PI voltage controller is correspondingly negative. This controller output is algebraically summed with a predefined nominal switching frequency that is intentionally selected to operate above the resonant frequency of the LLC tank. Consequently, the PI controller does not directly determine the

Table 23.2 MATLAB m-file for single-loop voltage controller design.

```
Kpv=1; Kiv=1;
Gcv = Kpv*(s+Kiv)/s;

Vref = 200;
%sisotool Single loop-voltage
T=sisoinit(1);
T.F.Value = Vref;
T.G.Value = Gvd;
T.C.Value = Gcv;
sisotool(T);
```

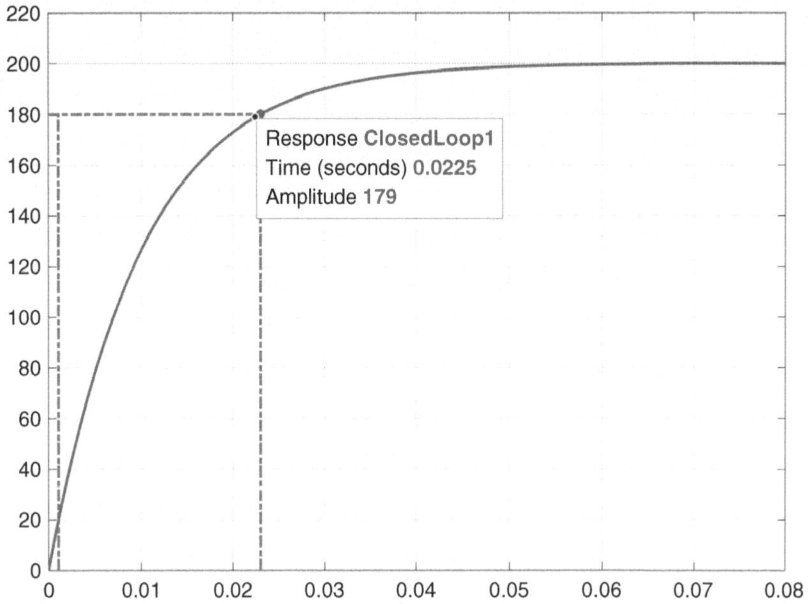

Figure 23.5 Step response of the output voltage in the LLC converter.

absolute switching frequency; instead, it generates a small frequency deviation around a fixed operating point.

This control strategy deliberately avoids operation near the resonant frequency, where the gain slope $dM/d\omega_s$ becomes small and the control authority of the LLC converter is weak. By confining the controller action to a narrow frequency range above resonance, stable voltage regulation and robust dynamic performance are ensured, even under load disturbances.

The resulting frequency command is converted into a phase-continuous sinusoidal carrier using an integrator-based oscillator. This carrier signal is then transformed into fixed-duty (50%) gate signals for the half-bridge switches, enabling stable and efficient operation of the LLC converter in the above-resonance region.

The complete PSIM schematic of the LLC converter with the proposed voltage control strategy is shown in Figure 23.6. The circuit parameters used in PSIM are identical to those employed in the MATLAB model.

The output-voltage waveforms corresponding to a reference step change are presented in Figure 23.7. The voltage response exhibits excellent agreement with the MATLAB simulation results, thereby validating the accuracy of the small-signal model and the correctness of the controller design.

The second waveform in Figure 23.7 shows the variation of the switching-frequency deviation $\Delta\omega_s$. The actual switching frequency applied to the converter

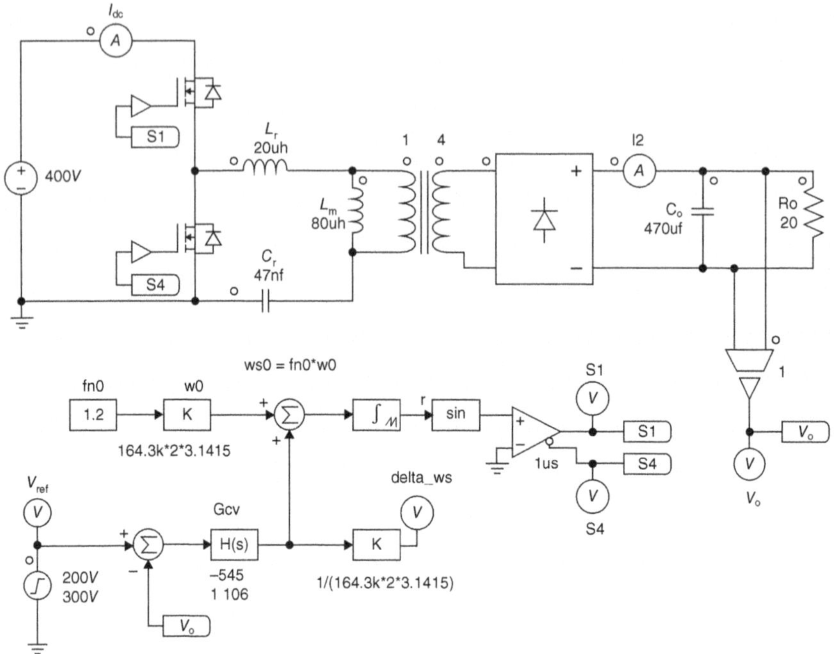

Figure 23.6 PSIM schematic for voltage control of the LLC resonant converter.

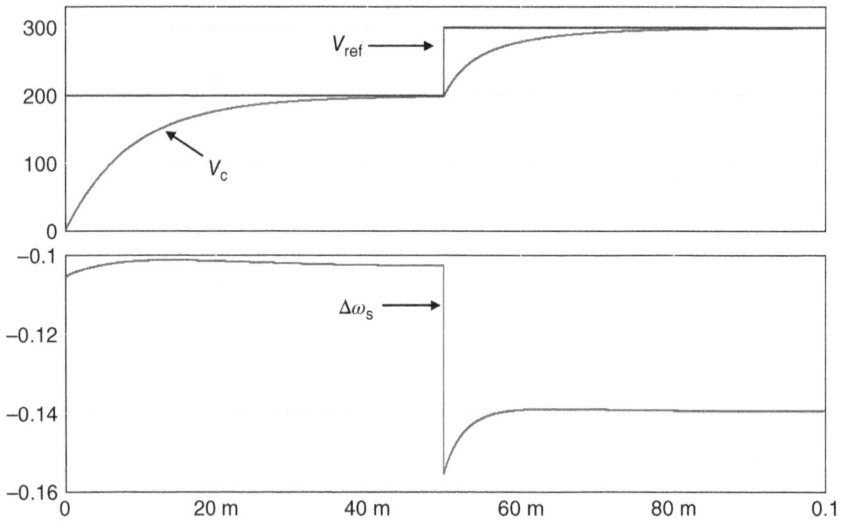

Figure 23.7 Step response of the output voltage and $\Delta\omega_s$.

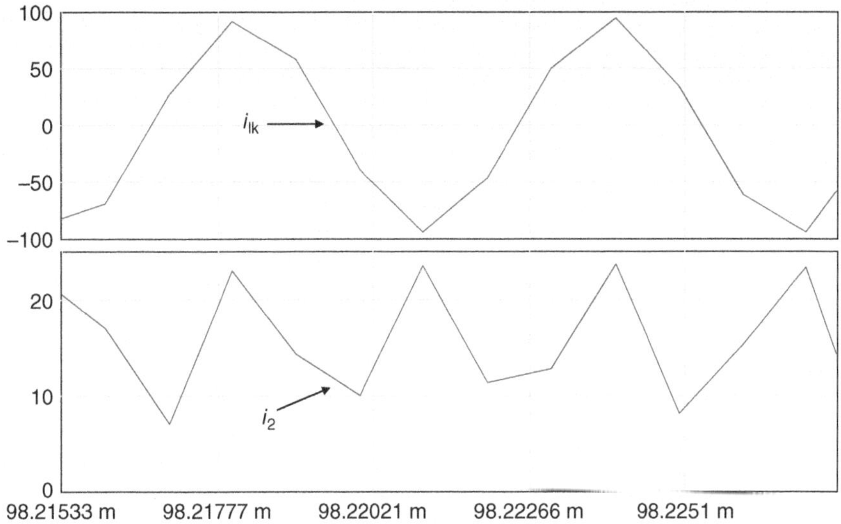

Figure 23.8 Primary and secondary current waveforms.

is obtained by subtracting $\Delta\omega_s$ from the nominal switching frequency. It can be observed that both the magnitude and the dynamic behavior of $\Delta\omega_s$ closely match the predictions of the analytical modeling equations, further confirming the validity of the proposed frequency-control approach.

Figure 23.8 illustrates the primary and secondary current waveforms of the converter. The upper plot shows the primary-side current i_{lk}, which clearly exhibits a bi-directional waveform with alternating positive and negative polarities. This behavior indicates that the energy transfer on the primary side is reversible within a switching period, which is a fundamental characteristic of phase-shift-based or transformer-isolated bi-directional converters. The current reversals occur as the applied voltage polarity across the leakage inductance changes, enabling controlled power transfer through the transformer.

The lower plot presents the secondary-side current i_2, which remains unidirectional due to the rectification process on the secondary side. Although the secondary current follows a pulsating waveform synchronized with the switching action, its magnitude reflects the instantaneous power delivered to the load. The difference in current characteristics between the primary and secondary sides highlights the role of the transformer and rectifier in shaping current flow while maintaining galvanic isolation and controlled power conversion.

24

Discontinuous Current Mode

24.1 Operation of the DCM in Buck Converter

The buck converter is a fundamental DC–DC power-conversion topology widely employed in power-electronic systems. Under typical operating conditions—characterized by sufficiently large inductance and moderate to heavy load current—the converter operates in continuous current mode (CCM), where the inductor current remains nonzero throughout the entire switching period. However, under light-load conditions, at high switching frequencies, or when the inductance value is reduced, the buck converter may enter discontinuous current mode (DCM). In DCM, the inductor current falls to zero for a finite portion of each switching cycle.

In DCM operation, the switching period is divided into three distinct intervals. During the first interval, the power switch is turned on, and the inductor current increases linearly as energy is drawn from the input source. In the second interval, the switch is turned off and the inductor current decreases while freewheeling through the diode, transferring stored energy to the output. In the third interval, the inductor current reaches zero and remains at zero until the beginning of the next switching cycle, during which the load is supplied solely by the output capacitor. The existence of this zero-current interval is the defining characteristic of DCM and fundamentally alters the converter's steady-state and dynamic behavior.

Figure 24.1 illustrates the inductor current waveform of a buck converter operating in DCM. The current increases during the switch-on interval, decreases during the freewheeling interval, and remains at zero for a finite duration before the next switching cycle begins.

Unlike CCM, the steady-state output voltage of a buck converter operating in DCM depends not only on the duty ratio and input voltage, but also on the load resistance, inductance, and switching frequency. Consequently, the duty-to-output voltage relationship becomes nonlinear and load-dependent. Although DCM operation eliminates subharmonic oscillation and removes the

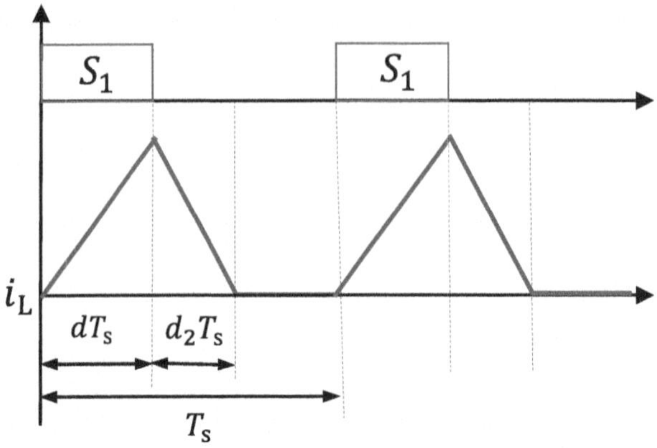

Figure 24.1 Inductor current waveform in DCM.

need for slope compensation in current-mode control, it results in higher peak and RMS currents, which increase conduction losses and impose greater stress on power components.

From a control and modeling perspective, DCM operation leads to a reduced-order dynamic model. Because the inductor current is zero for part of the switching period, the small-signal dynamics of the DCM buck converter are effectively first order and are dominated by the output capacitor and load. This characteristic simplifies voltage-loop controller design; however, the strong dependence of the plant gain on the operating point must be carefully accounted for to ensure stable and robust regulation over a wide load range.

For a buck converter, the boundary between continuous current mode and DCM is reached when the minimum inductor current becomes zero at the end of the freewheeling interval. At this boundary condition, the average inductor current is equal to one-half of the inductor current ripple. If the load current is reduced further, or if the inductance or switching period is insufficient to maintain continuous conduction, the converter transitions into DCM.

CCM–DCM Boundary Condition A buck converter operates in DCM when the inductor current reaches zero within a switching period and remains zero for a finite interval. The boundary between CCM and DCM occurs when the inductor current just reaches zero at the end of the freewheeling interval, without exhibiting a zero-current dwell time.

For an ideal buck converter operating at the CCM–DCM boundary, the average inductor current is equal to the output current, given by

$$I_L = I_o = \frac{V_o}{R}$$

The peak-to-peak inductor current ripple is expressed as

$$\Delta i_{\mathrm{L}} = \frac{(V_{\mathrm{in}} - V_{\mathrm{o}})DT_{\mathrm{s}}}{L}$$

where V_{in} is the input voltage, V_{o} is the output voltage, D is the duty ratio, T_{s} is the switching period, and L is the inductance.

At the boundary condition, the average inductor current equals one-half of the ripple current,

$$I_{\mathrm{o}} = \frac{\Delta i_{\mathrm{L}}}{2}$$

Substituting the ripple expression and using the steady-state relationship $V_{\mathrm{o}} = DV_{\mathrm{in}}$, the critical inductance that separates CCM and DCM is obtained as

$$L_{\mathrm{crit}} = \frac{(1 - D)RT_{\mathrm{s}}}{2}$$

Accordingly, DCM occurs when

$$L < L_{\mathrm{crit}} \Rightarrow \text{DCM operation}$$

This relationship clearly indicates that DCM operation is promoted by:

- Light load conditions (large R)
- Small inductance L
- High switching frequency (small T_{s})

Alternatively, the DCM condition can be expressed directly in terms of the load current as

$$I_{\mathrm{o}} < \frac{(V_{\mathrm{in}} - V_{\mathrm{o}})DT_{\mathrm{s}}}{2L}$$

When the load current falls below one half of the inductor ripple current, the inductor current becomes discontinuous, and the converter enters DCM.

24.2 State-Space Averaged Equation and Transfer Function

Consider an ideal buck converter operating in DCM with input voltage V_{in}, output capacitor voltage v_{c}, duty ratio d, switching period T_{s}, inductance L, output capacitance C, and load resistance R. In DCM, the inductor current starts from zero, rises linearly during the switch-on interval, falls to zero during the diode freewheeling interval, and then remains zero for the remainder of the switching period.

- ON interval $[0{\sim}dT_{\mathrm{s}}]$:

$$\frac{di_{\mathrm{L}}}{dt} = \frac{V_{\mathrm{in}} - v_{\mathrm{c}}}{L} \quad \Rightarrow \quad i_{\mathrm{pk}} = \frac{V_{\mathrm{in}} - v_{\mathrm{c}}}{L} dT_{\mathrm{s}}$$

- OFF (diode) interval $[dT_s \sim (d+d_2)T_s]$:

$$\frac{di_L}{dt} = \frac{-v_c}{L} \Rightarrow d_2 T_s = \frac{Li_{pk}}{v_c}$$

Therefore,

$$d_2 = \frac{V_{in} - v_c}{v_c} d$$

1) **Average inductor (output) current**

   Over the conduction interval $(d+d_2)T_s$, the inductor current is triangular $(0 \to i_{pk} \to 0)$. Hence, the averaged inductor current (equal to the average output current) is

$$i_o = \overline{i_L} = \frac{1}{T_s}\left(\frac{1}{2} i_{pk}(d+d_2)T_s\right) = \frac{1}{2} i_{pk}(d+d_2)$$

   Using $d + d_2 = d\left(1 + \frac{V_{in}-v_c}{v_c}\right) = d\frac{V_{in}}{v_c}$ and substituting i_{pk},

$$i_o = \frac{1}{2}\left(\frac{V_{in} - v_c}{L} dT_s\right)\left(d\frac{V_{in}}{v_c}\right) = \frac{T_s}{2L} d^2 \frac{V_{in}(V_{in} - v_c)}{v_c}$$

   At steady state $i_o = v_c/R$. Therefore:

$$\frac{v_c}{R} = \frac{T_s}{2L} d^2 \frac{V_{in}(V_{in} - v_c)}{v_c}$$

   Define the normalized output ratio $m \equiv v_c/V_{in}$ and

$$\alpha \equiv \frac{RT_s}{2L} d^2$$

   Then the DC equation becomes

$$m^2 = \alpha(1 - m) \Rightarrow m^2 + \alpha m - \alpha = 0$$

   Thus, the DCM conversion ratio is

$$\frac{V_c}{V_{in}} = m = \frac{-\alpha + \sqrt{\alpha^2 + 4\alpha}}{2}, \quad \alpha = \frac{RT_s}{2L} D^2$$

   This result highlights the load dependence of the DCM buck converter, in contrast to CCM where $V_c/V_{in} = D$.

2) **Averaged small-signal voltage equation**

   The capacitor dynamics are

$$C\frac{dv_c}{dt} = \overline{i_L} - \frac{v_c}{R}$$

   In DCM, the averaged inductor current over one switching period can be written as

$$\overline{i_L} = A\, d^2 \frac{V_{in}(V_{in} - v_c)}{v_c}, \quad A \equiv \frac{T_s}{2L}$$

Linearize about the operating point (V_c, D):

$$v_c = V_c + \hat{v}_c, d = D + \hat{d}$$

Then

$$C\frac{d\hat{v}_c}{dt} = \left.\frac{\partial \overline{i_L}}{\partial v_c}\right|_{DC} \cdot \hat{v}_c + \left.\frac{\partial \overline{i_L}}{\partial d}\right|_{DC} \cdot \hat{d} - \frac{1}{R}\hat{v}_c$$

The required partial derivatives are

$$\left.\frac{\partial \overline{i_L}}{\partial v_c}\right|_{DC} = -AD^2\left(\frac{V_{in}^2}{V_c^{\ 2}}\right), \quad \left.\frac{\partial \overline{i_L}}{\partial d}\right|_{DC} = 2AD\frac{V_{in}(V_{in} - V_c)}{V_c}$$

Substituting into the linearized capacitor equation and taking the Laplace transform yields

$$\left(Cs + \frac{1}{R} + AD^2\frac{V_{in}^2}{V_c^{\ 2}}\right)\hat{v}_c = 2AD\frac{V_{in}(V_{in} - V_c)}{V_c}\hat{d}$$

Hence, the duty-to-output voltage transfer function in DCM is

$$G_{vd}(s) = \frac{\hat{v}_c(s)}{\hat{d}(s)} = \frac{2AD\frac{V_{in}(V_{in}-V_c)}{V_c}}{Cs + \frac{1}{R} + AD^2\frac{V_{in}^2}{V_c^{\ 2}}}, \quad \left(A = \frac{T_s}{2L}\right) \tag{24.1}$$

Because the denominator is first order, $G_{vd}(s)$ can be written in the standard form,

$$G_{vd}(s) = \frac{\hat{v}_c(s)}{\hat{d}(s)} = \frac{K}{1 + s/\omega_p}$$

with

$$K = \frac{2AD\frac{V_{in}(V_{in}-V_c)}{V_c}}{\frac{1}{R} + AD^2\frac{V_{in}^2}{V_c^{\ 2}}}, \quad \omega_p = \frac{\frac{1}{R} + AD^2\frac{V_{in}^2}{V_c^{\ 2}}}{C}, \quad A = \frac{1}{2Lf_s}$$

3) **Averaged small-signal current equation**
Using the averaged current expression

$$\overline{i_L} = \frac{T_s}{2L}d^2\frac{V_{in}(V_{in} - v_c)}{v_c}$$

Linearize around (V_c, D) and retain the duty-perturbation term. With $i_L = I_L + \hat{i}_L$ and $d = D + \hat{d}$,

$$\hat{i}_L \approx \frac{\partial \overline{i_L}}{\partial d}\bigg|_{DC} \cdot \hat{d} = 2AD\frac{V_{in}(V_{in} - V_c)}{V_c}\hat{d}$$

Thus, the duty-to-averaged-inductor-current gain (DC in this simplified form) is

$$G_{\mathrm{id}}(s) \triangleq \frac{\hat{i}_{\mathrm{L}}(s)}{\hat{d}(s)} = \frac{V_{\mathrm{in}}(V_{\mathrm{in}} - V_{\mathrm{c}})}{LV_{\mathrm{c}}} DT_{\mathrm{s}} \tag{24.2}$$

This expression represents the operating-point-dependent gain from duty perturbation to the averaged inductor current in DCM.

24.3 Voltage Controller Design (Single Loop)

The proportional and integral gains, K_{pv} and K_{iv}, are determined using MATLAB's SISOTOOL based on the following frequency-domain design specifications:

- Bandwidth: 100 rad/sec
- Phase Margin: 90 deg

The MATLAB m-file used for the controller design is provided in Table 24.1.

Using the Automated Tuning feature in SISOTOOL, the optimized PI controller is obtained as

$$G_{\mathrm{cv}}(s) = \frac{0.00057446(s + 1620)}{s}$$

The corresponding closed-loop step response of voltage is presented in Figure 24.2, showing a fast transient response with negligible overshoot and confirming adequate stability margins.

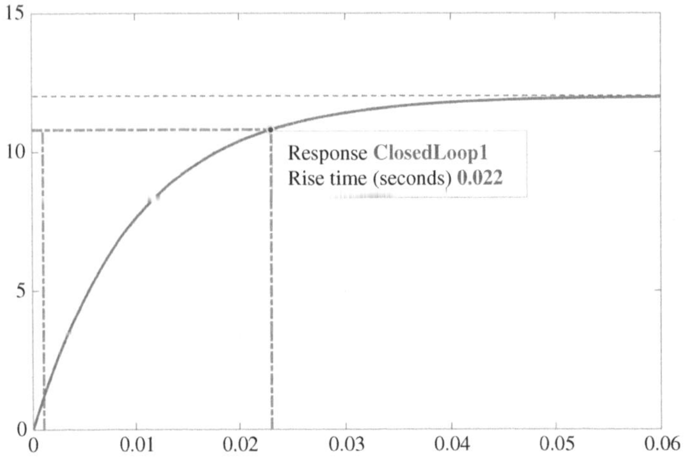

Figure 24.2 Closed-loop step response of the output voltage in DCM operation.

Table 24.1 MATLAB m-file for single-loop voltage controller (DCM buck converter).

```
%% ====Circuit Parameters for DCM Buck ====
Vin = 100;     % [V] input voltage
Vo = 12;       % [V] output voltage
fs = 20e3;     % [Hz] switching frequency
L  = 100e-6;   % [H] inductor
C  = 220e-6;   % [F] output capacitor
R  = 6;        % [Ohm] load resistance

A  = 1/(2*L*fs);      % A = Ts/(2L)
Ts = 1/fs;

% Steady-state duty for DCM buck
Io = Vo/R;
D  = sqrt( (Io*Vo) / (A*Vin*(Vin-Vo)) );

% Small-signal model (DCM buck)
den0 = (1/R) + A*D^2*(Vin^2/Vo^2);
K    = (2*A*D*(Vin*(Vin-Vo)/Vo)) / den0;
wp   = den0 / C;    % [rad/s] plant pole

s = tf('s');
G2d = K / (1 + s/wp);               % hV0/hd
G1d = Vin*(Vin-Vo)*D*Ts/(L*Vo);     % hiL/hd

Kpv=1; Kiv=1;
Gcv = Kpv*(s+Kiv)/s;

Vref = 12;
%sisotool Single loop - voltage
T=sisoinit(1);
T.F.Value = Vref;
T.G.Value = G2d;
T.C.Value = Gcv;
sisotool(T);
```

PSIM Verification The complete PSIM schematic of the DCM buck converter equipped with the proposed single-loop voltage controller is shown in Figure 24.3. The circuit parameters implemented in PSIM are identical to those used in the MATLAB-based small-signal model.

The output-voltage waveforms corresponding to a reference step change are shown in Figure 24.4. The PSIM response closely matches the MATLAB

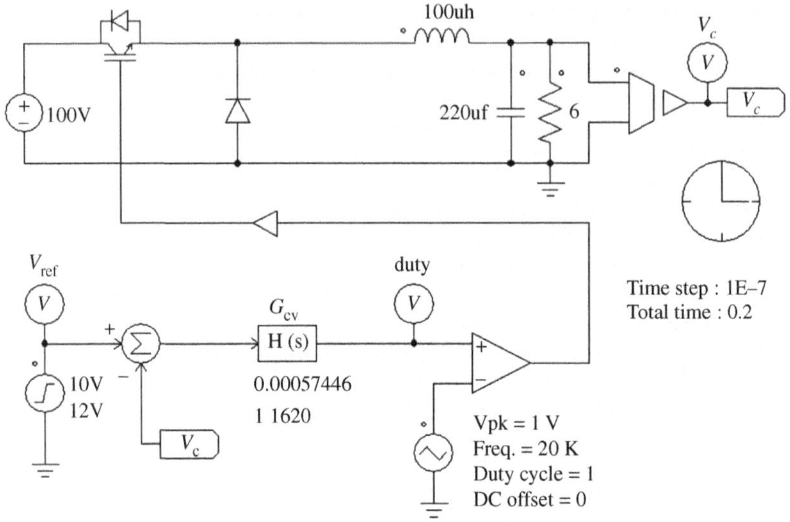

Figure 24.3 PSIM schematic for the DCM buck converter with single-loop voltage control.

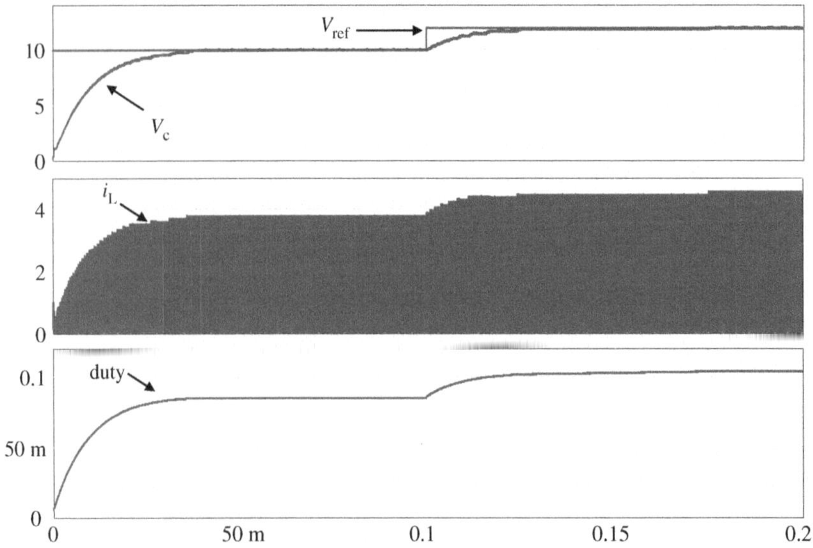

Figure 24.4 Step response of the output voltage and corresponding current waveforms.

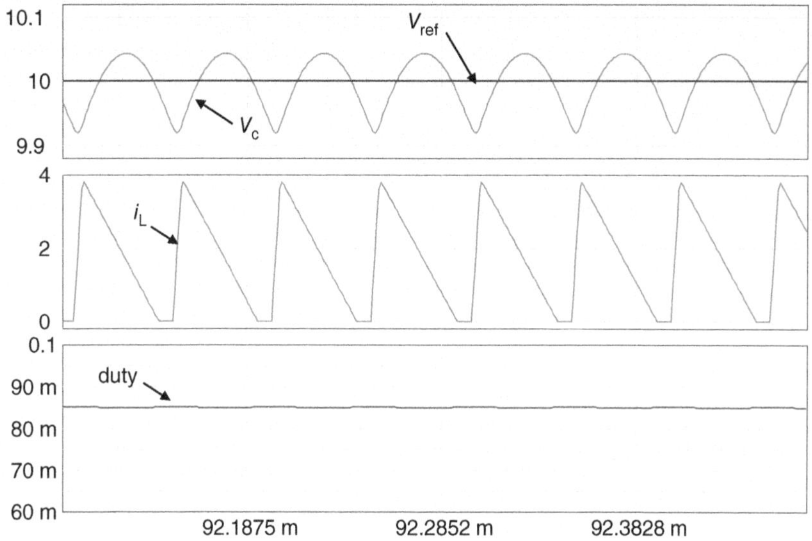

Figure 24.5 Magnified waveforms of the output voltage and current in DCM operation.

simulation, validating both the derived small-signal model and the controller design procedure. In addition, enlarged waveforms of the output voltage and current are presented in Figure 24.5, where the measured ripple and transient behavior agree well with the analytical predictions, further confirming the effectiveness of the proposed control approach.

25

Modular Multilevel Converter

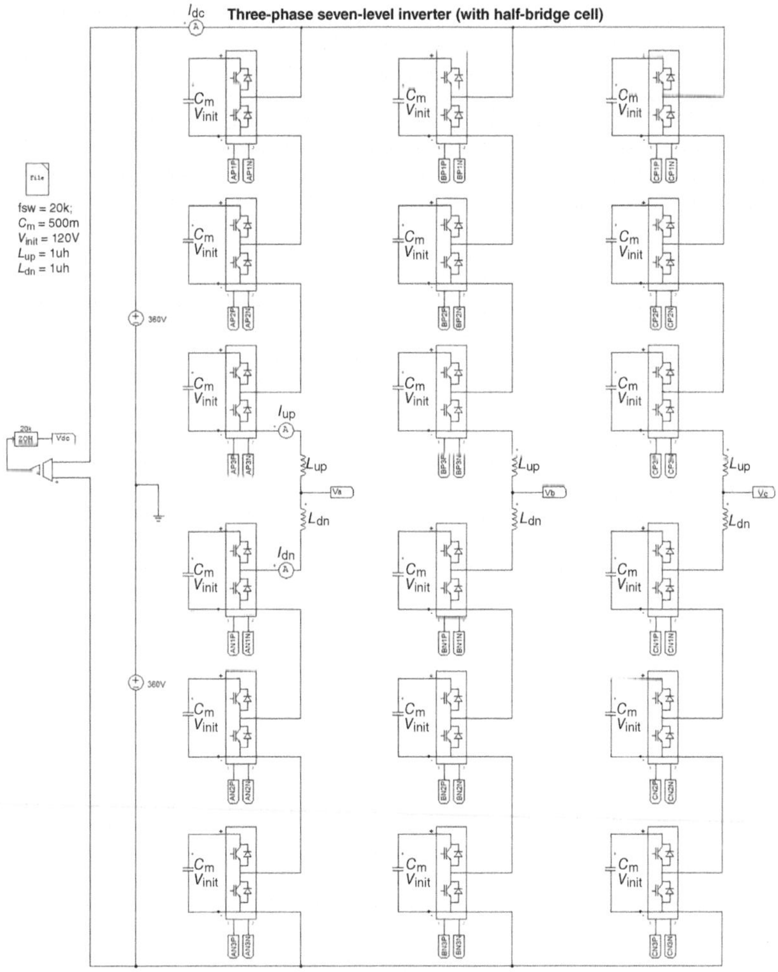

25.1 Modular Multilevel Converter (MMC)

A multilevel converter is a type of power conversion topology that synthesizes a stepped voltage waveform using power semiconductor switches such as IGBTs. The output voltage of a multilevel converter is typically generated by combining multiple voltage levels, which are obtained by connecting capacitors or other energy storage elements in series with the power switches. By constructing a staircase-like output waveform, multilevel converters can achieve higher output voltage levels compared to conventional two-level converters. In addition, they offer improved output waveform quality, reduced harmonic distortion, and enhanced conversion efficiency.

Multilevel converters can be broadly classified into two categories: *diode-clamped and flying-capacitor* (also referred to as capacitor-clamped) converters. In a diode-clamped multilevel converter, a series of capacitors is connected in parallel with the power switches to synthesize the stepped output voltage. In contrast, a flying-capacitor multilevel converter generates the output voltage by connecting a series of capacitors directly between the power switches. Figure 25.1 illustrates the configurations of multilevel inverters in both the diode-clamped and capacitor-clamped topologies. These configurations were previously introduced

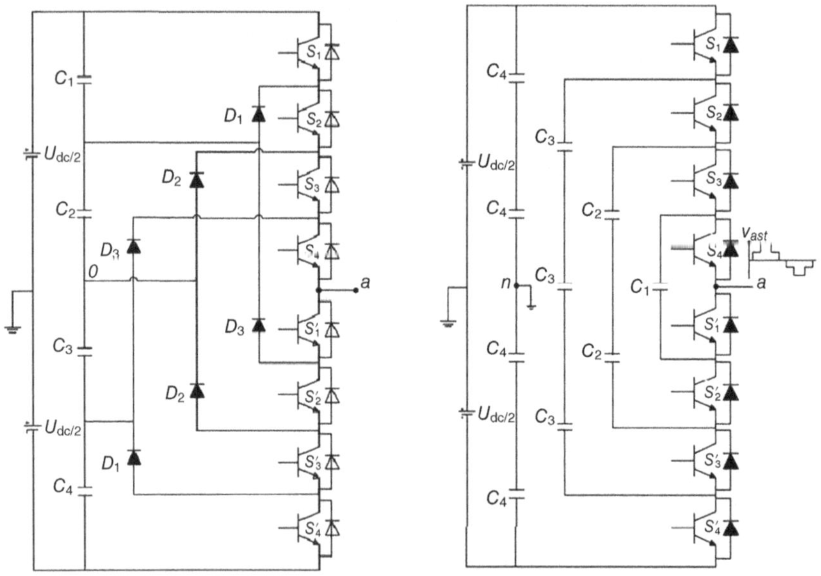

Figure 25.1 Configurations of multilevel inverter. (a) Five-level diode-clamped and (b) Five-level capacitor-clamped.

in Chapter 18 on neutral-point-clamped inverters and, therefore, will not be presented again in this chapter.

The modular multilevel converter (MMC) is a relatively recent addition to the family of multilevel power converters and has emerged as a transformative topology, particularly in high-voltage direct current (HVDC) transmission systems. The block diagram of a three-phase MMC is presented in Figure 25.2. Each phase arm consists of two multi-valves—an upper arm and a lower arm—each of which, for example, may include three series-connected power submodules. Based on the configuration of the cascaded submodules, MMCs can be classified into two main types: (i) Half-bridge and (ii) Full-bridge. The example shown in the figure illustrates the structure of a half-bridge MMC.

Modulation PWM signal Various modulation techniques have been explored for generating the output waveforms of MMCs. Among them, carrier phase-shifted sinusoidal PWM is one of the most widely adopted strategies. This method enables effective reduction of harmonic distortion while maintaining a relatively low switching frequency, making it well-suited for high-power MMC applications.

The three triangular carrier signals U_1, U_2, and U_3 are phase-shifted by 120° for the three upper and lower submodules of phase- a, as illustrated in Figure 25.3.

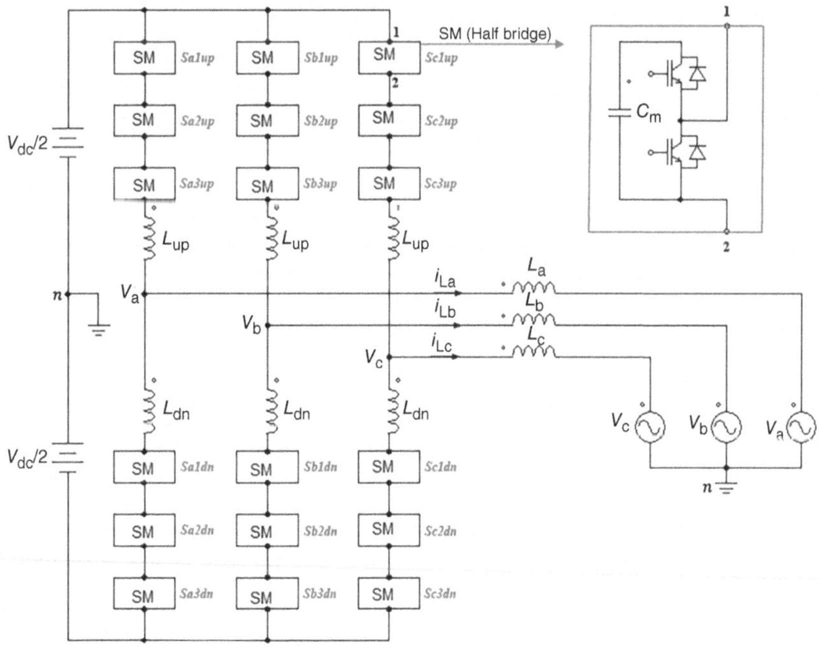

Figure 25.2 Three-phase MMC with three submodules per arm.

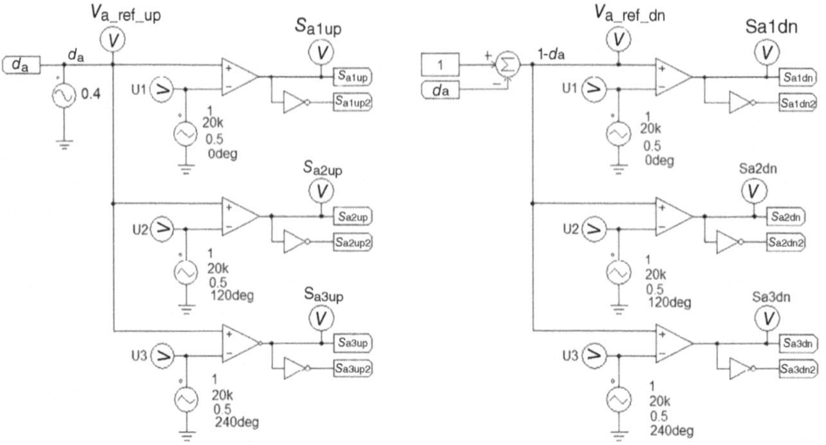

Figure 25.3 The *a*-phase PWM signal for up and down submodules.

The reference signals for the upper and lower arms of phase *a* are phase-shifted by 180° from each other. If the reference signal for the upper arm is denoted by d_a, then the corresponding reference signal for the lower arm is given by $1 - d_a$.

By comparing the sinusoidal reference waveform with the three carrier waveforms, the firing pulses for the three submodules in each phase are generated. Each transition in the multilevel output—from one level to the next—corresponds to the respective phase shift between the modulating and carrier signals.

The resulting PWM signals for phase-*a* are presented in Figure 25.4. The first plot shows the triangular carrier waveforms of each submodule along with the upper and lower reference signals. The second plot illustrates the seven-level sinusoidal output voltage V_a. The switching states of the upper and lower arms are shown in the third and fourth plots, respectively.

25.2 Controller Design

This section presents the design and implementation of a current controller for a three-phase, seven-level MMC configured as a power conditioning system (PCS) for grid-connected operation. In the considered topology, each phase leg of the MMC is interfaced with the three-phase utility grid through a dedicated filter inductor, which serves to attenuate switching harmonics and shape the injected grid current.

Accurate and robust current control is a critical requirement for MMC-based PCS applications, as it directly determines power quality, dynamic response, and

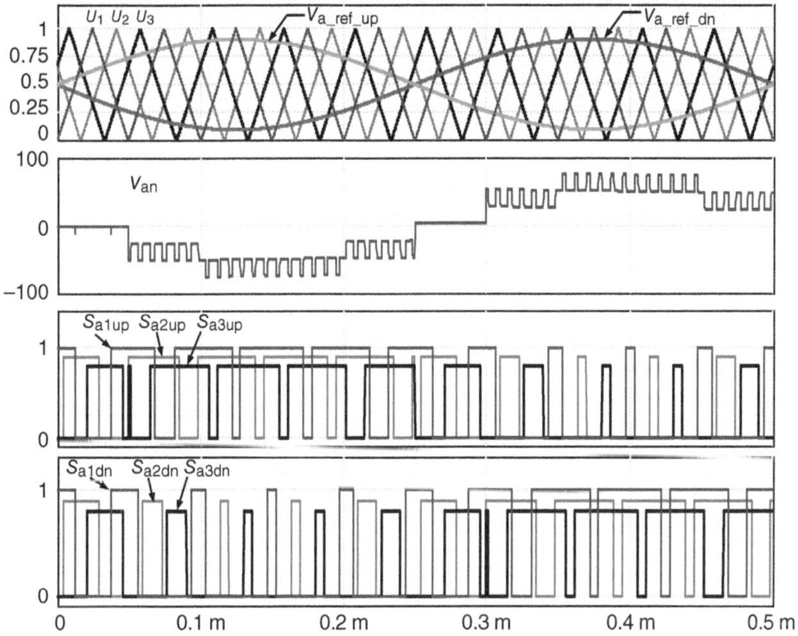

Figure 25.4 The *a*-phase waveforms of MMC.

compliance with grid codes. Owing to the multilevel structure and the presence of internal arm dynamics, MMC current control is inherently more complex than that of conventional two-level voltage-source converters. Consequently, a wide variety of current-control strategies—ranging from stationary-frame controllers to synchronous-frame and predictive methods—have been reported in the literature.

In order to maintain conceptual continuity with the control principles developed in earlier chapters of this book, and to facilitate intuitive understanding, this section focuses on two representative and practically relevant current-control approaches:

a) Instantaneous AC current control, implemented in the stationary reference frame, and
b) *dq*-axis (DQ) current control, implemented in the synchronous reference frame.

The instantaneous AC current control method offers a simple structure and straightforward implementation, making it well-suited for introductory analysis and rapid prototyping. In contrast, the *dq*-axis current control approach provides enhanced flexibility in regulating active and reactive power independently, enabling more systematic control of power factor and grid interaction. By presenting both methods within the same MMC framework, their respective characteristics, advantages, and limitations can be clearly compared.

25.2.1 Instantaneous AC Current Control

Figure 25.5 shows the PSIM schematic of an MMC current controller based on the instantaneous AC current control method. The controller is implemented using a C-block and directly reuses the *AC_3P_Control* module presented in Table 12.1 of Figure 12.3, without any modification.

In this control scheme, the phase currents are regulated independently by comparing each phase current with its corresponding reference and applying a proportional (*P*) controller. Bi-directional current flow is easily achieved by changing the sign of the reference current magnitude I_p. This approach offers a simple structure and high reliability, making it attractive for practical implementation. However, it has certain limitations: fault-current handling and power-factor control are not as intuitive as in more advanced control frameworks.

The PSIM schematic is divided into two main parts: the control stage and the power stage. Figure 25.5(a) illustrates the output stage connected to the grid together with the current-control C-block. This section is identical to the instantaneous AC current-control block introduced in Chapter 12.

The controller receives the phase control inputs d_a, d_b, and d_c, and generates the corresponding PWM signals for the upper (*UP*) and lower (*DN*) arm modules of each phase. The duty ratio is directly applied to the *UP* module, while its complementary value, (1-*duty*), is applied to the *DN* module. Through this mechanism, balanced modulation of the MMC arms is achieved, enabling effective regulation of the phase currents.

Figure 25.5(b) illustrates the power stage of the seven-level MMC. The converter is composed of three upper (*UP*) and three lower (*DN*) submodules per phase, forming a seven-level output voltage. Each phase's upper and lower modules are equipped with small inductors to suppress short-circuit currents, while the output of each phase bridge is connected to the grid through a filter inductor to attenuate switching harmonics and ensure high-quality grid current.

The corresponding simulation results are shown in Figure 25.6. In Figure 25.6(a), the magnitude of the current reference is set to +20 A, demonstrating power flow from the DC side to the grid. Under this condition, the grid voltage and current are synchronized, achieving unity power factor operation. In Figure 25.6(b), the current reference magnitude is set to −20 A, illustrating reverse power flow from the grid to the DC side. In this case, the phase relationship between the

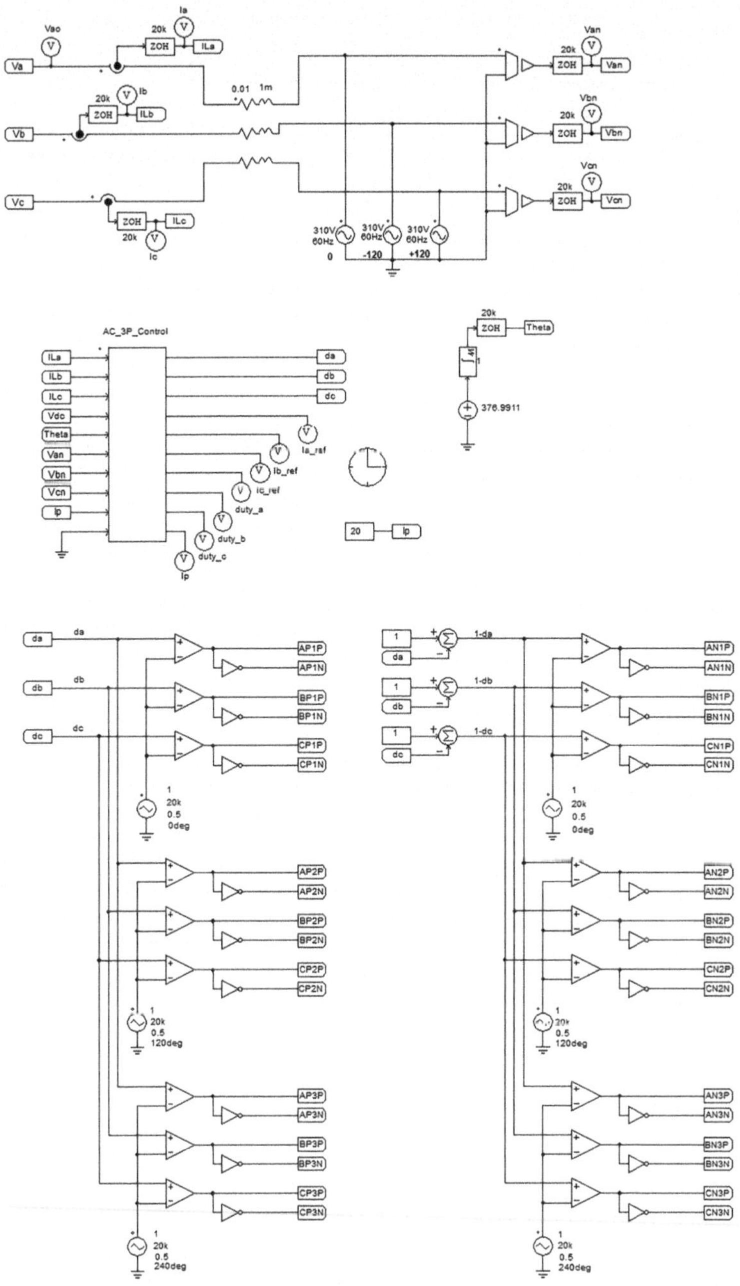

(a)

Figure 25.5 MMC instantaneous AC current control schematic. (a) MMC controller and PWM modules and (b) Three-phase, seven-level MMC power circuit.

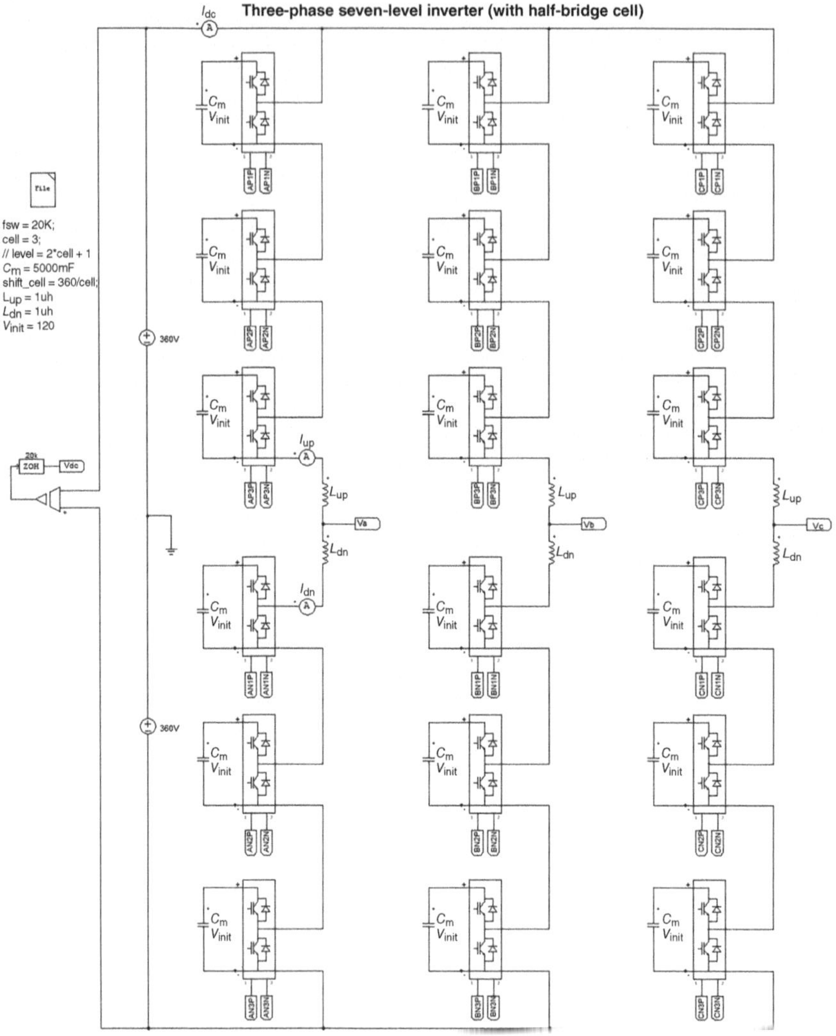

(b)

Figure 25.5 (*Continued*)

grid voltage and current is intentionally reversed to verify bi-directional power transfer.

In both operating modes, the current regulation is well maintained, and the MMC output voltage clearly exhibits seven discrete voltage levels, confirming proper modulation and effective current control.

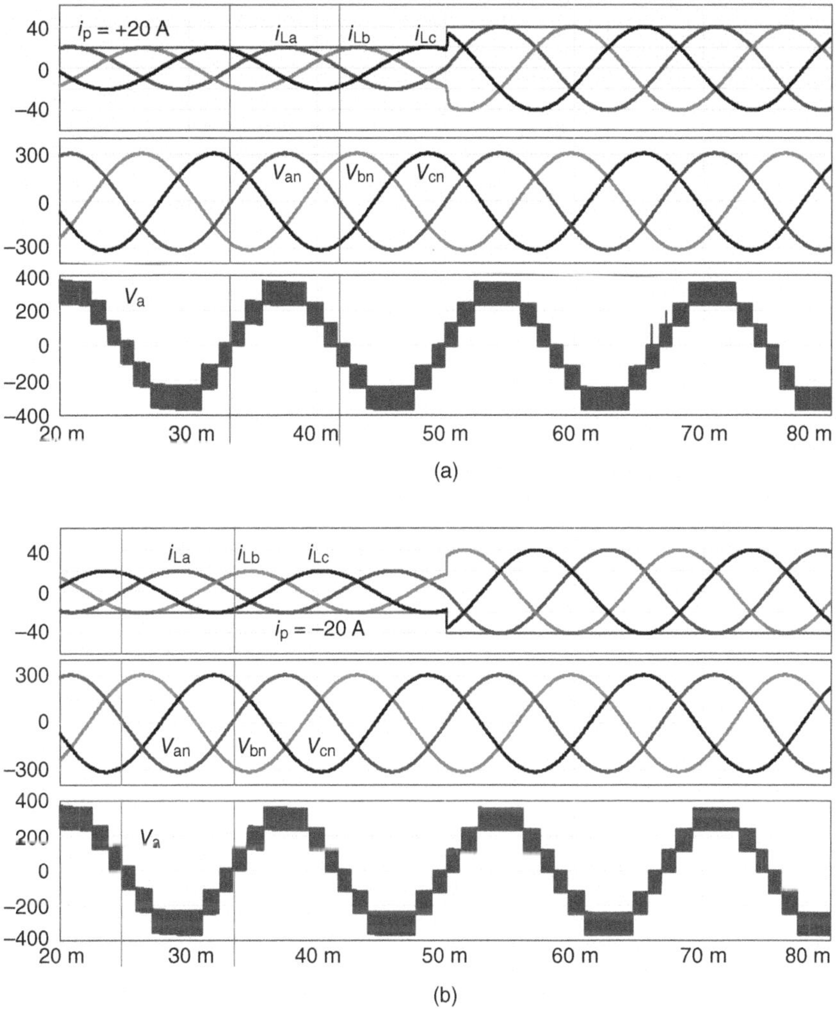

Figure 25.6 Bi-directional power flow simulation results of MMC. (a) Power flow from DC to grid ($I_p > 0$) and (b) Reverse power flow from grid to DC ($I_p < 0$).

25.2.2 DQ Current Control

The design of the MMC *dq* current controller is identical to that employed for the three-phase PCS presented in Chapter 12. Accordingly, the C-block *DQ_CURRENT* shown in Figure 12.41 is reused without modification. For brevity, the detailed simulation results are omitted, and readers are encouraged to verify the controller's performance through their own simulations.

Concluding Remarks

Throughout many years of teaching power electronics and control subjects in electrical engineering programs, the author has continually reflected on a fundamental question: how can controller design be taught in a way that is simple, systematic, and unified? For many graduate students, controller design is perceived as particularly challenging, as it often demands a strong mathematical background and is introduced through fragmented and system-specific methodologies. Bridging this gap between theoretical rigor and practical accessibility became the primary motivation for writing this book.

After approximately two years of preparation and one year of focused writing, this book has been completed. It began with the modest objective of presenting the fundamental principles of controller design in a form that could be readily understood and directly applied. As the manuscript evolved, its scope naturally expanded to encompass a wide range of power conversion systems that are commonly employed in commercial and industrial applications.

To ensure consistency and coherence throughout the book, special emphasis was placed on the unification of terminology, mathematical formulations, and simulation frameworks. All equations were derived by the author and systematically verified through simulation. Through this approach, the author hopes that readers will build confidence by clearly observing the direct correspondence between analytical results and simulation outcomes.

Although the explanations in this book are intentionally concise and straightforward, the underlying concepts are firmly grounded in rigorous and well-established theory. Accordingly, the material is intended to be valuable as both a textbook for university-level education and a practical reference for engineers working in real-world power electronics applications.

With the hope that this book will serve as a foundational resource in academia and a reliable reference in industry, the author concludes these remarks.

January 2026 Chungju City, Republic of Korea

Index

Wiley The manufacturer's authorized representative according to the EU General Product Safety Regulation is Wiley-VCH GmbH, Boschstr. 12, 69469 Weinheim, Germany, e-mail: Product_Safety@wiley.com.

Printed and bound by CPI Group (UK) Ltd, Croydon, CR0 4YY
08/10/2026
02229717-0001